Lecture Notes in Artificial Intelligence   3702

Edited by J. G. Carbonell and J. Siekmann

Subseries of Lecture Notes in Computer Science

Bernhard Beckert (Ed.)

Automated Reasoning with Analytic Tableaux and Related Methods

International Conference, TABLEAUX 2005
Koblenz, Germany, September 14-17, 2005
Proceedings

 Springer

Series Editors

Jaime G. Carbonell, Carnegie Mellon University, Pittsburgh, PA, USA
Jörg Siekmann, University of Saarland, Saarbrücken, Germany

Volume Editor

Bernhard Beckert
University of Koblenz-Landau
Department of Computer Science
AI Research Group
Universitätsstr. 1, 56070 Koblenz, Germany
E-mail: beckert@uni-koblenz.de

Library of Congress Control Number: 2005931930

CR Subject Classification (1998): I.2.3, F.4.1, I.2, D.1.6, D.2.4

ISSN   0302-9743
ISBN-10   3-540-28931-3 Springer Berlin Heidelberg New York
ISBN-13   978-3-540-28931-9 Springer Berlin Heidelberg New York

Springer is a part of Springer Science+Business Media

springeronline.com

© Springer-Verlag Berlin Heidelberg 2005
Printed in Germany

Typesetting: Camera-ready by author, data conversion by Scientific Publishing Services, Chennai, India
Printed on acid-free paper   SPIN: 11554554   06/3142   5 4 3 2 1 0

Preface

This volume contains the research papers presented at the International Conference on Automated Reasoning with Analytic Tableaux and Related Methods (TABLEAUX 2005) held September 14–17, 2005 in Koblenz, Germany. This conference was the fourteenth in a series of international meetings since 1992 (listed on page IX). It was part of the International Conference Summer Koblenz 2005, which included conferences covering a broad spectrum of topics and interesting fields of application for tableau-based methods: artificial intelligence (KI), multi-agent systems (MATES), automated reasoning and knowledge representation (FTP), and software engineering and formal methods (SEFM).

The Program Committee of TABLEAUX 2005 received 46 submissions from 16 countries. After each paper was reviewed by three referees, and an intensive discussion on the borderline papers was held during the online meeting of the Program Committee, 18 research papers and 7 system descriptions were accepted based on originality, technical soundness, presentation, and relevance. I wish to sincerely thank all the authors who submitted their work for consideration. And I would like to thank the Program Committee members and other referees for their great effort and professional work in the review and selection process. Their names are listed on the following pages.

In addition to the contributed papers, the program included four excellent keynote talks. I am grateful to Prof. Diego Calvanese (Free University of Bolzano, Italy), Prof. Ian Horrocks (University of Manchester, UK), Prof. Hans Jürgen Ohlbach (Ludwig Maximilian University, Munich, Germany), and Prof. Erik Rosenthal (University of New Haven, USA) for accepting the invitation to address the conference.

Three very good tutorials were part of TABLEAUX 2005: *Instance-Based Methods* (P. Baumgartner and G. Stenz), *Analytic Systems and Dialogue Games* (C. Fermüller), and *A Tutorial on Agda* (M. Benke). I would like to express my thanks to the tutorial presenters for their contribution.

It was a team effort that made the conference so successful. I am truly grateful to the Steering Committee members for their support and to Gernot Stenz for his effective work as Publicity Chair. And I particularly thank the local organizers for their hard work and help in making the conference a success: Gerd Beuster, Sibille Burkhardt, Ruth Götten, Vladimir Klebanov, Thomas Kleemann, Jan Murray, Oliver Obst, Alex Sinner, Christoph Wernhard, and Doris Wertz.

September 2005   Bernhard Beckert

Organization

Program and Conference Chair

Bernhard Beckert   University of Koblenz-Landau, Germany

Program Committee

Peter Baumgartner	MPI Saarbrücken, Germany
Marta Cialdea Mayer	Università Roma Tre, Italy
Roy Dyckhoff	University of St Andrews, Scotland
Christian Fermüller	Vienna University of Technology, Austria
Ulrich Furbach	University of Koblenz-Landau, Germany
Didier Galmiche	LORIA, Université Henri Poincaré, France
Martin Giese	RICAM, Austrian Acad. of Sci., Linz, Austria
Rajeev Goré	Australian Nat. Univ., Canberra, Australia
Jean Goubault-Larrecq	École Normale Supérieure de Cachan, France
Reiner Hähnle	Chalmers University, Gothenburg, Sweden
Ullrich Hustadt	University of Liverpool, UK
Christoph Kreitz	University of Potsdam, Germany
Reinhold Letz	Munich University of Technology, Germany
Carsten Lutz	Dresden University of Technology, Germany
Maarten Marx	University of Amsterdam, The Netherlands
Ugo Moscato	University of Milano-Bicocca, Italy
Neil V. Murray	University at Albany – SUNY, USA
Ilkka Niemelä	Helsinki University of Technology, Finland
Nicola Olivetti	University of Turin, Italy
Lawrence Paulson	University of Cambridge, UK
David A. Plaisted	University of North Carolina, USA
Peter H. Schmitt	University of Karlsruhe, Germany
Viorica Sofronie-Stokkermans	MPI Saarbrücken, Germany
Arild Waaler	University of Oslo, Norway
Calogero G. Zarba	LORIA and INRIA-Lorraine, France, and University of New Mexico, USA

Additional Referees

Pietro Abate
Wolfgang Ahrendt
Alessandro Avellone
Arnon Avron
Matthias Baaz
Mathieu Baudet
Maria Paola Bonacina
Claudio Castellini
Serenella Cerrito
Hans de Nivelle
Stéphane Demri
Mauro Ferrari
Camillo Fiorentini
Guido Fiorino
Lilia Georgieva

Laura Giordano
Michael Hansen
Keijo Heljanko
Ian Horrocks
Herman Ruge Jervell
Tommi Junttila
Yevgeny Kazakov
Thomas Kleemann
Lars Kristiansen
Martin Lange
Dominique
    Larchey-Wendling
Patrick Maier
Daniel Méry
George Metcalfe

Claudia Nalon
Andreas Nonnengart
Mario Ornaghi
Vincent Risch
Luca Roversi
Gernot Salzer
Ulrike Sattler
Niklas Sörensson
Gernot Stenz
Charles Stewart
Christoph Wernhard
Paul Wong
Hantao Zhang

Steering Committee

Neil Murray
    (President)

University at Albany – SUNY, USA

Jean Goubault-Larrecq
    (Vice-President)

École Normale Suprieure de Cachan, France

Bernhard Beckert
Peter Baumgartner
Roy Dyckhoff
Didier Galmiche
Reiner Hähnle

University of Koblenz-Landau, Germany
MPI Saarbrücken, Germany
University of St Andrews, Scotland
LORIA, Université Henri Poincaré, France
Chalmers University, Gothenburg, Sweden

Organization

TABLEAUX 2005 was organized by the Artificial Intelligence Group at the Institute for Computer Science of the University of Koblenz-Landau.

Sponsoring Institutions

German Research Foundation (DFG)
University of Koblenz-Landau
City of Koblenz
Government of the Rhineland-Palatinate
Griesson – de Beukelaer GmbH

Previous Meetings

1992	Lautenbach, Germany	1993	Marseille, France
1994	Abingdon, UK	1995	St. Goar, Germany
1996	Terrasini, Italy	1997	Pont-à-Mousson, France
1998	Oisterwijk, The Netherlands	1999	Saratoga Springs, USA
2000	St Andrews, UK	2001	Siena, Italy (part of IJCAR)
2002	Copenhagen, Denmark	2003	Rome, Italy
2004	Cork, Ireland (part of IJCAR)		

Table of Contents

System Descriptions

Query Processing in Peer-to-Peer Systems: An Epistemic Logic Approach

Diego Calvanese

Free University of Bolzano/Bozen, Faculty of Computer Science
calvanese@inf.unibz.it

Abstract

In peer-to-peer (P2P) systems, each peer exports information in terms of its own schema, and interoperation is achieved by means of mappings among the peer schemas. Peers are autonomous systems and mappings are dynamically created and changed. One of the challenges in these systems is processing queries posed to one peer taking into account the mappings. Obviously, query processing strongly depends on the semantics of the overall system.

In this talk, we overview the various approaches that have been proposed for modeling P2P systems, considering several central properties of such systems, such as modularity, generality, and computational issues related to query processing. We argue that an approach based on epistemic logic is superior with respect to all the above properties to previously proposed approaches based on first-order logic. Specifically, the epistemic logic approach naturally captures the modular nature of P2P systems, and moreover query answering can be performed efficiently (i.e., polynomially) in the size of the data stored in the various peers. This holds independently of the topology of the mappings among peers, and hence respecting one of the fundamental assumptions in P2P systems: namely, that peers are autonomouns entities that can establish mappings to other peers without requiring the intervention of any centralized authority.

B. Beckert (Ed): TABLEAUX 2005, LNAI 3702, p. 1, 2005.

Description Logics in Ontology Applications

Ian Horrocks

School of Computer Science, University of Manchester,
Oxford Road, Manchester M13 9PL, UK
horrocks@cs.man.ac.uk

Abstract. Description Logics (DLs) are a family of logic based knowledge representation formalisms. Although they have a range of applications (e.g., configuration and information integration), they are perhaps best known as the basis for widely used ontology languages such as OWL (now a W3C recommendation). This decision was motivated by a requirement that key inference problems be decidable, and that it should be possible to provide reasoning services to support ontology design and deployment. Such reasoning services are typically provided by highly optimised implementations of tableaux decision procedures; these have proved to be effective in applications in spite of the high worst case complexity of key inference problems. The increasing use of DL based ontologies in areas such as e-Science and the Semantic Web is, however, already stretching the capabilities of existing DL systems, and brings with it a range of research challenges.

1 Introduction

Description Logics (DLs) are a family of class (concept) based knowledge representation formalisms. They are characterised by the use of various constructors to build complex concepts from simpler ones, an emphasis on the decidability of key reasoning tasks, and by the provision of sound, complete and (empirically) tractable reasoning services.

Although they have a range of applications (e.g., reasoning with database schemas and queries [1,2,3]), DLs are perhaps best known as the basis for ontology languages such as OIL, DAML+OIL and OWL [4]. The decision to base these languages on DLs was motivated by a requirement not only that key inference problems (such as class satisfiability and subsumption) be decidable, but that "practical" decision procedures and "efficient" implemented systems also be available.

That DLs were able to meet the above requirements was the result of extensive research within the DL community over the course of the preceding 20 years or more. This research mapped out a complex landscape of languages, exploring a range of different language constructors, studying the effects of various combinations of these constructors on decidability and worst case complexity, and devising decision procedures, the latter often being tableaux based algorithms. At the same time, work on implementation and optimisation techniques demonstrated that, in spite of the high worst case complexity of key inference problems

B. Beckert (Ed): TABLEAUX 2005, LNAI 3702, pp. 2–13, 2005.

(usually at least ExpTime), highly optimised DL systems were capable of providing practical reasoning support in the typical cases encountered in realistic applications.

With the added impetus provided by the OWL standardisation effort, DL systems are now being used to provide computational services for a rapidly expanding range of ontology tools and applications [5,6,7,8,9,10]. The increasing use of DL based ontologies in areas such as e-Science and the Semantic Web is, however, already stretching the capabilities of existing DL systems, and brings with it a range of research challenges.

2   Ontologies and Ontology Reasoning

In Computer Science, an ontology is usually taken to mean a conceptual model (of some domain), typically realised as a hierarchical vocabulary of terms, together with formal specifications of the meaning of each term. These specifications are often given with reference to other (simpler) terms in the ontology. For example, in a medical terminology ontology, the meaning of the term *Gastritis* might be specified as an *InflammatoryProcess* whose *outcome* is *InflammationOf-Stomach*, where *InflammatoryProcess*, *outcome* and *InflammationOfStomach* are all terms from the ontology. Such vocabularies may be used, e.g., to facilitate data sharing and reuse (often by annotating data using terms from a shared ontology), to structure data, or simply to explicate and investigate knowledge of a domain.

Ontologies play a major role in the Semantic Web (where they are used to annotate web resources) [11,12], and are widely used in, e.g., knowledge management systems, e-Science, and bio-informatics and medical terminologies [13,14,15,16]. They are also of increasing importance in the Grid, where they may be used, e.g., to support the discovery, execution and monitoring of Grid services [17,18,19].

Given the formal and compositional nature of ontologies, it is natural to use logics as the basis for ontology languages—this allows for the precise definition of the meaning of compositional operators (such as "and" and "or"), and of relationships between terms (such as "subclass" and "instance"). The effective use of logic based ontology languages in applications will, however, critically depend on the provision of efficient reasoning support. On the one hand, such support is required by ontology engineers in order to help them to design and maintain sound, well-balanced ontologies [20]. On the other hand, such support is required by applications in order to exploit the formal specification of meaning captured in ontologies: querying ontologies and ontology structured data, is equivalent to computing logical entailments [21].

3   Ontology Languages and Description Logics

The OWL recommendation actually consists of three languages of increasing expressive power: OWL Lite, OWL DL and OWL Full. Like OWL's predecessor

DAML+OIL, OWL Lite and OWL DL are basically very expressive description logics with an RDF syntax. OWL Full provides a more complete integration with RDF, but its formal properties are less well understood, and key inference problems would certainly be *much* harder to compute.[1] For these reasons, OWL Full will not be considered here.

More precisely, OWL DL is based on the $\mathcal{SHOIQ}$ DL [23]; it restricts the form of number restrictions to be unqualified (see [24]), and adds a simple form of Datatypes (often called concrete domains in DLs [25]). Following the usual DL naming conventions, the resulting logic is called $\mathcal{SHOIN}(\mathbf{D})$, with the different letters in the name standing for (sets of) constructors available in the language: $\mathcal{S}$ stands for the basic $\mathcal{ALC}$ DL (equivalent to the propositional modal logic $\mathbf{K}_{(m)}$) extended with transitive roles [22], $\mathcal{H}$ stands for role hierarchies (equivalently, inclusion axioms between roles), $\mathcal{O}$ stands for nominals (classes whose extension is a single individual) [26], $\mathcal{N}$ stands for unqualified number restrictions and $(\mathbf{D})$ stands for datatypes) [27]. OWL Lite is equivalent to the slightly simpler $\mathcal{SHIF}(\mathbf{D})$ DL (i.e., $\mathcal{SHOIQ}$ without nominals, and with only functional number restrictions).

These equivalences allow OWL to exploit the considerable existing body of description logic research, e.g.:

- to define the semantics of the language and to understand its formal properties, in particular the decidability and complexity of key inference problems [28];
- as a source of sound and complete algorithms and optimised implementation techniques for deciding key inference problems [29,22,27];
- to use implemented DL systems in order to provide (partial) reasoning support [30,31,32].

3.1 $\mathcal{SHOIN}$ Syntax and Semantics

The syntax and semantics of $\mathcal{SHOIN}$ are briefly introduced here (we will ignore datatypes, as adding a datatype component would complicate the presentation and has little affect on reasoning [33]).

Definition 1. *Let* $\mathbf{R}$ *be a set of* role names *with both transitive and normal role names* $\mathbf{R}_+ \cup \mathbf{R}_P = \mathbf{R}$, *where* $\mathbf{R}_P \cap \mathbf{R}_+ = \emptyset$. *The set of* $\mathcal{SHOIN}$-roles *(or* roles *for short) is* $\mathbf{R} \cup \{R^- \mid R \in \mathbf{R}\}$. *A* role inclusion axiom *is of the form* $R \sqsubseteq S$, *for two roles* R *and* S. *A* role hierarchy *is a finite set of role inclusion axioms.*

An interpretation *$\mathcal{I} = (\Delta^{\mathcal{I}}, \cdot^{\mathcal{I}})$ consists of a non-empty set $\Delta^{\mathcal{I}}$, called the* domain *of $\mathcal{I}$, and a function $\cdot^{\mathcal{I}}$ which maps every role to a subset of $\Delta^{\mathcal{I}} \times \Delta^{\mathcal{I}}$ such that, for $P \in \mathbf{R}$ and $R \in \mathbf{R}_+$,*

$$\langle x, y \rangle \in P^{\mathcal{I}} \text{ iff } \langle y, x \rangle \in P^{-\mathcal{I}},$$
$$\text{and if } \langle x, y \rangle \in R^{\mathcal{I}} \text{ and } \langle y, z \rangle \in R^{\mathcal{I}}, \text{ then } \langle x, z \rangle \in R^{\mathcal{I}}.$$

---

[1] Inference in OWL Full is clearly undecidable as OWL Full does not include restrictions on the use of transitive properties which are required in order to maintain decidability [22].

An interpretation $\mathcal{I}$ satisfies a role hierarchy $\mathcal{R}$ iff $R^{\mathcal{I}} \subseteq S^{\mathcal{I}}$ for each $R \sqsubseteq S \in \mathcal{R}$; such an interpretation is called a model *of $\mathcal{R}$.*

Definition 2. *Let N_C be a set of* concept names *with a subset $N_I \subseteq N_C$ of* nominals. *The set of $\mathcal{SHOIN}$-concepts (or concepts for short) is the smallest set such that*

1. *every concept name $C \in N_C$ is a concept,*
2. *if C and D are concepts and R is a role, then $(C \sqcap D)$, $(C \sqcup D)$, $(\neg C)$, $(\forall R.C)$, and $(\exists R.C)$ are also concepts (the last two are called universal and existential restrictions, resp.), and*
3. *if R is a simple role[2] and $n \in \mathbb{N}$, then $\leqslant nR$ and $\geqslant nR$ are also concepts (called atmost and atleast number restrictions).*

The interpretation function $\cdot^{\mathcal{I}}$ of an interpretation $\mathcal{I} = (\Delta^{\mathcal{I}}, \cdot^{\mathcal{I}})$ maps, additionally, every concept to a subset of $\Delta^{\mathcal{I}}$ such that

$$(C \sqcap D)^{\mathcal{I}} = C^{\mathcal{I}} \cap D^{\mathcal{I}}, \qquad (C \sqcup D)^{\mathcal{I}} = C^{\mathcal{I}} \cup D^{\mathcal{I}}, \qquad \neg C^{\mathcal{I}} = \Delta^{\mathcal{I}} \setminus C^{\mathcal{I}},$$
$$\sharp o^{\mathcal{I}} = 1 \qquad \text{for all } o \in N_I,$$
$$(\exists R.C)^{\mathcal{I}} = \{x \in \Delta^{\mathcal{I}} \mid \text{There is a } y \in \Delta^{\mathcal{I}} \text{ with } \langle x, y \rangle \in R^{\mathcal{I}} \text{ and } y \in C^{\mathcal{I}}\},$$
$$(\forall R.C)^{\mathcal{I}} = \{x \in \Delta^{\mathcal{I}} \mid \text{For all } y \in \Delta^{\mathcal{I}}, \text{ if } \langle x, y \rangle \in R^{\mathcal{I}}, \text{then } y \in C^{\mathcal{I}}\},$$
$$\leqslant nR^{\mathcal{I}} = \{x \in \Delta^{\mathcal{I}} \mid \sharp\{y \mid \langle x, y \rangle \in R^{\mathcal{I}}\} \leqslant n\},$$
$$\geqslant nR^{\mathcal{I}} = \{x \in \Delta^{\mathcal{I}} \mid \sharp\{y \mid \langle x, y \rangle \in R^{\mathcal{I}}\} \geqslant n\},$$

where, for a set M, we denote the cardinality of M by $\sharp M$.

For C and D (possibly complex) concepts, $C \stackrel{.}{\sqsubseteq} D$ is called a general concept inclusion *(GCI), and a finite set of GCIs is called a* TBox.

An interpretation $\mathcal{I}$ satisfies a GCI $C \stackrel{.}{\sqsubseteq} D$ if $C^{\mathcal{I}} \subseteq D^{\mathcal{I}}$, and $\mathcal{I}$ satisfies a TBox $\mathcal{T}$ if $\mathcal{I}$ satisfies each GCI in $\mathcal{T}$; such an interpretation is called a model *of $\mathcal{T}$.*

A concept C is called satisfiable *with respect to a role hierarchy $\mathcal{R}$ and a TBox $\mathcal{T}$ if there is a model $\mathcal{I}$ of $\mathcal{R}$ and $\mathcal{T}$ with $C^{\mathcal{I}} \neq \emptyset$. Such an interpretation is called a* model *of C w.r.t. $\mathcal{R}$ and $\mathcal{T}$. A concept D* subsumes *a concept C w.r.t. $\mathcal{R}$ and $\mathcal{T}$ (written $C \sqsubseteq_{\mathcal{R},\mathcal{T}} D$) if $C^{\mathcal{I}} \subseteq D^{\mathcal{I}}$ holds in every model $\mathcal{I}$ of $\mathcal{R}$ and $\mathcal{T}$. Two concepts C, D are* equivalent *w.r.t. $\mathcal{R}$ and $\mathcal{T}$ (written $C \equiv_{\mathcal{R},\mathcal{T}} D$) iff they are mutually subsuming w.r.t. $\mathcal{R}$ and $\mathcal{T}$. (When $\mathcal{R}$ and $\mathcal{T}$ are obvious from the context, we will often write $C \sqsubseteq D$ and $C \equiv D$.) For an interpretation $\mathcal{I}$, an individual $x \in \Delta^{\mathcal{I}}$ is called an* instance *of a concept C iff $x \in C^{\mathcal{I}}$.*

Note that, as usual, subsumption and satisfiability can be reduced to each other, and reasoning w.r.t. *general TBoxes* and role hierarchies can be reduced to reasoning w.r.t. role hierarchies only [22,27].

3.2 Practical Reasoning Services

Most modern DL systems use *tableaux* algorithms to test concept satisfiability. These algorithms work by trying to construct (a tree representation of) a model

---

[2] A role is simple if it is neither transitive nor has any transitive subroles. Restricting number restrictions to simple roles is required in order to yield a decidable logic [22].

of the concept, starting from an individual instance. Tableaux expansion rules decompose concept expressions, add new individuals (e.g., as required by $\exists R.C$ terms),[3] and merge existing individuals (e.g., as required by $\leqslant nR.C$ terms). Non-determinism (e.g., resulting from the expansion of disjunctions) is dealt with by searching the various possible models. For an unsatisfiable concept, all possible expansions will lead to the discovery of an obvious contradiction known as a *clash* (e.g., an individual that must be an instance of both A and $\neg A$ for some concept A); for a satisfiable concept, a complete and clash-free model will be constructed [34].

Tableaux algorithms have many advantages. It is relatively easy to design provably sound, complete and terminating algorithms, and the basic technique can be extended to deal with a wide range of class and role constructors. More-over, although many algorithms have a higher worst case complexity than that of the underlying problem, they are usually quite efficient at solving the relatively easy problems that are typical of realistic applications.

Even in realistic applications, however, problems can occur that are much too hard to be solved by naive implementations of theoretical algorithms. Modern DL systems, therefore, include a wide range of optimisation techniques, the use of which has been shown to improve typical case performance by several orders of magnitude [29,35,36,32,37,38]. Key techniques include lazy unfolding, absorption and dependency directed backtracking.

Lazy Unfolding. In an ontology, or DL Tbox, large and complex concepts are seldom described monolithically, but are built up from a hierarchy of named concepts whose descriptions are less complex. The tableaux algorithm can take advantage of this structure by trying to find contradictions between concept names before adding expressions derived from Tbox axioms. This strategy is known as *lazy unfolding* [29,36].

The benefits of lazy unfolding can be maximised by lexically *normalising* and *naming* all concept expressions and, recursively, their sub-expressions. An ex-pression C is normalised by rewriting it in a standard form (e.g., disjunctions are rewritten as negated conjunctions); it is named by substituting it with a new con-cept name A, and adding an axiom $A \equiv C$ to the Tbox. The normalisation step allows lexically equivalent expressions to be recognised and identically named, and can even detect syntactically "obvious" satisfiability and unsatisfiability.

Absorption. Not all axioms are amenable to lazy unfolding. In particular, so called *general concept inclusions* (GCIs), axioms of the form $C \sqsubseteq D$ where C is non-atomic, must be dealt with by explicitly making every individual in the model an instance of $D \sqcup \neg C$. Large numbers of GCIs result in a very high degree of non-determinism and catastrophic performance degradation [36].

Absorption is another rewriting technique that tries to reduce the number of GCIs in the Tbox by absorbing them into axioms of the form $A \sqsubseteq C$, where A is a concept name. The basic idea is that an axiom of the form $A \sqcap D \sqsubseteq D'$

---

[3] Cycle detection techniques known as *blocking* may be required in order to guarantee termination.

can be rewritten as $A \sqsubseteq D' \sqcup \neg D$ and absorbed into an existing $A \sqsubseteq C$ axiom to give $A \sqsubseteq C \sqcap (D' \sqcup \neg D)$ [39]. Although the disjunction is still present, lazy unfolding ensures that it is only applied to individuals that are already known to be instances of A.

Dependency Directed Backtracking. Inherent unsatisfiability concealed in sub-expressions can lead to large amounts of unproductive backtracking search known as thrashing. For example, expanding the expression $(C_1 \sqcup D_1) \sqcap \ldots \sqcap (C_n \sqcup D_n) \sqcap \exists R.(A \sqcap B) \sqcap \forall R.\neg A$ could lead to the fruitless exploration of 2^n possible expansions of $(C_1 \sqcup D_1) \sqcap \ldots \sqcap (C_n \sqcup D_n)$ before the inherent unsatisfiability of $\exists R.(A \sqcap B) \sqcap \forall R.\neg A$ is discovered. This problem is addressed by adapting a form of dependency directed backtracking called *backjumping*, which has been used in solving constraint satisfiability problems [40].

Backjumping works by labelling concepts with a dependency set indicating the non-deterministic expansion choices on which they depend. When a clash is discovered, the dependency sets of the clashing concepts can be used to identify the most recent non-deterministic expansion where an alternative choice might alleviate the cause of the clash. The algorithm can then jump back over intervening non-deterministic expansions *without* exploring any alternative choices. Similar techniques have been used in first order theorem provers, e.g., the "proof condensation" technique employed in the HARP theorem prover [41].

4   Research Challenges for Ontology Reasoning

The development of the OWL language, and the successful use of reasoning systems in tools such as the Protégé editor [42], has demonstrated the utility of logic and automated reasoning in the ontology domain. The increasing use of DL based ontologies in areas such as e-Science and the Semantic Web is, however, already stretching the capabilities of existing DL systems, and brings with it a range of challenges for future research.

Scalability. Practical ontologies may be very large—tens or even hundreds of thousands of classes. Dealing with large-scale ontologies already presents a challenge to the current generation of DL reasoners, in spite of the fact that many existing large-scale ontologies are relatively simple. In the 40,000 concept Gene Ontology (GO), for example, much of the semantics is currently encoded in class names such as "heparin-metabolism"; enriching GO with more complex definitions, e.g., by explicitly modelling the fact that heparin-metabolism is a kind of "metabolism" that "acts-on" the carbohydrate "heparin", would make the semantics more accessible, and would greatly increase the value of GO by enabling new kinds of query such as "what biological processes act on glycosaminoglycan" (heparin is a kind of glycosaminoglycan) [43]. However, adding more complex class definitions can cause the performance of existing reasoners to degrade to the point where it is no longer acceptable to users. Similar problems have been encountered with large medical terminology ontologies, such as the GALEN ontology [44].

Moreover, as well as using a conceptual model of the domain, many applications will also need to deal with very large volumes of instance data—the Gene Ontology, for example, is used to annotate millions of individuals, and practitioners want to answer queries that refer both to the ontology and to the relationships between these individuals, e.g., "what DNA binding products interact with insulin receptors". Answering this query requires a reasoner not only to identify individuals that are (perhaps only implicitly) instances of DNA binding products and of insulin receptors, but also to identify which pairs of individuals are (perhaps only implicitly) instances of the interactsWith role. For existing ontology languages it is possible to use DL reasoning to answer such queries, but dealing with the large volume of GO annotated gene product data is far beyond the capabilities of existing DL systems [45].

Several different approaches to this problem are already under investigation. One of these involves the use of a hybrid DL-DB architecture in which instance data is stored in a database, and query answering exploits the relatively simple relational structure encountered in typical data sets in order minimise the use of DL reasoning and maximise the use of database operations [46]. Another technique that is under investigation is to use reasoning techniques based on the encoding of $\mathcal{SHIQ}$ ontologies in Datalog [47]. On the one hand, theoretical investigations of this technique have revealed that data complexity (i.e., the complexity of answering queries against a fixed ontology and set of instance data) is significantly lower than the complexity of class consistency reasoning (i.e., NP-complete for $\mathcal{SHIQ}$, and even polynomial-time for a slight restriction of $\mathcal{SHIQ}$) [48]; on the other hand, the technique would allow relatively efficient Datalog engines to be used to store and reason with large volumes of instance data.

Expressive Power. OWL is a relatively rich ontology language, but many applications require even greater expressive power than that which is provided by the existing OWL standard. For example, in ontologies describing complex physically structured domains such as biology [43] and medicine [44], it is often important to describe aggregation relationships between structures and their component parts, and to assert that certain properties of the component parts transfer to the structure as a whole (a femur with a fractured shaft is a fractured femur) [49]. The importance of this kind of knowledge can be gauged from the fact that various "work-arounds" have been described for use with ontology languages that cannot express it directly [50].

It may not be possible to satisfy all expressive requirements while staying within a decidable fragment of first order logic. Recent research has, therefore, studied the use in ontology reasoning of semi-decision procedures such as resolution based theorem provers for full first order logic [51]. There have also been studies of languages that combine a DL with some other logical formalism, often Datalog style rules, with the connection between the two formalisms being restricted so as to maintain decidability [52,47,53]

Extended Reasoning Services. Finally, in addition to solving problems of class consistency/subsumption and instance checking, explaining how such inferences are derived may be important, e.g., to help an ontology designer to rectify problems identified by reasoning support, or to explain to a user why an application behaved in an unexpected manner.

Work on developing practical explanation systems is at a relatively early stage, with different approaches still being developed and evaluated. One such technique involves exploiting standard reasoning services to identify a small set of axioms that still support the inference in question, the hope being that presenting a much smaller (than the complete ontology) set of axioms to the user will help them to understand the "cause" of the inference [54]. Another (possibly complementary) technique involves explaining the steps by which the inference was derived, e.g., using a sequence of simple natural deduction style inferences [55,56].

As well as explanation, so-called "non-standard inferences" could also be important in supporting ontology design; these include matching, approximation, and difference computations. Non-standard inferences are the subject of ongoing research [57,58,59,60]; it is still not clear if they can be extended to deal with logics as expressive as those that underpin modern ontology languages, or if they will scale to large applications ontologies.

5   Summary

Description Logics are a family of class based knowledge representation formalisms characterised by the use of various constructors to build complex classes from simpler ones, and by an emphasis on the provision of sound, complete and (empirically) tractable reasoning services. They have been used in a wide range of applications, but perhaps most notably (at least in recent times) in providing a formal basis and reasoning services for (web) ontology languages such as OWL.

The effective use of logic based ontology languages in applications will, however, critically depend on the provision of efficient reasoning services to support both ontology design and deployment. The increasing use of DL based ontologies in areas such as e-Science and the Semantic Web is, however, already stretching the capabilities of existing DL systems, and brings with it a range of challenges for future research. The extended ontology languages needed in some applications may demand the use of more expressive DLs, and even for existing languages, providing efficient reasoning services is extremely challenging.

Some applications may even call for ontology languages based on larger fragments of FOL. The development of such languages, and reasoning services to support them, extends these challenges to the whole logic based Knowledge Representation community.

Acknowledgements

I would like to acknowledge the contribution of those who provided me with inspiration and guidance, and the many collaborators with whom I have been privileged to work. These include Franz Baader, Sean Bechhofer, Dieter Fensel, Carole Goble, Frank van Harmelen, Carsten Lutz, Alan Rector, Ulrike Sattler, Peter F. Patel-Schneider, Stephan Tobies and Andrei Voronkov.

References

1. Calvanese, D., De Giacomo, G., Lenzerini, M., Nardi, D., Rosati, R.: Description logic framework for information integration. In: Proc. of the 6th Int. Conf. on Principles of Knowledge Representation and Reasoning (KR'98). (1998) 2–13
2. Calvanese, D., De Giacomo, G., Lenzerini, M.: On the decidability of query containment under constraints. In: Proc. of the 17th ACM SIGACT SIGMOD SIGART Symp. on Principles of Database Systems (PODS'98). (1998) 149–158
3. Horrocks, I., Tessaris, S., Sattler, U., Tobies, S.: How to decide query containment under constraints using a description logic. In: Proc. of the 7th Int. Workshop on Knowledge Representation meets Databases (KRDB 2000), CEUR (http://ceur-ws.org/) (2000)
4. Horrocks, I., Patel-Schneider, P.F., van Harmelen, F.: From $\mathcal{SHIQ}$ and RDF to OWL: The making of a web ontology language. J. of Web Semantics **1** (2003) 7–26
5. Knublauch, H., Fergerson, R., Noy, N., Musen, M.: The protégé OWL plugin: An open development environment for semantic web applications. In McIlraith, S.A., Plexousakis, D., van Harmelen, F., eds.: Proc. of the 2004 International Semantic Web Conference (ISWC 2004). Number 3298 in Lecture Notes in Computer Science, Springer (2004) 229–243
6. Liebig, T., Noppens, O.: Ontotrack: Combining browsing and editing with reasoning and explaining for OWL Lite ontologies. In McIlraith, S.A., Plexousakis, D., van Harmelen, F., eds.: Proc. of the 2004 International Semantic Web Conference (ISWC 2004). Number 3298 in Lecture Notes in Computer Science, Springer (2004) 244–258
7. Rector, A.L., Nowlan, W.A., Glowinski, A.: Goals for concept representation in the GALEN project. In: Proc. of the 17th Annual Symposium on Computer Applications in Medical Care (SCAMC'93), Washington DC, USA (1993) 414–418
8. Visser, U., Stuckenschmidt, H., Schuster, G., Vögele, T.: Ontologies for geographic information processing. Computers in Geosciences (to appear)
9. Oberle, D., Sabou, M., Richards, D.: An ontology for semantic middleware: extending daml-s beyond web-services. In: Proceedings of ODBASE 2003. (2003)
10. Wroe, C., Goble, C.A., Roberts, A., Greenwood, M.: A suite of DAML+OIL ontologies to describe bioinformatics web services and data. Int. J. of Cooperative Information Systems (2003) Special Issue on Bioinformatics.
11. Berners-Lee, T., Hendler, J., Lassila, O.: The semantic Web. Scientific American **284** (2001) 34–43
12. The DAML Services Coalition: DAML-S: Web service description for the semantic web. In: Proc. of the 2003 International Semantic Web Conference (ISWC 2003). Number 2870 in Lecture Notes in Computer Science, Springer (2003)
13. Uschold, M., King, M., Moralee, S., Zorgios, Y.: The enterprise ontology. Knowledge Engineering Review **13** (1998)
14. Stevens, R., Goble, C., Horrocks, I., Bechhofer, S.: Building a bioinformatics ontology using OIL. IEEE Transactions on Information Technology in Biomedicine **6** (2002) 135–141

15. Rector, A., Horrocks, I.: Experience building a large, re-usable medical ontology using a description logic with transitivity and concept inclusions. In: Proceedings of the Workshop on Ontological Engineering, AAAI Spring Symposium (AAAI'97), AAAI Press, Menlo Park, California (1997)

16. Spackman, K.: Managing clinical terminology hierarchies using algorithmic calculation of subsumption: Experience with SNOMED-RT. J. of the Amer. Med. Informatics Ass. (2000) Fall Symposium Special Issue.

17. Emmen, A.: The grid needs ontologies—onto-what? (2002) http://www.hoise.com/primeur/03/articles/monthly/AE-PR-02-03-7.html.

18. Tuecke, S., Czajkowski, K., Foster, I., Frey, J., Graham, S., Kesselman, C., Vanderbilt, P.: Grid service specification (draft). GWD-I draft , GGF Open Grid Services Infrastructure Working Group (2002) http://www.globalgridforum.org/.

19. Foster, I., Kesselman, C., Nick, J., Tuecke, S.: The physiology of the grid: An open grid services architecture for distributed systems integration (2002) http://www.globus.org/research/papers/ogsa.pdf.

20. Wolstencroft, K., McEntire, R., Stevens, R., Tabernero, L., Brass, A.: Constructing Ontology-Driven Protein Family Databases. Bioinformatics 21 (2005) 1685–1692

21. Horrocks, I., Tessaris, S.: Querying the semantic web: a formal approach. In Horrocks, I., Hendler, J., eds.: Proc. of the 2002 International Semantic Web Conference (ISWC 2002). Number 2342 in Lecture Notes in Computer Science, Springer-Verlag (2002) 177–191

22. Horrocks, I., Sattler, U., Tobies, S.: Practical reasoning for expressive description logics. In Ganzinger, H., McAllester, D., Voronkov, A., eds.: Proc. of the 6th Int. Conf. on Logic for Programming and Automated Reasoning (LPAR'99). Number 1705 in Lecture Notes in Artificial Intelligence, Springer (1999) 161–180

23. Horrocks, I., Sattler, U.: A tableaux decision procedure for $\mathcal{SHOIQ}$. In: Proc. of the 19th Int. Joint Conf. on Artificial Intelligence (IJCAI 2005). (2005) To appear.

24. Baader, F., Calvanese, D., McGuinness, D., Nardi, D., Patel-Schneider, P.F., eds.: The Description Logic Handbook: Theory, Implementation and Applications. Cambridge University Press (2003)

25. Baader, F., Hanschke, P.: A schema for integrating concrete domains into concept languages. In: Proc. of the 12th Int. Joint Conf. on Artificial Intelligence (IJCAI'91). (1991) 452–457

26. Blackburn, P., Seligman, J.: Hybrid languages. J. of Logic, Language and Information 4 (1995) 251–272

27. Horrocks, I., Sattler, U.: Ontology reasoning in the $\mathcal{SHOQ}$(D) description logic. In: Proc. of the 17th Int. Joint Conf. on Artificial Intelligence (IJCAI 2001). (2001) 199–204

28. Donini, F.M., Lenzerini, M., Nardi, D., Nutt, W.: The complexity of concept languages. Information and Computation 134 (1997) 1–58

29. Baader, F., Franconi, E., Hollunder, B., Nebel, B., Profitlich, H.J.: An empirical analysis of optimization techniques for terminological representation systems or: Making KRIS get a move on. Applied Artificial Intelligence. Special Issue on Knowledge Base Management 4 (1994) 109–132

30. Horrocks, I.: The FaCT system. In de Swart, H., ed.: Proc. of the 2nd Int. Conf. on Analytic Tableaux and Related Methods (TABLEAUX'98). Volume 1397 of Lecture Notes in Artificial Intelligence., Springer (1998) 307–312

31. Patel-Schneider, P.F.: DLP system description. In: Proc. of the 1998 Description Logic Workshop (DL'98), CEUR Electronic Workshop Proceedings, http://ceur-ws.org/Vol-11/ (1998) 87–89

32. Haarslev, V., Möller, R.: RACER system description. In: Proc. of the Int. Joint Conf. on Automated Reasoning (IJCAR 2001). Volume 2083 of Lecture Notes in Artificial Intelligence., Springer (2001) 701–705
33. Pan, J.Z.: Description Logics: Reasoning Support for the Semantic Web. PhD thesis, University of Manchester (2004)
34. Horrocks, I., Sattler, U., Tobies, S.: Practical reasoning for very expressive description logics. J. of the Interest Group in Pure and Applied Logic 8 (2000) 239–264
35. Bresciani, P., Franconi, E., Tessaris, S.: Implementing and testing expressive description logics: Preliminary report. In: Proc. of the 1995 Description Logic Workshop (DL'95). (1995) 131–139
36. Horrocks, I.: Using an expressive description logic: FaCT or fiction? In: Proc. of the 6th Int. Conf. on Principles of Knowledge Representation and Reasoning (KR'98). (1998) 636–647
37. Patel-Schneider, P.F.: DLP. In: Proc. of the 1999 Description Logic Workshop (DL'99), CEUR Electronic Workshop Proceedings, http://ceur-ws.org/Vol-22/ (1999) 9–13
38. Horrocks, I., Patel-Schneider, P.F.: Optimizing description logic subsumption. J. of Logic and Computation 9 (1999) 267–293
39. Horrocks, I., Tobies, S.: Reasoning with axioms: Theory and practice. In: Proc. of the 7th Int. Conf. on Principles of Knowledge Representation and Reasoning (KR 2000). (2000) 285–296
40. Baker, A.B.: Intelligent Backtracking on Constraint Satisfaction Problems: Experimental and Theoretical Results. PhD thesis, University of Oregon (1995)
41. Oppacher, F., Suen, E.: HARP: A tableau-based theorem prover. J. of Automated Reasoning 4 (1988) 69–100
42. Protégé: http://protege.stanford.edu/ (2003)
43. Wroe, C., Stevens, R., Goble, C.A., Ashburner, M.: A methodology to migrate the Gene Ontology to a description logic environment using DAML+OIL. In: Proc. of the 8th Pacific Symposium on Biocomputing (PSB). (2003)
44. Rogers, J.E., Roberts, A., Solomon, W.D., van der Haring, E., Wroe, C.J., Zanstra, P.E., Rector, A.L.: GALEN ten years on: Tasks and supporting tools. In: Proc. of MEDINFO2001. (2001) 256–260
45. Horrocks, I., Li, L., Turi, D., Bechhofer, S.: The instance store: DL reasoning with large numbers of individuals. In: Proc. of the 2004 Description Logic Workshop (DL 2004). (2004) 31–40
46. Bechhofer, S., Horrocks, I., Turi, D.: The OWL instance store: System description. In: Proc. of the 20th Int. Conf. on Automated Deduction (CADE-20). Lecture Notes in Artificial Intelligence, Springer (2005) To appear.
47. Hustadt, U., Motik, B., Sattler, U.: Reducing SHIQ-description logic to disjunctive datalog programs. In: Proc. of the 9th Int. Conf. on Principles of Knowledge Representation and Reasoning (KR 2004). (2004) 152–162
48. Motik, B., Sattler, U., Studer, R.: Query answering for OWL-DL with rules. In: Proc. of the 2004 International Semantic Web Conference (ISWC 2004). (2004) 549–563
49. Rector, A.: Analysis of propagation along transitive roles: Formalisation of the galen experience with medical ontologies. In: Proc. of DL 2002, CEUR (http://ceur-ws.org/) (2002)
50. Schulz, S., Hahn, U.: Parts, locations, and holes - formal reasoning about anatomical structures. In: Proc. of AIME 2001. Volume 2101 of Lecture Notes in Artificial Intelligence., Springer (2001)

51. Tsarkov, D., Riazanov, A., Bechhofer, S., Horrocks, I.: Using Vampire to reason with OWL. In McIlraith, S.A., Plexousakis, D., van Harmelen, F., eds.: Proc. of the 2004 International Semantic Web Conference (ISWC 2004). Number 3298 in Lecture Notes in Computer Science, Springer (2004) 471–485

52. Eiter, T., Lukasiewicz, T., Schindlauer, R., Tompits, H.: Combining answer set programming with description logics for the semantic web. In: Proc. of the 9th Int. Conf. on Principles of Knowledge Representation and Reasoning (KR 2004), Morgan Kaufmann, Los Altos (2004) 141–151

53. Rosati, R.: On the decidability and complexity of integrating ontologies and rules. J. of Web Semantics **3** (2005) 61–73

54. Schlobach, S., Cornet, R.: Explanation of terminological reason-ing: A preliminary report. In: Proc. of the 2003 Description Logic Workshop (DL 2003). (2003)

55. McGuinness, D.L.: Explaining Reasoning in Description Logics. PhD thesis, Rutgers, The State University of New Jersey (1996)

56. Borgida, A., Franconi, E., Horrocks, I.: Explaining $\mathcal{ALC}$ subsumption. In: Proc. of the 14th Eur. Conf. on Artificial Intelligence (ECAI 2000). (2000)

57. Baader, F., Küsters, R., Borgida, A., McGuinness, D.L.: Matching in description logics. J. of Logic and Computation **9** (1999) 411–447

58. Brandt, S., Turhan, A.Y.: Using non-standard inferences in description logics — what does it buy me? In: Proc. of KI-2001 Workshop on Applications of Description Logics (KIDLWS'01). Volume 44 of CEUR (http://ceur-ws.org/). (2001)

59. Küsters, R.: Non-Standard Inferences in Description Logics. Volume 2100 of Lecture Notes in Artificial Intelligence. Springer Verlag (2001)

60. Brandt, S., Küsters, R., Turhan, A.Y.: Approximation and difference in description logics. In: Proc. of the 8th Int. Conf. on Principles of Knowledge Representation and Reasoning (KR 2002). (2002) 203–214

Automated Reasoning in the Context of the Semantic Web

Hans Jürgen Ohlbach

Institute for Informatics, Ludwig-Maximilians University, Munich
ohlbach@lmu.de

"The Semantic Web is specifically a web of machine-readable information whose meaning is well-defined by standards" (Tim Berners-Lee in the foreword of the book "Spinning the Web"). This is a very simplified definition of the Semantic Web. The crucial part is the last word "standards". Since machine readable information in the web can be almost anything, the standards must also be about almost anything. Taken to the extreme, it requires a standardised model of the whole world, physical as well as conceptual, against which the information is interpreted. The world model must contain concrete data, for example the location of my office in Munich, as well as abstract relationships, for example, that an office is a room.

Abstract relationships like this are part of the Ontology layer of the Semantic Web tower. Many scientific communities in various sciences are currently busy building ontologies for their domains. These ontologies are in principle generalised and simplified world models. The preferred tool to build them is OWL, which is based on the Description Logic SHIQ. The two layers above the Ontology layer,

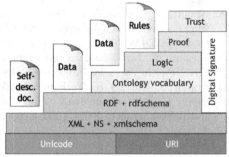

Semantic Web Tower

'Logic' and 'Proof', are the places where automated reasoning comes into play. Not much has been done so far in these areas. Instead there are many works about meta systems for the Semantic Web, in particular XML, XML query and transformation languages, rule systems (eg. RuleML), web service architectures, web service combination etc. These systems are meta systems in the sense that they can be used to represent and manipulate facts about the world, but they have not much built in knowledge about the world.

The next steps in these developments pose two grand challenges. The first challenge is to built more sophisticated world models which combine concrete data, algorithms, specialised logical calculi and constraint systems with abstract ontologies. The second challenge is to enrich the meta systems with the world models. Both require combination methods in the style of theory resolution or concrete domains for Description Logics.

Our group is involved in the EU Network of Excellence REWERSE (www. rewerse.net) where we develop the XML query language Xcerpt, together with 'geotemporal' and 'geospatial' world models.

B. Beckert (Ed): TABLEAUX 2005, LNAI 3702, p. 14, 2005.

Formal Versus Rigorous Mathematics: How to Get Your Papers Published*

Erik Rosenthal

Department of Mathematics, University of New Haven,
West Haven, CT 06516, USA
`erosenthal@newhaven.edu`

Abstract. This talk will consider rigorous mathematics and the nature of proof. It begins with an historical perspective and follows the development of formal mathematics. The talk will conclude with examples demonstrating that understanding the relationship between formal mathematics and rigorous proof can assist with both the discovery and the quality of real proofs of real results.

Keywords: *rigor, formal mathematics, nature of proof.*

1   Introduction

There was a symposium honoring Woody Bledsoe on his 70th birthday in Austin, Texas in 1991,[1] during which I had the pleasure of sharing a meal with Alan Robinson [15]. One of the topics that interrupted the repast was mathematical rigor and the role played in it by formalism. Alan's thoughtful comments provided me with an insight — perhaps even an epiphany — into the very nature of mathematics. I believe it made me both a better mathematician and a better teacher of mathematics. The purpose of this talk is to convey that insight; I hope I can do justice to Alan's wisdom.

Section 2 explores the history of mathematical rigor, ending with a discussion of modern standards of proof, and Section 3 discusses Alan's insight; both sections are expository. Section 4 looks at examples from [10] and is more technical.

There are many thousands of references for the material discussed in Section 2. The selected few are not necessarily the best sources; they are simply the ones I found useful. Let me call particular attention to the websites in the bibliography. My favorite is MacTutor [18], which is maintained by the University of St Andrews in Scotland.[2]

---

* I would like to thank David Rosenthal, Jeffrey Rosenthal, Peter Rosenthal, and Donald Sarason, who made numerous valuable suggestions. Neil Murray, never more than a phone call away when advice and insight were required, deserves special mention. Finally, there is no way to properly thank my wife Jean, who listened and made suggestions, did research, and provided much needed hugs at crucial moments.

[1] Woody died in 1995; his loss was deeply felt.

[2] I apologize to those readers who dislike footnotes; more than a few sprinkle these pages with their fine print.

B. Beckert (Ed): TABLEAUX 2005, LNAI 3702, pp. 15–32, 2005.

2 What Is Mathematical Rigor?

Our goal is to obtain some understanding of how formal definitions can assist with the discovery (creation?) of rigorous and (hopefully!) sound proofs in mathematics. To achieve that goal, we will explore the notion of rigorous proof. We begin with a simple question:

2.1 What Is Mathematics?

When students ask that question, I sometimes answer, "the study of functions." Not the study of sets? No. Sets serve as the language of mathematics, but we study functions: differentiable functions in calculus, continuous functions in topology, matrices in linear algebra, homomorphisms in group theory, binary operators in logic. It is useful for students to understand this, and it is a workable and, as it were, functional definition.[3] But that answer misses something. For one thing, some of what mathematicians investigate is not about functions. More importantly, it is a safe bet that no budding mathematician answered the question, "What do you want to do when you grow up?" with, "I want to study functions."

Humans seem to love puzzles, perhaps because we evolved to solve the problem, *How do you capture dinner if it is bigger, faster, and stronger than you are?*[4] Puzzles come in all sizes and shapes, whether it be finding the odd coin with a balance scale, determining the murderer in a whodunit, or answering an open question in mathematics. *Mathematical reasoning* may thus be defined as the process of using logical deduction to discover truth. Picking out the counterfeit coin, Nero Wolfe exposing the villain at a dinner party, and Andrew Wiles' proof of Fermat's Last Theorem are all examples of mathematical reasoning. Let us therefore define mathematics as that body of knowledge that uses mathematical reasoning to study, well, functions (and a few other things); in particular, to prove theorems about functions.

2.2 A Brief History of Rigor

Mathematics has been studied for millennia, and mathematical discoveries have been made independently in many parts of the world. The earliest developments had very practical roots, with counting leading the way.[5] It is in Classical Greece,

---

[3] For the record, though not necessarily relevant to this discussion, it is an abomination to define a function as a set of ordered pairs. That definition creates a misconception about functions, which in fact are assignments of elements between sets. The ordered pair definition yields the *graph of a function*, a related but nonetheless different animal altogether. The terms *set*, *element*, and *function* are the basic building blocks of mathematics and thus must be undefined. We provide intuition into their meaning with synonyms such as *collection, point, assignment*, but we cannot give precise definitions.

[4] The solution to that particular puzzle is *the American Express Card*.

[5] This may be evidence that lower back problems have plagued humans since before mathematics; otherwise, we would count with base twenty.

as far as we know, that mathematics evolves from a collection of practical techniques to a systematic body of knowledge [11]. Aristotle (384 BC – 328 BC) wrote *Analytica Posteriora* [2] around 340 BC; it formally introduced the notion of logical demonstration. Euclid's *Elements* [9], written around 280 BC, is most likely the first book on pure mathematics; it is certainly the most influential mathematics book in history. This is also the beginning of rigorous mathematics. Although there are a couple of errors in *Elements*, Euclid (325 BC – 265 BC) adheres to the standards of rigor set forth by Aristotle, and modern mathematicians use those same standards. That is not to say that mathematicians have never strayed from the path of rigor in the intervening years. Quite the contrary: As we shall see, when Europe emerged from the Middle Ages, rigor was not of primary importance to mathematicians.

The fall of the Roman Empire a few centuries after Euclid's treatise ushered in the Dark Ages,[6] and we leap ahead two millenia to 1450 and the invention of the printing press [8]. It is interesting to note that while, not surprisingly, the *Bible* was the first book printed with the new press, the first technical book was Euclid's *Elements*, published in 1482.[7] The Renaissance and the printing press are inextricably intertwined: The early Renaissance enabled the invention of the printing press, and the printing press in turn enabled the Renaissance to flourish.

The arts and natural philosophy[8] blossomed, the latter largely in support of the new technologies spurred by the new mercantilism. In the seventeenth and eighteenth centuries, the Renaissance becomes the Age of Reason, and the growth of natural philosophy continues.

Natural philosophy in this period is largely concerned with applications to technology, and mathematics is not considered a separate discipline. The effect is that rigor is less important than results. Below we will examine some weaknesses in the development of the calculus, but let us look at one example from Leonhard Euler (1707 – 1783) [6], probably the most prolific of all mathematicians.

Consider

$$\sum_{n=1}^{\infty} \frac{1}{n^2}.$$

This series is related to Bernoulli numbers and to the Riemann Zeta function. The Bernoulli brothers tried but failed to determine its sum; Euler succeeded. Euler's approach, based on two simple observations, was brilliant if outrageous.

---

[6] This is a Euro-centric point of view: Serious scholarship continued in many parts of the world. For example, algebra was developed in the Middle East, and Euclid's *Elements* came to Europe via the Middle East. But our concern is with mathematical rigor, and so historical rigor will have to play second fiddle after Rome burns.

[7] The 1482 edition, *Elementa Geometria*, a Latin translation from Arabic, was published in Venice.

[8] *Natural philosophy* included both mathematics and the sciences.

He observed first that if $a_1, a_2, ..., a_n$ are all the roots of a polynomial (repeated as necessary if there are multiplicities greater than one), then, for some constant C, the polynomial is the product

$$C(a_1 - x)(a_2 - x) \cdots (a_n - x).$$

If none of the roots is zero, then

$$p(x) = \left(1 - \frac{x}{a_1}\right)\left(1 - \frac{x}{a_2}\right) \cdots \left(1 - \frac{x}{a_n}\right)$$

is the unique polynomial with those roots and with $p(0) = 1$.

He next observed that the Maclaurin series for $\sin x$ is

$$x - \frac{x^3}{3!} + \frac{x^5}{5!} - \frac{x^7}{7!} + \ldots.$$

From this he concluded that[9]

$$\frac{\sin x}{x} = 1 - \frac{x^2}{3!} + \frac{x^4}{5!} - \frac{x^6}{7!} + \ldots.$$

Also, the roots of this function are precisely the non-zero roots of the sine function: $\{\pm\pi, \pm 2\pi, \pm 3\pi \ldots\}$. Treating $\frac{\sin x}{x}$ as he would any polynomial[10] and noting that it evaluates to 1 at 0,[11] it must also be the product

$$\left(1 - \frac{x}{\pi}\right)\left(1 + \frac{x}{\pi}\right)\left(1 - \frac{x}{2\pi}\right)\left(1 + \frac{x}{2\pi}\right)\left(1 - \frac{x}{3\pi}\right)\left(1 + \frac{x}{3\pi}\right) \cdots$$

$$= \left(1 - \frac{x^2}{\pi^2}\right)\left(1 - \frac{x^2}{4\pi^2}\right)\left(1 - \frac{x^2}{9\pi^2}\right) \cdots.$$

Multiplying out this latter product, we obtain

$$1 - x^2\left(\frac{1}{\pi^2} + \frac{1}{4\pi^2} + \frac{1}{9\pi^2} + \ldots\right) + \text{(terms containing higher powers of x)}.$$

Equating the coefficient of x^2 here with the coefficient in the Maclaurin expansion yields

$$\sum_{1}^{\infty} \frac{1}{\pi^2 n^2} = \frac{1}{3!},$$

whence

$$\sum_{1}^{\infty} \frac{1}{n^2} = \frac{\pi^2}{6}.$$

---

[9] He apparently was not terribly concerned about the possibility of dividing by 0.
[10] Why not?
[11] It made sense to Euler. In fact, he was correct.

How extraordinary: What does this series have to do with the circumference of a circle?

Euler may not have been all that rigorous, but he was brilliant.

The rigor of Aristotle was not restored until Gauss and Cauchy — more on that later.

2.3   The Tirade of Lord Bishop George Berkeley

Bishop George Berkeley (1685 – 1753) published *The Analyst* [3] in 1734. The title page includes, "A DISCOURSE Addressed to an Infidel Mathematician. WHEREIN It is examined whether the Object, Principles, and Inferences of the modern Analysis are more distinctly conceived, or more evidently deduced, than Religious Mysteries and Points of Faith."

The infidel in question may have been Isaac Newton (1643 – 1727), but Newton had died seven years earlier, and he had become quite religious at the end of his life. Another possibility is Edmond Halley (1656 – 1742). More likely, Berkeley had no particular infidel in mind; rather, he was simply responding to "reasoned" attacks on religious faith. The primary argument in *The Analyst* is that Analysis — the Calculus — requires as much faith as religion. He was quite the mathematician himself: His criticisms are remarkably well thought out, reminiscent of a careful referee of a paper submitted to a journal. It is worth our while to understand the logic of his argument.

Calculus [7] is concerned with two problems: finding the slope of tangent lines to a curve, and finding the area under a curve. The solution to the first is called *differentiation* — Newton called it *fluxions* for *rate of flow* — and the solution to the second is called *integration* — Newton called it *fluents* because he thought of it as flow. Newton was not the first to consider either of these processes: Archimedes (287 BC – 212 BC) integrated and Pierre de Fermat (1601 – 1665) differentiated. The primary reason that Newton and Leibniz (Gottfried Wilhelm Leibniz (1646 – 1716)) are considered to have discovered Calculus is that they recognized what we now call *the Fundamental Theorem of Calculus*, which states that differentiation and integration are inverse processes.[12]

We need to understand how Newton calculated his fluxions to understand the gaps in rigor discovered by Berkeley. To find the slope of a tangent line, Newton looked at secant lines that are close to the tangent line — see Figure 1. Assume we want the tangent at x_0. Choose x_1 close to x_0 and look at the secant between the points $(x_0, f(x_0))$ and $(x_1, f(x_1))$. Newton called the increment from x_0 to x_1 an *infinitesimal*, a non-zero, non-negative number that is less than any positive number. Then the infinitesimal is $\mathcal{O} = x_1 - x_0$, and the slope of the secant is

---

[12] This is an extraordinarily important result: Most scientific advances of the last three centuries depend on it. There is little doubt that the technology used to build those pesky electronic devices that students use to avoid understanding the Fundamental Theorem could not have been invented without it.

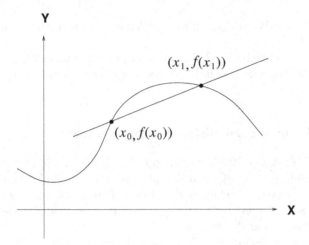

Fig. 1. Estimating Slopes of Tangent Lines

$$\frac{f(x_1) - f(x_0)}{x_1 - x_0} = \frac{f(x_0 + \mathcal{O}) - f(x_0)}{\mathcal{O}},$$

which is a good estimate of the slope of the tangent if x_1 is close enough to x_0.

Newton dealt with specific functions rather than the general notion of function, so let us see how he would find the slope of $f(x) = x^2$. For that function, the slope of the secant line is

$$\frac{f(x_0 + \mathcal{O}) - f(x_0)}{\mathcal{O}} = \frac{(x_0 + \mathcal{O})^2 - x_0^2}{\mathcal{O}} = \frac{2x_0\mathcal{O} + \mathcal{O}^2}{\mathcal{O}} = \frac{\mathcal{O}}{\mathcal{O}}(2x_0 + \mathcal{O}) .$$

Since the infinitesimal $\mathcal{O}$ is not 0, we can cancel, so the slope of the secant is $2x_0 + \mathcal{O}$. Now, to calculate the slope of the tangent line, let $\mathcal{O}$ be 0, and voila: The slope is $2x_0$.

Wait one second, Bishop Berkeley screamed. Referring to infinitesimals as "the ghosts of departed quantities," he argued that if $\mathcal{O}$ is 0, you cannot cancel — indeed, the expression for the slope of the secant is undefined. On the other hand, if $\mathcal{O}$ is not 0, how can you set it equal to zero?

2.4   Mathematicians' Response to Bishop Berkeley

Berkeley's complaint resounded in the mathematical community. By the end of the eighteenth century, led by Karl Friedrich Gauss (1777 – 1855) and Augustin Louis Cauchy (1789 – 1857), mathematicians restored Aristotle's standards of rigor. The gaps in the development of the Calculus were filled in by Bernhard Bolzano (1781 – 1848), Cauchy, Karl Weierstrass (1815 – 1897), and Bernhard

Riemann (1826 – 1866) [7]. The key to their success was to eliminate infinitesimals by making *limits* rigorous. The limit concept provides a perfect example of the central theme of this talk — of what Alan Robinson was trying to convey to me in Austin — see Section 3.

There were very rich developments in mathematics in the nineteenth century following the "rigorization" of limits. Another great advance was the set theory[13] [5] of Georg Cantor (1845 – 1918). There seemed to be no limits[14] to mathematics; the achievements of the nineteenth century were surely unimaginable to Newton and Leibniz.

Then a young Bertrand Russell (1872 – 1970) wrote a letter [17] to Gottlob Frege (1848 – 1925), a leading logician, in 1902.

That letter contained the now famous *Russell's paradox*: Let $\mathcal{K}$ be the set defined by $\mathcal{K} = \{A | A \notin A\}$. A small difficulty arises with the question, is $\mathcal{K} \in \mathcal{K}$? Russell's paradox formalizes *the liar's paradox — This sentence is false —* but is far more serious for mathematics since the liar's paradox can be attributed to inconsistencies in natural language. Note that *the truth teller's paradox — This sentence is true —* is also a problem. The liar's paradox is neither true nor false, and the truth teller's paradox is both true and false. It is interesting to observe that the same duality exists with Russell's paradox: If $\mathcal{M} = \{A | A \in A\}$, then it appears to be logically valid to answer the question, is $\mathcal{M} \in \mathcal{M}$? with both yes and no. This is not surprising: The paradox arises from self-reference, and that exists in both forms.

Ernst Zermelo (1871 – 1953) axiomatized set theory in 1908, and Abraham Fraenkel (1891 – 1965) and Thoralf Skolem (1887 – 1963) made significant additional contributions in 1922. The result was what we now call *Zermelo-Fraenkel set theory* [5]. As far as we know, the Zermelo-Fraenkel axioms are consistent; in particular, Russell's paradox does not arise.

But Russell's paradox raised fundamental questions about mathematics. Many mathematicians were concerned, and David Hilbert (1862 – 1943), one of the most prominent mathematicians of his day, put forward a proposal that has come to be known as *Hilbert's Program* [20] to investigate the foundations of mathematics. It called for formalizing mathematics in axiomatic form and for a proof that the resulting axiomatization is consistent. Kurt Gödel (1906 – 1978) put an end to that dream with his extraordinary 1931 paper [14,16]: He proved that any logical system that includes Peano's axiomatization of the natural numbers could not be both consistent and complete. To make matters worse, he also proved that such a system could not prove itself consistent without proving itself inconsistent!

A common misconception is that Paul Cohen's 1963 result [4] that the continuum hypothesis is independent of the Zermelo-Fraenkel axioms is an example of

---

[13] Cantor's work on sets was surprisingly controversial. At times, he was bitterly attacked by other mathematicians. The worst came from Leopold Kronecker (1823 – 1891). Cantor suffered from depression; many say it was brought on by Kronecker.

[14] Pun intended?

Gödel's incompleteness theorem.[15] That any particular axiom system be incomplete is not surprising. One easy way to find an incomplete system is to remove one axiom from a given system (unless the removed axiom is not independent of the others). What Gödel's theorem tells us is that *every* consistent axiom system that contains the integers must be incomplete. Thus, for example, adding the continuum hypothesis to the Zermelo-Fraenkel axioms cannot create a complete system (unless it is inconsistent).

2.5 From Gödel to $\mathcal{NP}$

The comments in this section are based entirely on my own observations and intuition.

Mathematicians learned to live with Gödel's results — what choice did they have? One professor of mine dismissed the issue by saying, "Zermelo-Fraenkel set theory avoids the known paradoxes but not the rest." Most mathematicians were confident that their systems were in fact consistent, and the post World War II era saw a boom in mathematics. Partly in reaction to the Russell-Gödel period and partly in reaction to computers, at which mathematicians looked down their noses, "pure mathematics" became the thing. Although it was rarely stated explicitly, many mathematicians thought that real applications somehow lessened the value of their work.[16] They were especially disdainful of computer science.

Attitudes towards computer science began to change in the 1980's after the question, does $\mathcal{NP} = \mathcal{P}$ emerged. Although I cannot prove this, I am certain that a number of mathematicians said, "Those idiot computer people. I'll show 'em how to handle $\mathcal{NP}$." When they discovered that the question was a wee bit more difficult than they had expected, they (albeit begrudgingly) developed some respect for computer science. Slowly they began to recognize the substance inherent in all of computer science. Then they discovered $\tau_\epsilon\chi$ and it was all over.

Our discussion so far has included an extensive examination of rigor without addressing the question:

2.6 What Is a Proof?

This question has been asked by mathematicians and philosophers alike (if there is a difference) and by students of mathematics. It should be noted that when members of the latter group ask the question, it is because they are baffled,

---

[15] The continuum hypothesis states that there are no uncountable sets that are smaller than the set of real numbers. This is not very satisfactory: Given the size difference between the reals and the integers, there ought to be such sets. But Cohen's result tells us that if there are any such subsets of the real numbers, their cardinality cannot be determined.

[16] Paul Halmos is purported [18] to have said, "Applied mathematics will always need pure mathematics just as anteaters will always need ants."

whereas when members of the former group ask the question, they think they know the answer.[17] The following dialog was recently overheard in the hallowed halls of the University of New Haven.

SM[18]: Professor, what is a proof?

MP[19]: My dear boy. A proof is a sequence of logically deduced steps leading inevitably from the hypothesis of the theorem to the conclusion.

SM: Yes, sir, but, well, today's lecture was wonderful...

MP: Thank you. We aim to please.

SM: Yes, sir. But, you see, I had some difficulty understanding the proof of Green's Theorem. I mean, I could not always tell why one step logically followed from previous steps.

MP: Naturally, these steps could be broken down into smaller steps, each of which can be so clearly inferred from the previous steps that there is no controversy. Indeed, Bertrand Russell successfully did just that once or twice.

SM: I see, sir. But why don't we do that with all theorems?

MP: Do you think a mathematician has the time to spend on such tedious twaddle? My dear boy, ...

Our good friend MP did capture the feelings of most mathematicians. His "tedious twaddle" is in fact the notion of proof that came out of the Hilbert Program [20]: A *Hilbert Proof* of a theorem is a sequence of statements, each of which is an hypothesis of the theorem, an axiom or definition, a (previously proved) theorem, or a consequence that is deduced from previous statements with the application inference rules. Hilbert also required a proof checker — a verification algorithm. Since the Hilbert Program was BC,[20] they probably did not expect the verification to be carried out by computers but rather by the next best thing.[21]

---

[17] Mathematicians, arrogant creatures that they are, think they know everything, except possibly for the solutions to a few open problems, such as the Riemann hypothesis or the question, does $\mathcal{NP} = \mathcal{P}$? Inexplicably, in light of the failures of so many really smart people, most actually expect, or at least hope, to solve some of these problems.

[18] Student of Mathematics.

[19] Mathematician/Philosopher.

[20] Before Computers.

[21] Graduate students.

I am unaware of any examples of Hilbert Proofs except for the most trivial.[22] In fact, the proofs that mathematicians actually write are not materially different from Euclid's. Define an *Aristotelian Proof* of a theorem to be a sequence of statements that begins with the hypothesis and ends with the conclusion, in which each statement is logically deduced from axioms, definitions, and the previous statements. The practical difference between *logically deduced* and *deduced with the application inference rules* is of course enormous. Modern mathematicians also require a verification algorithm; they call it *peer review*. The obvious question is, "Is peer review all that reliable?" Most mathematicians would answer, "Yes." They would add that this is especially true of important results. That is, some incorrect proofs may escape a referee, but not when the result is significant. Wiles' proof of Fermat's Last Theorem, for example, has been carefully checked by many mathematicians, and we can be sure of its correctness.

Sounds a bit like Bishop Berkeley's blind faith, doesn't it?

3   Alan's Insight: Limits

Cauchy was probably the first mathematician to treat limits rigorously. In doing so, he filled in the gaps that Berkeley had found. Like Newton and Leibniz, Cauchy started with the slope of a secant as an estimate of the slope of the tangent, but he avoided infinitesimals. The notation we use here is modern, but it captures Cauchy's thinking. First, use h in place of $\mathcal{O}$ for the increment from x_0 to x_1 (see Figure 1) to avoid the suggestion of infinitesimals. If

$$m(h) = \frac{f(x_0 + h) - f(x_0)}{h},$$

then $m(h)$ is the slope of the secant, and h can never be 0; i.e., the point $(x_1, f(x_1))$ on the curve must be different from the point $(x_0, f(x_0))$. But if h is near 0, i.e., if the two points are close to each other, then the slope of the secant should be close to the slope of the tangent. What Cauchy did was to provide a mathematical definition of *close*.

Let ϵ be the error in the slope estimate — i.e., the absolute value of the difference between the slope of the tangent and the slope of the secant. Cauchy's idea was that ϵ can be made arbitrarily small if h is close enough to (but not equal to!) 0. More generally, Cauchy defined

$$\lim_{x \to a} f(x) = L \tag{1}$$

to mean that $f(x)$ can be made as close to L as desired by making x close enough to (but not equal to) a.

Cauchy's presentation was rigorous, but the notation introduced by Weierstrass a few years later is better than Cauchy's. He defined expression (1) as follows:

---

[22] There are of course non-trivial examples of Hilbert Proofs generated by computer.

Given any $\epsilon > 0$, there exists a $\delta > 0$ such that $0 < |x - a| < \delta$ implies $|f(x) - L| < \epsilon$.

How elegant! The Weierstrass definition is so perfect that we still use his notation, and an enormous body of mathematics has been developed from it. It generalizes very easily. For example, to find the limit of a sequence $\{a_n\}$, we change the condition $|x - a| < \delta$ to a condition on the subscript: Define $\lim_{n \to \infty} a_n = L$ by

Given any $\epsilon > 0$, there exists an N such that $n > N$ implies $|g(x) - L| < \epsilon$.

Let us consider one example, $\lim_{x \to 0} x/x = 1$. Verifying this limit justifies the canceling done by Newton in most of his fluxion calculations. But this is almost trivial using the Weierstrass definition: Given $\epsilon > 0$, choose any positive δ. If $0 < x < \delta$, then $x/x = 1$, so $|x/x - 1| = 0 < \epsilon$.

Consider the journey that led to the Weierstrass definition. First came the idea of using slopes of secants to estimate the slope of the tangent, expecting that the limiting case would produce the actual slope of the tangent. That was followed by a clear intuitive definition of limit: $f(x)$ will be arbitrarily close to L if x is close enough to (but not equal to) a. Only then was the formal definition of limit possible. This brings us to the crux of the matter: What is the *real* definition of limit? For the answer, we return to lunch with Alan [15]:

Robinson's Clarity.[23] The ideas and concepts of mathematics are intuitive. Formal definitions are necessary for the rigorous development of mathematics, *but they are useful only if they capture the intuitive concepts.*

The real definition of limit is therefore the intuitive one, not the formal one: The Weierstrass epsilon-delta definition was crucial for the rich body of mathematics that followed, but *we use it because it captures the intuitive definition of limit precisely.*

Consider four possible definitions of $\lim_{x \to a} f(x)$ ($\ni$ is used for *such that*).

1. $(\forall \epsilon > 0)\, (\exists \delta > 0) \ni (0 < |x - a| < \delta) \Rightarrow (|f(x) - L| < \epsilon)$.
2. $(\exists \delta > 0) \ni (0 < |x - a| < \delta) \Rightarrow (f(x) = L)$.
3. $(\forall \epsilon > 0)\, (\exists \delta > 0) \ni (|x - a| < \delta) \Rightarrow (|f(x) - L| < \epsilon)$.
4. $(\forall \epsilon > 0)\, (\exists \delta > 0) \ni (\exists x, 0 < |x - a| < \delta) \Rightarrow (|f(x) - L| < \epsilon)$.

The first is the Weierstrass definition (using modern notation). In the second, the "epsilon condition" has been changed so that f is not just close to L, it is equal to L. This would be a perfectly workable definition, and it does yield $\lim_{x \to 0} x/x = 1$, solving Newton's cancellation difficulty. But it is not sufficiently general. In the third, the condition $x \neq a$ has been removed from the "delta condition." This is essentially the definition of continuous function and is quite useful, but it reintroduces the division-by-zero problem in the definition of

---

[23] The rigorous formulation of his intuition?

derivative. The last definition amounts to saying that f is frequently close to L if x is close to a. Some authors call such an L a *cluster point* of f, but we want limit to mean that $f(x)$ is *always* close to L if x is close enough to a.

Here is the point: Any of these definitions makes mathematical sense and would admit serious mathematical developments. Indeed, more than one PhD dissertation has been written about the last two. *But the first captures what we really mean by limit, and the others do not.* The formal definition comes from the intuitive definition, not the other way around.[24] At the same time, the intuitive definition is not enough; we do need the formal definition for rigorous proofs.

Perhaps Footnote 3, in which it was suggested that it might not be ideal to define a function from A to B as a subset of $A \times B$ in which each element of A appears exactly once, is relevant after all. This is a precise definition of the graph of a function, but it violates Robinson's Clarity because it does not capture what we really mean by function, namely an assignment of points in A to points in B. It is true that the "assignment definition" leaves function as an undefined term, but if we go with the "graph definition," then we misuse the word *function* every time we use it. Indeed, any mathematician who wanted to talk about the set of ordered pairs related to a function would say *graph of the function*, because the word function does not convey that meaning!

There is a practical side to Robinson's Clarity as well. It is not uncommon to have good intuition that a conjecture is a theorem but not be able to find a precise proof. Applying Robinson's Clarity sometimes provides the answer. In the next section, two theorems from [10] are described whose proofs were eventually[25] made rigorous using Robinson's Clarity.

We close this section by noting that many mathematicians believe there is a kind of converse to Robinson's Clarity: A good proof should provide intuition about the theorem.

4   How to Get Your Papers Published

The paper, *Completeness for Linear, Regular Negation Normal Form Inference Systems* [10], co-authored with Reiner Hähnle and Neil Murray, required three revisions[26] before it was accepted for publication. Both referees read it carefully and made numerous valuable suggestions for improvement.[27] The referees found gaps in several proofs. In this section, we will look at two of them and see how those gaps were filled with applications of Robinson's Clarity.

A significant part of that paper is concerned with completeness of inference systems that employ negation normal form,[28] so a definition of NNF is probably

---

[24] Unfortunately, the "other way around" happens far too often in the classroom.

[25] The original submitted proofs were vague, and their eventual rigor depended on good referees pointing out the gaps.

[26] The last was pretty minor, but three sounds more dramatic than two.

[27] One in particular was very thorough. If you happen to be that referee, please know that your efforts are very much appreciated.

[28] You may have guessed that from the title.

in order: A logical formula is in *negation normal form* (NNF) if $\wedge$ and $\vee$ are the only binary connectives, and if all negations are at the atomic level. Negation normal form differs from the more normalized *conjunctive* and *disjunctive* normal forms (CNF and DNF) in that NNF allows any degree of nesting of operators.

Several of the proofs employ generalizations of the Anderson-Bledsoe excess literal argument, which was developed for resolution. A key to their proof is the removal of a literal from one clause, say C. The induction hypothesis then provides a resolution proof of the resulting clause set. The removed literal is then restored to C and *to every clause in the proof that descends from C*. See [1] or [10] for the details. This process is straightforward in the clause setting, but quite another matter in NNF. The next theorem and proof, which are exactly what appeared in the original submitted version of the paper, generalize the Anderson-Bledsoe technique to NNF. The referees felt (correctly!) that removal and restoration of literals was less transparent in NNF. The fix was to nail down that process; i.e., to apply Robinson's Clarity. The reader may choose to skip ahead, rather than to slog through a gap-filled proof that depends on lemmas and definitions available only in the original paper.

Theorem 5.[29] The tableau method is a complete refutation procedure for unsatisfiable sets of (ground) NNF formulas.

Proof. Let $\mathcal{S}$ be an unsatisfiable set of ground NNF formulas; we proceed by induction on the number n of distinct atoms in $\mathcal{S}$.

If $n = 0$, then $\mathcal{S} = \mathit{false}$, and there is nothing to prove; so assume the theorem holds for all formulas with at most n distinct atoms, and let $\mathcal{S}$ be a set of formulas with $n + 1$ distinct atoms.

Suppose first that one of the formulas in $\mathcal{S}$ is the unit p. After the alpha rule is applied to the initial tableau, there is a single branch, and that branch is open and contains a node labeled p.

Otherwise, let p be an atom in $\mathcal{S}$, and remove from $\mathcal{S}$ all occurrences of $CE(p)$. Applying the Pure Rule (Lemma 5) to the resulting unlinked occurrences of $\bar{p}$ removes the d-extensions of all occurrences of $\bar{p}$; let $\mathcal{S}_p$ be the set of formulas produced. By Lemma 7, $\mathcal{S}_p$ is unsatisfiable, so by the induction hypothesis, there is a proof T_p for $\mathcal{S}_p$. Let T_p' be the tableau tree produced by applying each extension in T_p to the corresponding formulas in $\mathcal{S}$. If T_p' is closed, we are done. If not, then all open branches in T_p' must result from formulas containing p or $\bar{p}$.

A formula that contains p is a formula of $\mathcal{S}$ from which the c-extension of p was removed to produce the corresponding formula of $\mathcal{S}_p$. That c-extension must occur in a disjunction since a conjunction would be part of the c-extension. When such a disjunction is used for extension, an open branch containing p emerges — one that is simply not in T_p. The open branch will contain $CE(p)$, which is either the unit p or a conjunction that automatically produces the unit p on the branch by the alpha rule.

Consider now extensions by formulas containing $\bar{p}$. Recall that in the completeness proof for linear resolution for formulas in clause form (Theorem 2),

---

[29] The theorem number is from [10].

the clauses containing $\bar{p}$ were never used in the proof provided by the induction hypothesis. Such clauses are in effect removed by the Pure Rule since $DE(\bar{p})$ is precisely the clause containing $\bar{p}$. In the non-clausal case, $DE(\bar{p})$ is also deleted, but this in general is a *proper* subformula of the formula containing $\bar{p}$. In that case, $DE(\bar{p})$ is always an argument of a conjunction. Let $DE_c(\bar{p})$ be the conjunction, and let $DE'_c(\bar{p})$ be $DE_c(\bar{p})$ with $DE(\bar{p})$ removed. Thus, $DE_c(\bar{p}) = DE'_c(\bar{p}) \wedge DE(\bar{p})$. If $DE'_c(\bar{p})$ is used in an extension in T_p, then, when $DE_c(\bar{p})$ is extended in T'_p, $DE(\bar{p})$ is conjoined to $DE'_c(\bar{p})$. In particular, there are no new branches and therefore no open branches due to the formulas of $\mathcal{S}$ that contain $\bar{p}$. Hence, all open branches in T'_p contain nodes labeled p.

Similarly, by deleting occurrences of $CE(\bar{p})$ to produce the set of formulas $\mathcal{S}_{\bar{p}}$, the induction hypothesis provides a proof $T_{\bar{p}}$ of $\mathcal{S}_{\bar{p}}$ and a corresponding proof tree $T'_{\bar{p}}$ of $\mathcal{S}$. The leaves of all open branches of $T'_{\bar{p}}$ are labeled $\bar{p}$. Finally, to obtain a closed tableau from $T'_{\bar{p}}$ and T'_p, apply the steps of $T'_{\bar{p}}$ along each open branch (containing $\{p\}$) of T'_p. Observe that any branch not closed by (the steps of) $T'_{\bar{p}}$ or by T'_p has nodes labeled p and labeled $\bar{p}$ and thus can be closed.   □

The weakness in this proof stems from the structure of NNF formulas, which is considerably more complex than that of CNF or DNF formulas. In particular, the effect of removal and replacement of literals is transparent in the Anderson-Bledsoe proof but far from obvious in an NNF tableaux. As a result, a brief description of NNF is required for this discussion; more can be found in [10], and a detailed analysis can be found in [12]. We need two concepts here: *fundamental subgraph* and *full block*.

Given a logical formula $\mathcal{F}$ in NNF, think of the connectives as being n-ary, so that, for example, the n-ary representation of $(A \wedge B) \wedge C$ is $A \wedge B \wedge C$. Then a fundamental subgraph of $\mathcal{F}$ is one of the conjuncts of a conjunction or one of the disjuncts of a disjunction. To be clear, this means that a fundamental subgraph of a conjunction must itself be either a literal or a disjunction. A full block amounts to several fundamentals of a single conjunction (disjunction) conjoined (disjoined) together.

There were two keys to filling the gaps in the proof of Theorem 5. The first was to carefully describe the structure of NNF tableaux (i.e., tableaux developed for NNF formulas); the second was precise (and rigorous!) rules for removing and replacing full blocks from an NNF tableaux.

The structure of an NNF tableau differs from that of a CNF tableau in one crucial respect: Internal nodes in a CNF tableau are always branch points; this is not true for NNF tableaux because the nesting of operators necessitates the use of alpha rules to develop them. The set of tableau nodes produced by application of an alpha rule is called a *segment* in [10]. The meaning of segment is intuitively clear, sufficiently so that the term was not introduced in the submitted version. But segments are not completely transparent, primarily because "the set of tableau nodes produced by application of an alpha rule" depends on knowledge of how the tableau was developed. As one referee pointed out, such definitions should depend only on the structure of the tableau. The following is taken directly from the final version of the paper:

It will be necessary to deal with the set of nodes produced by an alpha extension, and so we define the *segment* of a node N on a branch Θ as follows: If N is a branch point or a leaf, then the segment of N is the set consisting of N plus the nodes on Θ above N but below (and not including) the first branch point above N. If N is not a branch point or a leaf, then the segment of N is the segment of the branch point or leaf immediately below N. Note that the head of the tableau is the segment of the first branch point (or of the leaf if there are none). Note also that each branch point must have at least two segments immediately below it.

This definition is not entirely trivial, but it is a perfect example of Robinson's Clarity: It captures precisely the intuitive notion of segment. It does depend on several terms — *branch, branch point, leaf, head* — all are carefully defined in the paper.

The Tableaux Removal-Replacement Rules provide another example of Robinson's Clarity. They make rigorous the process of removing a full block from a formula, constructing a tableau proof tree from the resulting formula, and then replacing the removed full block to construct a tableau for the original formula. These rules made filling in the gaps in Theorem 5 straightforward — see [10] for the final proof.

Tableaux Removal-Replacement Rules. Let $\mathcal{F}'$ be the formula produced when a full block $\mathcal{E}$ is removed from a formula $\mathcal{F}$. Our goal is to create from a sequence S' of tableau steps applied to $\mathcal{F}'$ a corresponding sequence S of tableau steps for $\mathcal{F}$. The full block $\mathcal{E}$ is either part of a conjunction or a disjunction. (Were $\mathcal{E}$ all of $\mathcal{F}$, $\mathcal{F}'$ would be empty and there would be nothing to do.) Specifically, let $\mathcal{G} = \mathcal{E} \wedge \mathcal{G}'$, where $\mathcal{G}$ is a fundamental subgraph of a disjunction, or let $\mathcal{G} = \mathcal{E} \vee \mathcal{G}'$, where $\mathcal{G}$ is a fundamental subgraph of a conjunction. Note that $\mathcal{F}'$ is $\mathcal{F}$ with $\mathcal{G}$ replaced by $\mathcal{G}'$; consider a sequence S' of tableau steps applied to $\mathcal{F}'$. Observe that any element of S' that leaves $\mathcal{G}'$ intact can be directly applied to $\mathcal{F}$ (using $\mathcal{G}$ in place of $\mathcal{G}'$). Whether $\mathcal{G}$ be a conjunction or a disjunction, $\mathcal{G}'$ may be a conjunction or a disjunction.

RR1. Suppose first that $\mathcal{G} = \mathcal{E} \wedge \mathcal{G}'$.

   a. If $\mathcal{G}'$ is a conjunction, it is unaffected by S' unless it is alpha extended. The corresponding step in S is to alpha extend $\mathcal{G}$. Any further step in S' is directly applicable and thus is identical in S.

   b. If $\mathcal{G}'$ is a disjunction or a single literal, then it is part of a larger disjunction $\mathcal{D}'$ in $\mathcal{F}'$ and is unaffected by S' unless $\mathcal{D}'$ is beta extended. Let $\mathcal{D}$ be the corresponding disjunction in $\mathcal{F}$: i.e., $\mathcal{D}$ contains $\mathcal{G}$ instead of $\mathcal{G}'$. Three steps are necessary in S. First, $\mathcal{D}$ is beta extended, and $\mathcal{G}$ becomes a single leaf instead of the several leaves created by the disjuncts of $\mathcal{G}'$. Secondly, there is an automatic alpha extension that creates a leaf segment with two (or more if $\mathcal{E}$ is a conjunction) nodes, labeled $\mathcal{E}$ and $\mathcal{G}'$. Thirdly, $\mathcal{G}'$ is beta extended. Observe that an extra branch point has been created, but there are no new leaves.

RR2. Suppose now that $G = \mathcal{E} \vee G'$.

    a. If G' is a disjunction, it is unaffected by S' unless it is beta extended. The corresponding step in S is to beta extend G, creating an extra leaf, labeled $\mathcal{E}$ (or extra leaves, labeled with the disjuncts of $\mathcal{E}$ if $\mathcal{E}$ is a disjunction). Any further step in S' is directly applicable and thus is identical in S.

    b. If G' is a conjunction or a single literal, then it is part of a larger conjunction C' in $\mathcal{F}'$ and is unaffected by S' unless C' is alpha extended. Two steps are necessary in S. First, the corresponding alpha extension creates G as a single node in a leaf segment instead of the several nodes created by the conjuncts of G'. Secondly, G is beta extended. Observe that an extra branch point has been created, and there is a new leaf, labeled $\mathcal{E}$ (or several, if $\mathcal{E}$ is a disjunction).

These rules appear as a remark in [10]; the remark amounts to a definition of the rules and a lemma that says they work. That they do indeed work may not be all that easy to see, but the point of Robinson's Clarity is not to make mathematics easy — that will never happen. But these rules do make the removal/replacement process precise, and it is easy to see how the rules operate. The proof of Theorem 5 that appeared is also not that easy, but formalizing the notion of segment and well defining the Removal-Replacement Rules led to a rigorous proof.

We conclude with a discussion of one of the more interesting results (and probably the most difficult) from [10].

Theorem 7. The tableau method restricted to regular u-connected tableaux, free of unit extensions, is a complete refutation procedure for unsatisfiable sets of (ground) NNF formulas.

Connectivity for the tableau method is analogous to linearity for resolution, the idea being always to use the most recent inference. Although regularity is also in the theorem, it will not be discussed here because connectivity is far more interesting. For a CNF tableau, connectivity is defined as follows: Two clauses C_1 and C_2 are *connected* if there is a link (complementary pair of literals) consisting of one literal from each clause. Two nodes N_1 and N_2 in a tableau proof tree are connected if they are on the same branch and the clauses labeling them are connected. A tableau is *weakly connected* if for any branch point B, at least one child N of B is connected to B or to a node above it, and, if B is not the first branch point, the connection is to a unit. The tableau is *connected* if it is weakly connected and if one child of each branch point is connected either to the branch point or to a node in the head.

This definition easily generalizes to NNF tableaux, but Theorem 7 does not hold for such connectivity. What is required is u-connectivity: A tableau is defined to be *u-connected* if for each branch point B there is a node N in one of the segments immediately below B that is either connected to the segment of B or has the following property: Each connection from N to a node above it is to a unit, and there is at least one such connection. An (alpha or beta) extension is *u-connected* if the tableau it produces is u-connected. Observe that a connected tableau with connections to the head need not be u-connected.

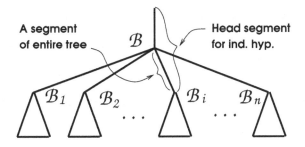

Fig. 2. Tableau Proof Tree After One Beta Extension

Notice the crucial role played by segments in this definition. Notice also that to be u-connected, if there are no nodes connected to the segment of the branch point B, then there must be a node with the property that *every* connection from it is to a unit. Keep in mind that in an NNF tableau, even if beta rules are never applied to units, alpha extensions can produce units anywhere in the tableau, not just in the head. In particular, the fact that u-connectivity is required for Theorem 7 does not weaken the result.

A rigorous proof of Theorem 7 cannot be obtained without Robinson's Clarity.[30] The proof is by induction, and the difficulty arises after the induction hypothesis is applied. The proof begins with a beta extension of a fundamental subgraph of the original formula, creating a branch point and several branches in the tableau — see Figure 2. The induction hypothesis provides a u-connected tableau for each branch. When each branch is treated separately, the head consists of the head of the entire tableau plus a part that is below the first branch point. Here is the problem: Suppose a node N in the ith branch is connected to the segment of B_i above the branch point B. When the tableaux for the branches are put together, that connection is no longer to the segment of B_1, and connectivity may be violated. Robinson's Clarity[31] provided the solution. What is proved is that when that happens, every connection from N is to a unit; i.e., *u-connectivity is not violated*. The details can be found in [10].

5   Conclusion

Use Robinson's Clarity.

References

1. Anderson, Robert, and Bledsoe, Woody, A linear format for resolution with merging and a new technique for establishing completeness, *J. ACM* **17,3** (1970), 525-534.
2. Aristotle, *Analytica Posteriora*, Athens, Greece, 340 BC.

---

[30] We did try in the submitted version of the paper, as the referees can attest.

[31] Not to mention a lot of blood, sweat, and, especially, tears.

3. Berkeley, George, *The Analyst: Or, a Discourse Addressed to an Infidel Mathematician*, Dublin, 1734.

4. Cohen, Paul J., *Set Theory and the Continuum Hypothesis* (4th edition), W. A. Benjamin, New York, 1996.

5. Devlin, Keith, *The Joy of Sets: Fundamentals of Contemporary Set Theory* (2nd edition), Springer, New York, 1994.

6. Dunham, William, *Euler: The Master of Us All*, The Mathematical Association of America, Washington, 1999.

7. Edwards, C.H. Jr., *The Historical Development of the Calculus*, Springer, New York, 1994.

8. Eisenstein, Elizabeth L., *The Printing Press as an Agent of Change*, Cambridge University Press, Cambridge, 1980.

9. Euclid, *Elements*, Athens, Greece, 280 BC.

10. Hähnle, Reiner, Murray, Neil V. and Rosenthal, Erik, Completeness for linear, regular negation normal form inference systems, *Theoretical Computer Science* **328,3** (2004), 325-354.

11. Heath, Thomas L., *History of Greek Mathematics: From Thales to Euclid*, Dover Publications, New York, 1981.

12. Murray, Neil V. and Rosenthal, Erik, Inference with path resolution and semantic graphs, *J. ACM* **34,2** (1987), 225–254.

13. Murray, Neil V. and Rosenthal, Erik, Dissolution: Making paths vanish, *J.ACM* **40,3** (1993), 504-535.

14. Nagel, Ernest and Newman, James R., *Gödel's Proof* (revised edition), New York University Press, New York, 2002.

15. Robinson, J. Alan and Rosenthal, Erik, Lunch at Woody's Place, *Ethereal Proceedings of the Symposium in Honor of Woody Bledsoe*, ed. Boyer, R.S., Austin, Texas, (November, 1991), $\aleph_3 - \aleph_{11}$.

16. Smullyan, Raymond M., *Gödel's Incompleteness Theorems* Oxford University Press, Oxford, 1992.

17. Van Heijenoort, Jean (Editor), *From Frege to Godel: A Source Book in Mathematical Logic, 1879 – 1931*, Harvard University Press, Cambridge, 1967.

Websites

18. The MacTutor History of Mathematics Archive,
    http://www-groups.dcs.st-and.ac.uk/ history/.

19. PRIME: The Platonic Realms Interactive Mathematics Encyclopedia,
    http://www.mathacademy.com/pr/prime/.

20. Stanford Encyclopedia of Philosophy,
    http://plato.stanford.edu/.

21. Wikipedia, the Free Encyclopedia,
    http://en.wikipedia.org/wiki/Main_Page.

Consistency of Variable Splitting in Free Variable Systems of First-Order Logic

Roger Antonsen[1] and Arild Waaler[1,2]

[1] Department of Informatics, University of Oslo, Norway
[2] Finnmark College, Alta, Norway

Abstract. We prove consistency of a sequent calculus for classical logic with explicit splitting of free variables by means of a semantical soundness argument. The free variable system is a mature formulation of the system proposed at TABLEAUX 2003 [1]. We also identify some challenging and interesting open research problems.

1 Introduction

Variable splitting is a technique for detecting variable independency in first-order free variable sequent calculi, matrix systems and systems of tableau. The underlying problem arises when a β-inference splits a branch into two new branches, such that a variable has two different contexts: one occurrence in each branch. Variable splitting provides us with criteria for deciding whether or not a substitution can assign different values to different occurrences of free variables. A novel feature of the technique is that it applies globally to the whole proof object: it considers the effect that an inference has on the whole proof by identifying different occurrences of the same variable all over the construction. The goal of the uniform variable splitting technique, first presented in [1], is to label variables differently (modulo a set of equations) exactly when they are independent.

Related ideas were first suggested for matrix systems by Bibel [2] under the heading "splitting by need". It was later developed by the authors of this paper in the context of sequent calculi. The overall motivation is to have a calculus which simultaneously has: (1) invariance under order of rule application (to enable goal-directed search, since rules then can be applied in any order), (2) introduction of free variables instead of arbitrary terms (to enable least-commitment search since instantiation can be postponed and in the end decided on the basis of a unification problem), and (3) a branchwise restriction of the search space (to enable early termination in cases of unprovability). For further discussion of these points, see [1].

In this paper we present a mature form of the system in [1] and show its consistency by means of a direct semantical soundness argument. The consistency proof is the main contribution of this paper. Although the basic ideas in [1] are easily recognized in the new system, the original system suffers from syntactical

B. Beckert (Ed): TABLEAUX 2005, LNAI 3702, pp. 33–47, 2005.

complexities beyond necessity and, more seriously, from an inconsistency due to a too weak admissibility condition on proofs [3]. These shortcomings have now been resolved.

At the end of the paper we compare the technique with a well-known technique for identifying rigid and universal variables, and raise some interesting and still unresolved research problems.

2   The Splitting Calculus

The set of first-order formulas (without equality) is defined in the standard way from function and relation symbols of all arities. If φ is, e.g., $\forall x \neg \psi$, we say that ψ is an *immediate subformula* of $\neg \psi$ and a *subformula* (but not an immediate subformula) of φ. A *signed formula* is a formula with a *polarity*, $\top$ or $\bot$. A signed formula has a *type* determined by its main connective and its polarity according to the table below. For a given signed formula, the polarities of its immediate subformulas are as specified in the table, from which the polarities of its subformulas are defined by recursion.

α	α_1	α_2		β	β_1	β_2		γ	γ_0		δ	δ_0
$(\varphi \vee \psi)^{\bot}$	$\varphi^{\bot}$	$\psi^{\bot}$		$(\varphi \wedge \psi)^{\bot}$	$\varphi^{\bot}$	$\psi^{\bot}$		$(\exists x \varphi)^{\bot}$	$\varphi^{\bot}$		$(\forall x \varphi)^{\bot}$	$\varphi^{\bot}$
$(\varphi \wedge \psi)^{\top}$	$\varphi^{\top}$	$\psi^{\top}$		$(\varphi \vee \psi)^{\top}$	$\varphi^{\top}$	$\psi^{\top}$		$(\forall x \varphi)^{\top}$	$\varphi^{\top}$		$(\exists x \varphi)^{\top}$	$\varphi^{\top}$
$(\varphi \rightarrow \psi)^{\bot}$	$\varphi^{\top}$	$\psi^{\bot}$		$(\varphi \rightarrow \psi)^{\top}$	$\varphi^{\bot}$	$\psi^{\top}$						
$(\neg \varphi)^{\top}$	$\varphi^{\bot}$											
$(\neg \varphi)^{\bot}$	$\varphi^{\top}$											

We need a fine-grained system to individuate different formula occurrences within derivations. We achieve this by means of a denotation system for all potential formula occurrences. Specifically, the different copies of γ-formulas will be indexed differently. The notational labor pays off in the formulation of the consistency argument.

For each signed formula φ a set of *positions* $\mathcal{P}(\varphi)$ and an *immediate descendant* relation $\ll_1$ is given such that $(\mathcal{P}(\varphi), \ll_1)$ is isomorphic to φ together with its subformula relation. For uniqueness, we assume that $\mathcal{P}(\varphi)$ and $\mathcal{P}(\psi)$ are disjoint if φ and ψ are distinct. If a is a position, $\mathrm{fml}(a)$ denotes its corresponding unsigned formula and $\mathrm{pol}(a)$ denotes the polarity of a.

While positions are used to individuate formulas, *indices* are introduced to distinguish different formula occurrences within a derivation. An *index* is a position a together with an *occurrence tag* t, written a^t. The polarity of an index is obtained by $\mathrm{pol}(a^t) = \mathrm{pol}(a)$, and the type of an index is the type of the underlying position.

Occurrence tags are used to keep track of copies generated by the γ-rules of the system. To give a smooth formulation of the γ-rules, we assume that tags form an infinite binary tree with root ε and that t' and t'' give the two distinct successor tags of t. Let i be an index a^t of type γ; since a is of type γ, it has a unique immediate ancestor b (i.e. $a \ll_1 b$). The successor functions are lifted to indices:

$$i' = a^{t'} \qquad \text{and} \qquad i'' = b^{t''}.$$

By means of this indexing system different occurrences of a formula with position a can be individuated by indices a^{t_1}, a^{t_2}, a^{t_3}, etc. A *root index* is of the form a^ε and is used to denote an "initial" formula occurrence in a way which is made precise below.

The language used to construct derivations is richer than the original language. The underlying first-order language is extended with *instantiation variables* of the form u_i, where i is an index, and *Skolem functions* of the form f_a, where a is a position. The set of indices $\mathcal{I}$ is defined by induction over positions along with a lifting of the fml function to a function from $\mathcal{I}$ and a lifting of the immediate descendant relation $\ll_1$ to a relation over $\mathcal{I}$. The set of indices corresponds exactly to the formulas generated by the rules of the calculus, so there is one case for each type.

- If $\mathsf{fml}(a)$ is closed, then a^ε is an index and $\mathsf{fml}(a^\varepsilon) = \mathsf{fml}(a)$.
- Suppose a^t is an index of type α or β, and that $a \ll_1 a_i$ ($i = 1, 2$). Then, a_i^t is also an index and $a^t \ll_1 a_i^t$. The formula $\mathsf{fml}(a_i^t)$ is the appropriate immediate subformula of $\mathsf{fml}(a^t)$.
- Suppose a^t is an index of type δ, that $a \ll_1 a_1$ and $\mathsf{fml}(a^t) = \mathsf{Q}x\varphi$. Then, a_1^t is an index, $\mathsf{fml}(a_1^t) = \varphi[x/f_a(\vec{u})]$ and $a^t \ll_1 a_1^t$, where $\vec{u}$ are the instantiation variables occurring in φ.
- Suppose i is an index of type γ with $\mathsf{fml}(i) = \mathsf{Q}x\varphi$. Then, i' is an index with $\mathsf{fml}(i') = \mathsf{Q}x\varphi$ and $i \ll_1 i'$; i'' is an index with $\mathsf{fml}(i'') = \varphi[x/u_i]$ and $i \ll_1 i''$.[1]

Lemma 1. $(\mathcal{I}, \ll_1)$ *defines a set of trees, in particular, for each index k there is at most one j such that $j \ll_1 k$.*

Proof. Suppose the contrary, i.e. two different indices are both $\ll_1$-related to k. The only case which could possibly give rise to this situation is that there are two distinct $\ll_1$-paths from a γ-type index i to k, one via i' and one via i''. But given that unrelated tags are selected for i' and i'', this situation cannot arise.   □

In light of this result, every index is situated in an *index tree*. If $i \ll_1 j$, we say that j is an *immediate ancestor* of i and that i is the *immediate descendant* of j; the descendant relation $\ll$ is defined from $\ll_1$ by transitive closure. The tree property ensures that if two indices, i and j, have a common descendant, they have a unique greatest common descendant, written $i \sqcap j$, with respect to $\ll$. Two different indices are β-*related*, written $i \triangle j$, if they are not $\ll$-related and $i \sqcap j$ is of type β. If two different indices are neither $\ll$-related nor β-related, we

---

[1] It is not common in the literature, see e.g. [4], to relate the different *copies* of γ-formulas (or the different copies of γ_0-formulas) by the descendant relation. However, defining the relation in this way gives a more precise control over the introduction of instantiation variables; this facilitates a smooth formulation of the splitting system.

call them α-*related* and write $i \circ j$. If i and j are β-related and aslo the immediate ancestors of a β-index a (i.e. $a \ll_1 i$ and $a \ll_1 j$), then i and j are *dual* indices and said to be of type β_0. The following simple observation illustrates these notions.

Lemma 2. *If b and c are duals and $a \circ b$, then $a \circ c$.*

Proof. Suppose that b and c are duals and $a \circ b$. First, a and c are not $\ll$-related: If $a \ll c$, then $a \ll b$, contradiction the assumption that $a \circ b$. If $c \ll a$, then a and b are β-related, also contradicting $a \circ b$. Second, a and c are not β-related: If they were, a and b would also be β-related (since $a \sqcap b = a \sqcap c$), contradicting $a \circ b$. So, $a \circ c$. $\square$

A *splitting set* is a set of indices of type β_0 such that no two indices are dual. A *colored index* is a pair iS of an index i and a splitting set S. A *sequent* is a finite set of colored indices. A *root sequent* is the lowermost sequent of a derivation tree and contains only root indices and empty splitting sets.[2]

Notational conventions. In rules and derivations we write the underlying formulas instead of colored indices; instead of writing iS, we simply write φ, or φS, where $\varphi = \mathrm{fml}(i)$. A sequent is often written as an ordered pair, $\Gamma \vdash \Delta$, where Γ consists of all colored indices with polarity $\top$ and Δ consists of all with polarity $\bot$. Γ^i denotes the set $\{\varphi(S \cup \{i\}) \mid \varphi S \in \Gamma\}$, where i has been added to all the splitting sets. In examples, natural numbers are used for indices, and indices for formula occurrences are indicated *below* the root sequent. Splitting sets are written as sequences of natural numbers displayed as superscripts, e.g. in $\forall x Pxu \to Pax^{589}$, where the splitting set is $\{5, 8, 9\}$.

The rules of the free variable system are given in Fig. 1, where α, β, γ and δ denote colored indices of the respective types. If γ is iS, then γ' is $i'S$ and γ_0'' is $i''S$; the other immediate ancestors are denoted α_1, α_2, β_1, β_2 and δ_0. A *derivation* is a tree regulated by these rules. As usual, an inference in a derivation is said to *expand* its principal colored index. Note that every instantiation variable and formula occurrence in a derivation is associated with a unique index.

$$\frac{\Gamma, \alpha_1, \alpha_2}{\Gamma, \alpha} \qquad \frac{\Gamma^i, \beta_1 \qquad \Gamma^j, \beta_2}{\Gamma, \beta} \qquad \frac{\Gamma, \gamma', \gamma_0''}{\Gamma, \gamma} \qquad \frac{\Gamma, \delta_0}{\Gamma, \delta}$$

Fig. 1. The rules of the splitting calculus. In the β-rule, β_1 is iS and β_2 is jS.

Every *branch* in a derivation can be identified with a set of indices corresponding to immediate ancestors of expanded β-indices. For example, if $(\varphi^i \vee \psi^j)^\top$ is

---

[2] Since there is only one available root index for a given signed formula, it follows that we cannot have two distinct formula occurrences of the same signed formula in a sequent. However, we can easily get this freedom by slightly modifying the occurrence tags.

expanded, one branch contains i, and the other contains j. A derivation is *balanced* if the following condition holds: If B and C are branches such that $i, j \in B$, $i \in C$ and $i \ll j$, then $j \in C$, i.e. if an index is expanded in one branch, it is also expanded in all other branches.

Example 1. In the following derivation extra γ-formulas are not displayed. Indices for the subformulas are shown below the root sequent; the important ones are 1, 2, 3 and 4. There are four branches: $\{1, 7\}$, $\{2, 7\}$, $\{3, 9\}$ and $\{4, 9\}$. The derivation is not balanced. In Ex. 3 we will consider what happens when non-atomic leaf occurrences are expanded into (A), (B), (C) and (D). Note that the derivation gets balanced when all leaf sequents are expanded one more step.

$$
\begin{array}{cccc}
(A) & (B) & (C) & (D) \\
\hline
Pa^7 \vdash Pu^1, Qa \wedge Qb^{17} & Pb^7 \vdash Pu^2, Qa \wedge Qb^{27} & Qu^3, Pa \vee Pb^{39} \vdash Qa^9 & Qu^4, Pa \vee Pb^{49} \vdash Qb^9
\end{array}
$$

$$
\begin{array}{cc}
\cline{1-1}
Pa \vee Pb^7 \vdash Pu, Qa \wedge Qb^7 & Qu, Pa \vee Pb^9 \vdash Qa \wedge Qb^9
\end{array}
$$

$$
Pa \vee Pb, Pu \to Qu \vdash Qa \wedge Qb
$$

$$
\underset{1 \quad 5 \quad 2 \quad \ 6 \quad \ 7 \quad 8 \quad 9 \qquad 3 \ \ 10 \ \ 4}{Pa \vee Pb, \forall x(Px \to Qx) \vdash Qa \wedge Qb}
$$

Let u_i be an instantiation variable occurring in the formula of a colored index with splitting set S. Then, $u_i T$, where T is $\{j \in S \mid i \not\ll j\}$, is a *colored variable*; when we need to emphasize the operation which gives rise to $u_i T$ we will denote it $\overline{u_i S}$. Note that the ancestors of i are not allowed to add to the splitting sets of u_j. If φ is a formula and S is a splitting set, then $\overline{\varphi S}$ is the formula φ where all instantiation variables u_i have been replaced by $\overline{u_i S}$; this operation is unproblematic since instantiation variables never occur bound. The function $\overline{(\cdot)}$ is extended to sets of formulas and sequents in the obvious way.

Lemma 3. *For a colored variable $u_i S$ obtained from the leaf sequent of a branch B, $a \in S$ if and only if $a \in B$ and $a \circ i$.*

Proof. Let $u_i S$ be a colored variable obtained from the leaf sequent of a branch B. Suppose $a \in S$. Since $S \subseteq B$, $a \in B$. Also, a is neither $\ll$-related to i nor β-related to i. First, $i \ll a$ is impossible by the definition of a colored variable, and $a \ll i$ is impossible, since β-rules do not change the splitting sets of the immediate ancestors of the β-index. Second, if a and i are β-related, then $u_i S$ could not be a colored variable. Conversely, suppose $a \in B$ and $a \circ i$. Since the colored variable $u_i S$ is obtained from the leaf sequent of B, and since $a \circ i$, there must be an index j, such that $j \ll i$, with a splitting set to which the index a is added by a β-rule application. Since splitting sets are only extended by rule applications, $a \in S$. $\qquad\square$

Convention. Unless an explicit exception is made, the derivation at hand is called π. The set of leaf sequents of π is called L. $\mathsf{Col}(L)$ denotes the set of colored instantiation variables from $\overline{L}$.

A *substitution* is a mapping from $\mathsf{Col}(L)$ to the set of ground terms.[3] A substitution σ *closes* a leaf sequent s if there is a subset $P\bar{s}S \vdash P\bar{t}T$ of s such that $\overline{P\bar{s}S}\sigma = \overline{P\bar{t}T}\sigma$; it closes L if it closes every sequent in L.

It is not sufficient to define a proof as a derivation along with a substitution which closes all its leaf sequents. We need to compensate for lack of balance (illustrated in Ex. 2) and check for a certain type of circularity caused by a *mutual splitting* (illustrated in Ex. 4).

Example 2. In the derivation in Ex. 1, $\mathsf{Col}(L) = \{u^1, u^2, u^3, u^4\}$. Note that $\sigma = \{u^1/a, u^2/b, u^3/a, u^4/b\}$ closes the derivation. However, the root sequent is falsifiable; a countermodel is: Pa, Qa true, Pb, Qb false.

We introduce the set $\mathsf{Bal}(L)$ of *balancing equations* for L. $\mathsf{Bal}(L)$ is a binary relation over $\mathsf{Col}(L)$ defined by all equations $uS \approx uT$ such that $S \cup T$ is a splitting set and $S \neq T$. $\mathsf{Bal}(L)$ is irreflexive and symmetric; its reflexive, transitive closure is an equivalence relation. This equivalence relation induces a partition of $\mathsf{Col}(L)$ (the set of its equivalence classes). Clearly, a substitution solves all equations in $\mathsf{Bal}(L)$ iff for every equivalence class in the partition it assigns the same value to all its members.

Example 3. The set of balancing equations for the derivation π in Ex. 1 is $\mathsf{Bal}(L) = \{u^1 \approx u^3, u^1 \approx u^4, u^2 \approx u^3, u^2 \approx u^4\}$. Since all members of $\mathsf{Col}(L)$ are equivalent, there is no substitution which closes all the leaf sequents of π and also solves $\mathsf{Bal}(L)$.

The example also illustrates the relationship between $\mathsf{Bal}(L)$ and extensions of π. Note that each of the four leaf sequents has a potential expansion:

(A) goes to (A_1) $Pa^{37} \vdash Pu^{13}, Qa^{17}$ and (A_2) $Pa^{47} \vdash Pu^{14}, Qb^{17}$.
(B) goes to (B_1) $Pb^{37} \vdash Pu^{23}, Qa^{27}$ and (B_2) $Pb^{47} \vdash Pu^{24}, Qb^{27}$.
(C) goes to (C_1) $Qu^{13}, Pa^{39} \vdash Qa^{19}$ and (C_2) $Qu^{23}, Pb^{39} \vdash Qa^{29}$.
(D) goes to (D_1) $Qu^{14}, Pa^{49} \vdash Qb^{19}$ and (D_2) $Qu^{24}, Pb^{49} \vdash Qb^{29}$.

Consider first π', the result of expanding (A) and (B). Note that the set of colored variables changes: $\mathsf{Col}(L') = \{u^{13}, u^{14}, u^{23}, u^{24}, u^3, u^4\}$. Both $u^{13} \approx u^3$ and $u^3 \approx u^{23}$ are balancing equations, while $u^{13} \approx u^{23}$ is not, since $\{1, 2, 3\}$ is not a splitting set. $\mathsf{Bal}(L')$ hence induces the equivalence classes $\{u^{13}, u^3, u^{23}\}$ and $\{u^{14}, u^4, u^{24}\}$. Note that no simultaneous solution of the equations in $\mathsf{Bal}(L')$ closes L'. Let π'' be the derivation obtained from π' by expanding (C) and (D). No substitution closes L''. Moreover, $\mathsf{Bal}(L'')$ is empty. This is because π'' is balanced; see Lemma 5.

Example 4. The following is a derivation for another sequent which is falsifiable. A countermodel is: Paa, Pbb true, Pab, Pba false. The variables u and v stand for u_5 and u_7. There are no balancing equations, and the substitution

---

[3] The groundedness assumption for substitutions simplifies the soundness argument. It can easily be skipped when focus is proof search and not consistency.

$\sigma = \{u^3/a, u^4/b, v^1/a, v^2/b\}$ solves all the equations and hence closes all leaf sequents.

$$
\begin{array}{cccc}
u^3 \approx a & u^4 \approx b & u^3 \approx a & u^4 \approx b \\
v^1 \approx a & v^1 \approx a & v^2 \approx b & v^2 \approx b \\
\hline
Pua^3 \vdash Pav^1 & Pua^4 \vdash Pbv^1 & Pub^3 \vdash Pav^2 & Pub^4 \vdash Pbv^2 \\
\end{array}
$$

$$
\cfrac{
\cfrac{
\cfrac{Pua \vdash Pav \land Pbv^1 \qquad Pub \vdash Pav \land Pbv^2}
{Pua \lor Pub \vdash Pav \land Pbv}}
{Pua \lor Pub \vdash \exists x(Pax \land Pbx)}}
{\forall x(\underset{5}{P}x\underset{1}{a} \lor \underset{6}{P}x\underset{2}{b}) \vdash \exists x(\underset{7}{P}a\underset{3}{x} \land \underset{8}{P}b\underset{4}{x})}
$$

In order to avoid this situation, a relation $\prec_\sigma$ on indices is introduced for each substitution σ. To this end, let $S \sqcap T = \{s \sqcap t \,|\, s \in S, t \in T, s \bigtriangleup t\}$, for splitting sets S and T. The *distance* between S and T is the number of elements in $S \sqcap T$. Let $a \prec_\sigma i$ if there are $u_i S$ and $u_i T$ in $\mathsf{Col}(L)$ such that $\sigma(u_i S) \neq \sigma(u_i T)$ and $S \sqcap T = \{a\}$. Observe that this is a singleton set and that the distance between S and T is one. The transitive closure of $\ll \cup \prec_\sigma$ is denoted $\vartriangleleft_\sigma$.

Definition 1. *A substitution σ is* admissible *for π if σ solves all balancing equations and $\vartriangleleft_\sigma$ is irreflexive. The pair $\langle \pi, \sigma \rangle$ is a* proof *of the root sequent of π if σ is admissible for π and closes all leaf sequents.*

A main intuition is that a well-founded $\vartriangleleft_\sigma$-relation encodes an optimal order in which to apply rules in a calculus without splitting.

The substitution σ from Ex. 4 gives rise to a cyclic $\vartriangleleft_\sigma$. Because $\sigma(u^3) \neq \sigma(u^4)$ and $3 \sqcap 4 = 8$ we have $8 \prec_\sigma u$. Likewise, because $\sigma(v^1) \neq \sigma(v^2)$ and $1 \sqcap 2 = 6$ we have $6 \prec_\sigma v$. Together with $\ll$ we get the following cycle: $u \ll 6 \prec_\sigma v \ll 8 \prec_\sigma u$. Since $\vartriangleleft_\sigma$ is not irreflexive, σ is not admissible.

3   Balanced Derivations

Balanced derivations play a central role in the soundness argument. This section states some of their properties.

Lemma 4. *Let B and C be branches in a balanced derivation and let index a occur in C and not in B. Then there are dual indices $b \in B$ and $c \in C$ such that $c \ll a$.*

Proof. Suppose not. By induction over the inferences in the branch represented by B, from the root sequent and upward, it is easy to show that if no such β-inference $b \sqcap c$ is in B such that $c \ll a$, there must be an index d in the leaf sequent of B such that $d \ll a$. This follows since an index which can potentially generate a will always be copied into the branch as an extra formula of an inference. Since the derivation is balanced, a must be in B, which contradicts the assumption. $\square$

Lemma 5. *No balancing equation can arise from the leaf sequents of a balanced derivation.*

Proof. Suppose for contradiction that a balancing equation $u_i S \approx u_i T$ arises from a balanced derivation. Let B and C be branches whose leaf sequents give rise to $u_i S$ and $u_i T$, respectively. Assume w.l.o.g. that $a \in T \setminus S$. By Lemma 4, there are dual indices $b \in B$ and $c \in C$ such that $c \ll a$. Neither b nor c is $\ll$-related to i: It is not the case that $c \ll i$, since u_i occurs in the leaf sequent of B, and it is not the case that $b \ll i$, since u_i occurs in the leaf sequent of C. Furthermore, it is not the case that $i \ll b$ or $i \ll c$, since this would imply that $i \ll a$, which is impossible, since $a \in T$ and, by Lemma 3, $a \circ i$. By Lemma 3, $b \in S$ and $c \in T$. Since b and c are duals, $S \cup T$ is not a splitting set, in which case $u_i S \approx u_i T$ cannot be a balancing equation. $\qquad\square$

Theorem 1. *For every proof $\langle \pi, \sigma \rangle$ of a sequent, there is a proof $\langle \pi', \sigma' \rangle$ of the same sequent such that π' is balanced.*

Proof. We show that one imbalance can be eliminated; by repeating this elimination a balanced derivation is obtained. An imbalance means that an index k is extended in a branch B_1 and not in a branch B_2, i.e. it must be in the leaf sequent of B_2. Let π' be the derivation obtained by expanding k in B_2. First, observe that if the set of colored variables is unchanged, $\mathsf{Col}(L) = \mathsf{Col}(L')$, then $\langle \pi', \sigma \rangle$ is still a proof. (This is necessarily the case if k is an α-, γ- or δ-index and possibly the case if k is a β-index.) Otherwise, k is a β-index. We can w.l.o.g. suppose that k is the index of a $\vee$-formula with polarity $\top$:

$$\frac{\Gamma^i, \varphi U \qquad \Gamma^j, \psi U}{\Gamma, \varphi \vee \psi U}$$

We now construct σ' from σ. Call the colored variables in $\mathsf{Col}(L) \cap \mathsf{Col}(L')$ *unchanged*, those in $\mathsf{Col}(L') \setminus \mathsf{Col}(L)$ *new* and those in $\mathsf{Col}(L)$ *old*. For the unchanged variables, let σ' agree with σ, i.e. $\sigma'(uS) = \sigma(uS)$. A new variable uS' must be of the form $uS \cup \{i\}$ or $uS \cup \{j\}$, for some old variable uS. In this case, let $\sigma'(uS') = \sigma(uS)$.

Claim: $\langle \pi', \sigma' \rangle$ is still a proof.

All leaf sequents are closed by σ'. This follows from the construction of σ' and that σ satisfies the balancing equations in $\mathsf{Bal}(L)$. (First, note that σ' closes all untouched leaf sequents. To see that the new leaf sequents are closed by σ', there are two cases to consider. First, the new leaf sequents contain only new variables. Then, they are closed by virtue of the construction of σ'. Second, the new leaf sequents contain unchanged variables. If these are in φU or ψU, it is unproblematic. However, it is possible that these are new relative to the expanded leaf sequent, but that they nevertheless occur in other, i.e. untouched, leaf sequents.

Then, an argument is needed. Suppose that $uS \cup \{i\}$ is such an unchanged variable, where uS is the old variable. We must show that $\sigma'(uS \cup \{i\}) = \sigma(uS)$. But this holds, since $uS \cup \{i\} \approx uS$ is in $\mathsf{Bal}(L)$.) It remains to show that σ' is admissible. Let $uS_1 \approx uS_2$ be a balancing equation in $\mathsf{Bal}(L')$. If both variables are unchanged, then $uS_1 \approx uS_2$ is in $\mathsf{Bal}(L)$ and therefore solved by σ'. If at least one of them is new, assume w.l.o.g. that $uS_1 = uS_1' \cup \{i\}$ for an old variable uS_1'. If uS_2 is unchanged, then $uS_1' \approx uS_2$ is in $\mathsf{Bal}(L)$, solved by σ, and $uS_1 \approx uS_2$ is solved by σ'. If uS_2 is new, we can assume that $uS_2 = uS_2' \cup \{i\}$ for an old variable uS_2', since S_2 cannot contain j. But then, $uS_1' \approx uS_2'$ is in $\mathsf{Bal}(L)$, and $uS_1 \approx uS_2$ is solved by σ'.

Finally, we show that $\prec_{\sigma'}$ is a subset of $\prec_\sigma$, so $\lhd_{\sigma'}$ must be irreflexive too. Suppose $a \prec_{\sigma'} m$. Then, there are $u_m S_1$ and $u_m S_2$ from $\mathsf{Col}(L')$ containing dual indices, $a_1 \in S_1$ and $a_2 \in S_2$, with $a = a_1 \sqcap a_2$, such that $\sigma'(u_m S_1) \neq \sigma'(u_m S_2)$. If both variables are unchanged, then $a \prec_\sigma m$. If at least one of them is new, assume w.l.o.g. that $u_m S_1 = u_m S_1' \cup \{i\}$ for an old variable $u_m S_1'$. Observe that $u_m S_2$ cannot also be new (if it were, σ' would, by construction, assign the same value to $u_m S_1$ and $u_m S_2$). So, $u_m S_2$ must be unchanged. If $a_1 = i$, then $a_2 = j$ and $u_m S_1' \approx u_m S_2 \in \mathsf{Bal}(L)$, implying that $\sigma'(u_m S_1) = \sigma'(u_m S_2)$, which is not possible. So, $a_1 \neq i$, which implies that $\sigma(u_m S_1') \neq \sigma(u_m S_2)$ and $a \prec_\sigma m$. □

Let $S \downarrow a$ denote the set of indices from S which are ancestors of a, i.e. $\{j \in S \mid a \ll j\}$. Let uS and uT be two colored variables. A β-*change of* uS *toward* uT is the result of replacing S by $(S \setminus (S \downarrow b)) \cup (T \downarrow b)$, for some $b \in S \sqcap T$. Notice that this operation always reduces the distance by exactly one (unless it is already zero). The set of β-*variants of* uS *toward* uT is defined by induction over zero or more β-changes toward uT. Observe that if the distance between S and T is n, there are exactly 2^n β-variants of uS toward uT.

Lemma 6. *Let* π *be balanced. If* uS *and* uT *are two colored variables in* $\mathsf{Col}(L)$, *then all* β-*variants of* uS *toward* uT *are also in* $\mathsf{Col}(L)$.

Proof. Suppose uS' is obtained from uS by one β-change, i.e. that S' is $(S \setminus (S \downarrow b)) \cup (T \downarrow b)$. By construction and Lemma 3, both $S \downarrow b$ and $T \downarrow b$ contain only indices α-related to u. Then, since the derivation is balanced, there must be a branch containing a leaf sequent which gives rise to uS'. □

Lemma 7. *Let* π *be balanced, and let* uS *and* uT *be two colored variables in* $\mathsf{Col}(L)$. *Suppose* $\sigma(uS) \neq \sigma(uT)$. *Then, there are* β-*variants* uS' *and* uS'' *of* uS *toward* uT *with a distance of one to each other, such that* $\sigma(uS') \neq \sigma(uS'')$.

Proof. By the previous lemma, all β-variants of uS toward uT are in $\mathsf{Col}(L)$ and hence in the domain of σ. Suppose that σ assigns the same value to all adjacent β-variants. Then all β-variants of uS, including uT, must be assigned the same value by σ, contradicting the assumption. □

4   Soundness

A model $\mathcal{M}$ for the underlying object language is comprised of a non-empty domain $|\mathcal{M}|$ over which function and relation symbols receive interpretation of appropriate types. The interpretation of a function symbol g is denoted $g^{\mathcal{M}}$. Free variables are interpreted by a variable assignment ϑ mapping all variables in the language into $|\mathcal{M}|$. Any formula in the language is either satisfied or falsified in $\mathcal{M}$ under ϑ. A signed formula $\varphi^{\top}$ is satisfied in $\mathcal{M}$ under ϑ iff φ is. Dually, $\varphi^{\perp}$ is satisfied in $\mathcal{M}$ under ϑ iff φ is not.

When interpreting the extended language, some care must be taken with instantiation variables. Note that there are two different languages at play: one which takes variables as colored and one in which splitting sets are neglected. In the former case we will talk about *colored* variable assignments. A colored variable assignment θ will always appear in relation to a given derivation; its domain will then be the set of all colored variables that can be defined from the leaf sequents of the derivation. In the latter case we will talk about *pure* variable assignments. A pure variable assignment τ will always appear in relation to a given set of formulas; its domain will then be the set of all instantiation variables that occur in the set.

In both cases we will assume that models interpret a Skolem term $f_p(\vec{u})$ *canonically*, i.e. if $Qx\varphi$ is a signed δ-type formula indexed by p^t with instantiation variables $\vec{u}$ satisfied in $\mathcal{M}$ under ϑ, then so is $\varphi[x/f_p(\vec{u})]$.

These notions are straightforwardly lifted to indices. Let θ be a colored variable assignment and τ a pure variable assignment. A colored index iS is satisfied in $\mathcal{M}$ under θ iff $\overline{\mathsf{fml}(i)^{\mathsf{pol}(i)}\,S}$ is. An index i is satisfied in $\mathcal{M}$ under τ iff $\mathsf{fml}(i)^{\mathsf{pol}(i)}$ is. The assignments θ and τ *agree on* u_i in a sequent s if $\theta(u_iS) = \tau(u_i)$ for every u_iS in $\mathsf{Col}(s)$. The following is then immediate.

Lemma 8. *Let θ and τ agree on all variables in s with respect to a given model. Then a colored index iS in s is satisfied under θ iff i is satisfied under τ.*

A sequent s is *falsifiable* if there is a model and a colored variable assignment which satisfy every colored index in s.[4] A sequent is *valid* if it is not falsifiable. Recall that a substitution maps colored variables to ground terms. Given a model $\mathcal{M}$, a substitution σ *induces* a colored variable assignment $\sigma_{\mathcal{M}}$ by: $\sigma_{\mathcal{M}}(uS) = (\sigma(uS))^{\mathcal{M}}$. Although simple, the following observation is important.

Lemma 9. *Assume that σ closes a sequent s and let $\mathcal{M}$ be any model. Then s is not falsified in $\mathcal{M}$ under $\sigma_{\mathcal{M}}$.*

Theorem 2. *Let π be a balanced derivation with leaf sequents L and a root sequent s falsified by a model $\mathcal{M}$. Let σ be an admissible substitution. Then, there is a sequent in L falsified by $\mathcal{M}$ under $\sigma_{\mathcal{M}}$.*

---

[4] It might seem odd that a sequent is falsified when all its members are satisfied, but according to this definition a sequent $\Gamma \vdash \Delta$ is falsified exactly when all formulas in Γ are true and all formulas in Δ are false, i.e. when it has a countermodel.

Proof. We first construct inductively, for each $k \geq 0$, the following sets:

- J_k: a set of indices
- B_k: a $\ll$-closed sub-branch in π
- τ_k: a pure variable substitution

Let M_k be the set of $\lhd_\sigma$-minimal indices in J_k. Since σ is admissible, $\lhd_\sigma$ is irreflexive; hence M_k is always non-empty. The goal of the construction is to identify a unique leaf sequent s falsified under σ_M. J_k approximates the set of indices which comprises s, B_k approximates the branch whose leaf sequent is s while τ_k gives the correct assignment of values to variables occurring in formulas indexed by members of M_k.

The induction proceeds by selecting an index i in M_k and deleting it from J_k. If i is a β-index, one immediate ancestor is added to B_k and the other is removed from J_k along with all its ancestors. This crucial branch selection depends on the values assigned to the instantiation variables in $\mathsf{fml}(i)$ by τ_k. If i is a γ-index, the value of $\tau(u_i)$ is specified, ensuring the falsifiability of the final sequent. To make this precise let us say that a leaf sequent is a B_k-*leaf* if it is the leaf sequent of a branch B such that $B_k \subseteq B$. Claim k is the following:

(1) In every B_k-leaf s, τ_k and σ_M agrees on every colored variable $u_i S$ from s such that u_i is in the domain of τ_k.
(2) M satisfies every index in M_k under τ_k.

Initially, J_0 is the set of all indices expanded in π, $B_0 = \emptyset$ and $\tau_0 = \emptyset$. Claim 0, the base step, holds since (1) the domain of τ_0 is empty and (2) all $\lhd_\sigma$-minimal indices in J_0 are root indices and satisfied by assumption.

For the induction step, choose a $\lhd_\sigma$-minimal index i from J_k such that there is a j with $i \ll j$. If no such j exists, the construction is completed. There are four cases, depending on the type of i.

If i is an α-index, suppose that a and b are the immediate ancestors of i. Let

- $J_{k+1} = J_k \setminus \{i\}$
- $B_{k+1} = B_k$
- $\tau_{k+1} = \tau_k$.

Condition (1) holds for J_{k+1} since it (by the induction hypothesis) holds for J_k. For condition (2), suppose j is $\lhd_\sigma$-minimal in J_{k+1}. If j is in J_k, then it is satisfiable. If not, then $j = a$ or $j = b$ and satisfiable by assumption. Thus, condition (2) holds.

If i is a β-index, suppose that a and b are the immediate ancestors of i. Assume w.l.o.g. that a is satisfied under τ_k. Let

- $J_{k+1} = J_k \setminus (\{i\} \cup \{j \mid b \ll j\})$ ("Remove i, b and all ancestors of b"),
- $B_{k+1} = B_k \cup \{a\}$,
- $\tau_{k+1} = \tau_k$.

The arguments for conditions (1) and (2) are as for the α-case.

If i is a δ-index, suppose that a is the immediate ancestor of i. Let

- $J_{k+1} = J_k \setminus \{i\}$
- $B_{k+1} = B_k$
- $\tau_{k+1} = \tau_k$.

As for the two previous cases, (1) holds by the induction hypothesis, and (2) holds since $\mathcal{M}$ by assumption interprets all Skolem functions canonically.

The hard case is when i is a γ-index (recall that i'' is the index of the instantiation of $\mathrm{fml}(i)$ with u_i). By definition of satisfaction $\mathrm{fml}(i'')$ is satisfiable for all values of u_i. The case is trivial if there are no occurrences of u_i in it, so suppose there are. Note that the domain of τ_k consists only of variables u_j with $j \vartriangleleft_\sigma i$ but not u_i. To interpret i'' we must therefore decide on a value for u_i under τ_{k+1}; when this is done, τ_{k+1} assigns value to all instantiation variables in $\mathrm{fml}(i'')$. This is decided from the assignment $\sigma_\mathcal{M}$ and the set B_k. Let

$$S = \{j \in B_k \mid j \circ i\} \quad \text{and} \quad V = \{u_i T \in \mathrm{Col}(L) \mid S \subseteq T\}.$$

Observation 1: Let $u_i T \in V$. Then $B_k \cup T$ contains no β-related indices. *Proof.* We first show that (†) $B_k \cup T$ contains no duals. Suppose for a contradiction that there are duals b and c such that $b \in B_k$ and $c \in T$. By Lemma 3, $c \circ i$. By Lemma 2, $b \circ i$. By construction of S, $b \in S$, and by construction of V, $b \in T$. But then T is not a splitting set, since it contains the duals b and c. This completes the proof of (†). Suppose next that there are β-related indices $b' \in B_k$ and $c' \in T$. By definition there are duals $b \ll b'$ and $c \ll c'$. Since B_k by construction is $\ll$-closed, $b \in B_k$. Now note that c is neither α-related to i nor β-related to i: c is not α-related to i, for then c is in T by Lemma 3 and this contradicts (†); c is not β-related to i, for then c' would also be β-related to i and $u_i T$ would not be a colored variable. So, c must be $\ll$-related to i. But, $i \ll c$ is impossible since c' would then not be in T by Lemma 3. Hence $c \ll i$. However, since B_k contains the dual of c, $c \notin J_k$, and hence $i \notin J_k$. This contradicts the fact that i has been selected from J_k.

Observation 2: Let $u_i T_1 \in V$ and $u_i T_2 \in V$. Then all β-variants of $u T_1$ toward $u T_2$ are also in V. *Proof.* Suppose that T_1' is a β-variant of $u T_1$ toward $u T_2$ which is not in V. Then, $S \not\subseteq T_1'$ and there must be a $j \in S$ such that $j \notin T_1'$. By the definition of a β-change, $j \notin T_2$. This contradicts the assumption that $u_i T_2 \in V$, since this implies that $j \in S \subseteq T_2$.

Observation 3: $V\sigma_\mathcal{M}$ is singleton. *Proof.* Suppose not. Then, there are $u_i T_1$ and $u_i T_2$ from V such that $\sigma_\mathcal{M}(u_i T_1) \neq \sigma_\mathcal{M}(u_i T_2)$. By Lemma 7, there are β-variants $u_i T_1'$ and $u_i T_1''$ of $u_i T_1$ with a distance of one to each other, such that $\sigma_\mathcal{M}(u_i T_1') \neq \sigma_\mathcal{M}(u_i T_1'')$. By Observation 2 above, $u_i T_1'$ and $u_i T_1''$ are in V. Then, there are dual indices $a' \in T_1'$ and $a'' \in T_1''$. By Observation 1, neither a' nor a'' are β-related to any index in B_k, and hence (‡) $a = a' \sqcap a''$ is not β-related to any index in B_k. Since $u_i T_1'$ and $u_i T_1''$ have distance one and $T_1' \sqcap T_1'' = \{a\}$, it follows by definition of $\prec_\sigma$ that $a \prec_\sigma i$, and hence that $a \lhd_\sigma i$. Then, a must have been selected as $\lhd_\sigma$-minimal in the generation of a B_n for some $n < k$. To see this, note that $a \notin J_k$, since i has just been selected and a is $\lhd_\sigma$-smaller than i. So, if a was not selected as $\lhd_\sigma$-minimal for some $n < k$, there is an $m < k$ such that a β-index b was selected as $\lhd_\sigma$-minimal from J_m and $a \in J_m$, but $a \notin J_{m+1}$. But then a is β-related to something in B_k, which we by ($\ddagger$) know that it is not. Hence a has been selected as $\lhd_\sigma$-minimal for some $n < k$, which means that either $a' \in B_k$ or $a'' \in B_k$. Suppose w.l.o.g. that $a' \in B_k$. Since $a' \in T_1'$ and $u_i T_1'$ is a colored variable, Lemma 3 gives that $a' \circ i$. Then, by definition of S, $a' \in S$. Since $u_i T_1'' \in V$, $S \subseteq T_1''$ (by definition of V), hence $a' \in T_1''$. But this is impossible, since T_1'' then would contain the duals a' and a'', and hence not be a splitting set.

Hence, $V\sigma_\mathcal{M}$ is singleton; let $V\sigma_\mathcal{M} = \{d\}$. Define

- $J_{k+1} = J_k \setminus \{i\}$
- $B_{k+1} = B_k$
- $\tau_{k+1} = \tau_k \cup \{u_i/d\}$.

Condition (1) holds by construction, since $V\sigma_\mathcal{M}$ is singleton. To prove (2), note that i' is satisfied in $\mathcal{M}$ under τ_{k+1} since i is satisfied (by the induction hypothesis), and that i'' is satisfied since all instances of $\mathsf{fml}(i'')$ is satisfied in $\mathcal{M}$ under τ_{k+1}. This concludes the induction step.

We can now complete the proof of the theorem. Let B, J and τ be the limit objects of the induction, and let s be the leaf sequent of B. Note that s consists of colored indices iS such that $i \in J$. By (2) we know that all indices in J are satisfied under τ and, by (1), that τ agrees with $\sigma_\mathcal{M}$ on all instantiation variables in s. Conclude by Lemma 8.   □

Theorem 3 (soundness). *If a sequent is provable, it is valid.*

Proof. Let π be a derivation of a sequent closed by an admissible σ. By Theorem 1 we can assume that π is balanced. Suppose that the sequent is not valid and that it is falsified by a model $\mathcal{M}$. By Theorem 2, there is a leaf sequent falsified in $\mathcal{M}$ under $\sigma_\mathcal{M}$. But we know that σ closes every leaf sequent of π, contradicting Lemma 9.   □

5 Comparison and Future Work

It is common for variables in a free-variable system to be treated *rigidly*, which means that substitutions must be applied to *all* free variables occurring in the derivation simultaneously. On the other hand, some free variables occurrences can be substituted with any term, regardless of other occurrences of the same variable. These are called *universal*, a property which in general is undecidable to prove, see e.g. [5].

An easily detectable subclass of universal variables is the class of so called *local* variables, see e.g. [6,7]. Let φ be a γ-formula with index i in a derivation. The instantiation variable u_i is called *rigid* if φ has a subformula ψ of type β such that u_i occurs in both immediate subformulas of ψ. Otherwise, u_i is called *local*.

$$\frac{\dfrac{\dot{P}u^1 \vdash \dot{P}a, Qa}{Pu^1 \vdash Pa \vee Qa} \quad \dfrac{\dot{P}u^2 \vdash \dot{P}b, Qb}{Pu^2 \vdash Pb \vee Qb}}{Pu \vdash (Pa \vee Qa) \wedge (Pb \vee Qb)} \quad \dfrac{\dfrac{\dot{Q}u^1 \vdash Pa, \dot{Q}a}{Qu^1 \vdash Pa \vee Qa} \quad \dfrac{\dot{Q}u^2 \vdash Pb, \dot{Q}b}{Qu^2 \vdash Pb \vee Qb}}{Qu \vdash (Pa \vee Qa) \wedge (Pb \vee Qb)}$$

$$\frac{Pu \vee Qu \vdash (Pa \vee Qa) \wedge (Pb \vee Qb)}{\underset{u}{\forall x}(Px \vee Qx) \vdash (Pa \underset{1}{\vee} Qa) \wedge (Pb \underset{2}{\vee} Qb)}$$

Fig. 2. Comparison to universal variables

Example 5. In Fig. 2 u is a rigid variable. From the leaf sequents L, two different colored variables are obtained, which can be substituted with different terms. The substitution $\sigma = \{u^1/a, u^2/b\}$ gives a proof: $\mathsf{Bal}(L) = \emptyset$ and $\prec_\sigma = \{\langle 1 \sqcap 2, u \rangle\}$ gives an irreflexive ordering, $\lhd_\sigma$.

Also in Ex. 1, u is a rigid variable. If (A), (B), (C) and (D) are expanded, there are in total eight different occurrences of u in the leaf sequents, one in each leaf sequent, from which four different colored variables are obtained, which in turn can be instantiated differently. But notice for instance that u^{14} occurs in both (A$_2$) and (D$_1$) and that *these two* occurrences cannot be instantiated differently.

It is not possible to detect this independency with universal or local variables, since the variable u becomes rigid immediately after expanding a β-formula.

Open problem. It is evident from the proof in Fig. 2 that variable splitting can give rise to shorter proofs than proofs in a system which exploits local variables. However, how great is the benefit? And how do proof lengths in the variable splitting system compare to proof lengths in plain free variable systems (with neither of the two variable-techniques) and ground, Gentzen-style systems?

Open problem. The definition of proofs gives rise to a cycle check on the $\lhd_\sigma$-ordering. From the $\ll$-relation on indices, define a smaller, "liberalized" relation $\ll^-$ in the following way. For a γ-index i and a β-index b, let $i \ll^- b$ if the instantiation variable u_i occurs in both of the immediate subformulas of $\mathsf{fml}(b)$. The motivation for this is that in order to satisfy a β-formula one must make a choice of one of the immediate β-subformulas and this choice depends on the choice of values for the variables in it. Let $\lhd_\sigma^-$ be the transitive closure of $\ll^- \cup \prec_\sigma$. Is it sufficient to require that $\lhd_\sigma^-$ is irreflexive in order to characterize provability?

Open problem. It is possible to replace the reference to the reduction ordering $\lhd_\sigma$ with an appropriate unification problem? This requires a term language to represent $\prec_\sigma$.

Future work. In [8] incremental closure is defined for a variable-pure system of first-order tableaux. Significant steps toward the adaptation of this technique to the variable splitting system are taken in [9]. We work toward finalizing this technique for the variable splitting system. We also work on an implementation of the system to facilitate experimental comparison of our system with standard first-order systems of tableaux.

Acknowledgments. We thank Christian M. Hansen for valuable comments and discussions.

References

1. Waaler, A., Antonsen, R.: A free variable sequent calculus with uniform variable splitting. In: Automated Reasoning with Analytic Tableaux and Related Methods: International Conference, TABLEAUX, Rome, Italy. Number 2796 in LNCS, Springer-Verlag (2003) 214–229

2. Bibel, W.: Automated Theorem Proving 2. Edition. Vieweg Verlag (1987)

3. Antonsen, R.: Uniform variable splitting. In: Contributions to the Doctoral Programme of the Second International Joint Conference on Automated Reasoning (IJCAR 2004), Cork, Ireland, 04 July - 08 July, 2004. Volume 106., CEUR Workshop Proceedings (2004) 1–5 Online: http://ceur-ws.org/Vol-106/01-antonsen.pdf.

4. Wallen, L.A.: Automated deduction in nonclassical logics. MIT Press (1990)

5. Hähnle, R.: Tableaux and related methods. In Robinson, A., Voronkov, A., eds.: Handbook of Automated Reasoning. Volume I. Elsevier Science (2001) 100–178

6. Letz, R.: Clausal tableaux. In Bibel, W., Schmidt, P.H., eds.: Automated Deduction: A Basis for Applications. Volume I, Foundations: Calculi and Methods. Kluwer Academic Publishers, Dordrecht (1998)

7. Letz, R., Stenz, G.: Universal variables in disconnection tableaux. In: Automated Reasoning with Analytic Tableaux and Related Methods: International Conference, TABLEAUX, Rome, Italy. Number 2796 in LNCS, Springer-Verlag (2003) 117–133

8. Giese, M.: Incremental Closure of Free Variable Tableaux. In: Proc. Intl. Joint Conf. on Automated Reasoning, Siena, Italy. Number 2083 in LNCS, Springer-Verlag (2001) 545–560

9. Hansen, C.M.: Incremental Proof Search in the Splitting Calculus. Master's thesis, Dept. of Informatics, University of Oslo (2004)

10. Antonsen, R.: Free variable sequent calculi. Master's thesis, University of Oslo, Language, Logic and Information, Department of Linguistics (2003)

11. Smullyan, R.M.: First-Order Logic. Volume 43 of Ergebnisse der Mathematik und ihrer Grenzgebiete. Springer-Verlag, New York (1968)

12. Waaler, A.: Connections in nonclassical logics. In Robinson, A., Voronkov, A., eds.: Handbook of Automated Reasoning. Volume II. Elsevier Science (2001) 1487–1578

On the Dynamic Increase of Multiplicities in Matrix Proof Methods for Classical Higher-Order Logic

Serge Autexier

Saarland University & German Research Center for Artificial
Intelligence (DFKI GmbH), Saarbrücken, Germany
autexier@ags.uni-sb.de

Abstract. A major source of the undecidability of a logic is the number of instances—the so-called multiplicities—of existentially quantified formulas that are required in a proof. We consider the problem in the context of matrix proof methods for classical higher-order logic and present a technique which improves the standard practice of iterative deepening over the multiplicities. We present a mechanism that allows to adjust multiplicities on demand during matrix-based proof search and not only preserves existing substitutions and connections, but additionally adapts them to the parts that result from the increase of the multiplicities.

1   Introduction

A major source of the undecidability of a logic is the number of instances—the so-called multiplicities—of existentially quantified formulas that are required in a proof. Specifying bounds for the *multiplicities* is impossible without sacrificing completeness. Proof procedures based on matrices [5,2] address this problem by fixing the multiplicities of existential quantifiers, create an initial expansion tree (resp. indexed formula tree) with the fixed multiplicity and then search for an admissible substitution and a spanning set of connections. The standard overall procedure then consists of performing an iterative deepening over the multiplicities until eventually the kernel procedure is successful—in case the conjecture is valid. Restarting the kernel proof procedure implies that substitutions and connections found in previous runs must be established again. Hence, there is a need to integrate the adjustment of the multiplicities into the kernel matrix proof search procedure.

In this article we present a new technique to include the increase of multiplicities into the kernel matrix proof search procedure and to carry over any information about substitutions and connections to the new parts. This technique was developed for extensional higher-order logic as well as for some (first-order) modal logics with constant domains. For sake of brevity we present here only the version for higher-order logic and refer the interested reader to [3].

In order to be complete with respect to higher-order logic, the substitutions need to be applied on the expansion tree, since whole new expansion trees can

B. Beckert (Ed): TABLEAUX 2005, LNAI 3702, pp. 48–62, 2005.

result from instantiation of literals with head variables, e.g. $X(t)$. Aiming at a complete proof procedure requires that each time a variable is instantiated, the multiplicity of its quantifier must be increased. Increasing the multiplicity while adapting existing connections and substitutions may in turn require to increase the multiplicities of further quantifiers. We present both a mechanism to increase the multiplicities of quantifiers as well as a constructive mechanism to determine, for given a variable which shall be instantiated, the minimal set of quantifiers for which the multiplicities must be increased.

The paper is organized as follows: Sect. 2 recapitulates the basic definitions for classical higher-order logic and Smullyan's uniform notation. Sect. 3 presents a sound and complete extensional expansion proof technique, which integrates the formulation from [2,9,10] with *indexed formula trees* from [12]. The dynamic increase of multiplicities is presented in Sect. 4 and illustrated by an example.

2   Preliminaries

For the definition of higher-order logic, we use a simple higher-order type system $\mathcal{T}$ [4], composed of a base type ι for individuals, a type o for formulas, and where $\tau \to \tau'$ denotes the type of functions from τ to τ'. As usual, we assume that the functional type constructor $\to$ associates to the right. We annotate constants f_τ and variables x_τ with types τ from $\mathcal{T}$ to indicate their type. A higher-order signature $\Sigma = (\mathcal{T}, \mathcal{F}, \mathcal{V})$ consists of types $\mathcal{T}$, constants $\mathcal{F}$ and variables $\mathcal{V}$, both typed over $\mathcal{T}$. The typed λ-calculus is standard and defined as follows:

Definition 1 (λ-Terms). *Let $\Sigma = (\mathcal{T}, \mathcal{F}, \mathcal{V})$ be a higher-order signature. Then the typed λ-terms $\mathcal{T}_{\Sigma,\mathcal{V}}$ over Σ and $\mathcal{V}$ are: (Var) for all $x_\tau \in \mathcal{V}$, $x \in \mathcal{T}_{\Sigma,\mathcal{V}}$ is a variable term of type τ; (Const) for all $c_\tau \in \mathcal{C}$, $c \in \mathcal{T}_{\Sigma,\mathcal{V}}$ is a constant term of type τ; (App) if $t : \tau, t' : \tau \to \tau' \in \mathcal{T}_{\Sigma,\mathcal{V}}$ are typed terms, then $(t'\ t) \in \mathcal{T}_{\Sigma,\mathcal{V}}$ is an application term of type τ'; (Abs) if $x : \tau \in \mathcal{V}$ and $t : \tau' \in \mathcal{T}_{\Sigma,\mathcal{V}}$, then $\lambda x_\tau . t \in \mathcal{T}_{\Sigma,\mathcal{V}}$ is an abstraction term of type $\tau \to \tau'$.*

Definition 2 (Substitutions). *Let $\Sigma = (\mathcal{T}, \mathcal{F}, \mathcal{V})$ be a higher-order signature. A substitution is a type preserving function[1] $\sigma : \mathcal{V} \to \mathcal{T}_{\mathcal{F},\mathcal{V}}$ that is the identity function on $\mathcal{V}$ except for finitely many elements from $\mathcal{V}$. This allows for a finite representation of σ as $\{t_1/x_1, \ldots, t_n/x_n\}$ where $\sigma(y) = y$ if $\forall 1 \le i \le n, y \ne x_i$.*

As usual we do not distinguish between a substitution and its homomorphic extension. A substitution σ is *idempotent* if, and only if, its homomorphic extension to terms is idempotent, i.e. $\sigma(\sigma(t)) = \sigma(t)$ for all $t \in \mathcal{T}_{\Sigma,\mathcal{V}}$. Given a substitution σ we denote by $dom(\sigma)$ the set of all variables for which $\sigma(x) \ne x$, i.e. the *domain* of σ. Higher-order λ-terms usually come with a certain set of reduction and expansion rules. We use the β reduction rule and the η expansion rule (see [4]), which give rise to the $\beta\eta$ long normal form, which is unique up to renaming of bound variables (α-equal). Throughout the rest of this paper we

---

[1] i.e. for all variables $x : \tau$, $\sigma(x)$ also has type τ.

α	α_0	α_1
$(\varphi \vee \psi)^+$	φ^+	ψ^+
$(\varphi \Rightarrow \psi)^+$	φ^-	ψ^+
$(\varphi \wedge \psi)^-$	φ^-	ψ^-
$(\neg\varphi)^+$	φ^-	$-$
$(\neg\varphi)^-$	φ^+	$-$

β	β_0	β_1
$(\varphi \wedge \psi)^+$	φ^+	ψ^+
$(\varphi \vee \psi)^-$	φ^-	ψ^-
$(\varphi \Rightarrow \psi)^-$	φ^+	ψ^-

γ	$\gamma_0(t)$
$(\forall x \cdot \varphi)^-$	$(\varphi[x/t])^-$
$(\exists x \cdot \varphi)^+$	$(\varphi[x/t])^+$

δ	$\delta_0(c)$
$(\forall x \cdot \varphi)^+$	$(\varphi[x/c])^+$
$(\exists x \cdot \varphi)^-$	$(\varphi[x/c])^-$

Fig. 1. Uniform Notation

assume substitutions are idempotent, all terms are in $\beta\eta$ long normal form, and substitution application includes $\beta\eta$-long normalisation.

For the semantics of higher-order logic we use the extensional general models from [7] by taking into account the corrections from [1]. It is based on the notion of *frames* that is a τ-indexed family $\{\overline{D}_\tau\}_{\tau \in \mathcal{T}}$ of nonempty domains, such that $\overline{D}_o = \{\top, \bot\}$ and $\overline{D}_{\tau \to \tau'}$ is a collection of functions mapping $\overline{D}_\tau$ into $\overline{D}_{\tau'}$. The members of $\overline{D}_o$ are called *truth values* and the members of $\overline{D}_\iota$ are called *individuals*. Given a frame $\{\overline{D}_\tau\}_{\tau \in \mathcal{T}}$, an *assignment* is a function ρ on $\mathcal{V}$ such that for each variable x_τ holds $\rho(x_\tau) \in \overline{D}_\tau$. Given an assignment ρ, a variable x_τ and an element $e \in \overline{D}_\tau$ we denote by $\rho[e/x]$ that assignment ρ' such that $\rho'(x_\tau) = e$ and $\rho'(y_{\tau'}) = \rho(y_{\tau'})$, if $y_{\tau'} \neq x_\tau$.

Definition 3 (Satisfiability & Validity). *A formula φ is satisfiable if, and only if, there is a model M and an assignment ρ such that $M^\rho(\varphi) = \top$. φ is true in a model M if, and only if, for all variable assignments ρ holds $M^\rho(\varphi) = \top$. φ is valid if, and only if, it is true in all models.*

In order to ease the formulation of our technique for the modal logics from [12], we formulate expansion trees close to the representation used in [12]. [12] uses indexed formula trees, which rely on the concept of polarities and uniform notation [12,6,11]. Polarities are assigned to formulas and sub-formulas and are either positive ($+$) or negative ($-$). Intuitively, positive polarity of a sub-formula indicates that it occurs in the succedent of a sequent in a sequent calculus proof and negative polarity is for formulas occurring in the antecedent of a sequent.

Formulas annotated with polarities are called *signed formulas*. Uniform notation assigns *uniform types* to signed formulas which encode their "behavior" in a sequent calculus proof: there are two propositional uniform types α and β for signed formulas where the respective sequent calculus decomposition rule results in a split of the proof (β) or not (α). Furthermore there are two types γ and δ for quantification over object variables, respectively for signed formulas where in the respective sequent calculus quantifier elimination rule the *Eigenvariable*-condition must hold (δ) or not (γ). In Fig. 1 we give the list of signed formulas for each uniform type. The labels of the first column indicate the uniform type of the signed formula and the labels of the columns containing the signed subformulas are the so-called *secondary types* of the signed subformulas.

In the following we agree to denote by $\alpha^p(\alpha_1^{p_1}, \alpha_2^{p_2})$ a signed formula of polarity p, uniform type α, and sub-formulas α_i with respective polarities p_i according the tables in Fig. 1. We use an analogous notation for signed formulas of the other uniform types, i.e. for $\gamma^p x . F^p$ and $\delta^p x . F^p$. Furthermore we define $\overline{\alpha}^p(F^q) :=$ $\alpha^p(F^q)$ and for $n > 1$, $\overline{\alpha}^p(F_1^{p_1}, \ldots, F_n^{p_n}) := \alpha^p(F_1^{p_1}, \overline{\alpha}^{p_2}(F_2^{p_2}, \ldots, F_n^{p_n}))$.

In this paper we are mainly concerned with signed formulas. To ease the presentation we extend the notion of satisfiability to signed formulas. In order to motivate this definition consider a sequent $\psi_1, \ldots, \psi_n \vdash \varphi$. It represents the proof status that we have to prove φ from the ψ_i. In terms of polarities, all the ψ_i have negative polarity while φ has positive polarity. The ψ_i are the assumptions and thus we consider the models that satisfy those formulas and prove that those models also satisfy φ. Hence, we define that a model M satisfies a *negative* formula ψ_i^- if, and only if, M satisfies ψ_i. From there we derive the dual definition for positive formulas, namely that a model M satisfies a positive formula F^+, if, and only if, M *does not* satisfy F. Formally:

Definition 4 (Satisfiability of Signed Formulas). *Let F^p be a signed formula of polarity p, M a model and ρ an assignment. Then we define: $M^\rho \models F^+$ holds if, and only if, $M^\rho \not\models F$. Conversely, we define $M^\rho \models F^-$ holds if, and only if, $M^\rho \models F$. We extend that notion to sets of signed formulas $\mathcal{F}$ by: $M^\rho \models \mathcal{F}$, if, and only if, for all $F^p \in \mathcal{F}$, $M^\rho \models F^p$.*

Lemma 1. *Let M be a model, ρ a variable assignment, F^q, G^r signed formulas of polarities q, r and p also some polarity. Then $M^\rho \models \alpha^p(F^q, G^r)$ holds if, and only if, both $M^\rho \models F^q$ and $M^\rho \models G^r$ hold; $M^\rho \models \beta^p(F^q, G^r)$ holds if, and only if, either $M^\rho \models F^q$ or $M^\rho \models G^r$ holds; $M^\rho \models \gamma^q x . F^q$ holds if, and only if, for all $a \in M_\tau$, $M^{\rho[a/x]} \models F^q$ holds; $M^\rho \models \delta^q x . F^q$ holds if, and only if, there is an $a \in M_\tau$ such that $M^{\rho[a/x]} \models F^q$ holds.*

Remark 1 (Notational Conventions). We denote formulas by capital Latin letters $A, B, \ldots, F, G, \ldots$, and formulas with holes by Greek letters $\varphi(.)$, which denotes λ-abstractions $\lambda x . \varphi$ where x occurs *at least once* in φ.

Definition 5 (Literals in Signed Formulas). *Let $\varphi(F)^p$ be a signed formula of polarity p, and F occurs exactly once. We say that F is a literal in $\varphi(F)^p$ if using the rules from Fig. 1 we can assign a polarity to F, but not to its sub-terms.*

3   Extensional Expansion Proofs

This section introduces extensional expansion trees in the style of the indexed formula trees from [12]. First we define the extensional expansion tree obtained initially from a signed formula, which we denote by *initial extensional expansion tree*. In a second step we define the transformation rules that insert new subtrees to the subtree, perform instantiations, and draw connections. The first kind of rules are those dealing with the introduction of Leibniz' equality, extensionality introduction for equations and equivalences, expansion of positive equivalences as well as the introduction of cut.

Definition 6 (Initial Extensional Expansion Tree). *We denote extensional expansion trees (EET) for some signed formula F^p by $\triangle^{F^p}$; we say that F^p is*

the label *of* $\triangle^{F^p}$. *In addition to the label of an EET* $\triangle$, *we define the* deep
signed formula $DF(\triangle)$ *contained in* $\triangle$.

- *Any literal* F^p *is a leaf node of an EET of polarity* p *and label* F^p; *it has no uniform type and* $DF(F^p) := F^p$.
- *If* $\triangle^{F^p}$ *is an EET for* F^p, *then* $\varpi^{-p}(\triangle^{F^p})$ *is an EET for* $\alpha^{-p}(F^p)$ *of polarity* $-p$ *and uniform type* α. *The secondary type of* $\triangle^{F^p}$ *is* α_1 *and* $DF(\varpi^{-p}(\triangle^{F^p}))$ $:= \alpha^{-p}(DF(\triangle^{F^p}))$.
- *If* $\triangle^{F^p}, \triangle^{G^q}$ *are EETs respectively for* F^p *and* G^q, *then* $\varpi^r(\triangle^{F^p}, \triangle^{G^q})$ *is an EET for* $\alpha^r(F^p, G^q)$ *of polarity* r *and uniform type* α. *The secondary types of* $\triangle^{F^p}$ *and* $\triangle^{G^q}$ *are* α_1 *and* α_2, *respectively, and* $DF(\varpi^r(\triangle^{F^p}, \triangle^{G^q})) := \alpha^r(DF(\triangle^{F^p}), DF(\triangle^{G^q}))$.
- *If* $\triangle^{F^p}, \triangle^{G^q}$ *are EETs respectively for* F^p *and* G^q, *then* $\beta^r(\triangle^{F^p}, \triangle^{G^q})$ *is an EET for* $\beta^r(F^p, G^q)$ *of polarity* r *and uniform type* β. *The secondary types of* $\triangle^{F^p}$ *and* $\triangle^{G^q}$ *are* β_1 *and* β_2, *respectively, and* $DF(\beta^r(\triangle^{F^p}, \triangle^{G^q})) := \beta^r(DF(\triangle^{F^p}), DF(\triangle^{G^q}))$.
- *If* $\triangle^{\varphi(X_i)^p}, 1 \leq i \leq m$ *are EETs for* $\varphi(X_i)^p$, *then* $\varpi^p x(\triangle^{\varphi(X_1)^p}, \ldots, \triangle^{\varphi(X_m)^p})$ *is an EET for* $\gamma^p x . \varphi(x)^p$ *of polarity* p, *uniform type* γ *and multiplicity* m. *The secondary types of the* $\triangle^{\varphi(X_i)^p}$ *are* γ_0 *and* $DF(\varpi^p x(\triangle^{\varphi(X_1)^p}, \ldots, \triangle^{\varphi(X_m)^p})) := \overline{\alpha}^p(DF(\triangle^{\varphi(X_1)^p}), \ldots, DF(\triangle^{\varphi(X_m)^p}))$. *We say that the* X_i *are* γ-*variables respectively bound on* $\triangle^{\varphi(X_i)^p}$.
- *If* $\triangle^{F^p}$ *is an EET for* F^p, *then* $\delta^p x \triangle^{F^p}$ *is an EET for* $\delta^p x . F^p$ *of polarity* p *and uniform type* δ. *The secondary type of* $\triangle^{F^p}$ *is* δ_0 *and* $DF(\delta^p x \triangle^{F^p}) := DF(\triangle^{F^p})$. *We say that* x *is a* δ-*variable.*

Remark 2. We denote by $\triangle_{\triangle^F}$ an extensional expansion tree which contains $\triangle^F$ as a subtree (not necessarily direct). We say that an extensional expansion tree has a *singular* multiplicity, if all its γ-type subtrees have multiplicity 1. Furthermore, for a given (initial) extensional expansion tree we denote the set of subtrees of secondary type γ_0 by Γ_0, and analogously we define Δ_0. Two subtrees $\triangle_1$ and $\triangle_2$ of some extensional expansion tree $\triangle$ are α-*related*, if, and only if, the smallest subtree of $\triangle$ which contains both $\triangle_1$ and $\triangle_2$ has uniform type α.

Example 1. As an example for an initial extensional expansion tree we consider the formula about natural numbers
$((\forall u . 0 + u = u) \wedge \forall x . \forall y . s(x) + y = s(x + y)) \Rightarrow \exists z . s(z) + (s(z) + c) = s(s(c))$
where 0_ι is the zero of natural numbers, $s_{\iota \to \iota}$ denote the successor and $+_{\iota \times \iota \to \iota}$ denotes the sum over natural numbers. Fig. 2 shows the initial extensional expansion tree with singular multiplicity for the positive formula.

Fig. 2. Initial extensional expansion tree for the running example

During proof search the γ-variables bound on nodes of secondary type γ_0 are instantiated. Following [9,10,12] the admissibility of a substitution σ is realized as an acyclicity check of a directed graph obtained from the structure of the extensional expansion tree and dependencies between binding nodes of the instantiated γ-variable X and binding nodes of δ-variables occurring in $\sigma(X)$. To this end we introduce the notions of *structural orderings* $\prec$ induced by the structure of an extensional expansion tree and of *quantifier orderings* $\prec$ induced by the substitution.

Definition 7 (Structural Ordering & Quantifier Ordering). *Let $\triangle$ be an extensional expansion tree and σ an (idempotent) substitution for γ-variables bound on γ_0-type nodes in $\triangle$ by terms containing only γ- and δ-variables also bound in $\triangle$. The* structural ordering $\prec$ *is a binary relation among the nodes in $\triangle$ defined by: $\triangle_1 \prec_\triangle \triangle_2$ iff $\triangle_1$ dominates $\triangle_2$ in $\triangle$. The* quantifier ordering $\prec^\sigma$ *induced by σ is the binary relation defined by: $\triangle_0 \prec^\sigma \triangle_1$ iff there is an $X \in dom(\sigma)$ bound on $\triangle_1$ and in $\sigma(X)$ occurs a δ-variable bound on $\triangle_0$.*

Definition 8 (Reduction-Relation & Admissible Substitutions). *Let $\triangle$ be an extensional expansion tree and σ a substitution. Then the* reduction relation $\lhd^\sigma_\triangle$ *is the transitive closure of the union of $\prec_\triangle$ and $\prec^\sigma$, i.e. $\lhd^\sigma_\triangle := (\prec_\triangle \cup \prec^\sigma)^+$. σ is* admissible *with respect to $\triangle$, if and only if $\lhd^\sigma_\triangle$ is irreflexive.*

The notion of admissibility of substitutions is equivalent to the notion defined in [12] if substitutions σ are idempotent. With respect to [9,10] our notion corresponds to the *dependency relation* among the instances of γ-variables.

Following [12,2] we define *(horizontal) paths* on extensional expansion trees.

Definition 9 (Paths). *Let $\triangle$ be an extensional expansion tree. Then a* path *in $\triangle$ is a sequence $\ll \triangle_1, \ldots, \triangle_n \gg$ of α-related subtrees in $\triangle$. The set of deep signed formulas of a path p is given by $DF(p) := \{DF(\triangle) \mid \triangle \in p\}$. The sets $\mathcal{P}(\triangle)$ of* paths through $\triangle$ *is the smallest set containing $\{\ll \triangle \gg\}$ and which is closed under the following operations: If $P \cup \{\ll \Gamma, \triangle' \gg\} \in \mathcal{P}(\triangle)$ then*

(α-**Dec**) *If $\triangle' := \alpha^p(\triangle_1, \triangle_2)$, then $P \cup \{\ll \Gamma, \triangle_1, \triangle_2 \gg\} \in \mathcal{P}(\triangle)$;*
(β-**Dec**) *If $\triangle' := \beta^p(\triangle_1, \triangle_2)$, then $P \cup \{\ll \Gamma, \triangle_1 \gg, \ll \Gamma, \triangle_2 \gg\} \in \mathcal{P}(\triangle)$;*
(γ-**Dec**) *If $\triangle' := \gamma^p x(\triangle_1, \ldots, \triangle_n)$, then $P \cup \{\ll \Gamma, \triangle_1, \ldots, \triangle_n \gg\} \in \mathcal{P}(\triangle)$.*
(δ-**Dec**) *If $\triangle' := \delta^p x(\triangle'')$, then $P \cup \{\ll \Gamma, \triangle'' \gg\} \in \mathcal{P}(\triangle)$.*

Definition 10 (Satisfiable & Unsatisfiable Ground Paths). *A path p is* ground *if all signed formulas in $DF(p)$ are ground. A ground path p is* satisfiable *if there exists a model M such that $M \models DF(p)$. Otherwise p is* unsatisfiable.

Corollary 1. *Let p be a path, such $DF(p)$ contains either $\top^+$, or $\bot^-$ or signed formulas F^- and F^+. Then any ground path for p is unsatisfiable, and so is p.*

Note the close relationship between the decomposition rules for paths for the different types of trees and the relationship between the satisfiability of the main signed formulas with respect to its constituent signed formulas: a path containing an α-type tree is replaced by a path containing both subtrees, while a path containing a β-type tree is decomposed into two paths each containing one of the subtrees. Analogously the decomposition of γ- and δ-type trees corresponds to the relationship between the satisfiability of the signed formula of the respective type to its constituent signed formulas. This relationship together with the acyclicity of the reduction ordering $\lhd_\triangle^\sigma$ entail that whenever we have obtained a set of paths where all paths are unsatisfiable, then the initial conjecture is valid. Indeed, if all paths are unsatisfiable, we can apply the model satisfiability rules backwards from the constituent signed formulas to the main signed formulas and obtain that there is no model M satisfying φ^+, i.e. $\forall M . M \not\models \varphi^+$, and thus it holds by definition $\forall M . M \models \varphi$. Hence φ is valid.

Analyzing the decomposition rules and the respective satisfiability relations in more detail, we observe that they are equivalence transformations which is necessary for our soundness and safeness results. *Safeness* means intuitively that no possible refutation is lost by such a transformation. The decomposition rules have this property and this allows to switch freely between the granularity of the decomposition of paths. However, the definition of further transformations on extensional expansion trees requires a weaker property which only requires the preservation of satisfiable paths during the transformation, which is the general condition for *soundness*. Formally:

Definition 11 (Soundness & Safeness). *Let $\triangle$, $\triangle'$ be two extensional expansion trees with respective admissible substitutions σ and σ', and $\triangle'$ results from $\triangle$ by some transformation. The transformation is sound if, and only if, if there is a satisfiable path in $\triangle$ then so there is in $\triangle'$. The transformation is safe if, and only if, if there is a satisfiable path in $\triangle'$ then so there is one in $\triangle$.*

Definition 12 (Connections). *Let $\triangle$ be an extensional expansion tree and p a path through $\triangle$. A connection of p is a pair of subtrees $(\triangle^{F^+}, \triangle^{F^-})$ of opposite polarities, of the same label F (α-equal) and which are both in p. A set C of connections is spanning for $\triangle$, if, and only if, there is a set of paths P through $\triangle$, such that each path in P either contains a connection from C or $\top^+$ or $\bot^-$.*

We denote a proof state of an extensional expansion proof by $\triangle; \sigma \rhd C$: It consists of an extensional expansion tree $\triangle$, an admissible substitution σ and a set of connections C. For a given closed formula we denote by $\triangle^{F^+,\mu}; id \rhd \emptyset$ the initial proof state, where $\triangle^{F^+,\mu}$ is the initial extensional expansion tree for F^+, μ the multiplicities of the γ-type subtrees and id denotes the empty substitution.

The extensional expansion proof procedure is defined by a set of transformation rules $\frac{\pi'}{\pi}$, which are to be read bottom-up, i.e. the problem of finding a spanning set of connections for π is reduced to find a spanning set of connections for π'. A rule is sound (resp. safe), if the transformation on the extensional expansion tree together with the substitution is sound (resp. safe) and the set of connections are indeed connections for the respective expansion trees. The rules are given in Fig. 3 and consist of (1) an axiom rule, (2) a rule to insert a connection, (3) the introduction of Leibniz' equality for equations and equivalences, (4) two rules for functional and boolean extensionality introduction, (5) a rule to expand positive equivalences into a conjunction of implications, (6) a substitution rule, and (7) a rule to introduce cuts.

$$\frac{\overline{}}{\triangle;\sigma \triangleright \mathcal{C}}\ Axiom$$

If $\mathcal{C}$ is a spanning set of connections for $\triangle$

$$\frac{\triangle';\sigma' \triangleright \{(\triangle^{F^-},\triangle^{F^+})\} \cup \mathcal{C}}{\triangle;\sigma \triangleright \mathcal{C}}\ Connect$$

If $\triangle^{F^-}$ and $\triangle^{F^+}$ are α-related in $\triangle$

$$\frac{\triangle_{\alpha(s=t^p,\triangle^{\forall q \cdot q(s) \Rightarrow q(t)^p})};\sigma \triangleright \mathcal{C}}{\triangle_{s=t^p};\sigma \triangleright \mathcal{C}}\ Leibniz_=$$

$$\frac{\triangle_{\alpha(F \Leftrightarrow G^p,\triangle^{\forall q \cdot q(F) \Rightarrow q(G)^p})};\sigma \triangleright \mathcal{C}}{\triangle_{F \Leftrightarrow G^p};\sigma \triangleright \mathcal{C}}\ Leibniz_\Leftrightarrow$$

$$\frac{\triangle_{\alpha((F \Leftrightarrow G)^+,\triangle_{(F \Rightarrow G) \wedge (G \Rightarrow F)^+})};\sigma \triangleright \mathcal{C}}{\triangle_{(F \Leftrightarrow G)^+};\sigma \triangleright \mathcal{C}}\ \Leftrightarrow\text{-}Elim$$

$$\frac{\sigma'(\triangle);\sigma' \circ \sigma \triangleright \sigma'(\mathcal{C})}{\triangle;\sigma \triangleright \mathcal{C}}\ Subst$$

if $\sigma' \circ \sigma$ admissible wrt. $\triangle$.

$$\frac{\triangle_{\alpha(s=t^p,\lambda x \cdot s = \lambda x \cdot t)};\sigma \triangleright \mathcal{C}}{\triangle_{s=t^p};\sigma \triangleright \mathcal{C}}\ f\text{-}Ext$$

if x local to $s=t$ in $\triangle_{s=t^p}$.

$$\frac{\triangle_{\alpha(A \Leftrightarrow B^p,\lambda x \cdot A = \lambda x \cdot B)};\sigma \triangleright \mathcal{C}}{\triangle_{A \Leftrightarrow B^p};\sigma \triangleright \mathcal{C}}\ b\text{-}Ext$$

if x local to $A \Leftrightarrow B$ in $\triangle_{A \Leftrightarrow B^p}$.

$$\frac{\triangle_{\alpha(\triangle^{FP},\triangle^{\gamma^p \overrightarrow{z}' \cdot \beta^p(A^-,A^+)})};\rho \circ \sigma \triangleright \mathcal{C}}{\triangle_{\triangle^{FP}};\sigma \triangleright \mathcal{C}}\ Cut\ A$$

where $\overrightarrow{x}'$ are new variables for the free γ- and δ-variables of A not bound above $\triangle^{FP}$ and $\rho := [\overrightarrow{x}'/\overrightarrow{x}]$

Fig. 3. The Extensional Expansion Proof Rules $\mathcal{EEP}$

To describe the transformation rules, we write $\alpha(\triangle, \triangle')$ to denote an expansion tree constructed from $\triangle$ and $\triangle'$ of uniform type α and *which has the same label and polarity than its first subtree* $\triangle$. For instance, $\alpha(s = t^p, \triangle^{\forall P \cdot P(s) \Rightarrow P(t)^p})$ is a subtree of uniform type α, *polarity p* and *label* $s = t^p$.

Axiom. The axiom rule terminates a proof, if the derived set of connections $\mathcal{C}$ is spanning for $\triangle$. The soundness is a direct consequence of Definition 12 and Corollary 1 and the admissibility of the substitution σ.

Insertion of Connections. Assume $\triangle^{F^+}$ and $\triangle^{F^-}$ are α-related subtrees in $\triangle$. Then there is a path through $\triangle$ which contains both $\triangle^{F^+}$ and $\triangle^{F^-}$, and hence we can make a connection between them. A connection between non-leaf trees $\triangle^{F^+}$ and $\triangle^{F^-}$ can always be propagated to connections between the subtrees. This possibly requires to increase the multiplicity of some γ-type node and applying a substitution in order to represent the connection, as is illustrated by the following two EETs for the same formula $\forall x \cdot F(x)$, but with opposite polarities: $\gamma^- x(F(a)^-)$ and $\delta^+ z(F(z)^+)$. Note that we have already instantiated the γ-variable x in the first tree by some a. Assuming the trees are α-related, we can make a connection between them[2]. In order to propagate the connection into the subtrees, we have to add an additional child to the first tree and instantiate it with the δ-variable z from the second tree, which results in $\gamma^- x(F(a)^-, F(z)^-)$. The new inserted subtree is maximal with respect to the $\lhd$ and remains maximal after subtitution, since its root–the binding position of the substituted variable– must be greater than the binding position of the substituted Eigenvariable. Hence the new substitution is admissible with respect to the new EET. For the special

---

[2] Note that whenever on one side we have a quantifer of type γ, the counter-part is a δ-type formula.

nodes $\alpha(s = t, \triangle')$ the connections are propagated into $\triangle'$ [3] . The connection rule we use always propagates connections to the leaf-nodes and therefore may change $\triangle$ and extend the overall substitution. The rule is both sound and safe.

$\Leftrightarrow$-*Elimination, Introduction of Leibniz' Equality.* These rules are only necessary since we the logical symbols "=" and "$\Leftrightarrow$" are part of your logical signature. The first replaces an equivalence $A \Leftrightarrow B$ by $(A \Rightarrow B) \wedge (B \Rightarrow A)$ and the latter replace equations $s = t$ and equivalences $A \Leftrightarrow B$ by Leibniz Definition of equality, i.e. $\forall q . q(s) \Rightarrow q(t)$ and $\forall q . q(A) \Rightarrow q(B)$. Due to the presence of the latter rules, the $\Leftrightarrow$-elimination rule can be safely restricted to positive formulas.

Functional and Boolean Extensionality. The functional and Boolean extensionality rules *f-Ext* and *b-Ext* require the variable x to be local with respect to the equation (resp. equivalence). This intuitively means that if x is a γ-variable, then it does not occur in a part which is β-related to the equation (resp. equivalence). If x is a δ-variable, then it does not occur in a part which is α-related[4].

The functional extensionality rule creates for a literal leaf node $s = t^p$, for which a variable x is local, a leaf node for $\lambda x . s = \lambda x . t^p$ and inserts it to $s = t^p$ by some node of uniform type α and label $s = t^p$. The Boolean extensionality rule is analogous for some literal leaf node $F \Leftrightarrow G^p$. The substitution remains admissible and the set of connections is still valid for the new expansion tree. Any path p which contained $s = t^p$ in the old expansion tree becomes $\ll p, \lambda x . s = \lambda x . t^p \gg$. It is easy to see that p is satisfiable if, and only if, $\ll p, \lambda x . s = \lambda x . t^p \gg$ is satisfiable.

Cut. The rule allows to insert a cut over some formula A at some arbitrary subtree $\triangle^{F^p}$ of the expansion tree. Assume $\overrightarrow{x}$ are the γ- and δ-variables that occur free in A and which are not bound by some binding node above $\triangle^{F^p}$, $\overrightarrow{x}'$ some new variables and $\rho := \{\overrightarrow{x}'/\overrightarrow{x}\}$. It creates an initial expansion tree for the signed formula $\gamma^p \overrightarrow{x}' . \beta^p(A^-, A^+)$ and inserts it on $\triangle^{F^p}$ by some α-type node of polarity p and label F^p. The admissibility of the substitution is trivially preserved since ρ creates a dependency only to subtrees adjacent to $\triangle^{F^p}$. Furthermore, any path p which contained F^p is split into two paths $\ll p, A^- \gg$ and $\ll p, A^+ \gg$. Hence, p is satisfiable, if, and only if, either $\ll p, A^- \gg$ or $\ll p, A^+ \gg$ is.

Remark 3 (Cut Elimination). Despite the admissibility of cut for the extensional expansion proofs from [10] on which we build upon, we present a set of rules which includes cut. The cut rule is only required in order to simulate the extensionality rule from [10]. The reason why we prefer the rules with cut is that the extensionality rule from [10] is less intuitive than our rules which are closer to the way extensionality is typically defined. We not only keep our rules for readability, but also because we use the extensional expansion trees presented here in the context of interactive theorem (see [3]), where intuitiveness of the rules matters much more than in automatic theorem proving. However, it seems to be straightforward to adapt the techniques presented in this paper to the extensional expansion trees from [10].

---

[3] This may require to add Leibniz' definition of equality for leaf nodes.
[4] For a complete formalization we refer to [3].

We denote the transformation rules shown in Fig. 3 by $\mathcal{EEP}$. Each application of a rule of name R is an *extensional expansion proof step* $\pi \vdash_R \pi'$, and $\vdash^*$ denotes its reflexive transitive closure. An extensional expansion proof with respect to $\mathcal{EEP}$ for some proof state π is a sequence of proof steps starting from π using rules from $\mathcal{EEP}$ and terminated with the *Axiom*-rule, i.e. $\pi \vdash^* \ldots \vdash_{Axiom}$. From [10] we can obtain the following soundness and completeness result:

Theorem 1 (Soundness & Completeness of $\mathcal{EEP}$). *F is a valid formula if, and only if, there exists a multiplicity μ for which there exists an extensional expansion proof with respect to $\mathcal{EEP}$ for $\triangle^{F^+,\mu}; id \rhd \emptyset$.*

Proof (Sketch). We proved that each rule preserves the existence of satisfiable paths, which proves the soundness of $\mathcal{EEP}$. Completeness of $\mathcal{EEP}$ follows from [10]: in contrast to [10] "$=$" and "$\Leftrightarrow$" are part of our basic set of logical symbols, while [10] replaces them by Leibniz' definition of equality resp. by a conjunction of implications. Therefore we need the rules $Leibniz_=$, $Leibniz_\Leftrightarrow$, and $\Leftrightarrow$-*Elim*. Our axiom rule is like in [10] and our connection rule is a macro for introducing many connections at once between leaf nodes. The functional and Boolean extensionality rules are different from those used in [10], but the later are admissible using our extensionality rules in combination with cut. Hence $\mathcal{EEP}$ is complete. $\qquad\square$

Example 2. We apply the $\mathcal{EEP}$-calculus on the initial extensional expansion tree from Fig. 2.

Using the rules (1) *Subst* $\sigma := \{0/X, s(0)+c/U, 0/Z, s(0)+c/Y\}$ (2) *Leibniz$_=$* on $\sigma(s(X)+Y = s(X+Y))$, (3) instantiating the set variable—introduced in 2—by $\lambda n \,.\, n = s(s(c))$ and (4) drawing a connection—indicated by the dashed line—between $s(0)+(s(0)+c) = s(s(c))^-$ and $s(0)+(s(0)+c) = s(s(c))^-$ results in the EET shown above. The proof is blocked in this situation since the initial multiplicities were insufficient.

4   Increase of Multiplicities

Proof search in extensional expansion proofs ([10]) proceeds by fixing the multiplicity of nodes of primary type γ, and subsequently searching for an admissible substitution and a spanning set of connections. To overcome the problem of guessing the right multiplicity beforehand there is a need to dynamically adjust multiplicities on the fly. An example is the instantiation of some γ-variable X bound on some γ_0-type subtree $\triangle_X$ of parent node $\triangle_\gamma$ of label $\gamma^p x \,.\, \varphi(x)$: in order to design a complete proof procedure, we must be able to "copy" that γ-variable before instantiating it. Furthermore, adding a new initial indexed formula tree $\triangle'_{X'}$ for $\varphi(X')$ for some new γ-variable X' and attaching it to $\triangle_\gamma$, prevents to carry over all proof information onto $\triangle'_{X'}$: instantiations, elimination of positive equivalences, Leibniz' equality introductions, extensionality introductions, and cuts are lost, as well as all connections that involved a subtree of $\triangle_X$ are not present for sub-nodes in $\triangle'_{X'}$. Finally, substitution information is not carried over onto $\triangle'_{X'}$. For instance, consider some γ-variable Y different from X which has been instantiated with a term containing X: there is no copy of Y which could be instantiated with the equivalent term containing X'. Thus, the proof information already established for $\triangle_X$ would have to be redone for $\triangle'_{X'}$.

In the following we present a mechanism to dynamically increase the multiplicity of some subtree which carries over all proof information to the new subtrees. First, note that the increase of multiplicity of some node may entail the increase of multiplicity of some other node. This is the case for example if a γ-variable to be copied occurs in the instance of some other γ-variable. Intuitively we need a notion of a *self contained set of subtrees with respect to a substitution*, in the sense that all γ- and δ-variables x which occur in the instance of some γ-variable Y, the set contains both the subtrees binding x and Y.

We formalize this intuition by the notion of a *convex set of subtrees with respect to some admissible substitution σ*.

Definition 13 (Convex Set of Subtrees). *Let $\triangle$ be an extensional expansion tree, σ an admissible substitution, and $\mathcal{K}$ a set of independent[5] subtrees of $\triangle$. $\mathcal{K}$ is convex with respect to σ if, and only if, for all $\triangle' \in \mathcal{K}$ and for all γ- and δ-variables x bound in $\triangle'$ we have: if x occurs in some instance $\sigma(Y)$ for some Y, then there exists some $\triangle'' \in \mathcal{K}$ in which Y is bound.*

A trivial example for such a set is the set that consists of the whole extensional expansion tree $\triangle$.

A convex set $\mathcal{K}$ of subtrees from some extensional expansion tree $\triangle$ has the property that it is not smaller with respect to $\lhd^\sigma_\triangle$ than any other part of $\triangle$ which is not in $\mathcal{K}$. In other words, there is no γ-variable bound outside $\mathcal{K}$ which is instantiated with some variable bound in $\mathcal{K}$. Consider the restriction $\sigma_\mathcal{K}$ of σ to those γ-variables that are bound in $\mathcal{K}$. Copying the subtrees from $\mathcal{K}$ yields a new set of subtrees $\mathcal{K}'$ and a renaming ρ of all γ- and δ-variables bound in $\mathcal{K}$. Then the renamed substitution $\sigma_{\mathcal{K}'} := \{\rho(\sigma_\mathcal{K}(x))/\rho(x) \mid x \in dom(\sigma_\mathcal{K})\}$ does not introduce any additional dependencies to parts not in $\mathcal{K}'$. Furthermore, if the original substitution was admissible, then so is $\sigma_{\mathcal{K}'} \circ \sigma$. In the following lemma we formalize this observation:

Lemma 2 (Maximality of Convex Sets of Subtrees). *Let $\triangle$ be an extensional expansion tree, σ an admissible substitution, and $\mathcal{K}$ a convex set of subtrees of $\triangle$. For all $x \in dom(\sigma)$, if x is not bound in $\mathcal{K}$, then all variables that occur in $\sigma(x)$ are also not bound in $\mathcal{K}$.*

Proof. Assume some x not bound in $\mathcal{K}$ and y bound in $\mathcal{K}$ that occurs in $\sigma(x)$. Then, by Definition 13, x should be bound in $\mathcal{K}$, which is a contradiction. □

---

[5] i.e. no nested subtrees.

Given a convex set of subtrees from Γ_0, we increase their parents multiplicities by first copying these subtrees and thereby renaming the variables if necessary. Note that we cannot just create extensional expansion trees, since we want to preserve the applications of rules, for instance Leibniz' equality introductions and extensionality introductions, during the copying process. From this copying process we obtain a renaming ρ of the copied γ- and δ-variables and an isomorphic function ι between the original subtrees and their copies. We agree that ρ is a total function, which is the identity function for all variables not bound in the copied subtrees, and ι is a total function which is the identity function on all nodes, except those occurring in the copied subtrees.

The new subtrees are also of secondary type γ_0 and are inserted as further children on the respective parent node of primary type γ, which increases their multiplicities. The renaming ρ and the node mapping ι are used in order to carry over the substitution information by enlarging the substitution σ by defining $\sigma' := \{\rho(\sigma(x))/\rho(x) \mid x \in dom(\rho)\}$ and use $\sigma' \circ \sigma$ as the new overall substitution.

Finally, the information about established connections in $\mathcal{C}$ is carried over by enlarging $\mathcal{C}$ using ι, i.e. we apply ι to the following connections $\mathcal{C}$ as follows: $\iota(\mathcal{C}) := \{(\iota(c), \iota(c')) \mid (c, c') \in \mathcal{C}, c \text{ or } c' \in dom(\iota)\}$.

Definition 14 (Multiplicity Increase). *Let $\triangle; \sigma \triangleright \mathcal{C}$ be an extensional expansion proof state and $\triangle_{X_1}, \ldots, \triangle_{X_n} \in \Gamma_0$ such that $\{\triangle_{X_1}, \ldots, \triangle_{X_n}\}$ is a convex set with respect to σ. Then the increase of multiplicities of their parent subtrees $\mathfrak{V}^p x_i(\ldots, \triangle_{X_i}, \ldots), 1 \leq i \leq n$ is defined by the rule*

$$\frac{\triangle_{\mathfrak{V}^p x_1(\ldots, \triangle_{X_1}, \triangle'_{X_1}, \ldots), \ldots, \mathfrak{V}^p x_n(\ldots, \triangle_{X_n}, \triangle'_{X_n}, \ldots)}; \sigma' \circ \sigma \triangleright \mathcal{C} \cup \iota(\mathcal{C})}{\triangle_{\mathfrak{V}^p x_1(\ldots, \triangle_{X_1}, \ldots), \ldots, \mathfrak{V}^p x_n(\ldots, \triangle_{X_n}, \ldots)}; \sigma \triangleright \mathcal{C}} \; \mu\text{-}Inc$$

where (1) $\triangle'_{X_i}$ are the copies of $\triangle_{X_i}$ and attached as additional subtrees to the respective γ-type parents, (2) ρ and ι are respectively the re-namings and the mapping of subtrees obtained from the copying, and (3) $\sigma' := [\rho(\sigma(X))/\rho(X) \mid$ for all γ-variables $X \in dom(\rho)]$.

We now introduce a constructive mechanism to determine the minimal set of nodes whose multiplicities need to be increased when increasing the multiplicity of some given node. Intuitively, if we have to copy a subtree $\triangle_m$ which contains the binding subtree of some γ-variable that occurs in the instance of some further γ-variable bound on some $\triangle'$, then we must copy $\triangle'$ as well.

Definition 15 (Determining Nodes to Increase Multiplicities). *Let $\triangle$ be an extensional expansion tree, σ an admissible substitution, and $\triangle_m$ a subtree of $\triangle$ of secondary type γ_0. The subtrees to copy in order to increase the multiplicity of $\triangle_m$'s parent are given by the $\mu(\triangle_m)$ that is inductively defined:*

$$\mu(\triangle_m) = \{\triangle_m\} \cup \left(\bigcup_{\triangle' \mid \triangle_m \prec_\triangle \triangle'} \mu(\triangle') \right) \cup \left(\bigcup_{\triangle' \in Inst_\triangle(\triangle_m)} \mu(\triangle') \right)$$

where $Inst_\triangle(\triangle_m)$ is the set of binding nodes of γ-variables X, such that y occurs in $\sigma(X)$ and y is bound on $\triangle_m$. If $\triangle_m$ is not a binding node, then $Inst_\triangle(\triangle_m) = \emptyset$. We denote by $\mu(\triangle_m)_{min}$ the minimal subtrees of $\mu(\triangle_m)$ with respect to $\triangleleft_\triangle^\sigma$.

Lemma 3. *Let $\triangle$ be an extensional expansion tree of secondary type γ_0 and σ an admissible substitution. Then $\mu(\triangle)_{min}$ (1) is a convex set of subtrees with respect to σ and (2) it contains only subtrees of secondary type γ_0.*

Proof (Sketch). For (1) we prove that $\mu(\triangle)$ is always a convex set of *dependent* subtrees by deriving a contradiction from assuming there is some $\triangle_x \in \mu(\triangle)$ binding some γ- or δ-variable x, x occurs in $\sigma(y)$, and y is bound on $\triangle_y \notin \mu(\triangle)$. Then $\mu(\triangle_m)_{min}$ is a convex set of subtrees is an easy consequence. (2) The elements of $\mu(\triangle_m)$ are all subtrees of $\triangle_m$ and all subtrees of trees binding some γ-variable Y. Since $\triangle_m$ has type γ_0, all minimal subtrees are of type γ_0.   □

Example 3. We illustrate the dynamic increase of multiplicities using our running example. We redo the same proof as before, but always increase the multiplicities of the quantifiers of the substituted variables. We use the function from Definition 15 to determine the set of convex subtrees in each case. Using the rules (1) Increase of multiplicities for $\eth^- u$ and $\eth^- x$ and (2) *Subst* $\sigma := \{0/X, Y/U\}$ results in the EET

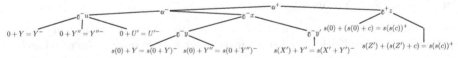

In order to apply the substitution $\sigma' := \{s(0) + c/Y, 0/Z\}$ in a safe way, we have to increase the multiplicities of the parents of Y and Z. However, $s(0) + Y = s(0 + Y)^-$ is the binding node of Y and $\mu(s(0) + Y = s(0 + Y)^-)_{min} := \{s(0) + Y = s(0 + Y)^-, 0 + Y = Y^-\}$. This indicates that if we want to increase the multiplicity of $\eth^- y$ we also have to increase the multiplicity of $\eth^- u$, since U has been substituted by Y already. Increasing the multiplicities of $\eth^- y, \eth^- u$ and $\eth^- z$ and applying the renaming $\rho := \{U''/U, Y''/Y, Z'/Z\}$ and only then applying σ' results in the extensional expansion tree

The overall substitution now is $\{0/X, s(0) + c/U, Y''/U'', s(0) + c/Y, 0/Z\}$. Applying, as before, *Leibniz$_=$*, then increase the multiplicity of $\eth^- p$ before substituting P with $\lambda n \cdot n = s(s(c))$ and drawing the same connection we obtain

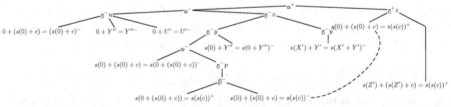

Now the proof is not blocked and can be continued. However, due to lack of space we do not present the extensional expansion tree of the final proof state.

Lemma 4 (Soundness & Safeness of Multiplicity Increase). *The increase of multiplicities is sound and safe.*

Proof. We first prove that the increase of multiplicities preserves the admissibility of the substitution and then that the resulting EET contains satisfiable paths, if, and only if, the former EET does.

(1) By Lemma 2 the convex set $\{\triangle_{x_1}, \ldots, \triangle_{x_n}\}$ is maximal with respect to $\triangleleft^{\triangle}_{\sigma}$. Thus there are disjoint substitutions σ_1, σ_2 such that the overall substitution $\sigma := \sigma_1 \uplus \sigma_2$ and $dom(\sigma_2)$ contains exactly all γ-variables from $dom(\sigma)$ which are bound in some $\triangle_{X_i}$. If σ is irreflexive, then $\triangleleft^{\sigma_2}_{\triangle_{X_i}, 1 \leq n}$ is irreflexive and for any $x \in dom(\sigma_2)$, $\sigma_2(x)$ contains only γ- and δ-variables bound in some $\triangle_{X_i}$. Furthermore, $\sigma' := \{\rho(\sigma_2(x))/\rho(x) \mid x \in dom(\sigma_2)\}$ and $\triangleleft^{\sigma'}_{\triangle_{X'_i}, 1 \leq n}$ is irreflexive since $\triangle_{X_i}$ and $\triangle'_{X'_i}$ are isomorphic. Since σ_2 was disjoint from σ_1 and ρ is a renaming of all γ- and δ-variables in σ_2, so is σ' from both σ_1 and σ_2. Hence $\triangleleft^{\sigma' \circ \sigma}_{\triangle}$ is irreflexive wrt. the new EET.

(2) For each $\triangle_{X_i}$ the increase of multiplicities adds its renamed copy $\triangle'_{X'_i}$ to each path that contained $\triangle_{X_i}$. Thus, if $\mathsf{DF}(\triangle_{X_i}) = F^p$, then $\mathsf{DF}(\triangle'_{X'_i}) = \rho(F^p)$. Since ρ is simply a renaming of variables, it trivially holds that F^p is satisfiable if, and only if, $\rho(F^p)$ is satisfiable. $\square$

Theorem 2 (Soundness & Completeness of $\mathcal{EEP} + \mu\text{-}Inc$). *$F$ is a valid formula if, and only if, there exists an extensional expansion proof using $\mathcal{EEP}$ augmented by μ-Inc for $\triangle^{F^+, \mu_1}; id \triangleright \emptyset$, where μ_1 denotes the singular multiplicity.*

Proof (Sketch). The proof is based on the fact that the only rule which is unsafe is the substitution rule. Combining the instantiation of some γ-variable X by some term t with the increase of the multiplicities of the subtrees determined by $\mu(\triangle_X)_{min}$ before application of the substitution we can make it a safe transformation. Let $\triangle$ be some $\triangle \in \mu(\triangle_X)_{min}$; any path $\ll \Gamma, \triangle \gg$ has the deep signed formulas $\{\varphi(X)\} \cup \mathsf{DF}(\Gamma)$. After increase of the multiplicities and application of the substitution, this path becomes $\ll \Gamma', \Gamma, \triangle, \triangle' \gg$, where $\triangle'$ is the "copy" of $\triangle$ and Γ' results from increasing the multiplicities of the other elements of $\triangle \in \mu(\triangle_X)_{min}$. The deep signed formulas now are $\{\varphi(t), \varphi(X')\} \cup \mathsf{DF}(\Gamma) \cup \mathsf{DF}(\Gamma')$, where X' is the new γ-variable for X. If this set of formulas is satisfiable, then so is $\{\varphi(X)\} \cup \mathsf{DF}(\Gamma)$. Hence the transformation is safe. $\square$

5   Related Work and Conclusion

Designing a complete matrix proof search procedure requires to search for the right multiplicities. We presented a mechanism to increase the multiplicities which includes the search for the multiplicities into the kernel matrix proof search procedure. It supports a demand-driven increase of the multiplicities which not only preserves existing substitutions and connections, but moreover inherits this information to the new subtrees that result from the increase of multiplicities.

As an easy consequence the transformation rules that result from the addition of the multiplicity increasing rule are sound and complete. Although the technique was developed also for some first-order classical modal logics with constant domains, this paper described the technique only for higher-order logic matrix calculi. The definition of the technique for the first-order modal logics is based on indexed formula trees from [12] and essentially consists of extending the notion of convex set of subtrees to take into account the modal quantifiers and treat them analogous to the variable quantifiers. Details of it can be found in [3].

Issar [8] presented a respective procedure for the JForms introduced in [2], but restricts it to the first-order fragment. It increases the multiplicities on demand for some problematic path for which no connection can be found. The increase of multiplicities is realized by copying and renaming a part of the JForm, but restricting the effect to the problematic path in order to control the increase of the number of new paths. However, any information about substitutions and connections that existed for the copied parts are not adapted during copying. The procedure is complete as substitutions are not applied to the JForm, which is probably a major reason why the technique is restricted to first-order logic.

References

1. Peter B. Andrews. General models, descriptions, and choice in type theory. *The Journal of Symbolic Logic*, 37(2):385–397, June 1972.
2. Peter B. Andrews. Theorem proving via general matings. *Journal of the Association for Computing Machinery*, 28(2):193–214, April 1981.
3. Serge Autexier. *Hierarchical Contextual Reasoning*. PhD thesis, Computer Science Department, Saarland University, Saarbrücken, Germany, 2003.
4. Henk P. Barendregt. *The Lambda Calculus – Its Syntax and Semantics*. North Holland, 1984.
5. Wolfgang Bibel. On matrices with connections. *Journal of the Association for Computing Machinery*, 28(4):633–645, October 1981.
6. Melvin Fitting. Tableau methods of proof for modal logics. *Notre Dame Journal of Formal Logic*, XIII:237–247, 1972.
7. Leon Henkin. Completeness in the theory of types. *The Journal of Symbolic Logic*, 15:81–91, 1950.
8. Sunil Issar. Path-focused duplication: A search procedure for general matings. In T. S. W. Dietterich, editor, *Proceedings of the 8th National Conference on Artificial Intelligence (AAAI 90)*, volume 1, pages 221–226, Menlo Park - Cambridge - London, July 1990. AAAI Press / MIT Press.
9. Dale A. Miller. *Proofs in Higher-Order Logic*. Phd thesis, Carnegie Mellon University, 1983.
10. Frank Pfenning. *Proof Transformation in Higher-Order Logic*. Phd thesis, Carnegie Mellon University, 1987.
11. R. M. Smullyan. *First-Order Logic*, volume 43 of *Ergebnisse der Mathematik*. Springer-Verlag, Berlin, 1968.
12. Lincoln Wallen. *Automated proof search in non-classical logics: efficient matrix proof methods for modal and intuitionistic logics*. MIT Press series in artificial intelligence, 1990.

A Tableau-Based Decision Procedure for Right Propositional Neighborhood Logic

Davide Bresolin and Angelo Montanari

Department of Mathematics and Computer Science, University of Udine, Italy
{bresolin, montana}@dimi.uniud.it

Abstract. Propositional interval temporal logics are quite expressive temporal logics that allow one to naturally express statements that refer to time intervals. Unfortunately, most such logics turned out to be (highly) undecidable. To get decidability, severe syntactic and/or semantic restrictions have been imposed to interval-based temporal logics that make it possible to reduce them to point-based ones. The problem of identifying expressive enough, yet decidable, new interval logics or fragments of existing ones which are *genuinely* interval-based is still largely unexplored. In this paper, we make one step in this direction by devising an original tableau-based decision procedure for the future fragment of Propositional Neighborhood Interval Temporal Logic, interpreted over natural numbers.

1 Introduction

Propositional interval temporal logics are quite expressive temporal logics that provide a natural framework for representing and reasoning about temporal properties in several areas of computer science. Among them, we mention Halpern and Shoham's Modal Logic of Time Intervals (HS) [6], Venema's CDT logic [10], Moszkowski's Propositional Interval Temporal Logic (PITL) [9], and Goranko, Montanari, and Sciavicco's Propositional Neighborhood Logic (PNL) [2] (an up-to-date survey of the field can be found in [4]). Unfortunately, most such logics turned out to be (highly) undecidable. To get decidability, severe syntactic and/or semantic restrictions have been imposed to make it possible to reduce them to point-based ones, thus leaving the problem of identifying expressive enough, yet decidable, new interval logics or fragments of existing ones which are *genuinely* interval-based largely unexplored. In this paper, we make one step in this direction by devising an original tableau-based decision procedure for the future fragment of PNL, interpreted over natural numbers.

Interval logics make it possible to express properties of *pairs* of time points (think of intervals as constructed out of points), rather than *single* time points. In most cases, this feature prevents one from the possibility of reducing interval-based temporal logics to point-based ones. However, there are a few exceptions where the logic satisfies suitable *syntactic* and/or *semantic restrictions*, and such a reduction can be defined, thus allowing one to benefit from the good computational properties of point-based logics [8].

B. Beckert (Ed): TABLEAUX 2005, LNAI 3702, pp. 63–77, 2005.

One can get decidability by making a suitable choice of the interval modalities. This is the case with the $\langle B \rangle \langle \overline{B} \rangle$ and $\langle E \rangle \langle \overline{E} \rangle$ fragments of HS. Given a formula ϕ and an interval $[d_0, d_1]$, $\langle B \rangle \phi$ (resp. $\langle \overline{B} \rangle \phi$) holds over $[d_0, d_1]$ if ϕ holds over $[d_0, d_2]$, for some $d_0 \leq d_2 < d_1$ (resp. $d_1 < d_2$), and $\langle E \rangle \phi$ (resp. $\langle \overline{E} \rangle \phi$) holds over $[d_0, d_1]$ if ϕ holds over $[d_2, d_1]$, for some $d_0 < d_2 \leq d_1$ (resp. $d_2 < d_1$). Consider the case of $\langle B \rangle \langle \overline{B} \rangle$ (the case of $\langle E \rangle \langle \overline{E} \rangle$ is similar). As shown by Goranko et al. [4], the decidability of $\langle B \rangle \langle \overline{B} \rangle$ can be obtained by embedding it into the propositional temporal logic of linear time LTL[F,P] with temporal modalities F (sometime in the future) and P (sometime in the past). The formulae of $\langle B \rangle \langle \overline{B} \rangle$ are simply translated into formulae of LTL[F,P] by a mapping that replaces $\langle B \rangle$ by P and $\langle \overline{B} \rangle$ by F. LTL[F,P] has the finite model property and is decidable.

As an alternative, decidability can be achieved by constraining the classes of temporal structures over which the interval logic is interpreted. This is the case with the so-called Split Logics (SLs) investigated by Montanari et al. in [7]. SLs are propositional interval logics equipped with operators borrowed from HS and CDT, but interpreted over specific structures, called split structures. The distinctive feature of split structures is that every interval can be 'chopped' in at most one way. The decidability of various SLs has been proved by embedding them into first-order fragments of monadic second-order decidable theories of time granularity (which are proper extensions of the well-known monadic second-order theory of one successor S1S).

Another possibility is to constrain the relation between the truth value of a formula over an interval and its truth value over subintervals of that interval. As an example, one can constrain a propositional variable to be true over an interval if and only if it is true at its starting point (*locality*) or can constrain it to be true over an interval if and only it it is true over all its subintervals (*homogeneity*). A decidable fragment of PITL extended with quantification over propositional variables (QPITL) has been obtained by imposing the *locality* constraint. By exploiting such a constraint, decidability of QPITL can be proved by embedding it into quantified LTL. (In fact, as already noted by Venema, the locality assumption yields decidability even in the case of quite expressive interval logics such as HS and CDT.)

A major challenge in the area of interval temporal logics is thus to identify *genuinely* interval-based decidable logics, that is, logics which are not explicitly translated into point-based logics and not invoking locality or other semantic restrictions. In this paper, we propose a tableau-based decision procedure for the future fragment of (strict) Propositional Neighborhood Logic, that we call Right Propositional Neighborhood Logic (RPNL$^-$ for short), interpreted over natural numbers. While various tableau methods have been developed for linear and branching time point-based temporal logics, not much work has been done on tableau methods for interval-based temporal logics. One reason for this disparity is that operators of interval temporal logics are in many respects more difficult to deal with [5]. As an example, there exist straightforward inductive definitions of the basic operators of point-based temporal logics, while inductive definitions of interval modalities turn out to be much more complex. In [3,5], Goranko et al.

propose a general tableau method for CDT, interpreted over partial orders. It combines features of the classical tableau method for first-order logic with those of explicit tableau methods for modal logics with constraint label management, and it can be easily tailored to most propositional interval temporal logics proposed in the literature. However, it only provides a semi-decision procedure for unsatisfiability. By combining syntactic restrictions (future temporal operators) and semantic ones (the domain of natural numbers), we succeeded in devising a tableau-based decision procedure for RPNL$^-$. Unlike the case of the $\langle B \rangle \langle \overline{B} \rangle$ and $\langle E \rangle \langle \overline{E} \rangle$ fragments, in such a case we cannot abstract way from the left endpoint of intervals: there can be contradictory formulae that hold over intervals that have the same right endpoint but a different left one. The proposed tableau method partly resembles the tableau-based decision procedure for LTL [11]. However, while the latter takes advantage of the so-called fix-point definition of temporal operators which makes it possible to proceed by splitting every temporal formula into a (possibly empty) part related to the current state and a part related to the next state, and to completely forget the past, our method must also keep track of universal and (pending) existential requests coming from the past.

The rest of the paper is organized as follows. In Section 2, we introduce the syntax and semantics of RPNL$^-$. In Section 3, we give an intuitive account of the proposed method. In Section 4, we present our decision procedure, we prove its soundness and completeness, and we address complexity issues. In Section 5, we show our procedure at work on a simple example. Conclusions provide an assessment of the work and outline future research directions.

2   The Logic RPNL$^-$

In this section, we give syntax and semantics of RPNL$^-$ interpreted over natural numbers or over a prefix of them. To this end, we introduce some preliminary notions. Let $\mathbb{D} = \langle D, < \rangle$ be a strict linear order, isomorphic to the set $\mathbb{N}$ of natural numbers or to a prefix of them. A strict *interval* on $\mathbb{D}$ is an ordered pair $[d_i, d_j]$ such that $d_i, d_j \in D$ and $d_i < d_j$. The set of all strict intervals on $\mathbb{D}$ will be denoted by $\mathbb{I}(\mathbb{D})^-$ (notice that every interval $[d_i, d_j] \in \mathbb{I}(\mathbb{D})^-$ contains only a finite number of points). The pair $\langle \mathbb{D}, \mathbb{I}(\mathbb{D})^- \rangle$ is called an *interval structure*.

RPNL$^-$ is a propositional interval temporal logic based on the neighborhood relation between intervals. Its formulae consist of a set AP of propositional letters $p, q, \ldots$, the Boolean connectives $\neg$ and $\vee$, and the future temporal operator $\langle A \rangle$. The other Boolean connectives, as well as the logical constants $\top$ (true) and $\bot$ (false), are defined in the usual way. Furthermore, we introduce the temporal operator $[A]$ as a shorthand for $\neg \langle A \rangle \neg$. The *formulae* of RPNL$^-$, denoted by $\varphi, \psi, \ldots$, are recursively defined by the following grammar:

$$\varphi = p \mid \neg \varphi \mid \varphi \vee \varphi \mid \langle A \rangle \varphi.$$

We denote by $|\varphi|$ the size of φ, that is, the number of symbols in φ (in the following, we shall use $|\ |$ to denote the cardinality of a set as well). Whenever there are no ambiguities, we call an RPNL$^-$ formula just a formula. A formula of the form $\langle A \rangle \psi$ or $\neg \langle A \rangle \psi$ is called a *temporal formula* (from now on, we identify $\neg \langle A \rangle \psi$ with $[A] \neg \psi$), while a formula devoid of temporal operators is called a *state formula* (state formulae are formulae of propositional logic).

A *model* for an RPNL$^-$ formula is a tuple $\mathbf{M}^- = \langle \langle \mathbb{D}, \mathbb{I}(\mathbb{D})^- \rangle, \mathcal{V} \rangle$, where $\langle \mathbb{D}, \mathbb{I}(\mathbb{D})^- \rangle$ is an interval structure and $\mathcal{V} : \mathbb{I}(\mathbb{D})^- \longrightarrow 2^{AP}$ is the *valuation function* that assigns to every interval the set of propositional letters true on it.

Let $\mathbf{M}^- = \langle \langle \mathbb{D}, \mathbb{I}(\mathbb{D})^- \rangle, \mathcal{V} \rangle$ be a model and let $[d_i, d_j] \in \mathbb{I}(\mathbb{D})^-$. The semantics of RPNL$^-$ is defined recursively by the *satisfiability relation* $\models$ as follows:

- for every propositional letter $p \in AP$, $\mathbf{M}^-, [d_i, d_j] \models p$ iff $p \in \mathcal{V}([d_i, d_j])$;
- $\mathbf{M}^-, [d_i, d_j] \models \neg \psi$ iff $\mathbf{M}^-, [d_i, d_j] \not\models \psi$;
- $\mathbf{M}^-, [d_i, d_j] \models \psi_1 \vee \psi_2$ iff $\mathbf{M}^-, [d_i, d_j] \models \psi_1$, or $\mathbf{M}^-, [d_i, d_j] \models \psi_2$;
- $\mathbf{M}^-, [d_i, d_j] \models \langle A \rangle \psi$ iff $\exists d_k \in D$, $d_k > d_j$, such that $\mathbf{M}^-, [d_j, d_k] \models \psi$.

Let d_0 be the initial element of D and let d_1 be its successor. The satisfiability problem for an RPNL$^-$ formula φ with respect to the *initial interval* $[d_0, d_1]$ of the structure is defined as follows: φ is satisfiable in a model $\mathbf{M}^- = \langle \langle \mathbb{D}, \mathbb{I}(\mathbb{D})^- \rangle, \mathcal{V} \rangle$ if and only if $\mathbf{M}^-, [d_0, d_1] \models \varphi$.

3   The Proposed Solution

In this section we give an intuitive account of the proposed tableau-based decision procedure for RPNL$^-$. To this end, we introduce the main features of a model building process that, given a formula φ to be checked for satisfiability, generates a model for it (if any) step by step. Such a process takes into consideration one element of the temporal domain at a time and, at each step, it progresses from one time point to the next one. For the moment, we completely ignore the problem of termination. In the next section, we shall show how to turn this process into an effective procedure.

Let $D = \{d_0, d_1, d_2, \ldots\}$ be the temporal domain, which we assumed to be isomorphic to $\mathbb{N}$ or to a prefix of it. The model building process begins from time point d_1 by considering the initial interval $[d_0, d_1]$. It associates with $[d_0, d_1]$ the set $C_{[d_0, d_1]}$ of all and only the formulae which hold over it.

Next, it moves from d_1 to its immediate successor d_2 and it takes into consideration the two intervals ending in d_2, namely, $[d_0, d_2]$ and $[d_1, d_2]$. As before, it associates with $[d_1, d_2]$ (resp. $[d_0, d_2]$) the set $C_{[d_1, d_2]}$ (resp. $C_{[d_0, d_2]}$) of all and only the formulae which hold over $[d_1, d_2]$ (resp. $[d_0, d_2]$). Since $[d_1, d_2]$ is a right neighbor of $[d_0, d_1]$, if $[A] \psi$ holds over $[d_0, d_1]$, then ψ must hold over $[d_1, d_2]$. Hence, for every formula $[A] \psi$ in $C_{[d_0, d_1]}$, it puts ψ in $C_{[d_1, d_2]}$. Moreover, since every interval which is a right neighbor of $[d_0, d_2]$ is also a right neighbor of $[d_1, d_2]$, and vice versa, for every formula ψ of the form $\langle A \rangle \xi$ or $[A] \xi$, ψ holds over $[d_0, d_2]$ if and only if it holds over $[d_1, d_2]$. Accordingly, it requires that $\psi \in C_{[d_0, d_2]}$ if and only if $\psi \in C_{[d_1, d_2]}$. Let us denote by $\text{REQ}(d_2)$ the set of formulae of the form $\langle A \rangle \psi$ or $[A] \psi$ which hold over an interval ending in d_2 (by analogy, let $\text{REQ}(d_1)$ be the set of formulae of the form $\langle A \rangle \psi$ or $[A] \psi$ which hold

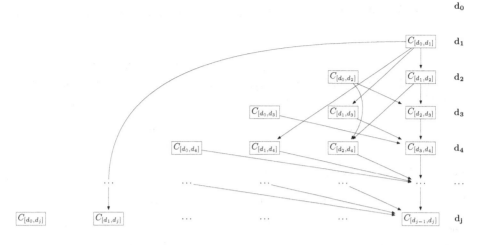

Fig. 1. The layered structure

over an interval ending in d_1, that is, the formulae $\langle A \rangle \psi$ or $[A]\psi$ which hold over $[d_0, d_1]$).

Next, the process moves from d_2 to its immediate successor d_3 and it takes into consideration the three intervals ending in d_3, namely, $[d_0, d_3]$, $[d_1, d_3]$, and $[d_2, d_3]$. As at the previous steps, for $i = 0, 1, 2$, it associates the set $C_{[d_i, d_3]}$ with $[d_i, d_3]$. Since $[d_1, d_3]$ is a right neighbor of $[d_0, d_1]$, for every formula $[A]\psi \in$ REQ(d_1), $\psi \in C_{[d_1, d_3]}$. Moreover, $[d_2, d_3]$ is a right neighbor of both $[d_0, d_2]$ and $[d_1, d_2]$, and thus for every formula $[A]\psi \in$ REQ(d_2), $\psi \in C_{[d_2, d_3]}$. Finally, for every formula ψ of the form $\langle A \rangle \xi$ or $[A]\xi$, we have that $\psi \in C_{[d_0, d_3]}$ if and only if $\psi \in C_{[d_1, d_3]}$ if and only if $\psi \in C_{[d_2, d_3]}$.

Next, the process moves from d_3 to its successor d_4 and it repeats the same operations; then it moves to the successor of d_4, and so on.

The layered structure generated by the process is graphically depicted in Figure 1. The first layer correspond to time point d_1, and for all $i > 1$, the i-th layer corresponds to time point d_i. If we associate with each node $C_{[d_i, d_j]}$ the corresponding interval $[d_i, d_j]$, we can interpret the set of edges as the neighborhood relation between pairs of intervals. As a general rule, given a time point $d_j \in D$, for every $d_i < d_j$, the set $C_{[d_i, d_j]}$ of all and only the formulae which hold over $[d_i, d_j]$ satisfies the following conditions:

- since $[d_i, d_j]$ is a right neighbor of every interval ending in d_i, for every formula $[A]\psi \in$ REQ(d_i), $\psi \in C_{[d_i, d_j]}$;
- since every right neighbor of $[d_i, d_j]$ is also a right neighbor of all intervals $[d_k, d_j]$ belonging to layer d_j, for every formula ψ of the form $\langle A \rangle \xi$ or $[A]\xi$, $\psi \in C_{[d_i, d_j]}$ if and only if it belongs to all sets $C_{[d_k, d_j]}$ belonging to the layer.

As we shall show in the next section, the layers of the structure depicted in Figure 1 will become the (macro)nodes of the tableau for φ, whose edges will connect ordered pairs of nodes corresponding to consecutive layers.

4   A Tableau-Based Decision Procedure for RPNL⁻

4.1   Basic Notions

Let φ be an RPNL⁻ formula to check for satisfiability and let AP be the set of its propositional variables. For the sake of brevity, we use $(A)\psi$ as a shorthand for both $\langle A \rangle \psi$ and $[A]\psi$.

Definition 1. *The* closure $\mathrm{CL}(\varphi)$ *of a formula* φ *is the set of all subformulae of* φ *and of their single negations (we identify* $\neg\neg\psi$ *with* ψ*).*

Lemma 1. *For every formula* φ*,* $|\mathrm{CL}(\varphi)| \leq 2 \cdot |\varphi|$*.*

Lemma 1 can be easily proved by induction on the structure of φ.

Definition 2. *The set of temporal requests of a formula* φ *is the set* $\mathrm{TF}(\varphi)$ *of all temporal formulae in* $\mathrm{CL}(\varphi)$*, that is,* $\mathrm{TF}(\varphi) = \{(A)\psi \in \mathrm{CL}(\varphi)\}$*.*

We are now ready to introduce the key notion of atom.

Definition 3. *Let* φ *be a formula of RPNL⁻. A* φ-atom *is a pair* (A, C)*, with* $A \subseteq \mathrm{TF}(\varphi)$ *and* $C \subseteq \mathrm{CL}(\varphi)$*, such that:*
- *for every* $(A)\psi \in \mathrm{TF}(\varphi)$*, if* $(A)\psi \in A$*, then* $\neg(A)\psi \notin A$*;*
- *for every* $\psi \in \mathrm{CL}(\varphi)$*,* $\psi \in C$ *iff* $\neg\psi \notin C$*;*
- *for every* $\psi_1 \vee \psi_2 \in \mathrm{CL}(\varphi)$*,* $\psi_1 \vee \psi_2 \in C$ *iff* $\psi_1 \in C$ *or* $\psi_2 \in C$*;*
- *for every* $[A]\psi \in A$*,* $\psi \in C$*.*

Temporal formulae in A *are called* active requests*, while formulae in* C *are called* current formulae*.*

We denote the set of all φ-atoms by A_φ. We have that $|A_\varphi| \leq 2^{2|\varphi|}$. As we shall later show, the proposed tableau method identifies any interval $[d_i, d_j]$ with an atom (A, C), where A includes all universal formulae $[A]\psi \in \mathrm{REQ}(d_i)$ as well as those existential formulae $\langle A \rangle \psi \in \mathrm{REQ}(d_i)$ which do not hold over any interval $[d_i, d_k]$, with $d_k < d_j$, and C includes all formulae $\psi \in \mathrm{CL}(\varphi)$ which hold over $[d_i, d_j]$. Moreover, for all $[A]\psi \in A$, $\psi \in C$, while for any $\langle A \rangle \psi \in A$, it may happen that $\psi \in C$, but this is not necessarily the case.

Atoms are connected by the following binary relation.

Definition 4. *Let* X_φ *be a binary relation over* A_φ *such that, for every pair of atoms* $(A, C), (A', C') \in A_\varphi$*,* $(A, C)X_\varphi(A', C')$ *if (and only if):*
- $A' \subseteq A$*;*
- *for every* $[A]\psi \in A$*,* $[A]\psi \in A'$*;*
- *for every* $\langle A \rangle \psi \in A$*,* $\langle A \rangle \psi \in A'$ *iff* $\neg\psi \in C$*.*

In the next section we shall show that for any pair $i < j$, the relation X_φ connects the atom associated with the interval $[d_i, d_j]$ to the atom associated with the interval $[d_i, d_{j+1}]$.

4.2   Tableau Construction, Fulfilling Paths, and Satisfiability

To check the satisfiability of a formula φ, we build a graph, called a *tableau* for φ, whose nodes are *sets of atoms*, associated with a layer of Figure 1 (namely, with a set of intervals that end at the same point), and whose edges represent the relation between a layer and the next one, that is, between a point and its immediate successor. We shall take advantage of such a construction to reduce the problem of finding a model for φ to the problem of finding a path in the tableau that satisfies suitable properties.

Definition 5. *A* node *is a set of φ-atoms $\mathcal{N}$ such that, for any pair $(A, C), (A', C') \in \mathcal{N}$ and any $(A)\psi \in \mathrm{TF}(\varphi)$, $(A)\psi \in C \Leftrightarrow (A)\psi \in C'$.*

We denote by $\mathcal{N}_\varphi$ the set of all nodes that can be built from A_φ, by $Init(\mathcal{N}_\varphi)$ the set of all *initial nodes*, that is, the set $\{\{(\emptyset, C)\} \in \mathcal{N}_\varphi\}$, and by $Fin(\mathcal{N}_\varphi)$ the set of all *final nodes*, that is, the set $\{\mathcal{N} \in \mathcal{N}_\varphi : \forall(A, C) \in \mathcal{N}, \forall\langle A\rangle\psi \in \mathrm{CL}(\varphi)(\langle A\rangle\psi \notin C)\}$. Furthermore, for any node $\mathcal{N}$, we denote by $\mathrm{REQ}(\mathcal{N})$ the set $\{(A)\psi : \exists(A, C) \in \mathcal{N}((A)\psi \in C)\}$ (or, equivalently, $\{(A)\psi : \forall(A, C) \in \mathcal{N}((A)\psi \in C)\}$). From Definition 5, it follows that $|\mathcal{N}_\varphi| \leq 2^{2^{2 \cdot |\varphi|}}$.

Definition 6. *The* tableau *for a formula φ is a directed graph $T_\varphi = \langle \mathcal{N}_\varphi, E_\varphi \rangle$, where for any pair $\mathcal{N}, \mathcal{M} \in \mathcal{N}_\varphi$, $(\mathcal{N}, \mathcal{M}) \in E_\varphi$ if and only if $\mathcal{M} = \{(A_\mathcal{N}, C_\mathcal{N})\} \cup \mathcal{M}'_\mathcal{N}$, where*
  1. *$(A_\mathcal{N}, C_\mathcal{N})$ is an atom such that $A_\mathcal{N} = \mathrm{REQ}(\mathcal{N})$;*
  2. *for every $(A, C) \in \mathcal{N}$, there exists $(A', C') \in \mathcal{M}'_\mathcal{N}$ such that $(A, C)X_\varphi(A', C')$;*
  3. *for every $(A', C') \in \mathcal{M}'_\mathcal{N}$, there exists $(A, C) \in \mathcal{N}$ such that $(A, C)X_\varphi(A', C')$.*

Definition 7. *Given a finite path $\pi = \mathcal{N}_1 \ldots \mathcal{N}_n$ in T_φ, an* atom path *in π is a sequence of atoms $(A_1, C_1), \ldots, (A_n, C_n)$ such that:*
  - *for every $1 \leq i \leq n$, $(A_i, C_i) \in \mathcal{N}_i$;*
  - *for every $1 \leq i < n$, $(A_i, C_i)X_\varphi(A_{i+1}, C_{i+1})$.*

Given a node $\mathcal{N}$ and an atom $(A, C) \in \mathcal{N}$, we say that the atom (A', C') is a descendant *of (A, C) if and only if there exists a node $\mathcal{M}$ such that $(A', C') \in \mathcal{M}$ and there exists a path π from $\mathcal{N}$ to $\mathcal{M}$ such that there is an atom path from (A, C) to (A', C') in π.*

The search for a φ model can by reduced to the search for a suitable path in T_φ.

Definition 8. *A* pre-model *for φ is a (finite or infinite) path $\pi = \mathcal{N}_1 \mathcal{N}_2 \mathcal{N}_3 \ldots$ in T_φ such that:*
  - *$\mathcal{N}_1(= \{(\emptyset, C)\}) \in Init(\mathcal{N}_\varphi)$ and $\varphi \in C$;*
  - *if π is finite and $\mathcal{N}_n$ is the last node of π, then $\mathcal{N}_n \in Fin(\mathcal{N}_\varphi)$.*

Let $\mathbf{M}^- = \langle\langle \mathbb{D}, \mathbb{I}(\mathbb{D})^-\rangle, \mathcal{V}\rangle$ be a model for φ. For every interval $[d_i, d_j] \in \mathbb{I}(\mathbb{D})^-$, we define an atom $(A_{[d_i, d_j]}, C_{[d_i, d_j]})$ such that:
$$A_{[d_i, d_j]} = \{[A]\psi \in \mathrm{REQ}(d_i)\} \cup$$
$$\{\langle A\rangle\psi \in \mathrm{REQ}(d_i) : \forall d_i < d_l < d_j(\mathbf{M}^-, [d_i, d_l] \models \neg\psi)\};$$
$$C_{[d_i, d_j]} = \{\psi \in \mathrm{CL}(\varphi) : \mathbf{M}^-, [d_i, d_j] \models \psi\}.$$

For every $j \geq 1$ (and $j < |D|$, if $|D|$ is finite), we have that:

$$(A_{[d_i,d_j]}, C_{[d_i,d_j]}) X_\varphi (A_{[d_i,d_{j+1}]}, C_{[d_i,d_{j+1}]}).$$

For every $d_j \in D$, with $j \geq 1$, let $\mathcal{N}_j = \{(A_{[d_i,d_j]}, C_{[d_i,d_j]}) : i < j\}$ and $\pi_{\mathbf{M}^-} = \mathcal{N}_1 \mathcal{N}_2 \mathcal{N}_3 \ldots$. We have that for all $j \geq 1$, $\mathcal{N}_j$ is a node; moreover, $\pi_{\mathbf{M}^-}$ is a pre-model for φ.

Conversely, for every pre-model for φ $\pi = \mathcal{N}_1 \mathcal{N}_2 \mathcal{N}_3 \ldots$ in T_φ, we build an interval structure $\langle \mathbb{D}, \mathbb{I}(\mathbb{D})^- \rangle$ and a set of interpretations $\mathbf{M}_\pi$ such that $\mathbf{M}^- = \langle \langle \mathbb{D}, \mathbb{I}(\mathbb{D})^- \rangle, \mathcal{V} \rangle \in \mathbf{M}_\pi$ if and only if:

- $\langle D \setminus \{d_0\}, < \rangle$ and π are isomorphic;
- for every $p \in AP$ and $[d_i, d_j] \in \mathbb{I}(\mathbb{D})^-$, $p \in \mathcal{V}([d_i, d_j])$ if and only if $\mu([d_i, d_j]) = (A, C)$ and $p \in C$, where $\mu : \mathbb{I}(\mathbb{D})^- \rightarrow A_\varphi$ maps every interval $[d_i, d_j]$ into an atom $\mu([d_i, d_j])$ in such a way that:

  1. $\mathcal{N}_1 = \{\mu([d_0, d_1])\}$;
  2. for every $d_i \in D$, with $d_1 < d_i$, $\mu([d_{i-1}, d_i]) = (A, C) \in \mathcal{N}_i$, with $A = \mathrm{REQ}(\mathcal{N}_{i-1})$;
  3. for every $d_i, d_j \in D$, with $d_i < d_{j-1}$, $\mu([d_i, d_j]) = (A, C) \in \mathcal{N}_j$, with $\mu([d_i, d_{j-1}]) X_\varphi (A, C)$;

Intuitively, μ assigns to every interval $[d_i, d_j]$ an atom belonging to the j-th node $\mathcal{N}_j$ in such a way that the interval relations depicted in Figure 1 are respected.

However, being π a pre-model for φ does not imply that there exists a model satisfying φ in $\mathbf{M}_\pi$, because formulae of the form $\langle A \rangle \psi$ are not necessarily satisfied by interpretations $\langle \langle \mathbb{D}, \mathbb{I}(\mathbb{D})^- \rangle, \mathcal{V} \rangle \in \mathbf{M}_\pi$. To overcome this problem, we restrict our attention to fulfilling paths.

Definition 9. *Let* $\pi = \mathcal{N}_1 \mathcal{N}_2 \mathcal{N}_3 \ldots$ *be a pre-model for* φ *in* T_φ. π *is a* fulfilling path *for* φ *in* T_φ *if and only if, for every* $\mathcal{N}_i$ *and every* $(A, C) \in \mathcal{N}_i$, *if* $\langle A \rangle \psi \in A$, *then there exist* $\mathcal{N}_j$, *with* $i \leq j$, *and* $(A', C') \in \mathcal{N}_j$, *such that* (A', C') *is a descendant of* (A, C) *in* π *and* $\psi \in C'$.

Theorem 1. *For any formula* φ, φ *is satisfiable if and only if there exists a fulfilling path for* φ *in* T_φ.

Proof. Let φ be a satisfiable formula and $\mathbf{M}^-$ be a model for it. It is easy to show that the pre-model $\pi_{\mathbf{M}^-} = \mathcal{N}_1 \mathcal{N}_2 \ldots$ is a fulfilling path for φ in T_φ. Let $(A_{[d_i,d_j]}, C_{[d_i,d_j]}) \in \mathcal{N}_j$ be an atom such that $\langle A \rangle \psi \in A_{[d_i,d_j]}$. By definition of $\pi_{\mathbf{M}^-}$, we have that $\langle A \rangle \psi \in \mathrm{REQ}(d_i)$ and, for all $d_i < d_l < d_j$, $\mathbf{M}^-, [d_i, d_l] \models \neg \psi$. Since $\mathbf{M}^-$ is a model for φ, there must exist an interval $[d_i, d_k]$, with $d_k \geq d_j$, satisfying ψ. Hence, by definition, $\psi \in C_{[d_i,d_k]}$.

As for the converse, let $\pi = \mathcal{N}_1 \mathcal{N}_2 \ldots$ be a fulfilling path for φ, and let $\mathbf{M}_\pi$ be the corresponding set of interpretations. We show that there exists a *fulfilling interpretation* $\mathbf{M}^- = \langle \langle \mathbb{D}, \mathbb{I}(\mathbb{D})^- \rangle, \mathcal{V} \rangle \in \mathbf{M}_\pi$, that is, an interpretation such that, for every interval $[d_i, d_j] \in \mathbb{I}(\mathbb{D})^-$, if $\mu([d_i, d_j]) = (A, C)$ and $\langle A \rangle \psi \in C$, then there exists an interval $[d_j, d_k] \in \mathbb{I}(\mathbb{D})^-$ such that $\mu([d_j, d_k]) = (A', C')$, with $\psi \in C'$. We define such an interpretation by induction on the index of the nodes in π, that is, we first show how to fulfill the $\langle A \rangle$-formulae in REQ($\mathcal{N}_1$) (base case) and then we show how to fulfill those in REQ($\mathcal{N}_j$) provided that we have already fulfilled those in REQ($\mathcal{N}_i$) for $i = 1, \ldots, j - 1$ (inductive step).

We begin from the initial node $\mathcal{N}_1 = \{(\emptyset, C)\}$. By the definition of μ, we have that $\mu([d_0, d_1]) = (\emptyset, C)$. Let $\langle A \rangle \psi_1, \langle A \rangle \psi_2, \ldots, \langle A \rangle \psi_n$ be the ordered list of the $\langle A \rangle$-formulae in REQ($\mathcal{N}_1$), if any (assume that they have been totally ordered on the basis of some syntactical criterion). We start with $\langle A \rangle \psi_1$. By Definition 6, there exists $(A_2, C_2) \in \mathcal{N}_2$ such that $\langle A \rangle \psi_1 \in A_2$. Since π is a fulfilling path for φ, by Definition 9 there exist $\mathcal{N}_k$, with $k \geq 2$, and $(A_k, C_k) \in \mathcal{N}_k$ such that (A_k, C_k) is a descendant of (A_2, C_2) in π and $\psi_1 \in C_k$. Let $(A_2, C_2) \ldots (A_k, C_k)$ be the atom path that leads from (A_2, C_2) to (A_k, C_k) in π. By putting $\mu([d_1, d_i]) = (A_i, C_i)$ for every $2 \leq i \leq k$, we meet the fulfilling requirement for $\langle A \rangle \psi_1$. Then we move to formula $\langle A \rangle \psi_2$. Two cases may arise: if there exists (A_i, C_i), with $2 \leq i \leq k$, such that $\psi_2 \in C_i$, then we are already done. Otherwise, we have that, for every (A_i, C_i), with $2 \leq i \leq k$, $\neg \psi_2 \in C_i$, and thus, by Definition 4, $\langle A \rangle \psi_2 \in A_i$ for $2 \leq i \leq k$. Since π is fulfilling, there exist $\mathcal{N}_h$, with $h > k$, and $(A_h, C_h) \in \mathcal{N}_h$ such that (A_h, C_h) is a descendant of (A_k, C_k) in π and $\psi_2 \in C_h$. Let $(A_k, C_k) \ldots (A_h, C_h)$ be the atom path that leads from (A_k, C_k) to (A_h, C_h) in π. As before, by putting $\mu([d_1, d_i]) = (A_i, C_i)$ for every $k + 1 \leq i \leq h$, we meet the fulfilling requirement for $\langle A \rangle \psi_2$. By repeating this process for the remaining formulae $\langle A \rangle \psi_3, \ldots, \langle A \rangle \psi_n$, we meet the fulfilling requirement for all $\langle A \rangle$-formulae in REQ($\mathcal{N}_1$).

Consider now a node $\mathcal{N}_j \in \pi$, with $j > 1$, and assume that, for every $i < j$, the fulfilling requirements for all $\langle A \rangle$-formulae in REQ($\mathcal{N}_i$) have been met. We have that for any pair of atoms $(A, C), (A', C') \in \mathcal{N}_j$ and any $\langle A \rangle \psi \in \mathrm{CL}(\varphi)$, $\langle A \rangle \psi \in C$ iff $\langle A \rangle \psi \in C'$ iff $\langle A \rangle \psi \in$ REQ($\mathcal{N}_j$). Moreover, no one of the intervals over which μ has been already defined can fulfill any $\langle A \rangle \psi \in$ REQ($\mathcal{N}_j$) since their left endpoints strictly precede d_j. We proceed as in the case of $\mathcal{N}_1$: we take the ordered list of $\langle A \rangle$-formulae in REQ($\mathcal{N}_j$) and we orderly fulfill them.

As for the intervals $[d_i, d_j]$ which are not involved in the fulfilling process, it suffices to define $\mu([d_i, d_j])$ so that it satisfies the general constraints on μ.

To complete the proof, it suffices to show that a fulfilling interpretation $\mathbf{M}^- = \langle \langle \mathbb{D}, \mathbb{I}(\mathbb{D})^- \rangle, \mathcal{V} \rangle$ is a model for φ. We show that for every $[d_i, d_j] \in \mathbb{I}(\mathbb{D})^-$ and every $\psi \in \mathrm{CL}(\varphi)$, $\psi \in C$, with $\mu([d_i, d_j]) = (A, C)$, if and only if $\mathbf{M}^-, [d_i, d_j] \models \psi$. We prove this by induction on the structure of ψ.

- If ψ is the propositional letter p, then $p \in C \overset{\mathcal{V} \ \mathrm{def.}}{\Longleftrightarrow} p \in \mathcal{V}([d_i, d_j]) \Leftrightarrow \mathbf{M}^-, [d_i, d_j] \models p$.

- If ψ is the formula $\neg \xi$, then $\neg \xi \in C \overset{\mathrm{atom \ def.}}{\Longleftrightarrow} \xi \notin C \overset{\mathrm{ind. \ hyp.}}{\Longleftrightarrow} \mathbf{M}^-, [d_i, d_j] \not\models \xi \Leftrightarrow \mathbf{M}^-, [d_i, d_j] \models \neg \xi$.

- If ψ is the formula $\xi_1 \vee \xi_2$, then $\xi_1 \vee \xi_2 \in C \overset{\mathrm{atom \ def.}}{\Longleftrightarrow} \xi_1 \in C$ or $\xi_2 \in C \overset{\mathrm{ind. \ hyp.}}{\Longleftrightarrow} \mathbf{M}^-, [d_i, d_j] \models \xi_1$ or $\mathbf{M}^-, [d_i, d_j] \models \xi_2 \Leftrightarrow \mathbf{M}^-, [d_i, d_j] \models \xi_1 \vee \xi_2$.

- Let ψ be the formula $\langle A \rangle \xi$. Suppose that $\langle A \rangle \xi \in C$. Since $\mathbf{M}^-$ is a fulfilling interpretation, there exists an interval $[d_j, d_k] \in \mathbb{I}(\mathbb{D})^-$ such that $\mu([d_j, d_k]) = (A', C')$ and $\xi \in C'$. By the inductive hypothesis, we have that $\mathbf{M}^-, [d_j, d_k] \models$

ξ, and thus $\mathbf{M}^-, [d_i, d_j] \models \langle A \rangle \xi$. As for the opposite implication, we assume by contradiction that $\mathbf{M}^-, [d_i, d_j] \models \langle A \rangle \xi$ and $\langle A \rangle \xi \notin C$. By atom definition, this implies that $\neg \langle A \rangle \xi = [A] \neg \xi \in C$. By definition of μ, we have that $\mu([d_j, d_k]) = (A', C')$ and $[A] \neg \xi \in A'$ for every $d_k > d_j$, and thus $\neg \xi \in C'$. By the inductive hypothesis, this implies that $\mathbf{M}^-, [d_j, d_k] \models \neg \xi$ for every $d_k > d_j$, and thus $\mathbf{M}^-, [d_i, d_j] \models [A] \neg \xi$, which contradicts the hypothesis that $\mathbf{M}^-, [d_i, d_j] \models \langle A \rangle \xi$.

Since π is a fulfilling path for φ, $\varphi \in C_{[d_0, d_1]}$, and thus $\mathbf{M}^-, [d_0, d_1] \models \varphi$.   □

4.3   Maximal Strongly Connected Components and Decidability

In the previous section, we reduced the satisfiability problem for RPNL$^-$ to the problem of finding a fulfilling path in the tableau for the formula φ to check. However, fulfilling paths may be infinite, and thus we must show how to finitely establish their existence.

Let $\mathcal{C}$ be a subgraph of T_φ. We say that $\mathcal{C}$ is a *strongly connected component* (*SCC* for short) of T_φ if for any two different nodes $\mathcal{N}, \mathcal{M} \in \mathcal{C}$ there exists a path in $\mathcal{C}$ leading from $\mathcal{N}$ to $\mathcal{M}$.

Let $\pi = \mathcal{N}_1 \mathcal{N}_2 \mathcal{N}_3 \ldots$ be an infinite fulfilling path in T_φ. Let $Inf(\pi)$ be the set of nodes that occurs infinitely often in π. It is not difficult to see that the subgraph defined by $Inf(\pi)$ is an SCC. We show that the search for a fulfilling path can be reduced to the search for a suitable SCC in T_φ. More precisely, we show that it is suffices to consider the *maximal strongly connected components* (*MSCC* for short) of T_φ, namely, SCC which are not properly contained in any other SCC.

Definition 10. *Let $\mathcal{C}$ be an SCC in T_φ. $\mathcal{C}$ is self-fulfilling if for every node $\mathcal{N} \in \mathcal{C}$, every atom $(A, C) \in \mathcal{N}$, and every formula $\langle A \rangle \psi \in A$, there exists a descendant (A', C') of (A, C) in $\mathcal{C}$ such that $\psi \in C'$.*

Let $\pi = \mathcal{N}_1 \mathcal{N}_2 \ldots$ be a fulfilling path for φ in T_φ. π starts from an initial node $\mathcal{N}_1$. If π is finite, it reaches a final node that belongs to a self-fulfilling SCC. If π is infinite, it reaches the SCC defined by $Inf(\pi)$ which is self-fulfilling as well.

The following lemma proves that being self-fulfilling is a monotone property.

Lemma 2. *Let $\mathcal{C}$ and $\mathcal{C}'$ be two non empty SCCs such that $\mathcal{C} \subseteq \mathcal{C}'$. If $\mathcal{C}$ is self-fulfilling, then $\mathcal{C}'$ is self-fulfilling too.*

Proof. Let $\mathcal{C} \subset \mathcal{C}'$. Suppose that there exist a node $\mathcal{N} \in \mathcal{C}'$ and an atom $(A, C) \in \mathcal{N}$ such that $\langle A \rangle \psi \in C$. Since $\mathcal{C}'$ is a SCC, there exists a path in $\mathcal{C}'$ that connects the node $\mathcal{N}$ to a node $\mathcal{M}$ in $\mathcal{C}$.

By Definition 6, there exists an atom $(A', C') \in \mathcal{M}$ which is a descendant of (A, C). Two cases may arise:

- if $\langle A \rangle \psi \notin A'$, then there exist a node $\mathcal{M}'$ in the path from $\mathcal{N}$ to $\mathcal{M}$ and a descendant (A'', C'') of (A, C) in $\mathcal{M}'$ with $\psi \in C''$;

- if $\langle A \rangle \psi \in A'$, since C is self-fulfilling, there exists (A'', C'') which is a descendant of (A', C') (and thus of (A, C)) with $\psi \in C''$.

In both cases, the formula $\langle A \rangle \psi$ gets fulfilled.   □

On the basis of Lemma 2, we define a simple algorithm searching for fulfilling paths, that progressively removes from T_φ *useless* MSCCs, that is, MSCCs that cannot participate in a fulfilling path. We call *transient state* an MSCC consisting of a single node $\mathcal{N}$ devoid of self loops, i.e., such that the edge $(\mathcal{N}, \mathcal{N}) \notin E_\varphi$.

Definition 11. *Let C be an MSCC in T_φ. C is* useless *if one of the following conditions holds:*

1. *C is not reachable from any initial node;*
2. *C is a transient state which has no outgoing edges and is not a final node;*
3. *C has no outgoing edges and it is not self-fulfilling.*

Algorithm 1. *Satisfiability checking procedure.*

$\langle \mathcal{N}_0, E_0 \rangle \leftarrow T_\varphi$
$i \leftarrow 0$
while $\langle \mathcal{N}_i, E_i \rangle$ is not empty and contains useless MSCC **do**
    let $C = \langle \mathcal{N}, E \rangle$ be a useless MSCC
    $i \leftarrow i + 1$
    $\mathcal{N}_i \leftarrow \mathcal{N}_{i-1} \setminus \mathcal{N}$
    $E_i \leftarrow E_{i-1} \cap (\mathcal{N}_i \times \mathcal{N}_i)$
if $\exists \mathcal{N} \in Init(\mathcal{N}_i)$ such that $\mathcal{N} = \{(\emptyset, C)\}$ with $\varphi \in C$ **then**
    return *true*
else
    return *false*

Let us denote by $\langle \mathcal{N}^*, E^* \rangle$ the structure computed by Algorithm 1. The correctness of the algorithm is based on the following lemma.

Lemma 3. *π is a fulfilling path for φ in $\langle \mathcal{N}_i, E_i \rangle$ if and only if it is a fulfilling path for φ in $\langle \mathcal{N}_{i+1}, E_{i+1} \rangle$.*

Proof. A fulfilling path for φ starts from an initial node and reaches either a final node that belongs to a self-fulfilling SCC (finite path) or the self-fulfilling SCC defined by $Inf(\pi)$ (infinite path). By Lemma 2, we know that being self-fulfilling is monotone and thus by removing useless MSCC from $\langle \mathcal{N}_i, E_i \rangle$ we cannot remove any fulfilling path.   □

Theorem 2. *For any formula φ, φ is satisfiable if and only if Algorithm 1 returns* true.

Proof. By Theorem 1, we have that φ is satisfiable if and only if there exists a fulfilling path in $T_\varphi = \langle \mathcal{N}_0, E_0 \rangle$. By Lemma 3, this holds if and only if there exists a fulfilling path in $\langle \mathcal{N}^*, E^* \rangle$, that is, there exists a finite path $\pi = \mathcal{N}_1 \mathcal{N}_2 \ldots \mathcal{N}_k$ in $\langle \mathcal{N}^*, E^* \rangle$ such that:

- $\mathcal{N}_1 = \{(\emptyset, C)\}$ is an initial node with $\varphi \in C$;
- $\mathcal{N}_k$ belongs to a self-fulfilling MSCC.

Since $\langle \mathcal{N}^*, E^* \rangle$ does not contain any useless MSCC, this is equivalent to the fact that there exists an initial node $\mathcal{N} = \{(\emptyset, C)\}$ in $\mathcal{N}^*$, with $\varphi \in C$. Thus, the algorithm correctly returns *true* if and only if φ is satisfiable. $\qquad\square$

As for computational complexity, we have:

- $|\mathcal{N}_\varphi| \leq 2^{2^{2|\varphi|}}$ and thus $|T_\varphi| = O(2^{2^{|\varphi|}})$;
- the decomposition of T_φ into MSCCs can be done in time $O(|T_\varphi|)$;
- the algorithm takes time $O(|A_\varphi| \cdot |T_\varphi|)$.

Hence, checking the satisfiability of a formula φ has an overall time bound of $O(2^{2^{|\varphi|}})$, that is, doubly exponential in the size of φ.

4.4   Improving the Complexity: An EXPSPACE Algorithm

In this section we describe an improvement of the proposed solution that exploits nondeterminism to find a fulfilling path in the tableau and thus to decide the satisfiability of a given formula φ, which is based on the definition of *ultimately periodic pre-model*.

Definition 12. *An infinite pre-model for φ $\pi = \mathcal{N}_1 \mathcal{N}_2 \ldots$ is ultimately periodic, with prefix l and period $p > 0$, if and only if, for all $i \geq l$, $\mathcal{N}_i = \mathcal{N}_{i+p}$.*

Theorem 3. *Let T_φ be the tableau for a formula φ. There exists an infinite fulfilling path in T_φ if and only if there exists an infinite fulfilling path that is ultimately periodic with prefix $l \leq |\mathcal{N}_\varphi|$ and period $p \leq |\mathcal{N}_\varphi|^2$.*

Proof. Let π be an infinite fulfilling path in T_φ. Consider now the SCC defined by $Inf(\pi)$ and the path σ connecting the initial node of π to it. An ultimately periodic fulfilling path satisfying the conditions of the theorem can be built as follows:

- let $\sigma = \mathcal{N}_1 \mathcal{N}_2 \ldots \mathcal{N}_n$. If $n > |\mathcal{N}_\varphi|$, take a path σ' from $\mathcal{N}_1$ to $\mathcal{N}_n$ of length $|\sigma'| \leq |\mathcal{N}_\varphi|$. Otherwise, take $\sigma' = \sigma$.
- Since $\mathcal{N}_n \in Inf(\pi)$, take a path σ_{loop} from $\mathcal{N}_n$ to $\mathcal{N}_n$. To guarantee the condition of fulfilling, we constrain σ_{loop} to contain all nodes in $Inf(\pi)$. By the definition of SCC, this loop exists and its length is less than or equal to $|\mathcal{N}_\varphi|^2$.

The infinite path $\pi' = \sigma' \sigma_{loop} \sigma_{loop} \sigma_{loop} \ldots$ is an ultimately periodic fulfilling path with prefix $l \leq |\mathcal{N}_\varphi|$ and period $p \leq |\mathcal{N}_\varphi|^2$. $\qquad\square$

The following algorithm exploits Theorem 3 to nondeterministically guess a fulfilling path satisfying the formula.

First, the algorithm guesses two numbers $l \leq |\mathcal{N}_\varphi|$ and $p \leq |\mathcal{N}_\varphi|^2$. If $p = 0$, it searches for a finite pre-model of length l. Otherwise, it takes l as the prefix and p as the period of the ultimately periodic pre-model. Next, the algorithm guesses the first node $\mathcal{N}_1$ of the pre-model, taking $\mathcal{N}_1 = \{(\emptyset, C)\}$ with $\varphi \in C$. Subsequently, it guesses the next node $\mathcal{N}_2$, incrementing a counter and checking that the edge $(\mathcal{N}_1, \mathcal{N}_2)$ is in E_φ. The algorithm proceeds in this way, incrementing the counter for every node it adds to the pre-model.

When the counter reaches l, two cases are possible: if $p = 0$, then the current node is the last node of a finite pre-model, and the algorithm checks if it is a self-fulfilling final node. If $p > 0$, the algorithm must guess the period of the pre-model. To this end, it keeps in $\mathcal{N}_p$ the current node (that is, the first node of the period), and it guesses the other nodes of the period by adding a node and by incrementing the counter at every step. Furthermore, for every atom $(A, C) \in \mathcal{N}_p$ and for every formula $\langle A \rangle \psi \in A$, it checks if the formula gets fulfilled in the period. When the counter reaches p, it checks if there exists an edge from the current node to $\mathcal{N}_p$ and if all $\langle A \rangle$-formulae in $\mathcal{N}_p$ has been fulfilled.

By Theorem 3, it follows that the algorithm returns *true* if and only if φ is satisfiable. Furthermore, the algorithm only needs to store:

- the numbers l, p and a counter ranging over them;
- the initial node $\mathcal{N}_1$;
- the current node and the next guessed node of the pre-model;
- the first node of the period $\mathcal{N}_p$;
- the set of $\langle A \rangle$-formulae that needs to be fulfilled.

Since the counters are bounded by $|\mathcal{N}_\varphi|^2 = O(2^{2^{|\varphi|}})$, and since the number of nodes is bounded by $O(2^{2^{|\varphi|}})$, the algorithm needs an amount of space which is exponential in the size of the formula.

Theorem 4. *The satisfiability problem for RPNL$^-$ is in EXPSPACE.*

5   The Decision Procedure at Work

In this section we apply the proposed decision procedure to the satisfiable formula $\varphi = \langle A \rangle p \land [A] \langle A \rangle p$ (which does not admit finite models). We show only a portion of the entire tableau, which is sufficiently large to include a fulfilling path for φ and thus to prove that φ is satisfiable.

Let $\mathbf{M}^- = \langle \langle \mathbb{D}, \mathbb{I}(\mathbb{D})^- \rangle, \mathcal{V} \rangle$ be a model that satisfies φ. Since $\mathbf{M}^-, [d_0, d_1] \models \varphi$ we have that $\mathbf{M}^-, [d_0, d_1] \models [A] \langle A \rangle p$ and $\mathbf{M}^-, [d_0, d_1] \models \langle A \rangle p$. It is easy to see that this implies that, for every interval $[d_i, d_j] \in \mathbb{I}(\mathbb{D})^-$, $\mathbf{M}^-, [d_i, d_j] \models [A] \langle A \rangle p$ and $\mathbf{M}^-, [d_i, d_j] \models \langle A \rangle p$. For this reason, we can consider (when searching for a fulfilling path for φ) only atoms obtained by combining one the following set of active requests with one of the following set of current formulae:

$$A_0 = \emptyset; \qquad\qquad C_0 = \{\varphi, [A]\langle A \rangle p, \langle A \rangle p, p\};$$
$$A_1 = \{[A]\langle A \rangle p\}; \qquad C_1 = \{\varphi, [A]\langle A \rangle p, \langle A \rangle p, \neg p\}.$$
$$A_2 = \{\langle A \rangle p, [A]\langle A \rangle p\};$$

As an example, consider the initial node $\mathcal{N}_1 = \{(\emptyset, C_0)\}$. Figure 2 depicts a portion of T_φ which is reachable from $\mathcal{N}_1$. An edge reaching a boxed set of nodes means that there is an edge reaching every node in the box, while an edge leaving from a box means that there is an edge leaving from every node in the box.

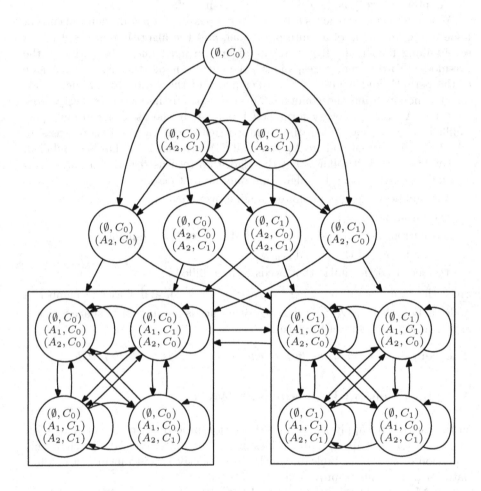

Fig. 2. A portion of the tableau for $\langle A \rangle p \wedge [A] \langle A \rangle p$

The only atoms with $\langle A \rangle$-formulae in their set of active requests are (A_2, C_0) and (A_2, C_1), since $\langle A \rangle p \in A_2$. The atom (A_2, C_0) immediately fulfills $\langle A \rangle p$, since $p \in C_0$. The atom (A_2, C_1) does not fulfill $\langle A \rangle p$, but we have that $(A_2, C_1) X_\varphi$ (A_2, C_0).

Consider now the two boxed set of nodes of Figure 2. They define an SCC $\mathcal{C}$ such that every node in $\mathcal{C}$ that contains the atom (A_2, C_1) has a descendant that contains the atom (A_2, C_0), which fulfills the request $\langle A \rangle p$. This means that $\mathcal{C}$ is a self-fulfilling SCC which is reachable from the initial node $\mathcal{N}_1$ and thus our decision procedure correctly concludes that the formula φ is satisfiable.

6   Conclusions and Further Work

In this paper we proposed an original tableau-based decision procedure for RPNL$^-$ interpreted over $\mathbb{N}$ (or over a prefix of it). We also provided an EX-PSPACE upper bound to the complexity of the satisfiability problem for RPNL$^-$. We do not know yet whether it is EXPSPACE-complete or not. As for possible extensions of the proposed method, we generalized it to branching time (where every timeline is isomorphic to $\mathbb{N}$) [1]. To this end, the definition of T_φ must be modified to take into account that every point of the tree (and, thus, every node) may have many immediate successors. Furthermore, the decision algorithm, instead of searching for fulfilling paths, has to test whether, for every formula $\langle A \rangle \psi$ that belongs to the set C of current formulae of an atom (A, C) of a node $\mathcal{N}$, there exists a successor $\mathcal{M}$ of $\mathcal{N}$ that fulfills its request. If this is the case, a model satisfying the formula can be obtained simply by taking the tree obtained through the unfolding of the tableau, starting from the initial node. The extension of the method to full PNL$^-$ turns out to be more difficult. In such a case the definition of nodes and edges of the tableau, as well as the definition of fulfilling path, must be revised to take into account past operators.

References

1. D. Bresolin, A. Montanari, A tableau-based decision procedure for Right Propositional Neighborhood Logic, Research Report 02, Dipartimento di Matematica e Informatica, Università di Udine, Italy, 2005.
2. V. Goranko, A. Montanari, G. Sciavicco, Propositional interval neighborhood temporal logics, Journal of Universal Computer Science 9 (9) (2003) 1137–1167.
3. V. Goranko, A. Montanari, G. Sciavicco, A general tableau method for propositional interval temporal logics, in: Proc. of the International Conference TABLEAUX 2003, Vol. 2796 of LNAI, Springer, 2003, pp. 102–116.
4. V. Goranko, A. Montanari, G. Sciavicco, A road map of interval temporal logics and duration calculi, Journal of Applied Non-Classical Logics 14 (1-2) (2004) 9–54.
5. V. Goranko, A. Montanari, G. Sciavicco, P. Sala, A general tableau method for propositional interval temporal logics: theory and implementation, Journal of Applied Logic, 2005 (to appear).
6. J. Halpern, Y. Shoham, A propositional modal logic of time intervals, Journal of the ACM 38 (4) (1991) 935–962.
7. A. Montanari, G. Sciavicco, N. Vitacolonna, Decidability of interval temporal logics over split-frames via granularity, in: Proc. of the European Conference on Logic in Artificial Intelligence 2002, LNAI 2424, Springer, 2002, 259–270.
8. A. Montanari, Propositional interval temporal logics: some promising paths, Proc. of the 12th International Symposium on Temporal Representation and Reasoning (TIME), IEEE Computer Society Press, 2005, 201-203.
9. B. Moszkowski, Reasoning about digital circuits, Ph.D. thesis, Dept. of Computer Science, Stanford University, Tech. Rep. STAN-CS-83-970, Stanford, CA, 1983.
10. Y. Venema, A modal logic for chopping intervals, Journal of Logic and Computation 1 (4) (1991) 453–476.
11. P. Wolper, The tableau method for temporal logic: An overview, Logique et Analyse 28 (1985) 119–136.

Cyclic Proofs for First-Order Logic with Inductive Definitions

James Brotherston

Laboratory for Foundations of Computer Science,
Division of Informatics, University of Edinburgh, James Clerk Maxwell Building,
King's Buildings, Mayfield Road, Edinburgh EH9 3JZ, Scotland, UK
J.Brotherston@sms.ed.ac.uk

Abstract. We consider a cyclic approach to inductive reasoning in the setting of first-order logic with inductive definitions. We present a proof system for this language in which proofs are represented as finite, locally sound derivation trees with a "repeat function" identifying cyclic proof sections. Soundness is guaranteed by a well-foundedness condition formulated globally in terms of *traces* over the proof tree, following an idea due to Sprenger and Dam. However, in contrast to their work, our proof system does not require an extension of logical syntax by ordinal variables.

A fundamental question in our setting is the strength of the cyclic proof system compared to the more familiar use of a non-cyclic proof system using explicit induction rules. We show that the cyclic proof system subsumes the use of explicit induction rules. In addition, we provide machinery for manipulating and analysing the structure of cyclic proofs, based primarily on viewing them as generating regular infinite trees, and also formulate a finitary trace condition sufficient (but not necessary) for soundness, that is computationally and combinatorially simpler than the general trace condition.

1 Introduction

Induction is essential to computer science, and mathematics in general, as the fundamental principle by which to reason about many of the structures that are ubiquitous within the fields. Indeed, any inductively defined structure comes equipped with corresponding recursion and induction principles that in, e.g., a functional programming language, are invoked when defining functions by recursion on that structure and in proving mathematical properties of these functions, respectively. Similarly, structures defined by mutual induction give rise to mutual recursion and induction principles.

The default approach to inductive reasoning is to directly employ the induction principles associated with the structures under consideration; however, there has been recent interest [3,4,5,7,9,15,16,17,18,19,20] in various forms of cyclic reasoning by which repeating sections of proof (cycles) are identified and various guardedness conditions are imposed on proofs to ensure their soundness. These conditions can be seen as encoding a termination argument or, more

B. Beckert (Ed): TABLEAUX 2005, LNAI 3702, pp. 78–92, 2005.
© Springer-Verlag Berlin Heidelberg 2005

pertinently for our purposes, as ensuring well-foundedness of an underlying inductive argument. Forms of cyclic, or circular, reasoning have been employed in local model checking [3], theorem proving tools and frameworks [4,7,9,16], in Turchin's supercompilation [19] and in program verification based on automata [20]. It has also been studied in the context of tableau-style proof systems for the μ-calculus by Sprenger and Dam [17,18] and Schöpp and Simpson [15], following an approach proposed by Dam and Gurov [5].

Our aim is to study cyclic reasoning in the relatively simple, yet expressive context of first-order logic extended with ordinary inductive definitions[1]. Similar formalisms typically form the basis of mechanised theorem proving tools [2,8,10,13]. The contribution of this paper is twofold. Firstly, we present a sound, powerful cyclic proof system that employs only the standard syntax of first-order logic. It seems the system is most likely to be of use for proving properties of mutually inductive definitions, for which the usual induction rules are often highly complex. We examine the relationship of the cyclic proof system to a standard non-cyclic proof system; specifically, we show that the cyclic proof system is at least as powerful as the standard one. We conjecture that it is in fact no more powerful, but whether or not this is true remains open at the time of writing. (Although Sprenger and Dam give an equivalence result for cyclic and non-cyclic μ-calculus proof systems [18], the approach taken there is not obviously applicable to our systems due to significant differences in the setting, which we discuss later). Secondly, we present machinery for analysing the structure of cyclic proofs that, as well as being useful for our system for first-order logic with inductive definitions, should be more generally applicable to other cyclic proof systems based on analogous trace conditions.

The remainder of this paper is structured as follows. In section 2 we recall the syntax and semantics of first-order logic with inductive definitions (FOL_{ind}). In section 3 we introduce Gentzen-style sequent calculus rules for this language; in particular, we give rules for the case-split and induction principles associated with any inductively defined predicate. In section 4 we define two types of proof: the usual *structural proofs* familiar from sequent calculus, which use induction rules explicitly; and *cyclic proofs* represented as finite proof trees with a "repeat function" identifying cyclic proof sections, together with a global condition stipulating the existence of a *trace* on every infinite path through the proof, inspired by a similar condition proposed by Sprenger and Dam [17]. This condition guarantees the soundness of cyclic proofs. We prove that cyclic proofs subsume structural proofs by giving a translation from structural to cyclic proofs. In section 5 we present machinery, based on viewing cyclic proofs as generating infinite trees, for transforming cyclic proofs into ones having convenient structural properties. In section 6, we examine the global trace condition (which is the most general possible) imposed on cyclic proofs and formulate a localised *trace manifold* proof condition which, although less general, contains more structural

---

[1] However, there is no difficulty in extending our approach to more complex formalisms such as iterated inductive definitions [11].

information. Finally, in section 7 we summarise our progress and outline our aims for our ongoing work in this area.

Due to space constraints, the proofs of many results in this paper have been omitted and the proofs that appear are only sketched. Full proofs will appear in the author's forthcoming PhD thesis.

2 Syntax and Semantics of FOL$_{\text{ind}}$

We assume a fixed first-order signature Σ. Further, we assume that the predicate symbols of Σ are separated into two distinct types: *ordinary* and *inductive* predicate symbols. For convenience, we use vector notation to denote finite sequences of variables and terms. For example, if P is a predicate symbol of arity k, we shall write $Pt(\boldsymbol{x})$ for $P(t_1(x_1, \ldots, x_n), \ldots, t_k(x_1, \ldots, x_n))$.

We extend the standard syntax and semantics of first-order logic by introducing a simple definitional mechanism for (mutually) inductive predicates, based on Martin-Löf's schema for *ordinary productions* [11]:

Definition 2.1 (Inductive definition set). An *inductive definition set* Φ is a finite set of *productions* of the form:

$$\frac{P_1 t_1(\boldsymbol{x}) \; \ldots \; P_m t_m(\boldsymbol{x})}{Pt(\boldsymbol{x})}$$

where P is an inductive predicate symbol and $P_1, \ldots, P_m$ may be either ordinary or inductive predicate symbols.

Example 2.2. We define the predicates N, E and O via the productions:

$$\frac{}{N0} \quad \frac{Nx}{Nsx} \quad \frac{}{E0} \quad \frac{Ex}{Osx} \quad \frac{Ox}{Esx}$$

In structures in which all "numerals" s^k0 for $k \geq 0$ are interpreted as distinct elements, the predicates N, E and O correspond to the properties of being a natural, even and odd number respectively.

We next define the interpretation of inductively defined predicates in a Σ-structure M with domain D. Given a definition set Φ for predicates $P_1, \ldots, P_n$ of arities $k_1, \ldots, k_n$ respectively, writing $Pow(A)$ for the powerset of A, one can construct a monotone operator φ_Φ with domain and codomain $(Pow(D^{k_1}), \ldots, Pow(D^{k_n}))$ whose least fixed point is the n-tuple of least subsets of (tuples of) D closed under the rules in Φ. This least fixed point can be approached in stages, by defining an ordinal-indexed sequence (φ_Φ^α) by $\varphi_\Phi^\alpha = \bigcup_{\beta < \alpha} \varphi_\Phi(\varphi_\Phi^\beta)$. Note that we thus have $\varphi_\Phi^0 = \emptyset$. The ith component of φ_Φ^α is then called the αth *approximant*[2]

---

[2] One can show that for any predicate P defined by our schema, $P^\alpha = P^\omega$ for any $\alpha > \omega$, so it is sufficient to index approximants by natural numbers. However, the sequence of approximants does not necessarily close at ω if more complex schemas are used. We retain the ordinal notation for modularity.

of P_i, written as P_i^α. For a full exposition of the construction of the operator φ_Φ (which is standard), see e.g. Aczel [1].

We consider standard first-order formulas over Σ. The semantics of formulas are defined as usual except for the semantics of atomic formulas, which are now defined as follows, where ρ is an environment interpreting free variables in M and P^M is the interpretation of the ordinary predicate symbol P in M:

$$M \models_\rho Pt \iff \begin{cases} \rho(t) \in \bigcup_\alpha P^\alpha & \text{if } P \text{ is an inductively defined predicate symbol} \\ \rho(t) \in P^M & \text{otherwise} \end{cases}$$

3 Sequent Calculus Proof Rules for FOL$_{\text{ind}}$

We shall consider proof systems for FOL$_{\text{ind}}$ presented in the sequent calculus style originally due to Gentzen [6], which is well-established as a convenient formalism for proof-theoretic reasoning. Our rules for inductively defined predicates are essentially sequent-calculus adaptations of Martin-Löf's natural deduction rules [11]; McDowell and Miller have considered a similar system [12]. We write sequents of the form $\Gamma \vdash \Delta$, where Γ, Δ are finite multisets of formulas. We use the standard sequent calculus rules for the propositional connectives and quantifiers, as well as the following structural rules:

$$\frac{\Gamma' \vdash \Delta'}{\Gamma \vdash \Delta} \; \Gamma' \subseteq \Gamma, \Delta' \subseteq \Delta \, (\text{Wk}) \qquad \frac{}{\Gamma \vdash \Delta} \; \Gamma \cap \Delta \neq \emptyset \; (\text{Axiom})$$

$$\frac{\Gamma, F, F \vdash \Delta}{\Gamma, F \vdash \Delta} \, (\text{ContrL}) \qquad \frac{\Gamma \vdash F, F, \Delta}{\Gamma \vdash F, \Delta} \, (\text{ContrR})$$

$$\frac{\Gamma \vdash F, \Delta \quad \Gamma, F \vdash \Delta}{\Gamma \vdash \Delta} \, (\text{Cut}) \qquad \frac{\Gamma \vdash \Delta}{\Gamma[\theta] \vdash \Delta[\theta]} \, (\text{Subst})$$

We also give proof rules governing the inductively defined predicates. First, for each ordinary production in Φ we obtain a right-introduction rule for an inductively defined predicate as follows:

$$\frac{P_1 t_1(x) \dots P_m t_m(x)}{Pt(x)} \quad \Rightarrow \quad \frac{\Gamma \vdash P_1 t_1(t_1'), \Delta \dots \Gamma \vdash P_m t_m(t_m'), \Delta}{\Gamma \vdash Pt(t'), \Delta} \, (PR_r)$$

Example 3.1. The right introduction rules for the predicate N defined in Example 2.2 are:

$$\frac{}{\Gamma \vdash N0, \Delta} \, (NR_1) \qquad \frac{\Gamma \vdash Nt, \Delta}{\Gamma \vdash Nst, \Delta} \, (NR_2)$$

Definition 3.2 (Mutual dependency). Let Φ be a fixed inductive definition set. Define the binary relation *Prem* on *inductive* predicate symbols as the least relation satisfying: whenever there is a production in Φ containing P in the conclusion and Q among the premises, then $Prem(P, Q)$ holds. Also define

Prem* to be the reflexive-transitive closure of *Prem*. Then we say two predicate symbols P and Q are *mutually dependent* if both $Prem^*(P,Q)$ and $Prem^*(Q,P)$ hold.

We now describe the construction of the induction rule for an inductively defined predicate P. First, for all predicates P_i such that P_i and P are mutually dependent, we first associate a formula $F_i z$ with P_i, where P_i is a k-ary predicate and z is a vector of k variables. Then the induction rule schema is:

$$\frac{\text{minor premises} \quad \Gamma, Ft \vdash \Delta}{\Gamma, Pt \vdash \Delta} \text{(Ind } P\text{)}$$

where Fz is the formula associated with P. Then for all productions having in their conclusion a predicate P_i such that P_i and P are mutually dependent, we obtain a *minor premise* as follows[3]:

$$\frac{P_1 t_1(x) \dots P_m t_m(x)}{P_i t(x)} \quad \Rightarrow \quad \Gamma, G_1 t_1(x), \dots, G_m t_m(x) \vdash F_i t(x), \Delta$$

where $x \notin FV(\Gamma \cup \Delta)$ for all $x \in x$ and G_k is defined for each $k \in \{1, \dots, m\}$ by:

$$G_k = \begin{cases} F_k & \text{if } P_k \text{ and } P \text{ are mutually dependent} \\ P_k & \text{otherwise} \end{cases}$$

Example 3.3. The induction rule for the even number predicate E defined in Example 2.2 is the following, where F and G are associated with E and O respectively:

$$\frac{\Gamma \vdash F0, \Delta \quad \Gamma, Fx \vdash Gsx, \Delta \quad \Gamma, Gx \vdash Fsx, \Delta \quad \Gamma, Ft \vdash \Delta}{\Gamma, Et \vdash \Delta} \text{(Ind } E\text{)}$$

We also consider rules implementing a *case-splitting* principle on inductively defined predicates. (Although case-splitting is a much weaker principle than induction, and indeed is subsumed by it, we shall see in Section 4 that cyclic reasoning together with the use of case-split rules subsumes explicit use of the induction rules above.) The case-split rule schema for an inductively defined predicate P is:

$$\frac{\text{case distinctions}}{\Gamma, Py \vdash \Delta} \text{(Case } P\text{)}$$

where y is a vector of variables and for each production having predicate P in its conclusion, we obtain a *case distinction* as follows:

$$\frac{P_1 t_1(x) \dots P_m t_m(x)}{Pt(x)} \Rightarrow \Gamma[t(x)/y], Pt(x), P_1 t_1(x), \dots, P_m t_m(x) \vdash \Delta[t(x)/y]$$

---

[3] Martin-Löf [11] uses a definition of "linkage" between predicate symbols to generate the minor deductions (corresponding to our minor premises) of his induction rules. However, his definition can produce redundant minor deductions in some cases. Our use of Definition 3.2 is intended to avoid this.

subject to the restriction that $x \notin FV(\Gamma \cup \Delta)$ for all $x \in \boldsymbol{x}$. In such rules, the formulas $P_1 \boldsymbol{t_1}(\boldsymbol{x}), \ldots, P_m \boldsymbol{t_m}(\boldsymbol{x})$ are said to be *case-descendents* of the *principal formula* $P\boldsymbol{y}$ in the rule conclusion.

Example 3.4. The case-split rule for the natural number predicate N is:

$$\frac{\Gamma[0/y] \vdash \Delta[0/y] \qquad \Gamma[sx/y], Nsx, Nx \vdash \Delta[sx/y]}{\Gamma, Ny \vdash \Delta} \text{(Case } N)$$

4   Structural and Cyclic Proofs

We now proceed to give definitions of *structural proofs* familiar from traditional sequent calculus proof systems, which are finite trees labelled with sequents and proof rules. We do this with more formality than is usual, based on the notion of a *rule graph*, in order to later facilitate a formal comparison between structural proofs and cyclic proofs. Note that we write $f : X \rightharpoonup Y$ to denote a partial function and $f : X \to Y$ to denote a total function from X to Y.

Definition 4.1 (Rule graph). Let *Seqs* denote the set of all well-formed sequents in some language and *Rules* denote some set of rules. Also let $n \in \mathbb{N}$ be the maximum number of premises of any $R \in Rules$. Then a *rule graph* is given by (V, s, r, p), where:

- V is a set of vertices, $s : V \to Seqs$, $r : V \rightharpoonup Rules$, and $p : V \rightharpoonup V^n$ (we write $p_j(v)$ for the jth component of $p(v)$);
- for all $v \in V$, $p_j(v)$ is defined just in case $r(v)$ is a rule with m premises, $1 \le j \le m$ and:
$$\frac{s(p_1(v)) \quad \cdots \quad s(p_m(v))}{s(v)}$$

is an instance of rule $r(v)$.

A rule graph G can be seen as a conventional graph whose vertex set is V and whose edge set is $E = \{(v, p_j(v)) \mid v \in V \text{ and } p_j(v) \text{ is defined}\}$.

A *path* in a rule graph is a (possibly infinite) sequence $v_0 j_0 v_1 j_1 v_2 j_2 \ldots$ such that for each $i \ge 0$, $v_{i+1} = p_{j_i}(v_i)$. (We often write paths simply as $v_0 v_1 v_2 \ldots$)

Definition 4.2 (Derivation tree). A rule graph $\mathcal{D} = (V, s, r, p)$ is a *derivation tree* if there is a distinguished node $v_0 \in V$ such that for all $v \in V$, there is a unique path in $\mathcal{D}$ from v_0 to v. v_0 is called the *root* of the tree, written $root(\mathcal{D})$.

Definition 4.3 (Structural proof). A *structural proof* of $\Gamma \vdash \Delta$ is a finite derivation tree $\mathcal{D} = (V, s, r, p)$ such that:

- the codomain of s is the set of all well-formed sequents of FOL_{ind};
- $s(root(\mathcal{D})) = \Gamma \vdash \Delta$;
- the codomain of r comprises the rules described in Section 3 above;

– r is a total function on V, i.e. every node in $\mathcal{D}$ is the conclusion of some rule instance. (Note that we consider axioms to be proof rules with 0 premises.)

Definition 4.4 (Satisfaction). Let M be a fixed Σ-structure and let ρ be an environment interpeting free variables in M. We write $\Gamma \models_\rho \Delta$ to mean that if $M \models_\rho J$ for all $J \in \Gamma$ then there is a $K \in \Delta$ such that $M \models_\rho K$. We write $\Gamma \models \Delta$ if $\Gamma \models_\rho \Delta$ for all ρ.

Theorem 4.5 (Soundness of structural proof). *If there is a structural proof of $\Gamma \vdash \Delta$ then $\Gamma \models \Delta$.* □

Proof. Routine.

Proposition 4.6. *For any inductively defined predicate P, the rule (Case P) is subsumed by the rule (Ind P) in structural proofs. Specifically, any instance of the rule (Case P) in a structural proof can be replaced by an equivalent derivation containing an instance of (Ind P) and instances of the standard sequent calculus rules for first-order logic, only.*

The well-known and important *cut elimination* theorem for sequent calculus for ordinary first-order logic due to Gentzen [6] says that any use of the (Cut) rule (which corresponds to the use of auxiliary lemmas) in a derivation can be avoided. This result should generalise to our structural proof system for FOL$_{\text{ind}}$. However, to our knowledge no proof for such a classical sequent calculus has so far appeared in the literature , although a proof for a related intuitionistic proof system appears in McDowell and Miller [12].

4.1   A Cyclic Proof System for FOL$_{\text{ind}}$

We now proceed to define a cyclic proof system for FOL$_{\text{ind}}$, in which the proof structures are finite derivation trees together with a function assigning to every unexpanded node in the proof tree (called a *bud*) an interior node with an identical sequent labelling (the *companion* of the bud). These structures (called *pre-proofs*) can then be viewed as cyclic rule graphs:

Definition 4.7 (Bud /companion nodes). Let $\mathcal{D} = (V, s, r, p)$ be a derivation tree. A *bud node* of $\mathcal{D}$ is a vertex $B \in V$ such that $r(B)$ is undefined, i.e. B is not the conclusion of any proof rule instance in $\mathcal{D}$.

A node $C \in V$ is said to be a *companion* for a bud node B if $r(C)$ is defined and $s(C) = s(B)$.

We remark that we do not require buds to have ancestor nodes as companions, i.e. C need not appear on the unique path in $\mathcal{D}$ from $root(\mathcal{D})$ to B.

Definition 4.8 (Cyclic pre-proof). A *cyclic pre-proof* (or simply *pre-proof*) of $\Gamma \vdash \Delta$ is a pair $(\mathcal{D} = (V, s, r, p), \mathcal{R})$, where $\mathcal{D}$ is a finite derivation tree and:

– the codomain of s is the set of all well-formed sequents of FOL_{ind};
– $s(root(\mathcal{D})) = \Gamma \vdash \Delta$;
– the codomain of r comprises the rules described in Section 3 above *except* for the induction rules;
– $\mathcal{R} : V \rightharpoonup V$ is a function assigning a companion to every bud node in $\mathcal{D}$.

We shall see shortly that our embargo on the explicit use of induction rules in the cyclic proof system is of no consequence; all induction rules turn out to be derivable within the system.

Definition 4.9 (Pre-proof graph). Let $\mathcal{P} = (\mathcal{D}, \mathcal{R})$ be a pre-proof, where $\mathcal{D} = (V, s, r, p)$. Then the *graph* of $\mathcal{P}$ is $\mathcal{G}_\mathcal{P} = (V', s, r, p)$, where V' is obtained from V by identifying each bud node B in $\mathcal{D}$ with its companion $\mathcal{R}(B)$.

We observe that the local soundness of our proof rules is not sufficient to guarantee that pre-proofs are sound, due to the (possible) cyclicity evident in their graph representations. In order to give a criterion for soundness, we follow Sprenger and Dam [17] in formulating the notion of a *trace* following a path:

Definition 4.10 (Trace). Let $\mathcal{P}$ be a pre-proof and let (v_i) be a path in $\mathcal{G}_\mathcal{P} = (V', s, r, p)$. A *trace following* (v_i) is a sequence (τ_i) such that, for all i:

– $\tau_i = P_i t_i \in \Gamma_i$, where $s(v_i) = \Gamma_i \vdash \Delta_i$ and P_i is inductively defined;
– if $r(v_i)$ is (Subst) then $\tau_i = \tau_{i+1}[\Theta]$, where Θ is the substitution associated with the rule instance;
– if $r(v_i)$ is (Case P) then either $\tau_{i+1} = \tau_i[t(x)/y]$, where $[t(x)/y]$ is the substitution associated with the case distinction at $s(v_{i+1})$, or τ_i is the principal formula Py of the rule instance and τ_{i+1} is a case-descendent of Py. In the latter case, i is said to be a *progress point* of the trace;
– if $r(v_i)$ is not (Subst) or (Case P), then $\tau_{i+1} = \tau_i$.

An *infinitely progressing trace* is a (necessarily infinite) trace having infinitely many progress points.

Informally, a trace follows (a part of) the construction of an inductively defined predicate occurring in the left hand side of the sequents occurring on a path in a pre-proof. These predicate constructions never become larger as we follow the trace along the path, and at progress points, they actually decrease. This property is encapsulated in the following lemma and motivates the subsequent definition of a *cyclic proof*:

Lemma 4.11. *Suppose we have a pre-proof of $\Gamma_0 \vdash \Delta_0$, but there exists ρ_0 such that $\Gamma_0 \nvDash_{\rho_0} \Delta_0$. Then there is an infinite path $(v_i)_{i \geq 0}$ in $\mathcal{G}_\mathcal{P}$ and an infinite sequence of environments $(\rho_i)_{i \geq 0}$ such that:*

1. *For all i, $\Gamma_i \nvDash_{\rho_i} \Delta_i$, where $s(v_i) = \Gamma_i \vdash \Delta_i$;*
2. *If there is a trace $(\tau_i)_{i \geq n}$ following some tail $(v_i)_{i \geq n}$ of $(v_i)_{i \geq 0}$, then the sequence $(\alpha_i)_{i \geq n}$ of ordinals defined by $\alpha_i = (least \; \alpha \; s.t. \; \rho_i(t_i) \in P_i^\alpha$ where $\tau_i = P_i t_i)$ is non-increasing. Furthermore, if j is a progress point of the trace then $\alpha_{j+1} < \alpha_j$.*

Proof. (Sketch) One constructs (v_i) and (ρ_i) inductively by assuming sequences $(v_i)_{0 \leq i \leq k}$ and $(\rho_i)_{0 \leq i \leq k}$ satisfying the two properties of the lemma and constructing v_{k+1} and ρ_{k+1}. The proof proceeds by a case analysis on the rule $r(v_k)$; the main interesting case is when $r(v)$ is a case-split rule.   □

Definition 4.12 (Cyclic proof). A pre-proof $\mathcal{P}$ is said to be a *cyclic proof* if, for every infinite path in $\mathcal{G}_\mathcal{P}$, there is an infinitely progressing trace following some tail of the path.

Theorem 4.13 (Soundness of cyclic proof). *If the sequent $\Gamma \vdash \Delta$ has a cyclic proof then $\Gamma \models \Delta$.*

Proof. (Sketch) If $\Gamma \not\models \Delta$ then we can apply Lemma 4.11 to construct infinite sequences (v_i) and (ρ_i) satisfying the two properties of the lemma. By Definition 4.12 there is an infinitely progressing trace following some tail of (v_i), so by the second property of the lemma we can construct an infinite descending chain of ordinals, which is a contradiction.   □

Importantly, the property of being a cyclic proof is decidable; the problem can be reduced to the problem of checking the language of a certain Büchi automaton for emptiness. A thorough analysis of automata-theoretic decision methods is given by Sprenger and Dam [17].

Example 4.14. The following is a cyclic proof of the statement $Ex \vee Ox \vdash Nx$, where N, E and O are as defined in Example 2.2 (we omit applications of weakening):

$$
\cfrac{
  \cfrac{}{\vdash N0}(NR_1) \qquad
  \cfrac{
    \cfrac{
      \cfrac{Ox \vdash Nx\ (\dagger)}{Oy \vdash Ny}(\text{Subst})
    }{Oy \vdash Nsy}(NR_2)
  }{Ex \vdash Nx\ (*)}(\text{Case } E)
  \qquad
  \cfrac{
    \cfrac{
      \cfrac{(*)\ Ex \vdash Nx}{Ey \vdash Ny}(\text{Subst})
    }{Ey \vdash Nsy}(NR_2)
  }{(\dagger)\ Ox \vdash Nx}(\text{Case } O)
}{Ex \vee Ox \vdash Nx}(\vee L)
$$

We use pairs of identical symbols $(*)$ and $(\dagger)$ above to indicate the pairing of buds with companions; we remark that this is an example of a proof in which bud nodes have non-ancestor nodes as companions. To see that this satisfies the cyclic proof condition, observe that any infinite path through the pre-proof necessarily has a tail consisting of repetitions of the "figure-of-8" loop in this proof, and there is a trace following this path: $(Ex, Oy, Oy, Ox \equiv Ox, Ey, Ey, Ex)$ (starting from $(*)$) that progresses at Ex and at Ox. We can thus concatenate copies of this trace to obtain an infinitely progressing trace as required.

Next, we address the fundamental question of the relative strengths of the standard structural proof system and the cyclic proof system. Our first result establishes that cyclic proofs are at least as powerful as structural proofs:

Theorem 4.15. *If there is a structural proof of $\Gamma \vdash \Delta$ then there is a cyclic proof of $\Gamma \vdash \Delta$.*

Proof. (Sketch) One shows that any use of an induction rule within a structural proof can be replaced with a derivation in our cyclic proof system. Each derivation contains a cyclic proof, constructed using the minor premises of the induction rule, which essentially is an explicit justification of local soundness. Moreover, as each of these cyclic proofs is self-contained, it follows easily that the result of uniformly substituting these derivations for induction rules is also a cyclic proof.

We illustrate the procedure by demonstrating the derivation of the rule (Ind E) given in Example 3.3. First of all, define the set of formulas $M_E = \{F0, \forall x.Fx \rightarrow Gsx, \forall x.Gx \rightarrow Fsx\}$. We start as follows:

$$
\cfrac{
\cfrac{
\cfrac{
\cfrac{
\cfrac{M_E, Ey \vdash Fy}{M_E, Et \vdash Ft}\ (\text{Subst})
}{\vdots}\ (\wedge\text{L})
}{
\cfrac{\bigwedge M_E, Et \vdash Ft}{\bigwedge M_E, \Gamma, Et \vdash Ft, \Delta}\ (\text{Wk})
}\ (\wedge\text{L})
}{\Gamma, Et \vdash Ft, \Delta}
\qquad
\cfrac{
\cfrac{
\cfrac{\Gamma \vdash \bigwedge M_E, \Delta}{\Gamma, Et \vdash \bigwedge M_E, Ft, \Delta}\ (\text{Wk})
}{}\ (\text{Cut})
\qquad \cfrac{\Gamma, Ft \vdash \Delta}{\Gamma, Et, Ft \vdash \Delta}\ (\text{Wk})
}{\Gamma, Et, Ft \vdash \Delta}\ (\text{Cut})
}{\Gamma, Et \vdash \Delta}
$$

minor premises

Obtaining the minor premises of (Ind E) from $\Gamma \vdash \bigwedge M_E, \Delta$ is straightforward. We continue by providing a cyclic proof of the sequent $M_E, Ey \vdash Fy$ occurring on the leftmost branch as follows (omitting applications of weakening):

$$
\cfrac{
\cfrac{
\cfrac{
\cfrac{
\cfrac{
\cfrac{
\cfrac{
\cfrac{\ }{Gsy \vdash Gsy}\ (\text{Ax}) \qquad M_E, Ey \vdash Fy\ (*)
}{M_E, Fy \rightarrow Gsy, Ey \vdash Gsy}\ (\rightarrow\text{L})
}{M_E, \forall x.Fx \rightarrow Gsx, Ey \vdash Gsy}\ (\forall\text{L})
}{M_E, Ey \vdash Gsy}\ (\text{ContrL})
}{M_E, Oy \vdash Gx}\ (\text{Case } O)
}{
\cfrac{\ }{Fsx \vdash Fsx}\ (\text{Ax}) \qquad \cfrac{}{M_E, Ox \vdash Gx}
}
}{M_E, Gx \rightarrow Fsx, Ox \vdash Fsx}\ (\rightarrow\text{L})
}{M_E, \forall x.Gx \rightarrow Fsx, Ox \vdash Fsx}\ (\forall\text{L})
}{M_E, Ox \vdash Fsx}\ (\text{ContrL})
$$

$$
\cfrac{
\cfrac{\ }{M_E \vdash F0}\ (\text{Ax}) \qquad \cfrac{M_E, Ox \vdash Fsx}{\ }\ (\text{Case } E)
}{M_E, Ey \vdash Fy\ (*)}
$$

To the single bud node of this derivation we assign the root sequent as companion (indicated by $(*)$). As there is a progressing trace from the companion to the bud, this is easily seen to be a cyclic proof.

We remark at this juncture that, although structural rules such as weakening or contraction are admissible in ordinary FOL sequent calculus, they clearly are essential to the cyclic proof system as we have formulated it here, as their removal can break the required syntactic identity between bud nodes and their companions.

Theorem 4.15 gives rise to the obvious question of whether its converse also holds:

Conjecture 4.16. *If there is a cyclic proof of $\Gamma \vdash \Delta$ then there is a structural proof of $\Gamma \vdash \Delta$.*

This is the main open question of our current research. An equivalence result for cyclic and non-cyclic proof systems for μ-calculus with explicit approxima- tions, given by Sprenger and Dam [18], gives some hope that it can be answered positively. However, this result is based on reducing cyclic reasoning with a syntax extended by ordinal variables to a principle of transfinite induction also expressed using ordinal variables. The problem of reducing cyclic reasoning on or- dinary syntax to induction over predicate definitions seems significantly harder. The most promising line of inquiry appears to be to establish a translation from cyclic to structural proofs. (A semantic proof would also be of interest, but we have no idea how to obtain one.)

5   Unfolding and Folding of Cyclic Proofs

In this section and the subsequent one we study the structure of cyclic proofs, which we view here as finite representations of infinite regular proof trees. As many different cyclic proofs can represent the same infinite tree, these trees give us a natural notion of equivalence on cyclic proofs. We give machinery based on infinite trees for transforming cyclic proofs into equivalent ones having simpler combinatorial structure. We hope that such concerns may be useful in eventually establishing a translation from cyclic to structural proofs.

Definition 5.1 (Associated tree). Let $\mathcal{P} = (\mathcal{D}, \mathcal{R})$ be a pre-proof with graph $\mathcal{G}_\mathcal{P} = (V', s, r, p)$. Define $Path(\mathcal{G}_\mathcal{P})$ to be the set of finite paths through $\mathcal{G}_\mathcal{P}$ starting from $v_0 = root(\mathcal{D})$. Then the *tree* of $\mathcal{P}$ is $\mathcal{T}_\mathcal{P} = (Path(\mathcal{G}_\mathcal{P}), s^*, r^*, p^*)$, where $s^*((v_i)_{0 \leq i \leq n}) = s(v_n)$, $r^*((v_i)_{0 \leq i \leq n}) = r(v_n)$, and:

$$p_j^*((v_i)_{0 \leq i \leq n}) = \begin{cases} ((v_i)_{0 \leq i \leq n}.p_j(v_n)) & \text{if } p_j(v_n) \text{ defined} \\ \text{undefined} & \text{otherwise} \end{cases}$$

Proposition 5.2. *For any pre-proof $\mathcal{P}$, $\mathcal{T}_\mathcal{P}$ is a derivation tree.*

We write $f(x) \simeq g(x)$, where f and g are partial functions, to mean that $f(x)$ is defined iff $g(x)$ is defined, and if $f(x)$ is defined then $f(x) = g(x)$.

Definition 5.3 (Rule graph homomorphism). Let $G = (V, s, r, p)$ and $H = (V', s', r', p')$ be rule graphs. A *rule graph homomorphism* from G to H is a map $f : V \to V'$ satisfying, for all $v \in V$, $s'(f(v)) = s(v)$, $r'(f(v)) \simeq r(v)$ and $p_j'(f(v))) \simeq f(p_j(v))$.

We say two rule graphs G and H are *isomorphic*, written $G \cong H$, if there exist rule graph homomorphisms $f : G \to H$ and $g : H \to G$ such that $f \circ g = g \circ f = id$, where id is the identity function.

Lemma 5.4. *For any pre-proof $\mathcal{P} = (\mathcal{D}, \mathcal{R})$, there is a surjective rule graph homomorphism $f_\mathcal{P} : \mathcal{T}_\mathcal{P} \to \mathcal{G}_\mathcal{P}$ such that $f_\mathcal{P}(root(\mathcal{T}_\mathcal{P})) = root(\mathcal{D})$.*

Lemma 5.5. *If $\mathcal{T}_\mathcal{P} \cong \mathcal{T}_{\mathcal{P}'}$ then $\mathcal{P}$ is a cyclic proof if and only if $\mathcal{P}'$ is.*

Theorem 5.6. *Let G be a rule graph, let T_1, T_2 be derivation trees, and let $f_1 : T_1 \to G$ and $f_2 : T_2 \to G$ be rule graph homomorphisms such that $f_1(root(T_1)) = f_2(root(T_2))$. Then $T_1 \cong T_2$.*

The following is a useful general theorem for extracting a pre-proof from a (possibly infinite) proof tree:

Theorem 5.7. *Let $T = (V, s, r, p)$ be a derivation tree with no bud nodes and with root node v_0, let G be a finite rule graph, and let $f : T \to G$ be a surjective rule graph homomorphism. Also, for each infinite branch $\pi = v_0 v_1 v_2 \ldots$ in T, let $m_\pi < n_\pi$ be numbers such that $f(v_{m_\pi}) = f(v_{n_\pi})$. Then we define $\mathcal{D} = (V', s', r', p')$ and $\mathcal{R}$ by "folding down" T as follows:*

- *$V' = \{v \in V \mid \text{for all infinite } \pi = v_0 v_1 v_2 \ldots \text{. if } \exists k. v = v_k \text{ then } k \leq n_\pi\}$*
- *$s'(v) = s(v)$ for all $v \in V' (\subset V)$*
- *if $v \in V'$ and $\exists \pi = v_0 v_1 v_2 \ldots$ s.t. $v = v_{n_\pi}$, then $r'(v)$ and $p'(v)$ are undefined, i.e. v is a bud node of $\mathcal{D}$ and we define $\mathcal{R}(v) = v_{m_\pi}$. (If there is more than one branch π meeting this criterion, we may choose any suitable v_{m_π}.) Otherwise we define $r'(v) = r(v)$ and $p'(v) = p(v)$.*

Then $\mathcal{P} = (\mathcal{D}, \mathcal{R})$ is a pre-proof and there are surjective rule graph homomorphisms $T \to \mathcal{G}_\mathcal{P} \to G$. Furthermore, the homomorphism from T to $\mathcal{G}_\mathcal{P}$ maps $v_0 = root(T)$ to $v_0 = root(\mathcal{D})$.

Our intended application of Theorem 5.7 is to obtain a cyclic proof $\mathcal{P}'$ with convenient structural properties from a given cyclic proof $\mathcal{P}$ by "folding down" $\mathcal{T}_\mathcal{P}$. This application is illustrated by our next result.

Definition 5.8 (Cycle normal form). *Let $\mathcal{P} = (\mathcal{D}, \mathcal{R})$ be a cyclic pre-proof. $\mathcal{P}$ is said to be in cycle normal form if, for every bud node B in $\mathcal{D}$, its companion $\mathcal{R}(B)$ is a (strict) ancestor of B.*

Theorem 5.9. *Let $\mathcal{P}$ be a cyclic proof and for each infinite branch $\pi = v_0 v_1 v_2 \ldots$ in $\mathcal{T}_\mathcal{P}$, let $m_\pi < n_\pi$ be numbers such that $f_\mathcal{P}(m_\pi) = f_\mathcal{P}(n_\pi)$. Then any pre-proof obtained from Theorem 5.7 by "folding down" $\mathcal{T}_\mathcal{P}$ is a cyclic proof, and furthermore is in cycle normal form.*

Proof. As there is a surjective rule graph homomorphism from $\mathcal{T}_\mathcal{P}$ to $\mathcal{G}_\mathcal{P}$ by Lemma 5.4, we can apply Theorem 5.7 to obtain a pre-proof $\mathcal{P}' = (\mathcal{D}', \mathcal{R}')$, and by the theorem, there is a surjective rule graph homomorphism $g : \mathcal{T}_\mathcal{P} \to \mathcal{G}_{\mathcal{P}'}$ such that $g(root(\mathcal{T}_\mathcal{P})) = root(\mathcal{D}')$. As there is also a surjective rule graph homomorphism $f_{\mathcal{P}'} : \mathcal{T}_{\mathcal{P}'} \to \mathcal{G}_{\mathcal{P}'}$ such that $f_{\mathcal{P}'}(root(\mathcal{T}_\mathcal{P})) = root(\mathcal{D}')$, again by Lemma 5.4, we can apply Theorem 5.6 to conclude $\mathcal{T}_\mathcal{P} \cong \mathcal{T}_{\mathcal{P}'}$, and since $\mathcal{P}$ is a cyclic proof, so is $\mathcal{P}'$ by Lemma 5.5.

To see that $\mathcal{P}'$ is in cycle normal form, we simply observe that the construction of Theorem 5.7 ensures $\mathcal{R}'(B)$ is always an ancestor of B for any bud node B. $\square$

We remark that we also have a direct proof of cycle-normalisation (the transformation of an arbitrary cyclic proof to one in cycle normal form), which will appear in the author's PhD thesis.

6   Trace-Based Proof Conditions

The general trace condition (cf. Definition 4.12) qualifying pre-proofs as cyclic proofs is both computationally and combinatorially complex. In order to simplify our analysis of cyclic proof structures, we consider the formulation of alternative trace conditions that are sufficient for pre-proofs to be cyclic proofs, and that also provide a greater degree of explicit structural information on pre-proofs.

We present one definition of such a condition — the existence of a so-called *trace manifold* for a pre-proof — which is apparently less general than Definition 4.12 and formulated with respect to a so-called *induction order* (a notion introduced by Schöpp [14] and crucially employed by Sprenger and Dam [18]). A trace manifold consists of finite trace segments together with conditions ensuring that for any infinite path, the segments can be "glued together" to yield an infinitely progressing trace on that path.

Definition 6.1 (Strong / weak connectivity). A directed graph $G = (V, E)$ is said to be *strongly connected* if for any $v, v' \in V$, there is a path in G from v to v'. G is said to be *weakly connected* if $(V, E \cup E^{-1})$ is strongly connected.

Definition 6.2 (Structural connectivity). Let $\mathcal{P} = (\mathcal{D}, \mathcal{R})$ be a pre-proof in cycle normal form and denote the set of bud nodes occuring in $\mathcal{D}$ by $\mathcal{B}$. Define the relation $\leq_{\mathcal{P}}$ on $\mathcal{B}$ by: $B_2 \leq_{\mathcal{P}} B_1$ if $\mathcal{R}(B_2)$ appears on the unique $\mathcal{D}$-path $\mathcal{R}(B_1) \ldots B_1$.

Definition 6.3 (Induction order). A partial order $\lhd$ on $\mathcal{B}$ is said to be an *induction order* for $\mathcal{P}$ if $\lhd$ is *forest-like*, i.e. $(B \lhd B_1 \wedge B \lhd B_2)$ implies $(B_1 = B_2 \vee B_1 \lhd B_2 \vee B_2 \lhd B_1)$, and every weakly $\leq_{\mathcal{P}}$-connected set $S \subseteq \mathcal{B}$ has a $\lhd$-greatest element.

Definition 6.4 (Trace manifold). Let $\mathcal{P} = (\mathcal{D}, \mathcal{R})$ be a pre-proof in cycle normal form, let $\mathcal{B} = \{B_1, \ldots, B_n\}$ be the set of bud nodes occurring in $\mathcal{P}$ and let $\lhd$ be an induction order for $\mathcal{P}$. A *trace manifold* with respect to $\lhd$ is a set of traces: $\{\tau_{ij} \mid B_i, B_j \in \mathcal{B}, B_i \lhd B_j\}$ satisfying:

- τ_{ij} follows the $\mathcal{D}$-path $\mathcal{R}(B_i) \ldots B_i$ in S;
- $\tau_{ij}(B_i) = \tau_{ij}(\mathcal{R}(B_i))$;
- $B_j \leq_{\mathcal{P}} B_i \lhd B_k$ implies $\tau_{jk}(\mathcal{R}(B_j)) = \tau_{ik}(\mathcal{R}(B_j))$;
- for each i, τ_{ii} has at least one progress point.

Lemma 6.5. *Let $\mathcal{P}$ be a pre-proof in cycle normal form and let $\lhd$ be an induction order for $\mathcal{P}$. If $\mathcal{P}$ has a trace manifold with respect to $\lhd$, then $\mathcal{P}$ is a cyclic proof.*

Proof. (Sketch) Definition 4.12 can be reformulated as quantified over strongly connected subgraphs of $\mathcal{G}_{\mathcal{P}}$. It can then be shown that, since $\mathcal{P}$ is in cycle normal form, such subgraphs can be characterised as $\mathcal{D}$-paths of the form $\mathcal{R}(B)\ldots B$ whose bud endpoints are weakly $\leq_{\mathcal{P}}$-connected. From this one can analyse the composition of any infinite path through $\mathcal{G}_{\mathcal{P}}$ and construct an infinitely progressing trace using the appropriate components given by the trace manifold.   □

We remark that, in fact, our translation from structural to cyclic proofs (cf. Theorem 4.15) transforms structural proofs into cyclic proofs with trace manifolds.

7   Conclusions and Future Work

We have formulated a cyclic proof system for first-order logic with ordinary inductive definitions that, importantly, uses only the standard syntax of sequent calculus for first-order logic and the standard sequent calculus proof rules, together with simple unfolding rules for inductively defined predicates. A global condition formulated in terms of *traces* over infinite paths in the proof is used to guarantee soundness. This approach essentially amounts to a postponement in the choice of induction principle; induction principles are not chosen within the proof itself, but rather implicitly via eventual satisfaction of the trace condition.

Cyclic proofs have been demonstrated to subsume the usual structural proofs that use explicit induction rules. Establishing the status of our Conjecture 4.16 — that cyclic proofs are no more powerful than structural proofs in terms of what can be proved — appears very difficult due to the complexity inherent in the trace condition on cyclic proofs and in our definition schema. We have developed tools for analysing the structure of cyclic proofs: in particular, a general theorem allowing cyclic proofs to be transformed via a folding operation on infinite trees. The useful result of cycle-normalisation is a simple corollary of this theorem.

We can also define cyclic proof more locally in terms of *trace manifolds* at the (possible) expense of some generality. We have observed that any structural proof can be transformed into a cyclic proof with a trace manifold. So structural provability $\Rightarrow$ trace manifold provability $\Rightarrow$ cyclic provability. Establishing whether either of these implications hold in reverse is clearly of interest but appears very difficult. The main problem with the latter — transforming an arbitrary cyclic proof into one having a trace manifold — is that as traces on two infinite paths can behave entirely differently, it is not obvious that a manifold need exist. It may be possible to use our "tree-folding" machinery of Section 5 in conjunction with a combinatorial argument about traces to establish this property, and we are looking into this presently. As for the former problem — translating cyclic proofs with trace manifolds into structural proofs — the main difficulty appears to lie in resolving the case-splits that represent trace progress points with cases of explicit induction rules.

Acknowledgements

The author wishes to thank, primarily, his supervisor, Alex Simpson. Thanks are also due to Alan Bundy, Lucas Dixon, Geoff Hamilton, Alberto Momigliano, Alan Smaill, and Rene Vestergaard for fruitful discussions, and to the anonymous referees.

References

1. Peter Aczel. An introduction to inductive definitions. In Jon Barwise, editor, *Handbook of Mathematical Logic*, pages 739–782. North-Holland, 1977.
2. Yves Bertot and Pierre Castéran. *Interactive Theorem Proving and Program Development*. EATCS: Texts in Theoretical Computer Science. Springer-Verlag, 2004.
3. Julian Bradfield and Colin Stirling. Local model checking for infinite state spaces. *Theoretical Computer Science*, 96:157–174, 1992.
4. Thierry Coquand. Infinite objects in type theory. In H. Barendregt and T. Nipkow, editors, *Types for Proofs and Programs*, pages 62–78. Springer, 1993.
5. Mads Dam and Dilian Gurov. μ-calculus with explicit points and approximations. *Journal of Logic and Computation*, 12(2):255–269, April 2002.
6. Gerhard Gentzen. Investigations into logical deduction. In M.E. Szabo, editor, *The Collected Papers of Gerhard Gentzen*, pages 68–131. North-Holland, 1969.
7. Eduardo Giménez. *A Calculus of Infinite Constructions and its application to the verification of communicating systems*. PhD thesis, Ecole Normale Supérieure de Lyon, 1996.
8. M.J.C. Gordon and T.F. Melham, editors. *Introduction to HOL: a theorem proving environment for higher order logic*. Cambridge University Press, 1993.
9. Geoff Hamilton. Poítin: Distilling theorems from conjectures. To appear.
10. Matt Kaufmann, Panagiotis Manolios, and J Strother Moore. *Computer-Aided Reasoning: An Approach*. Kluwer Academic Publishers, June 2000.
11. Per Martin-Löf. Haupstatz for the intuitionistic theory of iterated inductive definitions. In J.E. Fenstad, editor, *Proceedings of the Second Scandinavian Logic Symposium*. North-Holland, 1971.
12. Raymond McDowell and Dale Miller. Cut-elimination for a logic with definitions and induction. *Theoretical Computer Science*, 232:91–119, 2000.
13. Tobias Nipkow, Lawrence C. Paulson, and Markus Wenzel. *Isabelle/HOL: A Proof Assistant for Higher-Order Logic*, volume 2283 of *Lecture Notes in Computer Science*. Springer-Verlag, 2002.
14. Ulrich Schöpp. Formal verification of processes. Master's thesis, University of Edinburgh, 2001.
15. Ulrich Schöpp and Alex Simpson. Verifying temporal properties using explicit approximants: Completeness for context-free processes. In *Foundations of Software Science and Computation Structure: Proceedings of FoSSaCS 2002*, volume 2303 of *Lecture Notes in Computer Science*, pages 372–386. Springer-Verlag, 2002.
16. Carsten Schürmann. *Automating the Meta-Theory of Deductive Systems*. PhD thesis, Carnegie-Mellon University, 2000.
17. Christoph Sprenger and Mads Dam. A note on global induction mechanisms in a μ-calculus with explicit approximations. *Theoretical Informatics and Applications*, July 2003. Full version of FICS '02 paper.
18. Christoph Sprenger and Mads Dam. On the structure of inductive reasoning: circular and tree-shaped proofs in the μ-calculus. In *Proceedings of FOSSACS 2003*, volume 2620 of *Lecture Notes in Computer Science*, pages 425–440, 2003.
19. Valentin Turchin. The concept of a supercompiler. *ACM Transactions on Programming Languages and Systems*, 8:90–121, 1986.
20. M.Y. Vardi and P. Wolper. An automata-theoretic approach to automatic program verification. *Logic in Computer Science, LICS '86*, pages 322–331, 1986.

A Tableau-Based Decision Procedure for a Fragment of Graph Theory Involving Reachability and Acyclicity[*]

Domenico Cantone[1] and Calogero G. Zarba[2]

[1] University of Catania
[2] University of New Mexico

Abstract. We study the decision problem for the language **DGRA** (*directed graphs with reachability and acyclicity*), a quantifier-free fragment of graph theory involving the notions of reachability and acyclicity.

We prove that the language **DGRA** is decidable, and that its decidability problem is *NP*-complete. We do so by showing that the language enjoys a *small model property*: If a formula is satisfiable, then it has a model whose cardinality is polynomial in the size of the formula.

Moreover, we show how the small model property can be used in order to devise a tableau-based decision procedure for **DGRA**.

1 Introduction

Graphs arise naturally in many applications of mathematics and computer science. For instance, graphs arise as suitable data structures in most programs. In particular, when verifying programs manipulating pointers [3], one needs to reason about the reachability and acyclicity of graphs.

In this paper we introduce the language **DGRA** (*directed graphs with reachability and acyclicity*), a quantifier-free many-sorted fragment of directed graph theory. The language **DGRA** contains three sorts: node for nodes, set for sets of nodes, and graph for graphs. In the language **DGRA** graphs are modeled as binary relations over nodes or, alternatively, as sets of pairs of nodes. The language **DGRA** contains the set operators $\cup$, $\cap$, $\setminus$, $\{\cdot\}$ and the set predicates $\in$, and $\subseteq$. It also contains:

- a predicate *reachable*(a, b, G) stating that there is a nonempty path going from node a to node b in the graph G;
- a predicate *acyclic*(G) stating that the graph G is acyclic.

We prove that the language **DGRA** is decidable, and that its decidability problem is *NP*-complete. We do so by showing that the language enjoys a *small model property*: If a formula is satisfiable, then it has a model $\mathcal{A}$ whose cardinality is polynomial in the size of the formula.

---

[*] The second author was in part supported by grants NSF ITR CCR-0113611 and NSF CCR-0098114.

B. Beckert (Ed): TABLEAUX 2005, LNAI 3702, pp. 93–107, 2005.

More precisely, let φ be a satisfiable formula in the language **DGRA**, and let m and g be, respectively, the number of variables of sort node and graph occurring in φ. Then there exists a model $\mathcal{A}$ of φ such that its associated domain A_{node} has cardinality less than or equal to $m + m^2 \cdot g^2$.

At first sight, it seems that the small model property only suggests a brute force decision procedure for **DGRA**, consisting in enumerating all models up to a certain size. However, the bound on the cardinality of A_{node} can be cleverly exploited in order to devise a tableau-based decision procedure for **DGRA**.

Roughly speaking, the idea is as follows. Suppose that T is a tableau for the formula φ. We devise the tableau rules in such a way that at most $m^2 \cdot g^2$ fresh variables of sort node are added to any branch B of T. Furthermore, the tableau rules need to ensure that these fresh variables are to be interpreted as distinct from each other, and distinct from every old variable of sort node already occurring in φ.

We use the above intuition in order to devise a tableau calculus for **DGRA** that is terminating, sound, and complete. Consequently, we obtain a decision procedure for **DGRA** that is, at least potentially, more efficient than a naive brute force approach.

Organization of the paper. In Section 2 we define a notion of paths that will be used in the rest of the paper. In Section 3 we define the syntax and semantics of the language **DGRA**. In Section 4 we present our tableau calculus for **DGRA**. In Section 5 we show one example of our tableau calculus in action. In Section 6 we prove that our tableau calculus is terminating, sound, and complete, and therefore it yields a decision procedure for **DGRA**. In Section 7 we survey on related work. In Section 8 we draw final conclusions. For lack of space, we omit many proofs, which can be found in the extended version of this paper [6].

2   Paths

Definition 1 (Paths and cycles). Let A be a set. A (simple) PATH π over A is a sequence

$$\pi = \langle \nu_1, \ldots, \nu_n \rangle$$

such that

(a) $n \geq 2$;
(b) $\nu_i \in A$, for each $1 \leq i \leq n$;
(c) $\{\nu_1, \nu_n\} \cap \{\nu_2, \ldots, \nu_{n-1}\} = \emptyset$;
(d) $\nu_i \neq \nu_j$, for each $1 < i < j < n$.

A CYCLE is a path $\pi = \langle \nu_1, \ldots, \nu_n \rangle$ such that $\nu_1 = \nu_n$.   □

Note that, according to Definition 1, the sequence $\langle a, b, b, c \rangle$ is *not* a path.

We denote with $paths(A)$ the set of all paths over A. Let $\pi = \langle \nu_1, \ldots, \nu_n \rangle$ be a path in $paths(A)$, and let $R \subseteq A \times A$ be a binary relation. We write $\pi \subseteq R$ when $(\nu_i, \nu_{i+1}) \in R$, for each $1 \leq i < n$.

Sorts

node	nodes
set	sets of nodes
graph	graphs, modeled as sets of pairs of nodes

Symbols

	Function symbols	Predicate symbols
Sets	$\emptyset_{\mathsf{set}}$: set $\cup, \cap, \setminus$: set $\times$ set $\to$ set $\{\cdot\}$: node $\to$ set	$\in$: node $\times$ set $\subseteq$: set $\times$ set
Binary relations	$\emptyset_{\mathsf{graph}}$: graph $\cup, \cap, \setminus$: graph $\times$ graph $\to$ graph $\{(\cdot, \cdot)\}$: node $\times$ node $\to$ graph	$(\cdot, \cdot) \in \cdot$: node $\times$ node $\times$ graph $\subseteq$: graph $\times$ graph
Reachability		$reachable$: node $\times$ node $\times$ graph $acyclic$: graph

Fig. 1. The language **DGRA**

3   The Language DGRA

3.1   Syntax

The language **DGRA** (*directed graphs with reachability and acyclicity*) is a quantifier-free many-sorted language with equality [7]. Its sorts and symbols are depicted in Figure 1. Note that some symbols of the language are overloaded.

Definition 2. A **DGRA**-FORMULA is a well-sorted many-sorted formula constructed using:

- the function and predicate symbols in Figure 1;
- variables of sort τ, for $\tau \in \{\mathsf{node}, \mathsf{set}, \mathsf{graph}\}$;
- the equality predicate $=$;
- the propositional connectives $\neg$, $\wedge$, $\vee$, and $\to$.   □

Given a **DGRA**-formula φ, we denote with $vars_\tau(\varphi)$ the set of τ-variables occurring in φ. Moreover, we let $vars(\varphi) = vars_{\mathsf{node}}(\varphi) \cup vars_{\mathsf{set}}(\varphi) \cup vars_{\mathsf{graph}}(\varphi)$.

To increase readability, in the rest of the paper we will use the abbreviations depicted in Figure 2.

3.2   Semantics

Definition 3. Let V_τ be a set of τ-variables, for $\tau \in \{\mathsf{node}, \mathsf{set}, \mathsf{graph}\}$, and let $V = V_{\mathsf{node}} \cup V_{\mathsf{set}} \cup V_{\mathsf{graph}}$.

Syntactic sugar	Official formula
$a \notin x$	$\neg(a \in x)$
$G(a, b)$	$(a, b) \in G$
$\neg G(a, b)$	$\neg((a, b) \in G)$
$G^{+}(a, b)$	$reachable(a, b, G)$
$\neg G^{+}(a, b)$	$\neg reachable(a, b, G)$

Fig. 2. Syntactic sugar for the language **DGRA**

A **DGRA**-INTERPRETATION over V is a many-sorted interpretation satisfying the following conditions:

- each sort τ is mapped to a set A_τ such that:
  - $A_{\text{node}} \neq \emptyset$;
  - $A_{\text{set}} = \mathcal{P}(A_{\text{node}})$;
  - $A_{\text{graph}} = \mathcal{P}(A_{\text{node}} \times A_{\text{node}})$;
- each variable $u \in V$ of sort τ is mapped to an element $u^{\mathcal{A}} \in A_\tau$;
- the set symbols $\emptyset_{\text{set}}$, $\cup$, $\cap$, $\setminus$, $\{\cdot\}$, $\in$, and $\subseteq$ are interpreted according to their standard interpretation over sets of nodes;
- the binary relation symbols $\emptyset_{\text{graph}}$, $\cup$, $\cap$, $\setminus$, $\{(\cdot, \cdot)\}$, $(\cdot, \cdot) \in \cdot$, and $\subseteq$ are interpreted according to their standard interpretation over sets of pairs of nodes;
- $[reachable(a, b, G)]^{\mathcal{A}} = true$ if and only if there exists a path $\pi \in paths(A_{\text{node}})$ such that $\pi \subseteq G^{\mathcal{A}}$;
- $[acyclic(G)]^{\mathcal{A}} = true$ if and only if there is no cycle $\pi \in paths(A_{\text{node}})$ such that $\pi \subseteq G^{\mathcal{A}}$. □

If $\mathcal{A}$ is a **DGRA**-interpretation, we denote with $vars_\tau(\mathcal{A})$ the set of variables of sort τ that are interpreted by $\mathcal{A}$. Moreover, we let $vars(\mathcal{A}) = vars_{\text{node}}(\mathcal{A}) \cup vars_{\text{set}}(\mathcal{A}) \cup vars_{\text{graph}}(\mathcal{A})$. If $V \subseteq vars(\mathcal{A})$, we let $V^{\mathcal{A}} = \{u^{\mathcal{A}} \mid u \in V\}$.

Definition 4. A **DGRA**-formula φ is **DGRA**-SATISFIABLE if there exists a **DGRA**-interpretation $\mathcal{A}$ such that φ is true in $\mathcal{A}$. □

3.3   Examples

The following are examples of valid statements over graphs that can be expressed in the language **DGRA**:

$$(G^{+}(a, b) \ \wedge \ G^{+}(b, c)) \ \rightarrow \ G^{+}(a, c) \tag{1}$$

$$(G \subseteq H \ \wedge \ acyclic(H)) \ \rightarrow \ acyclic(G) \tag{2}$$

$$(G^{+}(a, b) \ \wedge \ H^{+}(b, a)) \ \rightarrow \ \neg acyclic(G \cup H) \tag{3}$$

$$\neg acyclic(\{(a, b)\}) \ \rightarrow \ a = b \tag{4}$$

In particular:

- (1) expresses the transitivity property of the reachability relation.
- (2) states that if a graph H is acyclic, then any of its subgraphs is also acyclic.
- (3) states that if it is possible to go from node a to node b in a graph G, and from node b to node a in a graph H, then the graph $G \cup H$ contains a cycle.
- (4) states that if a graph contains only the edge (a, b), and it is not acyclic, then a and b must be the same node.

3.4 Normalized Literals

Definition 5. A literal is FLAT if it is of the form $x = y$, $x \neq y$, $x = f(y_1, \ldots, y_n)$, $p(y_1, \ldots, y_n)$, and $\neg p(y_1, \ldots, y_n)$, where $x, y, y_1, \ldots, y_n$ are variables, f is a function symbol, and p is a predicate symbol. □

Definition 6. A **DGRA**-literal is NORMALIZED if it is a flat literal of the form:

$$a \neq b,$$

$$x = y \cup z, \qquad x = y \setminus z, \qquad x = \{a\},$$

$$G = H \cup L, \qquad G = H \setminus L, \qquad G = \{(a, b)\},$$

$$G^+(a, b), \qquad \neg G^+(a, b), \qquad acyclic(G).$$

where a, b are node-variables, x, y, z are set-variables, and G, H, L are graph-variables. □

Lemma 7. *The problem of deciding the **DGRA**-satisfiability of **DGRA**-formulae is equivalent to the problem of deciding the **DGRA**-satisfiability of conjunctions of normalized **DGRA** literals. Moreover, if the latter problem is in NP, so is the former.* □

3.5 The Small Model Property

Definition 8. Let $\mathcal{A}$ be a **DGRA**-interpretation, let $V \subseteq vars_{\text{node}}(\mathcal{A})$ be a set of node-variables, and let $k \geq 0$. We say that $\mathcal{A}$ is k-SMALL with respect to V if $\mathcal{A}_{\text{node}} = V^{\mathcal{A}} \cup A'$, for some set A' such that $|A'| \leq k$. □

Lemma 9 (Small model property). *Let Γ be a conjunction of normalized **DGRA**-literals, and let $V_\tau = vars_\tau(\Gamma)$, for each sort τ. Also, let $m = |V_{\text{node}}|$ and $g = |V_{\text{graph}}|$. Then the following are equivalent:*

1. *Γ is **DGRA**-satisfiable;*
2. *Γ is true in a **DGRA**-interpretation $\mathcal{A}$ that is $(m^2 \cdot g^2)$-small with respect to V_{node}.* □

The next two theorems show how Lemma 9 entails the decidability and NP-completeness of the language **DGRA**.

$$\frac{}{a = a} \; (E1) \qquad\qquad \frac{\begin{array}{c}\ell(a)\\ a = b\end{array}}{\ell(b)} \; (E2)$$

Note: In rule (E1), a is a node-variable already occurring in the tableau. In rule (E2), a and b are node-variables, and ℓ is a **DGRA**-literal.

Fig. 3. Equality rules

Theorem 10 (Decidability).*The problem of deciding the* **DGRA***-satisfiability of* **DGRA***-formulae is decidable.* $\qquad\square$

Theorem 11 (Complexity).*The problem of deciding the* **DGRA***-satisfiability of* **DGRA***-formulae is NP-complete.* $\qquad\square$

4   A Tableau Calculus for DGRA

In this section we show how the small model property can be used in order to devise a tableau-based decision procedure for **DGRA**. Our tableau calculus is based on the insight that if a **DGRA**-formula φ is true in a k-small **DGRA**-interpretation, then it is enough to generate only k fresh node-variables in order to prove the **DGRA**-satisfiability of φ.

Without loss of generality, we assume that the input of our decision procedure is a conjunction of normalized **DGRA**-literals. Thus, let Γ be a conjunction of normalized **DGRA**-literals, and let $V_\tau = vars_\tau(\Gamma)$, for each sort τ. Intuitively, a **DGRA***-tableau* for Γ is a tree whose nodes are labeled by normalized **DGRA**-literals.

Definition 12 (DGRA-tableaux). Let Γ be a conjunction of normalized **DGRA**-literals, and let $V_\tau = vars_\tau(\Gamma)$, for each sort τ. An INITIAL **DGRA**-TABLEAU for Γ is a tree consisting of only one branch B whose nodes are labeled by the literals in Γ.

A **DGRA**-TABLEAU for Γ is either an initial **DGRA**-tableau for Γ, or is obtained by applying to a **DGRA**-tableau for Γ one of the rules in Figures 3–6. $\qquad\square$

Definition 13. A branch B of a **DGRA**-tableau is CLOSED if at least one of the following two conditions hold:

(a) B contains two complementary literals $\ell, \neg\ell$;
(b) B contains literals of the form $acyclic(G)$ and $G^+(a, a)$.

A branch which is not closed is OPEN. A **DGRA**-tableau is CLOSED if all its branches are closed; otherwise it is OPEN. $\qquad\square$

$$\frac{\begin{array}{c} x = y \cup z \\ a \in x \end{array}}{a \in y \mid a \in z} \ (S1) \qquad \frac{\begin{array}{c} x = y \cup z \\ a \in y \end{array}}{a \in x} \ (S2) \qquad \frac{\begin{array}{c} x = y \cup z \\ a \in z \end{array}}{a \in x} \ (S3)$$

$$\frac{\begin{array}{c} x = y \setminus z \\ a \in x \end{array}}{\begin{array}{c} a \in y \\ a \notin z \end{array}} \ (S4) \qquad \frac{\begin{array}{c} x = y \setminus z \\ a \in y \end{array}}{a \in z \mid \begin{array}{c} a \notin z \\ a \in x \end{array}} \ (S5)$$

$$\frac{\begin{array}{c} x = \{a\} \\ b \in x \end{array}}{a = b} \ (S6) \qquad \frac{x = \{a\}}{a \in x} \ (S7)$$

Fig. 4. Set rules

$$\frac{\begin{array}{c} G = H \cup L \\ G(a,b) \end{array}}{H(a,b) \mid L(a,b)} \ (G1) \qquad \frac{\begin{array}{c} G = H \cup L \\ H(a,b) \end{array}}{G(a,b)} \ (G2) \qquad \frac{\begin{array}{c} G = H \cup L \\ L(a,b) \end{array}}{G(a,b)} \ (G3)$$

$$\frac{\begin{array}{c} G = H \setminus L \\ G(a,b) \end{array}}{\begin{array}{c} H(a,b) \\ \neg L(a,b) \end{array}} \ (G4) \qquad \frac{\begin{array}{c} G = H \setminus L \\ H(a,b) \end{array}}{L(a,b) \mid \begin{array}{c} \neg L(a,b) \\ G(a,b) \end{array}} \ (G5)$$

$$\frac{\begin{array}{c} G = \{(a,b)\} \\ G(c,d) \end{array}}{\begin{array}{c} a = c \\ b = d \end{array}} \ (G6) \qquad \frac{G = \{(a,b)\}}{G(a,b)} \ (G7)$$

Fig. 5. Graph rules

Given a **DGRA**-tableau T, we can associate to it a **DGRA**-formula $\phi(\mathsf{T})$ in disjunctive normal form as follows. For each branch B of T we let

$$\phi(\mathsf{B}) = \bigwedge_{\ell \in \mathsf{B}} \ell,$$

$$\frac{G(a,b)}{G^+(a,b)} \ (R1) \qquad \frac{\begin{array}{c} G^+(a,b) \\ G^+(b,c) \end{array}}{G^+(a,c)} \ (R2) \qquad \frac{}{G(a,b) \mid \neg G(a,b)} \ (R3)$$

$$\frac{G^+(a,b)}{\begin{array}{c} G(a,w) \\ G^+(w,b) \\ w \neq c_1 \\ \vdots \\ w \neq c_m \end{array}} \ (R4) \qquad\qquad \frac{}{\neg G^+(a,b)} \ (R5)$$

Note: Let Γ be a conjunction of normalized **DGRA**-literals, and let $V_\tau = vars_\tau(\Gamma)$, for each sort τ. Also, let $m = |V_{\text{node}}|$ and $g = |V_{\text{graph}}|$. Finally, let B be a branch of a **DGRA**-tableau form Γ.

Rule (R3) can be applied to B provided that:

(a) $a, b \in vars_{\text{node}}(\mathsf{B})$.

Rule (R4) can be applied to B provided that:

(b) B is saturated with respect to rule (R2);
(c) B does not contain literals of the form $G(a, d_1), G(d_1, d_2), \ldots, G(d_{k-1}, d_k), G(d_k, b)$;
(d) $vars_{\text{node}}(\mathsf{B}) = \{c_1, \ldots, c_n\}$;
(e) $|vars_{\text{node}}(\mathsf{B})| < m + m^2 \cdot g^2$.

Rule (R5) can be applied to B provided that:

(a) $a, b \in vars_{\text{node}}(\mathsf{B})$;
(b) B is saturated with respect to rule (R3);
(c) B does not contain literals of the form $G(a, d_1), G(d_1, d_2), \ldots, G(d_{k-1}, d_k), G(d_k, b)$;
(f) $|vars_{\text{node}}(\mathsf{B})| = m + m^2 \cdot g^2$.

Intuition behind rule (R4): Conditions (b) and (c) imply the existence of a w such that $G(a, w)$ and $G^+(w, b)$. Furthermore, w must be distinct from all the node-variables already occurring in B.

Intuition behind rule (R5): Conditions (b) and (c) imply the existence of a w such that $G(a, w)$ and $G^+(w, b)$. Furthermore, w must be distinct from all the node-variables already occurring in B. But since we are looking for "small" models, condition (f) tells us that we cannot add a fresh node-variables w to B. It must necessarily follow $\neg G^+(a, b)$.

Fig. 6. Reachability rules

where ℓ denotes a **DGRA**-literal. Then, we let

$$\phi(\mathsf{T}) = \bigvee_{\mathsf{B} \in \mathsf{T}} \phi(\mathsf{B}).$$

Definition 14. A **DGRA**-tableau T is SATISFIABLE if there exists a **DGRA**-interpretation $\mathcal{A}$ such that $\phi(\mathsf{T})$ is true in $\mathcal{A}$. □

Definition 15. A branch B of a **DGRA**-tableau is SATURATED if no application of any rule in Figures 3–6 can add new literals to B. A **DGRA**-tableau is SATURATED if all its branches are saturated. □

5 An Example

Figure 7 shows a closed **DGRA**-tableau for the following **DGRA**-unsatisfiable conjunction of normalized **DGRA**-literals:

$$\Gamma = \left\{ \begin{array}{l} acyclic(G), \\ acyclic(L), \\ G = H \setminus L, \\ H(a, a) \end{array} \right\}.$$

The inferences in the tableau can be justified as follows:

- Nodes 5 thru 7 are obtained by means of an application of rule $(G5)$.
- Node 8 is obtained by means of an application of rule (R1). The resulting branch is closed because it contains the literals $acyclic(L)$ and $L^+(a, a)$.
- Node 9 is obtained by means of an application of rule (R1). The resulting branch is closed because it contains the literals $acyclic(G)$ and $G^+(a, a)$.

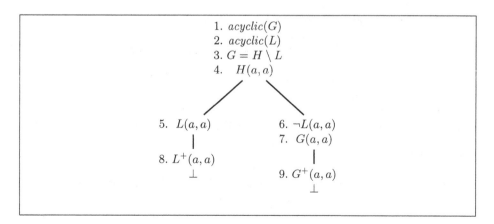

Fig. 7. A closed **DGRA**-tableau

6   Correctness

In this section we prove that our tableau calculus for **DGRA** is terminating, sound, and complete, and therefore it yields a decision procedure for **DGRA**. We follow standard arguments in the proofs of termination and completeness. Nonetheless, the proof of soundness is somewhat tricky, and it is based on the small model property.

6.1   Termination

Lemma 16 (Termination). *The tableau rules in Figure 3–6 are terminating.*□

PROOF. Let Γ be a conjunction of **DGRA**-literals, and let T be a saturated **DGRA**-tableaux. We want to show that T is finite.

Note that all rules in Figure 3–6 deduce only flat **DGRA**-literals. Furthermore, by inspecting rule (R4), it follows that the number of fresh variables that can be generated is bounded by $m^2 \cdot g^2$, where $m = |vars_{\mathsf{node}}(\Gamma)|$ and $g = |vars_{\mathsf{graph}}(\Gamma)|$.

Thus, if B is any branch of T, then B contains only flat literals constructed using a finite number of variables. It follows that the number of literals occurring in B is finite. Since all branches in T are finite, T is also finite.   ■

Note on complexity. Let Γ be a conjunction of normalized **DGRA**-literals, and let T be a saturated **DGRA**-tableau for Γ. By inspection of the proof of Lemma 16, it follows that the size of each branch in T is polynomially bounded by the size of Γ. This implies that our tableau-based decision procedure for **DGRA** is in *NP*, confirming the complexity result of Theorem 11.

6.2   Soundness

At first glance, it seems that our tableau calculus is not sound. "How can rule (R5) be sound?", may wonder the reader. Nonetheless, the following lemma shows that all the rules of our tableau calculus are sound in the sense that they preserve **DGRA**-satisfiability with respect to k-small **DGRA**-interpretations.

Lemma 17. *Let Γ be a conjunction of normalized **DGRA**-literals, and let $V_\tau = vars_\tau(\Gamma)$, for each sort τ. Also, let $m = |V_{\mathsf{node}}|$ and $g = |V_{\mathsf{graph}}|$. Finally, let T be a **DGRA**-tableau, and let T' be the result of applying to T one of the rules in Figures 3–6. Assume that there exists a **DGRA**-interpretation $\mathcal{A}$ such that:*

(α_1) $\phi(\mathsf{T})$ *is true in $\mathcal{A}$;*
(α_2) $\mathcal{A}$ *is $(m^2 \cdot g^2)$-small with respect to V_{node};*
(α_3) $c_i^{\mathcal{A}} \neq c_j^{\mathcal{A}}$, *whenever c_i is a fresh node-variable not occurring in $vars_{\mathsf{node}}(\Gamma)$, and c_j is a node-variable distinct from c_i.*

Then there exists a **DGRA**-*interpretation* $\mathcal{B}$ *such that:*

(β_1) $\phi(\mathsf{T}')$ *is true in* $\mathcal{B}$;
(β_2) $\mathcal{B}$ *is* $(m^2 \cdot g^2)$-*small with respect to* V_{node};
(β_3) $c_i^{\mathcal{B}} \neq c_j^{\mathcal{B}}$, *whenever* c_i *is a fresh* node-*variable not occurring in* $vars_{\mathsf{node}}(\Gamma)$, *and* c_j *is a* node-*variable distinct from* c_i. □

PROOF. We concentrate only on rules (R4) and (R5), since the proof goes straightforwardly for the other rules.

Concerning rule (R4), assume that the literal $G^+(a, b)$ is in B, and that conditions (b), (c), (d), and (e) in Figure 6 hold. Also, let $\mathcal{A}$ be a **DGRA**-interpretation $\mathcal{A}$ satisfying conditions (α_1), (α_2), and (α_3). By condition (b) and (c) and the fact that $(a^{\mathcal{A}}, b^{\mathcal{A}}) \in (G^{\mathcal{A}})^+$, it follows that there exists a node $\nu \in A_{\mathsf{node}}$ such that $(a^{\mathcal{A}}, \nu) \in G^{\mathcal{A}}$, $(\nu, b^{\mathcal{A}}) \in (G^{\mathcal{A}})^+$, and $\nu \neq c_i^{\mathcal{A}}$, for each $c_i \in vars_{\mathsf{node}}(\mathcal{B})$. Clearly, $\phi(\mathsf{T}')$ is true in the **DGRA**-interpretation $\mathcal{B}$ obtained from $\mathcal{A}$ by letting $w^{\mathcal{B}} = \nu$. By condition (e), $\mathcal{B}$ is $(m^2 \cdot g^2)$-small with respect to V_{node}. Moreover, condition (α_3) implies condition (β_3).

Concerning rule (R5), assume that conditions (a), (b), (c), and (f) in Figure 6 hold. Also, let $\mathcal{A}$ be a **DGRA**-interpretation $\mathcal{A}$ satisfying conditions (α_1), (α_2), and (α_3). By condition (f), it follows that $A_{\mathsf{node}} = vars_{\mathsf{node}}(\mathsf{B})$. But then, by conditions (b) and (c), we have $(a^{\mathcal{A}}, b^{\mathcal{A}}) \notin (G^{\mathcal{A}})^+$, and soundness of rule (R5) follows by letting $\mathcal{B} = \mathcal{A}$. ■

Lemma 18 (Soundness). *Let* Γ *be a conjunction of* **DGRA**-*literals. If there exists a closed* **DGRA**-*tableau for* Γ, *then* Γ *is* **DGRA**-*unsatisfiable.* □

PROOF. Let T be a closed **DGRA**-tableau for Γ, and suppose by contradiction that Γ is **DGRA**-satisfiable. Let $m = |vars_{\mathsf{node}}(\Gamma)|$ and $g = |vars_{\mathsf{graph}}(\Gamma)|$. By Lemmas 9 and 17, there exists an **DGRA**-interpretation $\mathcal{A}$ that is $(m^2 \cdot g^2)$-small with respect to $vars_{\mathsf{node}}(\Gamma)$, and such that $\phi(\mathsf{T})$ is true in $\mathcal{A}$. It follows that T is satisfiable. But this is a contradiction because T is closed, and closed **DGRA**-tableaux cannot be satisfiable. ■

6.3 Completeness

Lemma 19. *Let* Γ *be a conjunction of normalized* **DGRA**-*literals, and let* B *be a an open and saturated branch of a* **DGRA**-*tableau for* Γ. *Then* B *is satisfiable.* □

PROOF. Our goal is to define a **DGRA**-interpretation $\mathcal{A}$ satisfying B.

Let $V_\tau = vars_\tau(\Gamma)$, for each sort τ. Also, let W_{node} be the set of fresh node-variables introduced by applications of rule (R4), that is, $W_{\mathsf{node}} = vars_{\mathsf{node}}(\mathsf{B}) \setminus V_{\mathsf{node}}$. Finally, let $\sim$ be the equivalence relation over $V_{\mathsf{node}} \cup W_{\mathsf{node}}$ induced by the literals of the form $a = b$ occurring in B.

We let $\mathcal{A}$ be the unique **DGRA**-interpretation over $vars(\mathsf{B})$ defined by letting

$$A_{\mathsf{node}} = (V_{\mathsf{node}} \cup W_{\mathsf{node}})/\sim,$$

and

$$a^{\mathcal{A}} = [a]_\sim \,, \qquad\qquad\qquad\qquad \text{for each } a \in V_{\text{node}} \cup W_{\text{node}} \,,$$

$$x^{\mathcal{A}} = \{[a]_\sim \mid \text{the literal } a \in x \text{ is in B}\} \,, \qquad \text{for each } x \in V_{\text{set}} \,,$$

$$G^{\mathcal{A}} = \{([a]_\sim, [b]_\sim) \mid \text{the literal } G(a,b) \text{ is in B}\} \,, \quad \text{for each } G \in V_{\text{graph}} \,.$$

We claim that all literals occurring in B are true in $\mathcal{A}$.

Literals of the form $a = b$, $a \neq b$, $a \in x$, and $G(a,b)$. Immediate.

Literals of the form $a \notin x$. Let the literal $a \notin x$ be in B, and assume by contradiction that $a^{\mathcal{A}} \in x^{\mathcal{A}}$. Then there exists a node-variable b such that $a \sim b$, and the literal $b \in x$ is in B. By saturation with respect to the equality rules, the literal $a \in x$ is also in B, which implies that B is closed, a contradiction.

Literals of the form $\neg G(a,b)$. This case is similar to the case of literals of the form $a \notin x$.

Literals of the form $x = y \cup z$. Let the literal $x = y \cup z$ be in B. We want to prove that $x^{\mathcal{A}} = y^{\mathcal{A}} \cup z^{\mathcal{A}}$.

Assume first that $\nu \in x^{\mathcal{A}}$. Then there exists a node-variable a such that $\nu = a^{\mathcal{A}}$ and the literal $a \in x$ is in B. By saturation with respect to rule (S1), either the literal $a \in y$ is in B or the literal $a \in z$ is in B. In the former case, $\nu \in y^{\mathcal{A}}$; in the latter, $\nu \in z^{\mathcal{A}}$.

Vice versa, assume that $\nu \in y^{\mathcal{A}} \cup z^{\mathcal{A}}$ and suppose, without loss of generality, that $\nu \in y^{\mathcal{A}}$. Then there exists a node-variable a such that $\nu = a^{\mathcal{A}}$ and the literal $a \in y$ is in B. By saturation with respect to rule (S2), the literal $a \in x$ is in B. Thus, $\nu \in x^{\mathcal{A}}$.

Literals of the form $x = y \setminus z$, and $x = \{a\}$. These cases are similar to the case of literals of the form $x = y \cup z$.

Literals of the form $G = H \cup L$, $G = H \setminus L$, and $G = \{(a,b)\}$. These cases are similar to the cases of literals of the form $x = y \cup z$, $x = y \setminus z$, and $x = \{a\}$.

Literals of the form $G^+(a,b)$. Let the literal $G^+(a,b)$ be in B. If B contains literals of the form $G(a,d_1), G(d_1,d_2), \ldots, G(d_{k-1},d_k), G(d_k,b)$ then we clearly have $(a^{\mathcal{A}}, b^{\mathcal{A}}) \in (G^{\mathcal{A}})^+$. Otherwise, conditions (a), (b), (c), and (f) in Figure 6 hold, which implies that the literal $\neg G^+(a,b)$ is in B. It follows that B is closed, a contradiction.

Literals of the form $\neg G^+(a,b)$. Let the literal $\neg G^+(a,b)$ be in B, and assume by contradiction that $[G^+(a,b)]^{\mathcal{A}} = true$. Then there exist node-variables $c_1, \ldots, c_n$, with $n \geq 0$, such that the literals $G(a,c_1), G(c_1,c_2), \ldots, G(c_{n-1},c_n)$, and $G(c_n,b)$ are in B. By saturation with respect to rules (R1) and (R2), the literal $G^+(a,b)$ is in B, a contradiction.

Literals of the form $acyclic(G)$. Let the literal $acyclic(G)$ be in B, and assume by contradiction that $[acyclic(G)]^{\mathcal{A}} = false$. Then there exist node-variables $a_1, \ldots, a_n$, with $n \geq 1$, such that the literals $G(a_1, a_2)$, $G(a_2, a_3)$, $\ldots$, $G(a_{n-1}, a_n)$, and $G(a_n, a_1)$ are in B. By saturation with respect to rules (R1) and (R2), the literal $G^+(a_1, a_1)$ is in B, a contradiction.   ∎

Lemma 20 (Completeness). *Let Γ be a conjunction of normalized* **DGRA**-*literals. If Γ is* **DGRA**-*unsatisfiable then there exists a closed* **DGRA**-*tableau for Γ.*   □

PROOF. Assume, by contradiction, that Γ has no closed **DGRA**-tableau, and let T be a saturated **DGRA**-tableau for Γ. Since Γ has no closed **DGRA**-tableau, T must contain an open and saturated branch B. By Lemma 19, B is **DGRA**-satisfiable, which implies that Γ is also **DGRA**-satisfiable, a contradiction.   ∎

7 Related Work

7.1 Graph Theory

To our knowledge, the decision problem for graph theory was first addressed by Moser [9], who presented a decision procedure for a quantifier-free fragment of directed graph theory involving the operators of singleton graph construction, graph union, and graph intersection.

This result was extended by Cantone and Cutello [5], who proved the decidability of a more expressive quantifier-free fragment of graph theory. Cantone and Cutello's language can deal with both directed and undirected graphs, and it allows one to express the operators singleton, union, intersection, and difference, as well as some notions which are characteristic of graphs such as transitivity, completeness, cliques, independent sets, and the set of all self-loops. Cantone and Cutello's language does not deal with reachability and acyclicity.

Cantone and Cincotti [4] studied the decision problem for the language **UGRA** (*undirected graphs with reachability and acyclicity*). Intuitively, **UGRA** is the same as **DGRA**, except that it deals with undirected graphs. Unfortunately, due to a flaw in [4], it is still an open problem whether the language **UGRA** is decidable. Nonetheless, the ideas presented in [4] are very promising. Our proof of the small model property for **DGRA** is inspired by these ideas.

7.2 Static Analysis and Verification

Graph-based logics are of particular interest in the fields of static analysis and verification, where researchers use various abstractions based on graphs in order to represent the states of the memory of a program. We mention here four of such logics.

Benedikt, Reps, and Sagiv [2] introduced a logic of reachability expressions L_r. In this logic, one can express that it is possible to go from a certain node a to another node b by following a path that is specified by a regular expression R.

For instance, in L_r the expression $a\langle(R_1 \mid R_2)^*\rangle b$ asserts that it is possible to go from node a to node b by following 0 or more edges labeled by either R_1 or R_2.

Kuncak and Rinard [8] introduced the role logic RL, a logic that has the same expressivity of first-order logic with transitive closure. They also proved that a fragment RL^2 of role logic is decidable by reducing it to the two-variable logic with counting C^2.

Resink [11] introduced the local shape logic LSL. In this logic, it is possible to constrain the multiplicities of nodes and edges in a given graph. The logic LSL is equivalent to integer linear programming.

Ranise and Zarba [10] are currently designing together a logic for linked lists LLL, with the specific goal of verifying C programs manipulating linked lists. In this logic, graphs are specified by functional arrays, with the consequence that each node of a graph has at most one outgoing edge.

7.3   Description Logics

Baader [1] introduced description logic languages with transitive closure on roles. These languages are related to **DGRA** because sets of nodes are akin to concepts, roles are akin to graphs, and transitive closure of roles is akin to reachability. Therefore, we envisage a bright future in which advances in description logics will lead to advances in graph theory, and vice versa, advances in graph theory will lead to advances in description logics.

8   Conclusion

We presented a tableau-based decision procedure for the language **DGRA**, a quantifier-free fragment of directed graph theory involving the notions of reachability and acyclicity. We showed that the decidability of **DGRA** is a consequence of its *small model property*: If a formula is satisfiable, then it has a model whose cardinality is polynomial in the size of the formula. The small model property is at the heart of our tableau calculus, which can be seen as a search strategy of "small" models of the input formula.

We plan to continue this research by using (extensions of) **DGRA** in order to formally verify programs manipulating pointers. Finally, we want to study the decision problem for the language **UGRA** (*undirected graphs with reachability and acyclicity*) originally introduced in [4]. Although we do not know whether the language **UGRA** is decidable, we conjecture that decidability holds, at least in the case in which the acyclicity predicate is removed from the language.

Acknowledgments

This paper could not have existed without the exciting discussions with the following members of the research community: Aaron R. Bradley, Gianluca Cincotti, Jean-Christophe Filliâtre, Bernd Finkbeiner, Thomas In der Rieden,

Jean Goubault-Larrecq, Deepak Kapur, Yevgeny Kazakov, Dirk Leinenbach, Claude Marché, David Nowak, Silvio Ranise, Sriram Sankaranarayanan, Viorica Sofronie-Stokkermans, Uwe Waldmann, and Thomas Wies. We are also grateful to the three anonymous reviewers for pointing out a mistake in the submitted version of this paper, and for providing instructive references to the literature.

References

1. Franz Baader. Augmenting concept languages by transitive closure of roles: An alternative to terminological cycles. In John Mylopoulos and Raymond Reiter, editors, *International Joint Conference on Artificial Intelligence*, pages 446–451, 1991.
2. Michael Benedikt, Thomas W. Reps, and Shmuel Sagiv. A decidable logic for describing linked data structures. In S. Doaitse Swierstra, editor, *European Symposium on Programming*, volume 1576 of *Lecture Notes in Computer Science*, pages 2–19. Springer, 1999.
3. Rodney M. Burstall. Some techniques for proving correctness of programs which alter data structures. *Machine Intelligence*, 7:23–50, 1972.
4. Domenico Cantone and Gianluca Cincotti. The decision problem in graph theory with reachability related constructs. In Peter Baumgartner and Hantao Zhang, editors, *First-Order Theorem Proving*, Technical Report 5/2000, pages 68–90. Universität Koblenz-Landau, 2000.
5. Domenico Cantone and Vincenzo Cutello. A decidable fragment of the elementary theory of relations and some applications. In *International Symposium on Symbolic and Algebraic Computation*, pages 24–29, 1990.
6. Domenico Cantone and Calogero G. Zarba. A tableau-based decision procedure for a fragment of graph theory involving reachability and acyclicity. Technical report, Department of Computer Science, University of New Mexico, 2005.
7. Jean H. Gallier. *Logic for Computer Science: Foundations of Automatic Theorem Proving*. Harper & Row, 1986.
8. Viktor Kuncak and Martin C. Rinard. Generalized records and spatial conjunction in role logic. In Roberto Giacobazzi, editor, *Static Analysis*, volume 3148 of *lncs*, pages 361–376. Springer, 2004.
9. Louise E. Moser. A decision procedure for unquantified formulas of graph theory. In Ewing L. Lusk and Ross A. Overbeek, editors, *9th International Conference on Automated Deduction*, volume 310 of *Lecture Notes in Computer Science*, pages 344–357. Springer, 1988.
10. Silvio Ranise and Calogero G. Zarba. A decidable logic for pointer programs manipulating linked lists. Unpublished, 2005.
11. Arend Rensink. Canonical graph shapes. In David A. Schmidt, editor, *European Symposium on Programming*, volume 2986 of *Lecture Notes in Computer Science*, pages 401–415. Springer, 2004.

Embedding Static Analysis into Tableaux and Sequent Based Frameworks

Tobias Gedell

Department of Computing Science, Chalmers University of Technology,
SE-412 96 Göteborg, Sweden
gedell@cs.chalmers.se

Abstract. In this paper we present a method for embedding static analysis into tableaux and sequent based frameworks. In these frameworks, the information flows from the root node to the leaf nodes. We show that the existence of free variables in such frameworks introduces a bi-directional flow, which can be used to collect and synthesize arbitrary information.

We use free variables to embed a static program analysis in a sequent style theorem prover used for verification of Java programs. The analysis we embed is a reaching definitions analysis, which is a common and well-known analysis that shows the potential of our method.

The achieved results are promising and open up for new areas of application of tableaux and sequent based theorem provers.

1 Introduction

The aim of our work is to integrate static program analysis into theorem provers used for program verification. In order to do so, we must bridge the mismatch between the synthetic nature of static program analysis and analytic nature of tableaux and sequent calculi. One of the major differences is the flow of information.

In a program analysis, information is often synthesized by dividing a program into its subcomponents, calculating some information for each component and then merging the calculated information. This gives a flow of information that is directed bottom-up, with the subcomponents at the bottom.

Both tableaux and sequent style provers work in the opposite way. They take a theorem as input and, by applying the rules of their calculi, gradually divide it into branches, corresponding to logical case distinction, until all branches can be proved or refuted. In a ground calculus, there is no flow of information between the branches. Neither is there a need for that since the rules of the calculus only extend the proof by adding new nodes. Because of this, the information flow in a ground calculus is uni-directional—directed top-down, from the root to the leafs of the proof tree.

Tableaux calculi are often extended with *free variables* which are used for handling universal quantification (in the setting of sequent calculi, free variables correspond to *meta variables*, which are used for existential quantification) [Fit96]. Adding free variables breaks the uni-directional flow of information. When a branch chooses to instantiate a free variable, the instantiation has to be made visible at the point where the free variable was introduced. Therefore, some kind of information flow backwards in the

B. Beckert (Ed): TABLEAUX 2005, LNAI 3702, pp. 108–122, 2005.

proof has to exist. By exploiting this bi-directional flow we can collect and synthesize arbitrary information which opens up for new areas of application of the calculi.

We embed our program analysis in a sequent calculus using meta variables. The reason for choosing a program analysis is that logics for program verification could greatly benefit from an integration with program analysis. An example of this is the handling of loops in programs. Often a human must manually handle things like loops and recursive functions. Even for program constructs, which a verification system can cope with automatically, the system sometimes performs unnecessary work. Such a system could benefit from having a program analysis that could cheaply identify loops and other program constructs that can be handled using specialized rules of the program logics that do not require user interaction. An advantage of embedding a program analysis in a theorem prover instead of implementing it in an external framework, is that it allows for a closer integration of the analysis and prover.

The main contributions of this work are that:

- We show how synthesis can be performed in a tableau or sequent style prover, which opens up for new areas of application.
- We show how the rules of a program analysis can be embedded into a program logic and coexist with the original rules by using a tactic language.
- We give a proof-of-concept of our method. We do this by giving the full embedding of a program analysis in an interactive theorem prover.

The outline of this paper is as follows: In Section 2 we elaborate more on how we use the bi-directional flow of information; In Section 3 we briefly describe the theorem prover used for implementing our program analysis; In Section 4 we describe the program analysis; in Section 5 we present the embedding of the analysis in the theorem prover; in Section 6 we draw some conclusions; and in Section 7 we discuss future work.

2   Flow of Information

By using the mechanism for free variables we can send information from arbitrary nodes in the proof to nodes closer to the root. This is very useful to us since our program analysis needs to send information from the subcomponents of the program to the root node. In a proof, the subcomponents of the program correspond to leaf nodes. To show how it works, consider a tableau created with a destructive calculus where, at the root node, a free variable I is introduced. When I is instantiated by a branch closure, the closing substitution is applied to all branches where I occurs. This allows us to embed various analyses. One could, for example, imagine a very simple analysis that finds out whether a property P is true for any branch in a proof. In order to do so, we modify the closure rule. Normally, the closure rule tries to find two formulas φ and $\neg\psi$ in the same branch and a substitution that unifies φ and ψ. The new closure rule is modified to search for a closing substitution for a branch and if it finds one, check whether P is true for the particular branch. If it is, then the closing substitution is extended with an instantiation of the free variable I to a constant symbol c. We can now use this calculus

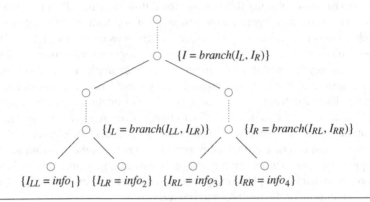

Fig. 1. Example tableau

to construct a proof as usual and when it is done, check whether I has been instantiated or not. If it has, then we know that P was true for at least one of the branches. Note that we are not interested in *what* I was instantiated to, just the fact that it *was* instantiated.

There is still a limit to how much information that can be passed to the root node. It is not possible to gather different information from each branch closure since they all use the same variable, I, to send their information. In particular, the reaching definitions analysis that we want to embed needs to be able to compute different information for each branch in the proof.

This can be changed by modifying the extension rule. When two branches are created in a proof, two new free variables, I_L and I_R, are introduced and I instantiated to $branch(I_L, I_R)$. I_L is used as the new I-variable for the left branch and I_R for the right branch. By doing this we ensure that each branch has its own variable for sending information. This removes the possibility of conflicting instantiations, since each I-variable will be instantiated at most once, either by extending or closing the branch to which it belongs.

When the tableau shown in Figure 1 has been closed, we get the instantiation of I, which will be the term $branch(branch(info_1, info_2), branch(info_3, info_4))$ that contains the information collected from all four branches. We have now used the tableau calculus to synthesize information from the leaf nodes.

We are now close to being able to implement our program analysis. The remaining problem is that we want to be able to distinguish between different types of branches. An example of this is found in Section 4.2 where different types of branches compute different collections of equations. We overcome this problem by, instead of always using the function symbol *branch*, allowing arbitrary function symbols when branching.

2.1 Non Destructive Calculi

In a non destructive constraint tableau, as described in [Gie01], it is possible to embed analyses using the same method.

In a constraint tableau, each node n has a *sink* object that contains all closing substitutions for the sub tableau having n as its top node. When adding a node to a branch, all closing substitutions of the branch are added to the node's sink object. The substitutions in the sink object are then sent to the sink object of the parent node. If the parent node is a node with more than one child, it has a *merger* object that receives the substitution and checks whether it is a closing substitution for all children. If it is, then it is propagated upwards to the sink object of the parent node, otherwise it is discarded. If the parent node only has one child, the substitution is directly sent to the node's parent node.

A tableau working like this is called non destructive since the free variables are never instantiated. Instead, a set of all possible closing instantiations is calculated for each branch and propagated upwards. When a closing substitution reaches the root node, the search is over since we know that it closes the entire tableau.

Using our method in a non destructive constraint tableau is easy. We modify the sink object of the root node to not only, when a closing substitution is found, tell us that the tableau is closable but also give us the closing substitution. The infrastructure with the sink objects could also make it easy to implement some of the extensions described in Section 7.

3   The KeY Prover

For the implementation, we choose an interactive theorem prover with a tactic programming language, the KeY system [ABB$^+$04]. The KeY system is a theorem prover for the Java Card language that uses a dynamic logic [Bec01]. The dynamic logic is a modal logic in which Java programs can occur as parts of formulas. An example of this is the formula,

$$\langle\{ \ i = 1; \ \}\rangle \ i > 0 \ ,$$

that denotes that after executing the assignment $i = 1$; the value of the variable i is greater than 0.

The KeY system is based on a non destructive sequent calculus with a standard semantics. It is well known that sequent calculi can be seen as the duality of tableaux calculi and we use this to carry over the method described in Section 2 to the sequent calculus used by KeY.

3.1   Tactic Programming Language

Theorem provers for program verification typically need to have a large set of rules at hand to handle all constructs in a language. Instead of hard-wiring these into the core of the theorem prover, one can opt for a more general solution and create a domain specific tactic language, which is used to implement the rules.

The rules written in the tactic language of KeY are called taclets [BGH$^+$04]. A taclet can be be seen as an implementation of a sequent calculus rule. In most theorem provers for sequent calculi, the rules perform some kind of pattern matching on sequents. Typically, the rules consist of a guard pattern and an action. If a sequent matches the guard pattern then the rule is applied and the action performed on the sequent. What it means for the pattern of a taclet to match a sequent is that there is a unifying substitution for the pattern and the sequent under consideration. The actions that can be performed include closing a proof branch, creating modified copies of sequents, and creating new proof branches.

We now have a look at the syntax of the tactic language and start with one of the simplest rules, the close_by_true rule.

```
close_by_true {
  find (==> true)
  close goal
};
```

The pattern matches sequents where *true* can be found on the right hand side. If *true* can be found on the right hand side, we know that we can close the proof branch under consideration, which is done by the close goal action.

If we, instead of closing the branch, want to create a modified copy of the sequent we use the replacewith action.

```
not_left {
  find (!b ==>)
  replacewith (==> b)
};
```

If we find a negated formula *b* on the left hand side we replace it with *b* on the right hand side.[1] The proof branch will remain open, but contain the modified sequent. We can also create new proof branches by using multiple replacewith actions.

So far, we have only considered sequents that do not contain embedded Java programs. When attaching programs to formulas, one has to choose a modality operator. There are a number of different modality operators having different semantics. The diamond operator <{p}>ϕ says that there is a terminating execution of the program p after which the formula ϕ holds. The box operator [{p}]ϕ says that after all terminating executions the formula ϕ holds. For our purpose, the modalities do not have any meaning since we are not trying to construct a proof in the traditional way. Regardless of this, the syntax of the taclet language forces us to have a modality operator attached to all programs. We, therefore, arbitrarily choose to use the diamond operator. In the future, it would be better to have a general-purpose operator with a free semantics that could be used for cases like this.

As an example of a taclet matching an embedded Java program, consider the following taclet, that matches an assignment of a literal to a variable attached to the formula *true* and closes the proof branch:

```
term_assign_literal {
  find (==> <{#var = #literal;}>(true))
  close goal
};
```

---

[1] Note that Γ and Δ are only implicitly present in the taclet.

4 Reaching Definitions Analysis

The analysis we choose to implement using our technique is *reaching definitions analysis* [NNH99]. This analysis is commonly used by compilers to perform several kinds of optimization such as, for example, loop optimization and constant computation [ASU86]. The analysis calculates which assignments may reach each individual statement in a program. Consider the following program, consisting of three assignments, where each statement is annotated with a label so that we can uniquely identify them.

$$a \overset{0}{=} 1; \quad b \overset{1}{=} 1; \quad a \overset{2}{=} 1;$$

Let us look at the statement annotated with 1. The statement executed before it (which we will call its previous statement) is the assignment $a \overset{0}{=} 1$; and since a has not yet been reassigned it still contains the value 1. We say that the assignment annotated with 0 *reaches* the statement annotated with 1. For each statement, we calculate the set of labels of the assignments that reach the statement before and after it has been executed. We call these sets the entry and exit sets. For this example, the label 0 will be in the entry set of the last assignment but not in its exit set, since the variable a is re-assigned. We do not just store the labels of the assignments in the sets, but also the name of the variable that is assigned. The complete entry and exit sets for our example program look as follows:

label	Entry	Exit
0	{}	{(a, 0)}
1	{(a, 0)}	{(a, 0), (b, 1)}
2	{(a, 0), (b, 1)}	{(b, 1), (a, 2)}

It is important to understand that the results of the analysis will be an *approximation*. It is undecidable to calculate the exact reaching information, which can easily be proven by using the halting problem. We will, however, ensure that the approximation is *safe*, which in this context means that if an assignment reaches a statement then the label of the assignment must be present in the entry set of that statement. The reverse may not hold, a label of an assignment being present in an entry set of a statement, does not necessarily mean that the assignment may reach that statement.

It is easy to see that for any program, a sound result of the analysis would be to let all entry and exit sets be equal to the set of all labels occurring in the program. This result would, of course, not be useful; what we want are as precise results as possible.

The analysis consists of two parts: a constraint-generation part and a constraint-solving part. The constraint-generation part traverses the program and generates a collection of equations defining the entry and exit sets. The equations are then solved by the constraint-solving part that calculates the actual sets.

4.1 Input Language

As input language, we choose a very simple language, the WHILE-language, which consists of assignments, block statements and if- and while-statements. We choose a simple language because we do not want to wrestle with a large language but instead show the concept of how the static program analysis can be implemented.

In the language, a program consists of a number of statements.

$$
\begin{array}{lll}
\text{Programs} & \textit{program} ::= \textit{stmt}^+ \\
\text{Statements} & \textit{stmt} & ::= \textit{var} \overset{lbl}{=} \textit{expr}; \\
& & |\quad \textbf{if}_{lbl}(\textit{term})\ \textit{stmt}\ \textbf{else}\ \textit{stmt} \\
& & |\quad \textbf{while}_{lbl}(\textit{term})\ \textit{stmt} \\
& & |\quad \{\textit{stmt}^*\}
\end{array}
$$

lbl ranges over the natural numbers and will be unique for each statement. We do not annotate block statements since they are just used to group multiple statements.

To simplify our analysis, we impose the restriction that all expressions $expr$ must be free from side-effects. Since removing side-effects from expressions is a simple and common program transformation, this restriction is reasonable to make.

4.2 Rules of the Analysis

We now look at the constraint-generation part of the analysis and start by defining the collections of equations that will be generated. These equations will characterize the reaching information in the analyzed program.

$$
\begin{array}{ll}
\text{Equations} \quad \Pi ::= \emptyset \\
\quad\quad |\quad \textbf{Entry}(lbl) = \Sigma \\
\quad\quad |\quad \textbf{Exit}(lbl) = \Sigma \\
\quad\quad |\quad \Pi \wedge \Pi
\end{array} \tag{1}
$$

$\emptyset$ is the empty collection of equations. $\textbf{Entry}(lbl) = \Sigma$ and $\textbf{Exit}(lbl) = \Sigma$ are equations defining the entry and exit sets of the statement annotated with lbl to be equal to the set expression Σ. We let $\wedge$ be the conjunction operator that merges two collections of equations.

The set expressions,

$$
\begin{array}{ll}
\text{Set expressions} \quad \Sigma ::= \emptyset \\
\quad\quad |\quad (\textit{var}, lbl) \\
\quad\quad |\quad \textbf{Entry}(lbl) \\
\quad\quad |\quad \textbf{Exit}(lbl) \\
\quad\quad |\quad \Sigma \cup \Sigma \\
\quad\quad |\quad \Sigma \text{-} \Sigma
\end{array} \tag{2}
$$

are used to build up the entry and exit sets. $\emptyset$ is the empty set (the overloading of this symbol will not cause any confusion). $(\textit{var}, lbl)$ is the set consisting of only a single reaching assignment. $\textbf{Entry}(lbl)$ and $\textbf{Exit}(lbl)$ refer to the values of the entry and exit sets of the statement annotated with lbl. $\cup$ and - are the union and difference operators.

The rules of the analysis are of the form $\ell_0 \vdash s \Downarrow \ell_1 : \Pi$, where s is the statement under consideration, ℓ_0 is the label of the statement executed before s (we will sometimes call this statement the *previous* statement), ℓ_1 the label of the last executed statement in s, and Π the equations characterizing the reaching information of the statement s.

The intuition behind this form is that we need to know the label of the statement executed before s because we will use its exit set when analyzing s. After we have analyzed s, we need to know the label of the last executed statement in s (which will often be s itself) because the statement executed after s needs to use the right exit set. Then, the most important thing to know is, of course, what equations were collected when analyzing s.

In the assignment rule,

ASSIGN

$$\frac{}{\ell_0 \vdash x \overset{\ell_1}{=} e; \Downarrow \ell_1 : \mathbf{Entry}(\ell_1) = \mathbf{Exit}(\ell_0) \wedge} ,$$
$$\mathbf{Exit}(\ell_1) = (x, \ell_1) \cup (\mathbf{Entry}(\ell_1) - \bigcup_{\ell \in lbl} (x, \ell))$$

we know that the reaching assignments in the entry set will be exactly those that were reaching after the previous statement was executed. This is expressed by the equation $\mathbf{Entry}(\ell_1) = \mathbf{Exit}(\ell_0)$. For the exit set, we know that all previous assignments of x will no longer be reaching. The assignments of all other variables will remain untouched. We therefore let the exit set be equal to the entry set from which we have first removed all previous assignments of x and then added the assignment (x, ℓ_1). This is expressed by the equation $\mathbf{Exit}(\ell_1) = (x, \ell_1) \cup (\mathbf{Entry}(\ell_1) - \bigcup_{\ell \in lbl}(x, \ell))$.

So far, we have not seen the need for including the label of the previous statement in the rules. This is illustrated by the rule for if-statements:

IF

$$\frac{\ell_0 \vdash s_0 \Downarrow \ell_2 : \Pi_0 \qquad \ell_0 \vdash s_1 \Downarrow \ell_3 : \Pi_1}{\ell_0 \vdash \mathbf{if}_{\ell_1}(e) \ s_0 \ \mathbf{else} \ s_1 \Downarrow \ell_1 : \Pi_0 \wedge \Pi_1 \wedge \mathbf{Entry}(\ell_1) = \mathbf{Exit}(\ell_0) \wedge}{\mathbf{Exit}(\ell_1) = \mathbf{Exit}(\ell_2) \cup \mathbf{Exit}(\ell_3)}$$

For an if-statement, the entry set will be equal to the exit set of the previous statement, which is expressed by the equation $\mathbf{Entry}(\ell_1) = \mathbf{Exit}(\ell_0)$. When analyzing the two branches s_0 and s_1, we use l_0 as the label of the previous statement since it is important that they, when referring to the exit set of the previous statement, use $\mathbf{Exit}(l_0)$ and not the exit set of the if-statement. From the two branches, we get the collections of the generated equations Π_0 and Π_1, along with the labels l_2 and l_3, which are the labels of the last executed statements in s_0 and s_1. Since we do not know which branch is going to be taken, we must approximate and assume that both branches can be taken. The exit set of the if-statement will therefore be equal to the union of the exit set of the last executed statements in s_0 and s_1, expressed by the equation $\mathbf{Exit}(\ell_1) = \mathbf{Exit}(\ell_2) \cup \mathbf{Exit}(\ell_3)$.

The rule for while-statements,

WHILE

$$\frac{\ell_1 \vdash s \Downarrow \ell_2 : \Pi_0}{\ell_0 \vdash \mathbf{while}_{\ell_1}(e) \ s \Downarrow \ell_1 : \Pi_0 \wedge \mathbf{Entry}(\ell_1) = \mathbf{Exit}(\ell_0) \cup \mathbf{Exit}(\ell_2) \wedge} ,$$
$$\mathbf{Exit}(\ell_1) = \mathbf{Entry}(\ell_1)$$

differs significantly from the rule for if-statements. For the entry set, we include the exit set of the last executed statement before the loop, but also the exit set of the last executed statement in the loop body. We must do this because there are two execution paths leading to the while loop. The first is from the statement executed before the loop, and the second from executing the loop body. For the exit set, we do not know if the body was executed or not. We could, therefore, let the exit set be equal to the union of the entry set of the while-statement and the exit set of the last executed statement in s. Since this is exactly what the entry set is defined to be, we just let the exit set be equal to the entry set. When analyzing the body of the loop we must once again approximate. The first time s is executed, it should use the exit set of l_0, since that was the last statement executed. The second time and all times after that, it should instead use the exit set of l_1, since the body of the while loop was the last statement executed. We approximate this by not separating the two cases and always use l_1 as the label of the previous statement.

We do not have a special rule for programs. Instead, we treat a program as a block statement and use the rules for sequential statements, which should not require much description:

$$\text{SEQ-EMPTY} \over {\ell_0 \vdash \{\} \Downarrow \ell_0 : \emptyset}$$

$$\text{SEQ} \over {\ell_0 \vdash s_1 \Downarrow \ell_1 : \Pi_1 \quad \cdots \quad \ell_{n-1} \vdash s_n \Downarrow \ell_n : \Pi_n} \over {\ell_0 \vdash \{s_1 \ldots s_n\} \Downarrow \ell_n : \Pi_1 \wedge \cdots \wedge \Pi_n}$$

5   Embedding the Analysis into the Prover

5.1   Encoding the Datatypes

In order to encode Σ, Π, and labels, we must declare the types we want to use. We declare **VarSet** which is the type of Σ, **Equations** which is the type of Π and **Label** which is the type of labels. The type of variable names, **Quoted**, is already defined by the system.

In the constructors for Σ, defined by (2), we have, for convenience, replaced the difference operator with the constructor **CutVar**. **CutVar**(s, x) denotes the set expression $s - \bigcup_{\ell \in lbl}(x, \ell)$. Our constructors are defined as function symbols by the following code:

```
VarSet Empty;
VarSet Singleton(Quoted, Label);
VarSet Entry(Label);
VarSet Exit(Label);
VarSet Union(VarSet, VarSet);
VarSet CutVar(VarSet, Quoted);
```

The constructors for Π, defined by (1), are defined analogously to the ones for Σ:

```
Equations None;
Equations EntryEq(Label, VarSet);
Equations ExitEq(Label, VarSet);
Equations Join(Equations, Equations);
```

The KeY system does not feature a unique labeling of statements so we need to annotate each statement ourselves. In order to generate the labels we define the **Zero** and **Succ** constructors with which we can easily enumerate all needed labels. The first label will be **Zero**, the second **Succ(Zero)**, the third **Succ(Succ(Zero))**, and so on.

```
Label Zero;
Label Succ(Label);
```

Since the rules of the analysis refer back to the exit set of the previous statement, there is a problem with handling the very first statement of a program (which does not have any previous statement). To solve this problem we define the label **Start** which we exclusively use as the label of the (non-existing) statement before the first statement. When solving the equations we let the exit set of this label, **Exit(Start)**, be the empty set.

```
Label Start;
```

Since one can only attach *formulas* to embedded Java programs, we need to wrap our parameters in a predicate. The parameters we need are exactly those used in our judgments,

$$\ell_0 \vdash s \Downarrow \ell_1 : \Pi \ .$$

We wrap the label of the previous statement, ℓ_0, the label of the last executed statement, ℓ_1, and the collection of equations, Π, in a predicate called *wrapper* (we do not need to include the statement s since the wrapper will be attached to it). In the predicate, we also include two labels needed for the generation of the labels used for annotating the program: the first unused label before annotating the statement and the first unused label after annotated the statement. The wrapper formula looks as follows:

```
wrapper(Label, Label, Equations, Label, Label);
```

5.2   Encoding the Rules

Before implementing the rules of our analysis as taclets, we declare the variables that we want to use in our taclets. These declarations should be fairly self explanatory.

```
program variable #x;
program simple expression #e;
program statement #s, #s0, #s1;
Equations pi0, pi1, pi2;
Label lbl0, lbl1, lbl2, lbl3, lbl4, lbl5;
Quoted name;
```

We now look at how the rules of the analysis are implemented and start with the rule for empty block statements. When implemented as a taclet we let it match an empty block statement, written as **<{ {} }>**, and a wrapper formula where the first argument is equal to the second argument, the collection of equations is empty, and the fourth argument is equal to the fifth. The formula pattern is written as **wrapper(lbl0, lbl0, None, lbl1, lbl1)**. The action that should be performed when this rule is applied is that the current proof branch should be closed. This is the case because the Seq-Empty rule has no premises. The complete taclet is written as follows:

```
rdef_seq_empty {
  find (==> <{{}}>(wrapper(lbl0, lbl0, None, lbl1, lbl1)))
  close goal
};
```

The rule for non-empty block statements is a bit more tricky. The rule handles an arbitrary number of statements in a block statement. This is, however, hard to express in the taclet language. Instead, we modify the rule to separate the statements into the head and the trailing list. This is equivalent to the original rule except that a block statement needs one application of the rule for each statement it contains. After being modified, the rule looks like this, where we let $\bar{s}_2$ range over lists of statements:

SEQ-MODIFIED
$$\frac{\ell_0 \vdash s_1 \Downarrow \ell_1 : \Pi_1 \qquad \ell_1 \vdash \{\bar{s}_2\} \Downarrow \ell_2 : \Pi_2}{\ell_0 \vdash \{s_1\ \bar{s}_2\} \Downarrow \ell_2 : \Pi_1 \wedge \Pi_2}$$

When implemented as a taclet, we let it match the head and the tail of the list, written as **<{.. #s1 ...}>**. In this pattern, **#s1** matches the head and the dots, **.. ...**[2], match the tail. We also let it match a wrapper formula containing the necessary labels together with the conjunction of the two collections of equations Π_1 and Π_2. For each premise, we create a proof branch by using the **replacewith** action. Note how the two last labels are threaded through the taclet:

```
rdef_seq {
  find (==> <{.. #s1 ...}>(wrapper(lbl0, lbl2, Join(pi1, pi2), lbl3, lbl5)))
  replacewith (==> <{#s1}>(wrapper(lbl0, lbl1, pi1, lbl3, lbl4)));
  replacewith (==> <{.. ...}>(wrapper(lbl1, lbl2, pi2, lbl4, lbl5)))
};
```

In the rule for assignments, we must take care of the annotation of the assignment. Since we know that the fourth argument in the wrapper predicate is the first free label, we bind lbl1 to it. We then use lbl1 to annotate the assignment. Since we have now used that label, we must increment the counter of the first free label. We do that by letting the fifth argument be the successor of lbl1. (Remember that the fifth argument in the wrapper predicate is the first free label after annotated the statement.) In the taclet we use a **varcond** construction to bind the name of the variable matching #x to name.

```
rdef_assign {
  find (==> <{#x = #e;}>
  (wrapper(lbl0, lbl1,
           Join(EntryEq(lbl1, Exit(lbl0)),
```

---

[2] The leading two dots match the surrounding context which for our analysis is known to always be empty. They are however still required by the KeY system.

```
              ExitEq (lbl1, Union(Singleton(name, lbl1),
                                   CutVar(Entry(lbl1), name)))),
           lbl1, Succ(lbl1))))
 varcond (name quotes #x)
 close goal
};
```

The taclet for if-statements is larger than the previously shown taclets, but since it introduces no new concepts, it should be easily understood:

```
rdef_if {
 find (==> <{if(#e) #s0 else #s1}>
  (wrapper(lbl0, lbl1,
           Join(Join(pi0, pi1),
               Join(EntryEq(lbl1, Exit(lbl0)),
                    ExitEq (lbl1, Union(Exit(lbl2), Exit(lbl3)))))),
           lbl1, lbl5)))
  replacewith (==> <{#s0}>(wrapper(lbl0, lbl2, pi0, Succ(lbl1), lbl4)));
  replacewith (==> <{#s1}>(wrapper(lbl0, lbl3, pi1, lbl4, lbl5)))
};
```

This is also the case with the taclet for while-statements and it is, therefore, left without further description:

```
rdef_while {
 find (==> <{while(#e) #s}>
 (wrapper(lbl0, lbl1,
          Join(pi0, Join(EntryEq(lbl1, Union(Exit(lbl0),Exit(lbl2))),
                        ExitEq (lbl1, Entry(lbl1))))),
          lbl1, lbl3)))
  replacewith (==> <{#s}>(wrapper(lbl1, lbl2, pi0, Succ(lbl1), lbl3)))
};
```

5.3 Experiments

We have tested the implementation of our analysis on a number of different programs. For all tested programs the analysis gave the expected entry and exit sets, which is not that surprising since there is a one-to-one correspondence between the rules of the analysis and the taclets implementing them.

As an example, consider the minimal program a = 1;, consisting of only an assignment. We embed this program in a formula, over which we existentially quantify the equations, s, the label of the last executed statement, lbl0, and the first free label after annotated the program, lbl1:

```
ex lbl0:Label. ex s:Equations. ex lbl1:Label.
 <{ a = 1; }>wrapper(Start, lbl0, s, Zero, lbl1)
```

When applying the rules of the analysis, the first thing that happens in that lbl0, s, and lbl1 are instantiated with meta variables. This is done by a built-in rule for existential quantification. The resulting formula is the following where L0, S and L1 are meta variables:

```
<{ a = 1; }>wrapper(Start, L0, S, Zero, L1)
```

We know that the KeY system will succeed in automatically applying the rules since the analysis is complete and, therefore, works for all programs. Being complete is an essential property for all program analyses and for our analysis it is easy to see that for any program there exists a set of equations which characterize the reaching information of the program.

When the proof has been created, we fetch the instantiation of all meta variables, which for our example are the following.

```
{
 S : Equations =
    Join(
      EntryEq(L0, Exit(Start)),
      ExitEq (L0, Union(Singleton(a, L0), CutVar(Entry(L0), a)))),
 L0 : Label = Zero,
 L1 : Label = Succ(L0)
}
```

We take these constraints and let a stand-alone constraint solver solve them. Recall that the analysis is divided into two parts. The first part, which is done by the KeY system, is to collect the constraints. The second part, which is done by the constraint solver, solves the constraints.

The constraint solver extracts the equations from the constraints and solves them yielding the following sets, which is the expected result:

```
Entry_0 = {}
Exit_0  = {(a, 0)}
```

6 Conclusions

It is interesting to see how well-suited an interactive theorem prover such as the KeY system is to embed the reaching definitions analysis in. One reason for this is that the rules of the dynamic logic are, in a way, not that different from the rules of the analysis. They are both syntax-driven, i.e., which rule to apply is decided by looking at the syntactic shape of the current formula or statement. This shows that theorem provers with free variables or meta variables can be seen as not just theorem provers for a specific logic but, rather, as generic frameworks for syntactic manipulation of formulas. Having this view, it is not that strange that we can be rather radical and disregard the usual semantic meaning of the tactic language, and use it for whatever purpose we want.

The key feature that allows us to implement our analysis is the machinery for meta variables, that we use to create a bi-directional flow of information. Using meta variables, we can let our analysis collect almost any type of information. We are, however, limited in what calculation we can do on the information. So far, we cannot do any calculation on the information while constructing the proof. We cannot, for example, do any simplification of the set expressions. One possible way of overcoming this would be to extend the constraint language to not just include syntactic constraints but also semantic constraints.

When it comes to the efficiency of the implementation of the constraint-generation part, it is a somewhat open issue. One can informally argue that the overhead of using the KeY system, instead of writing a specialized tool for the analysis, should be a constant factor. It might be the case that one needs to optimize the constraint solver to handle unification constraints in a way that is more efficient for the analysis. An optimized constraint solver should be able to handle all constraints, generated by the analysis, in a linear way.

7   Future Work

This work presented in this paper is a starting point and opens up for a lot of future work:

- Try different theorem provers to see how well the method presented in this paper works for other theorem provers.
- Further analyse the overhead of using a theorem prover to implement program analyses.
- Modify the calculus of the KeY prover to make use of the information calculated by the program analysis. We need to identify where the result of the analysis can help and how the rules of the calculus should be modified to use it. It is when this is done that the true potential of the integration is unleashed.
- Explore other analyses. We chose to implement the *reaching definitions analysis* because it is a well known and simple analysis that is well suited for illustrating our ideas. Now that we have shown that it is possible to implement a static program analysis in the KeY system, it is time to look for the analyses that would benefit the KeY system the most. Among the possible candidates for this are:
  - An analysis that calculates the possible side-effects of a method. For example what objects and variables that may change.
  - A path-based flow analysis helping the KeY system to resolve aliasing problems.
  - A flow analysis calculating the set of possible implementation classes of objects. This would help reducing the branching for abstract types like interfaces and abstract classes
  - A null pointer analysis that identifies object references which are not equal to null. This would help the system which currently has to always check whether a reference is equal to null before using it.

One limitation of the sequent calculus in the KeY prover is that the unification constraints, used for instantiating the meta variables, can only express syntactic equality. This is a limitation since it prevents the system from doing any semantic simplification of the synthesized information. If it was able to perform simplification of the information while it is synthesized, not only could it make the whole process more efficient, but also let it guide the construction of the proof to a larger extent. Useful extensions of the constraint language are for example the common set operations: test for membership, union, intersection, difference and so on. In a constraint tableaux setting, the simplification of these operations would then take place in the sink objects associated with each node in the proof.

A more general issue that is not just specific to the work presented in this paper is on which level static program analysis and theorem proving should be integrated. The level of integration can vary from having a program analysis run on a program and then give the result of the analysis together with the program to a theorem prover, to having a general framework in which program analysis and theorem proving are woven together. The former kind of integration is no doubt the easiest to implement but also the most limited. The latter is much more dynamic and allows for an incremental exchange of information between the calculus of the prover and program analysis.

Acknowledgments

We would like to thank Wolfgang Ahrendt, Martin Giese and Reiner Hähnle for the fruitful discussions and help with the KeY system. We thank the following people for reading the draft and providing valuable feedback: Richard Bubel, Jörgen Gustavsson, Kyle Ross, Philipp Rümmer and the anonymous reviewers.

References

[ABB$^+$04]   Wolfgang Ahrendt, Thomas Baar, Bernhard Beckert, Richard Bubel, Martin Giese, Reiner Hähnle and Wolfram Menzel, Wojciech Mostowski, Andreas Roth, Steffen Schlager, and Peter H. Schmitt. The KeY tool. *Software and System Modeling*, 2004. Online First issue, to appear in print.

[ASU86]   Alfred V. Aho, Ravi Sethi, and Jeffrey D. Ullman. *Compilers: Princiles, Techniques, and Tools.* Addison-Wesley, 1986.

[Bec01]   Bernhard Beckert. A dynamic logic for the formal verification of Java Card programs. In I. Attali and T. Jensen, editors, *Java on Smart Cards: Programming and Security. Revised Papers, Java Card 2000, International Workshop, Cannes, France*, LNCS 2041, pages 6–24. Springer, 2001.

[BGH$^+$04]   Bernhard Beckert, Martin Giese, Elmar Habermalz, Reiner Hähnle, Andreas Roth, Philipp Rümmer, and Steffen Schlager. Taclets: a new paradigm for constructing interactive theorem provers. *Revista de la Real Academia de Ciencias Exactas, Físicas y Naturales, Serie A: Matemáticas*, to appear, 2004. Special Issue on Computational Logic.

[Fit96]   Melvin C. Fitting. *First-Order Logic and Automated Theorem Proving.* Springer-Verlag, New York, second edition, 1996.

[Gie01]   Martin Giese. Incremental Closure of Free Variable Tableaux. In *Proc. Intl. Joint Conf. on Automated Reasoning, Siena, Italy*, number 2083 in LNCS, pages 545–560. Springer-Verlag, 2001.

[NNH99]   Flemming Nielson, Hanne Riis Nielson, and Chris Hankin. *Principles of Program Analysis.* Springer, 1999.

A Calculus for Type Predicates and Type Coercion

Martin Giese

Johann Radon Institute for Computational and Applied Mathematics,
Altenbergerstr. 69, A-4040 Linz, Austria
martin.giese@oeaw.ac.at

Abstract. We extend classical first-order logic with subtyping by type predicates and type coercion. Type predicates assert that the value of a term belongs to a more special type than the signature guarantees, while type coercion allows using terms of a more general type where the signature calls for a more special one. These operations are important e.g. in the specification and verification of object-oriented programs. We present a tableau calculus for this logic and prove its completeness.

1 Introduction

When mathematicians have shown that the ratio $\frac{p}{q}$ is in fact a natural number, they have no qualms about writing $(\frac{p}{q})!$, even though the factorial is not defined for a rational argument which $\frac{p}{q}$, syntactically speaking, is. This is very different from the common practice in strongly typed programming languages, like e.g. Java: here, an explicit *type coercion* or *cast* is needed in most cases to adjust the type of the fraction to the the required argument type of the factorial function, maybe $(rat\text{-}to\text{-}nat(\frac{p}{q}))!$. The compiled program might check at runtime that the value of the fraction is indeed a natural number, and signal an error otherwise. Typically, such a strongly typed programming language will also provide *type predicates*, which allow checking whether a value is indeed of a more special type than its syntax implies.

Previous work published about tableaux for logics with subtyping ('order-sorted logic'), e.g. [3,2] does not consider type coercions. There are no type predicates in the work of Schmitt and Wernecke, whereas Weidenbach does not consider static typing. We will amend these shortcomings in the present paper, by first defining in Sect. 2 a logic with static typing, subtypes, type coercions, type predicates, and equality, which is not too complicated. In Sect. 3, we give a tableau calculus for this logic, which is somewhat more complicated than might be expected. And in Sect. 4, we give the completeness proof for that calculus, which is actually the main contribution of this paper.

2 A Logic with Type Predicates and Type Coercions

2.1 Types

Let us first clarify an important distinction. In the definition of our logic, there are going to be two kinds of entities that will carry a type: terms and domain elements.

B. Beckert (Ed): TABLEAUX 2005, LNAI 3702, pp. 123–137, 2005.

- The type of a term is given by the signature, more precisely, a term's type is the type declared as return type of the term's outermost function symbol. The syntax is defined in such a way that a term of a certain type can be used wherever the syntax calls for a term of a supertype. To make it clear that we mean this notion of type, we will talk about the *static type of a term*.
- When we *evaluate* a term using some interpretation, we get an element of the domain. Every element of the domain has exactly one type. Our semantics is defined in such a way that the type of the value of a term will be a subtype of, or equal to the static type of the term. When we mean this notion of type, we will talk about the *dynamic type of a domain element*.

For example let us assume two types Z, signifying the integers, and Q, signifying the rationals. We want Z to be a subtype of Q. $zero, one, two : Z$ are constants of static type Z. The operation $div : Q, Q \rightarrow Q$ produces a term of static type Q from two terms of static type Q. Since Z is a subtype of Q, we can use two as argument for div. The static type of the term $div(two, two)$ is Q, since that is the return type of div. The domain of the standard model consists of the set of rational numbers, where the integers have dynamic type Z, and all other elements have dynamic type Q. In the standard model, the *value* of $div(two, two)$ is the domain element 1. The dynamic type of the value of $div(two, two)$ is therefore Z, which is all right since Z is a subtype of Q.

We assume as given a set of types $\mathcal{T}$ with a partial ordering $\sqsubseteq$. We require $\mathcal{T}$ to be closed under greatest lower bounds, i.e. for any two $A, B \in \mathcal{T}$, there is a $C \in \mathcal{T}$ with $C \sqsubseteq A$ and $C \sqsubseteq B$, such that for any other $D \in \mathcal{T}$ with $D \sqsubseteq A$ and $D \sqsubseteq B$, $D \sqsubseteq C$. It is well known that C is then uniquely defined, and we write $A \sqcap B$ for C.

In the approach of Weidenbach [2], types are just unary predicates and the relationship between types is given by axioms of a certain form. These unary predicates correspond to our type predicates, which limit the dynamic type of a domain element. There is no concept of static typing of terms and variables. Indeed, there is a straightforward modelling of type casts in the Weidenbach setting, but it requires working with ill-typed terms. In our calculus, only well-typed formulae are ever constructed.

2.2 Syntax

A signature for our logic consists of a set of function symbols and predicate symbols. Each function and predicate symbol has a (possibly empty) list of argument types and function symbols also have a return type. We write

$$f : A_1, \ldots, A_n \rightarrow A$$

resp.

$$p : A_1, \ldots, A_n$$

to express that the function f, resp. predicate p has argument types $A_1, \ldots, A_n$ and return type A. Constants are included in this definition as functions without arguments.

To simplify things, we do not allow overloading: any function or predicate symbol can have only one list of argument types and return type.

We also provide ourselves with a set of variables, each of which has a type. We write $v : A$ to say that v has type A.

The syntax of our logic is like that of usual classical first order logic, except that we have a series of function symbols and a series of predicate symbols with a predefined meaning, and for which we will use a special syntax.

Definition 1. *We inductively define the system of sets $\{T_A\}_{A \in \mathcal{T}}$ of terms of static type A to be the least system of sets such that*

- *$x \in T_A$ for any variable $x : A$,*
- *$f(t_1, \ldots, t_n) \in T_A$ for any function symbol $f : A_1, \ldots, A_n \to A$, and terms $t_i \in T_{A_i'}$ with $A_i' \sqsubseteq A_i$ for $i = 1, \ldots, n$,*
- *$(A)t \in T_A$ for any term $t \in T_{A'}$ where A' is an arbitrary type.*

We write the static type of t as $\sigma(t) := A$ for any term $t \in T_A$.

A term $(A)t$ is a *type coercion*, also called a *type cast* or *cast* for short. Its intended semantics is that whatever the static type of t, *if* the dynamic type of the value of t is $\sqsubseteq A$, then the value of $(A)t$ is the same as that of t, but the term has the static type A, permitting it to be used in a context where a term of that type is required.

Definition 2. *We inductively define the set of formulae F to be the least set such that*

- *$p(t_1, \ldots, t_n) \in F$ for any predicate symbol $p : A_1, \ldots, A_n$ and terms $t_i \in T_{A_i'}$ with $A_i' \sqsubseteq A_i$ for $i = 1, \ldots, n$,*
- *$t_1 \doteq t_2 \in F$ for any terms $t_1 \in T_{A_1}$ and $t_2 \in T_{A_2}$,*
- *$\neg \phi, \phi \lor \psi, \phi \land \psi, \ldots \in F$ for any $\phi, \psi \in F$.*
- *$\forall x.\phi, \exists x.\phi \in F$ for any $\phi \in F$ and any variable x.*
- *$t \sqsubseteq A \in F$ for any term t and type A.*

The formulae $t_1 \doteq t_2$ and $t_2 \doteq t_1$ are considered syntactically identical. An atom *is a formula of the shape $p(t_1, \ldots, t_n)$, or $t_1 \doteq t_2$, or $t \sqsubseteq A$. A* literal *is an atom or a negated atom. A closed formula is defined as usual to be a formula without free variables.*

The formula $t \sqsubseteq A$ is intended to express that the dynamic type of the value of t is a subtype of or equal to A.

2.3   Semantics

The terms and formulae of our logic will be evaluated with respect to a *structure* $\mathcal{S} = (\mathcal{D}, \mathcal{I})$, consisting of a *domain* and an *interpretation*.

The semantics of a term is going to be a value in the domain $\mathcal{D}$. Each domain element $x \in \mathcal{D}$ has a dynamic type $\delta(x) \in \mathcal{T}$.

While each domain element has a unique dynamic type, it may still occur as the value of terms of different static types. Our semantic definitions will be such, that the dynamic type of the value of a term is guaranteed to be a subtype of the static type of the term. We denote the set of valid domain elements for a certain static type by

$$\mathcal{D}_A := \{x \in \mathcal{D} \mid \delta(x) \sqsubseteq A\}$$

We will require each of these sets to be non-empty:

$$\mathcal{D}_A \neq \emptyset \quad \text{for all types } A \in \mathcal{T}.$$

This restriction spares us the trouble of handling quantification over empty domains.

An interpretation $\mathcal{I}$ assigns a meaning to every function and predicate symbol given by the signature. More precisely, for a function symbol

$$f : A_1, \ldots, A_n \to A \quad,$$

the interpretation yields a function

$$\mathcal{I}(f) : \mathcal{D}_{A_1} \times \ldots \times \mathcal{D}_{A_n} \to \mathcal{D}_A$$

and for a predicate symbol

$$p : A_1, \ldots, A_n$$

some set of tuples of domain elements.

$$\mathcal{I}(p) \subseteq \mathcal{D}_{A_1} \times \ldots \times \mathcal{D}_{A_n} \quad.$$

For type coercions, we require $\mathcal{I}$ to exhibit a particular behaviour, namely

$$\mathcal{I}((A))(x) = x \quad \text{if } \delta(x) \sqsubseteq A.$$

Otherwise, $\mathcal{I}((A))(x)$ may be an arbitrary element of $\mathcal{D}_A$. Finally, for type predicates, we require that

$$\mathcal{I}(\mathbb{E}A) = \mathcal{D}_A \quad.$$

A variable assignment β is simply a function that assigns some domain value $\beta(x) \in \mathcal{D}_A$ to every variable $x : A$.

β_x^d denotes the modification of the variable assignment β such that $\beta_x^d(x) = d$ and $\beta_x^d(y) = \beta(y)$ for all variables $y \neq x$.

For a structure $\mathcal{S} = (\mathcal{D}, \mathcal{I})$, we can now define the valuation function $\mathrm{val}_{\mathcal{S}}$ that takes a variable assignment and a term and yields the domain element that is the value of that term. Our definitions will ensure that $\mathrm{val}_{\mathcal{S}}(\beta, t) \in \mathcal{D}_A$ for any $t \in T_A$.

Definition 3. *We inductively define the valuation function* $\mathrm{val}_{\mathcal{S}}$ *by*

- $\mathrm{val}_{\mathcal{S}}(\beta, x) = \beta(x)$ *for any variable* x.
- $\mathrm{val}_{\mathcal{S}}(\beta, f(t_1, \ldots, t_n)) = \mathcal{I}(f)(\mathrm{val}_{\mathcal{S}}(\beta, t_1), \ldots, \mathrm{val}_{\mathcal{S}}(\beta, t_n))$.
- $\mathrm{val}_{\mathcal{S}}(\beta, (A)t) = \mathcal{I}((A))(\mathrm{val}_{\mathcal{S}}(\beta, t))$.

For a ground term t, we simply write $\mathrm{val}_\mathcal{S}(t)$.

Next, we define the validity relation $\mathcal{S}, \beta \models \phi$ that says whether a formula ϕ is valid in $\mathcal{S} = (\mathcal{D}, \mathcal{I})$ under some variable assignment β.

Definition 4. *We inductively define the validity relation* $\models$ *by*

- $\mathcal{S}, \beta \models p(t_1, \ldots, t_n)$ *iff* $(\mathrm{val}_\mathcal{S}(\beta, t_1), \ldots, \mathrm{val}_\mathcal{S}(\beta, t_n)) \in \mathcal{I}(p)$.
- $\mathcal{S}, \beta \models t_1 \doteq t_2$ *iff* $\mathrm{val}_\mathcal{S}(\beta, t_1) = \mathrm{val}_\mathcal{S}(\beta, t_2)$.
- $\mathcal{S}, \beta \models \neg\phi$ *iff not* $\mathcal{S}, \beta \models \phi$, *etc. for* $\phi \vee \psi$, $\phi \wedge \psi, \ldots$
- $\mathcal{S}, \beta \models \forall x.\phi$ *iff* $\mathcal{S}, \beta_x^d \models \phi$ *for every* $d \in \mathcal{D}_A$ *where* $x : A$, *i.e. A is the static type of the variable x. Similarly for* $\mathcal{S}, \beta \models \exists x.\phi$.
- $\mathcal{S}, \beta \models t \sqsubseteq A$ *iff* $\delta(\mathrm{val}_\mathcal{S}(\beta, t)) \sqsubseteq A$.

We write $\mathcal{S} \models \phi$, *for a closed formula* ϕ, *since* β *is then irrelevant. A closed formula* ϕ *is* unsatisfiable *iff* $\mathcal{S} \models \phi$ *for no structure* ϕ.

3   A Tableau Calculus

The main difficulty in finding a complete calculus for this logic is that it must be possible to prove e.g. a formula $s \doteq t$ containing casts in the presence of various type predicate literals $u \sqsubseteq A$, $v \sqsubseteq B$, if the latter imply the equality of s and t. The first idea would be to have a rule

$$\frac{\begin{array}{c} \phi[z/(A)t] \\ t \sqsubseteq A \end{array}}{\phi[z/t]} \ \textit{wrong}$$

that allows us to *remove* a cast if there is a type predicate that guarantees that the cast is successful. The problem here is that the formula $\phi[z/t]$ might no longer be well-typed, since the static type of t might be too general for the position where $(A)t$ stood – in fact, that is probably why the cast is there in the first place. One needs to go the opposite way: there must be a rule that allows to *add* casts wherever they are guaranteed not to change the semantics of a term. But this leads to difficulties in the completeness proof, since such a rule makes terms grow, defying the usual induction arguments employed in a Hintikka style completeness proof. We show how to solve this problem in Sect. 4.

In order to avoid the duplication of rules due to duality, we give a calculus for formulae in negation normal form, i.e. negation appears only before atoms $p(\ldots)$, or $t_1 \doteq t_2$, or $t \sqsubseteq A$.

The calculus is a 'ground', i.e. Smullyan style tableau calculus, without free variables. We expect it to be possible without difficulties to design a free variable version of the calculus, using order-sorted unification, in the style of [3,2].

See Fig. 1 for the rules of our calculus. Starting from an initial tableau containing only one branch with one formula, namely the one that we want to prove unsatisfiable, we build a tableau by applying these rules, adding the formulae beneath the line whenever the ones above the line are present. The β

$$\frac{\phi \wedge \psi}{\substack{\phi \\ \psi}} \; \alpha \qquad\qquad \frac{\phi \vee \psi}{\phi \quad \psi} \; \beta$$

$$\frac{\forall x.\phi}{\phi[x/t]} \; \gamma \qquad\qquad \frac{\exists x.\phi}{\phi[x/c]} \; \delta$$

with $t \in T_A$ ground, if $x : A$.   with $c : A$ a new constant, if $x : A$.

$$\frac{\substack{\phi[z/t_1] \\ t_1 \doteq t_2}}{\phi[z/t_2]} \; \text{apply-}\doteq \qquad\qquad \frac{\substack{\phi[z/t_1] \\ t_1 \doteq t_2}}{\phi[z/(A)t_2]} \; \text{apply-}\doteq'$$

$$\text{if } \sigma(t_2) \sqsubseteq \sigma(t_1). \qquad\qquad \text{with } A := \sigma(t_1).$$

$$\frac{t_1 \doteq t_2}{\substack{t_2 \sqsubseteq \sigma(t_1) \\ t_1 \sqsubseteq \sigma(t_2)}} \; \text{type-}\doteq \qquad\qquad \frac{\substack{t \sqsubseteq A \\ t \sqsubseteq B}}{t \sqsubseteq A \sqcap B} \; \text{type-}\sqcap$$

$$\frac{}{t \sqsubseteq \sigma(t)} \; \text{type-static}$$

$$\frac{\substack{\phi[z/t] \\ t \sqsubseteq A}}{\phi[z/(A)t]} \; \text{cast-add} \qquad\qquad \frac{\substack{(\neg)(A)t \sqsubseteq B \\ t \sqsubseteq A}}{(\neg)t \sqsubseteq B} \; \text{cast-type}$$

$$\text{where } A \sqsubseteq \sigma(t).$$

$$\frac{\phi[z/(A)t]}{\phi[z/t]} \; \text{cast-del}$$

$$\text{where } \sigma(t) \sqsubseteq A.$$

$$\frac{\substack{t \sqsubseteq A \\ \neg t \sqsubseteq B}}{\bot} \; \text{close-}\sqsubseteq \qquad\qquad \frac{\neg t \doteq t}{\bot} \; \text{close-}\doteq$$

$$\text{with } A \sqsubseteq B.$$

$$\frac{\phi, \neg\phi}{\bot} \; \text{close}$$

Fig. 1. The rules of our calculus

rule splits the current branch into two new branches, all other rules don't split. A tableau is closed, if $\bot$ is present on all branches.

Note in particular the cast-add rule, which can add a cast around a term, whenever there is a type predicate to guarantee that the value of the term remains the same with the cast.

The type-$\doteq$ rule might be a little unexpected at first. Remember that if $t_1 \doteq t_2$ holds, the terms have the same value $v \in \mathcal{D}$, so their values have the same dynamic type $\delta(v)$. On the other hand, the dynamic types of the values of terms are subtypes of their static types, so $\delta(v) \sqsubseteq \sigma(t_1)$ and $\delta(v) \sqsubseteq \sigma(t_2)$, which motivates the type-$\doteq$ rule. It is needed to deal with equalities between terms of unequal static types, see the proof of Lemma 4.

The rule cast-type allows us to remove the cast from $(A)t$, if $t \sqDoteq A$ is present, but only in the special case of the top-level of a type predicate literal. Such literals are a special case in our completeness proof, see Lemma 6.

Theorem 1. *A closed tableau can be constructed from a closed formula ϕ, iff ϕ is unsatisfiable.*

The proof of soundness, i.e. that a formula for which a closed tableau exists is indeed unsatisfiable, is rather standard, so we will not go into it here. The interesting part is the completeness proof, which is the main result of this paper.

4   Completeness

The standard approach for proving completeness fails because the cast 'normalization' rule cast-add, which is to be used to make semantically equal terms syntactically equal, makes terms grow. We can however work around this by using a modified signature and calculus.

Modified Signature. We show completeness in a modified calculus that uses an extended signature. We then show how a proof in the modified calculus can be transformed into one in the normal calculus. We modify the signature by adding a function symbol $f^A : \ldots \to A$ for any function symbol $f : \ldots \to B$ with $A \sqsubseteq B$. The semantics of $f^A(\ldots)$ will be defined to be the same as that of $f(\ldots)$, only with the static type A. Similarly, we add modified casts $(B)^A t$ with static type A for $A \sqsubseteq B$. We will say that $f^A(\ldots)$ and $(B)^A t$ carry a *superscript cast to A*. The idea of these superscript casts is to normalize terms such that each term carries a cast to the most specific type known for it. The modified calculus is just like the one of Fig. 1, except that rules cast-add and cast-type are replaced by those shown in Fig. 2, and the cast-strengthen rule is added.

$$\frac{\phi[z/t]}{\dfrac{t \sqDoteq A}{\phi[z/t^A]}} \text{ cast-add} \qquad \frac{(\neg)t^A \sqDoteq B}{\dfrac{t \sqDoteq A}{(\neg)t \sqDoteq B}} \text{ cast-type} \qquad \frac{\phi[z/t^A]}{\dfrac{t \sqDoteq B}{\phi[z/t^B]}} \text{ cast-strengthen}$$
$$\text{where } A \sqsubseteq \sigma(t). \qquad\qquad\qquad\qquad\qquad\qquad \text{where } B \sqsubseteq A.$$

Fig. 2. Alternative rules in the modified calculus

In these rules, t^A stands for either $f^A(\ldots)$ or $(B)^A t'$, depending on what kind of term t is. However, t itself may not have a superscript cast in the rules cast-add and cast-strengthen, since a term can have only one superscript cast. The rule cast-del of the original calculus can now be applied for casts with or without superscript casts.

To show completeness of this modified calculus, we need to show a *model lemma* that states that a (possibly infinite) open branch on which a certain set of relevant rule applications has been performed has a model. Showing from this that there is a closed tableau for every unsatisfiable formula ϕ is then standard.

The initial formula ϕ contains only a finite set of types as static types of terms or in type predicates $t \mathrel{\underline{\in}} A$. Some of the rules of our calculus can introduce finite intersections of types, but there will still always be only finitely many types in a tableau. In particular, $\sqsubseteq$ is a Noetherian ordering on $\mathcal{T}$.

Therefore, we can state the following definition:

Definition 5. *We call a tableau branch H type-saturated, if every possible application of type-$\sqcap$ and type-static for all ground terms t has been performed on H.*

Let H be a type-saturated branch. For any term t, we define the most specific known type $\kappa_H(t)$ w.r.t. H to be the unique least type A under the subtype ordering with $t \mathrel{\underline{\in}} A \in H$.

Existence and uniqueness of $\kappa_H(t)$ are guaranteed by the saturation w.r.t. the type-static and type-$\sqcap$ rules.

The purpose of our applications of cast-add, cast-strengthen, and cast-del is to bring terms to a normalized form, where the top operator of every subterm is superscripted with the most specific known type for the subterm, and there are no casts $(A)t$ which are redundant in the sense that $\sigma(t) \sqsubseteq A$.

Definition 6. *Let H be type-saturated. A term t is normalized w.r.t. H iff*

- $t = f^A(t_1, \ldots, t_n)$ *where all t_i are normalized, and* $A = \kappa_H(f(t_1, \ldots, t_n))$, *or*
- $t = (B)^A t'$ *where t' is normalized,* $A = \kappa_H((B)t')$, *and* $\sigma(t') \not\sqsubseteq B$.

We designate the set of ground terms normalized w.r.t. H by Norm_H. *An atom ϕ is normalized w.r.t. H, iff for some terms $t_i \in \mathrm{Norm}_H$,*

- $\phi = t_1 \doteq t_2$,
- $\phi = p(t_1, \ldots, t_n)$, *or*
- $\phi = f(t_1, \ldots, t_n) \mathrel{\underline{\in}} A$, *or*
- $\phi = (B)t_1 \mathrel{\underline{\in}} A$, *where* $\sigma(t_1) \not\sqsubseteq B$.

A normalized literal is a normalized atom or the negation of one. Note: normalized type predicate atoms do not have superscript casts on the top level operator.

Definition 7. *We call a tableau branch H saturated, if H is type-saturated, and for each of the following rules, if all premises are in H, then so is one of the conclusions:*

- *the α, β, and δ rules,*
- *the type-$\doteq$, cast-type, cast-strengthen, cast-del, and close-...rules on literals,*
- *the γ rule for every formula $\forall x.\phi \in H$ and every normalized ground term $t \in \mathrm{Norm}_H \cap T_A$ for $A \sqsubseteq B$ if $x : B$.*

- *the apply-$\dot{=}$ rule for equations $t_1 \dot{=} t_2$ with equal static types $\sigma(t_1) = \sigma(t_2)$,*
- *the cast-add rule, except on the top level of a type predicate literal, i.e.*
  $\phi[z/t] = t \sqsubseteq A$ for some A.

We do not apply cast-add on the top level of type literals $t \sqsubseteq A$: normalizing $t \sqsubseteq A$ to $t^A \sqsubseteq A$ would of course destroy the information of the literal. And rather than normalizing $t \sqsubseteq B$ to $t^A \sqsubseteq B$, we join the literals to $t \sqsubseteq A \sqcap B$.

Saturation w.r.t. cast-add and cast-strengthen implies that terms will be equipped with superscript casts for their most specific known types. Conversely, the following lemma guarantees that a superscript cast cannot appear without the corresponding type predicate literal, if we forbid spurious applications of the γ-rule for non-normalized terms.

Lemma 1. *Let H be a saturated branch obtained by a derivation in which the γ rule is only applied for terms t^A with $t \sqsubseteq A \in H$. Let $\phi \in H$ contain a subterm t^A, i.e. either $f^A(\dots)$ or $(B)^A t'$. Then H also contains the formula $t \sqsubseteq A$.*

Proof. We show this by induction on the length of the derivation of ϕ from formulae without superscript casts. As induction hypothesis, we can assume that the statement is true for the original formulae in the rule application which added ϕ to the branch. If ϕ results from a rule application where one of the original formulae already contained the term t^A, we are finished. Otherwise, ϕ might be the result of an application of the following rules:

- γ: The assumption of this lemma is that a quantifier can get instantiated with a term t^A only if $t \sqsubseteq A \in H$.
- apply-$\dot{=}$: If the superscript cast t^A does not occur in the term t_2, there must have been some $s^A[t_1]$ in ϕ which contains an occurrence of t_1 and $t^A = s^A[t_1]$. By the induction hypothesis, $s[t_1] \sqsubseteq A \in H$. But if H is saturated, then we must also have $s[t_2] \sqsubseteq A \in H$.
- cast-add: If ϕ is the result of adding ^{A} to t, then $t \sqsubseteq A$ is of course present. But t^A might also be $s^A[r^C]$ where where $r \sqsubseteq C \in H$ and ^{C} was added to r. Like for apply-$\dot{=}$, the induction hypothesis guarantees that $s[r] \sqsubseteq A \in H$. Due to saturation, we then also have $s[r^C] \sqsubseteq A \in H$
- cast-strengthen: The same argument as for cast-add applies.
- cast-del: The argument is again the same.

None of the other rules can introduce new terms t^A.   □

Domain for Saturated Branch. Given a saturated branch H, we will construct a model that makes all formulae in H true. We start by defining the domain as

$$\mathcal{D} := \mathrm{Norm}_H / \sim_H \quad,$$

where $\sim_H$ is the congruence induced by those (normalized) ground equations in H that have equal static type on both sides, i.e.

$$s \sim_H t \quad :\Leftrightarrow \quad s = t \text{ or } [s \dot{=} t \in H \text{ and } \sigma(s) = \sigma(t)]$$

We will denote the typical element of $\mathcal{D}$ by $[t]$ where t is a normalized term, and the $[]$ denote taking the congruence class.

$\sim_H$ is a congruence relation on normalized terms, meaning that

- $\sim_H$ is an equivalence relation on Norm_H.
- If $t \sim_H t'$, then $f^A(\ldots, t, \ldots) \sim_H f^A(\ldots, t', \ldots)$ for all argument positions of all function symbols f^A.
- If $t \sim_H t'$, then $(B)^A t \sim_H (B)^A t'$ for all casts $(B)^A$.

These properties follow from the saturation w.r.t. apply-$\doteq$.

We need to assign a dynamic type to every element $[t] \in \mathcal{D}$. We will simply take the static type of t. As t is normalized, this is the superscript on the outermost operator. The dynamic type is well-defined: the static type of all elements of the congruence class is the same, as we look only at equations of identical static types on both sides.

Interpretation for Saturated Branch. We now define the interpretation $\mathcal{I}$ of function and predicate symbols. We will give the definitions first, and then show that all cases are actually well-defined.

We define

$$\mathcal{I}(f)([t_1], \ldots, [t_n]) := [f^C(t_1, \ldots, t_n)] \qquad [A]$$

where $C := \kappa_H(f(t_1, \ldots, t_n))$. Similarly,

$$\mathcal{I}((B))([t']) := \begin{cases} [t'] & \text{if } \sigma(t') \sqsubseteq B, \\ [(B)^C t'] & \text{otherwise,} \end{cases} \qquad [B]$$

where $C := \kappa_H((B)t')$.

We want the function symbols with superscript casts to be semantically equivalent to the versions without, but we need to make sure that the static types are right. We use the interpretation of casts to ensure this.

$$\mathcal{I}(f^A)([t_1], \ldots, [t_n]) := \mathcal{I}((A))(\mathcal{I}(f)([t_1], \ldots, [t_n])) \quad . \qquad [C]$$

Similarly, for casts with superscript casts,

$$\mathcal{I}((B)^A)([t']) := \mathcal{I}((A))(\mathcal{I}((B))([t'])) \quad . \qquad [D]$$

For predicates, we define

$$([t_1], \ldots, [t_n]) \in \mathcal{I}(p) \quad \Leftrightarrow \quad p(t_1, \ldots, t_n) \in H \quad , \qquad [E]$$

as usual in Hintikka set constructions.

To show that these are valid definitions, we will need the following lemma:

Lemma 2. *Let H be a saturated branch and $t \sim_H t'$. Then $\kappa_H(t) = \kappa_H(t')$.*

Proof. To see that this is the case, note that due to the definition of $\sim_H$, we have either $t = t'$ or $t \doteq t' \in H$. If $t = t'$, the result follows immediately. Otherwise, due to the saturation w.r.t. apply-$\doteq$, it follows that $t \sqsubseteq A \in H$ iff $t' \sqsubseteq A \in H$ for all types A. Thus, the most specific known types for t and t' must be equal. $\square$

We will now show the validity of definitions $[A]$ to $[E]$.

Proof. For $[A]$, we need to show that

- $f^C(t_1, \ldots, t_n) \in \mathrm{Norm}_H$, if all $t_i \in \mathrm{Norm}_H$.
- The static type of $f^C(t_1, \ldots, t_n)$ is a subtype of the result type of f.
- $[f^C(t_1, \ldots, t_n)]$ is independent of the choice of particular $t_1, \ldots, t_n$.

Now, $f^C(t_1, \ldots, t_n)$ is indeed normalized w.r.t. H, if all t_i are normalized, since C is defined as $\kappa_H(f(t_1, \ldots, t_n))$, as required by the definition of normalized terms. As for the second point, the result type of f is of course the static type of $f(t_1, \ldots, t_n)$, which is a supertype of $C = \kappa_H(f(t_1, \ldots, t_n))$, which in turn is the static type of $f^C(t_1, \ldots, t_n)$. Finally, if $t_i \sim_H t'_i$ for $i = 1, \ldots, n$, then $\kappa(f(t_1, \ldots, t_n)) = \kappa(f(t'_1, \ldots, t'_n))$, due to Lemma 2, so C is independent of the choice of representatives. Also, $f^C(t_1, \ldots, t_n) \sim_H f^C(t'_1, \ldots, t'_n)$, since $\sim_H$ is a congruence relation.

For $[B]$, we need to make sure that the cast doesn't change its argument if its dynamic type is $\sqsubseteq B$. Since the dynamic type of $[t']$ is the static type of t', this is ensured by the first case. In the other case, one sees that $(B)^C t'$ is indeed a normalized term of static type $C \sqsubseteq B$. As for the choice of representatives, if $t' \sim_H t''$, then they have the same static type, so the same case applies for both. In the first case, $[t'] = [t'']$ trivially holds. In the second case C is well-defined due to Lemma 2, and $(B)^C t' \sim_H (B)^C t''$, again because $\sim_H$ is a congruence.

For $[C]$, and $[D]$, note that $\mathcal{I}((A))$ will deliver a value with dynamic type $\sqsubseteq A$ in both cases, as we required for superscript casts. These definitions do not require any choice of representatives.

Finally, for $[E]$, n-fold application of the saturation w.r.t. apply-$\doteq$ guarantees that if $t_i \sim_H t'_i$ for $i = 1, \ldots, n$, then $p(t_1, \ldots, t_n) \in H$ iff $p(t'_1, \ldots, t'_n) \in H$. Therefore, the definition of $\mathcal{I}(p)$ is independent of the choice of the t_i.

$\square$

A central property of our interpretation is that normalized ground terms get evaluated to their equivalence classes:

Lemma 3. *Let H be a saturated branch and let $\mathcal{S} = (\mathcal{D}, \mathcal{I})$ be the structure previously defined. For any normalized ground term $t \in \mathrm{Norm}_H$, $\mathrm{val}_{\mathcal{S}}(t) = [t]$.*

Proof. One shows this by induction on the term structure of t. Let $t = s^A$ be a normalized ground term, i.e. all its subterms are normalized, the top level operator carries a superscript cast to $A = \kappa_H(s)$, and the top level operator is not a redundant cast. By the induction hypothesis, $\mathrm{val}_{\mathcal{S}}(t_i) = [t_i]$ for all subterms

t_i of t. If the top level operator is a function term, $t = f^A(t_1, \ldots, t_n)$, $\mathrm{val}_{\mathcal{S}}(t) = \mathcal{I}((A))(\mathcal{I}(f)([t_1], \ldots, [t_n])) = \mathcal{I}((A))([f^A(t_1, \ldots, t_n)]) = [f^A(t_1, \ldots, t_n)] = [t]$. If $t = (B)^A t'$ is a cast, then $\mathrm{val}_{\mathcal{S}}(t) = \mathcal{I}((A))(\mathcal{I}((B))([t']))$. Since the cast is not redundant, this is equal to $\mathcal{I}((A))([(B)^A t']) = [(B)^A t'] = [t]$. □

Model Lemma. We now need to show that the constructed model satisfies all formulae in a saturated tableau branch. As usual, this is done by induction on some notion of formula complexity. However, for different kinds of formulae, the required notion will be a different one. We therefore split the whole proof into a series of lemmas, each of which requires a different proof idea.

First, we will show that all normalized literals are satisfied in $\mathcal{S} = (\mathcal{D}, \mathcal{I})$.

Lemma 4. *Let H be a saturated open branch and let $\mathcal{S} = (\mathcal{D}, \mathcal{I})$ be the structure defined previously. $\mathcal{S} \models \phi$ for any literal $\phi \in H$ that is normalized w.r.t. H.*

Proof. Lemma 3 tells us that all normalized subterms t_i of ϕ get evaluated to $[t_i]$. We will start with positive literals, i.e. non-negated atoms. For $\phi = p(t_1, \ldots, t_n)$, the result follows directly from our semantics and point $[E]$ in the definition of the interpretation $\mathcal{I}$. For a type predicate $f(t_1, \ldots, t_n) \in A$,

$$\mathrm{val}_{\mathcal{S}}(\beta, f(t_1, \ldots, t_n)) = [f^B(t_1, \ldots, t_n)]$$

for $B = \kappa_H(f(t_1, \ldots, t_n)) \sqsubseteq A$. Hence, the dynamic type of the value will be $\sqsubseteq A$, and therefore $\mathcal{S} \models f(t_1, \ldots, t_n) \in A$. Similarly, for $(B)t_1 \in A$,

$$\mathrm{val}_{\mathcal{S}}(\beta, (B)t_1) = \begin{cases} [t_1] & \text{if } \sigma(t_1) \sqsubseteq B, \\ [(B)^C t_1] & \text{otherwise,} \end{cases}$$

where $C = \kappa_H((B)t_1) \sqsubseteq A$. The first case is excluded, because $(B)t_1 \in A$ is normalized. In the second case, the dynamic type is clearly $C \sqsubseteq A$, so we are done.

For an equality atom $t_1 \doteq t_2 \in H$ with t_1 and t_2 of equal static types, $\mathrm{val}_{\mathcal{S}}(\beta, t_1) = \mathrm{val}_{\mathcal{S}}(\beta, t_2)$, since $t_1 \sim_H t_2$, so $\mathcal{S} \models t_1 \doteq t_2$. If t_1 and t_2 are of unequal static types A and B, respectively, since t_1 and t_2 are normalized, $t_1 = s_1^A$ and $t_2 = s_2^B$ for some terms s_1 and s_2. Also due to normalization, there must be literals $s_1 \in A \in H$ and $s_2 \in B \in H$. Due to saturation w.r.t. the type-$\doteq$ rule, we must also have literals $s_1 \in B \in H$ and $s_2 \in A \in H$. Now since H is type-saturated, H also contains $s_1 \in A \sqcap B$ and $s_2 \in A \sqcap B$ which, given that $A \neq B$, means that s_1^A and s_2^B are in fact *not* normalized which contradicts our assumptions.[1]

If ϕ is a negative literal, we again have to look at the different possible forms. If $\phi = \neg p(t_1, \ldots, t_n)$, assume that $\mathcal{S} \not\models \phi$. Then $\mathcal{S} \models p(t_1, \ldots, t_n)$, which implies that also $p(t_1, \ldots, t_n) \in H$, in which case H would be closed, which is a contradiction. For a negated type predicate $\phi = \neg t \in A$, t will be evaluated to an

---

[1] One sees here why the apply-$\doteq'$ rule that inserts a cast to make static types fit is not necessary, just convenient. It is not even necessary (but probably *very* convenient) to allow application of equations where the static type gets more special. Application for identical types is enough if we have the type-$\doteq$ and cast normalization rules.

object of dynamic type $C := \kappa_H(t)$. Assume $\mathcal{S} \not\models \phi$. Then $C \sqsubseteq A$. There must be a literal $t \sqsubseteq C \in H$, and therefore H is closed by the close-$\sqsubseteq$ rule. Finally, for a negated equation $\phi = \neg t_1 \doteq t_2$, assume $\mathcal{S} \not\models \phi$. Then $\mathcal{S} \models t_1 \doteq t_2$, and since t_1 and t_2 are normalized, $t_1 \sim_H t_2$. Then, either $t_1 = t_2$, implying that H is closed using close-$\doteq$, or there is an equation $t_1 \doteq t_2 \in H$. Saturation w.r.t. apply-$\doteq$ tells us that $\neg t_2 \doteq t_2$ must then also be in H, so H would again be closed, contradicting our assumptions. $\square$

Based on the previous lemma, we can show that almost all atoms are satisfied, with the exception of one special case which requires separate treatment.

Lemma 5. *Let H be a saturated open branch and $\mathcal{S}$ the same structure as before. $\mathcal{S} \models \phi$ for any literal $\phi \in H$, except if ϕ is a type predicate literal $\phi = (\neg)t^B \sqsubseteq A$ with a superscript cast on the top-level operator.*

Proof. We show this by induction on ϕ with an order $\succ$ defined as follows: $\phi \succ \psi$ iff $sct(\phi) \gg sct(\psi)$, where $sct(\phi)$ is the multiset of the superscript cast types of terms occurring in ϕ, except at the top-level of type predicate atoms, terms without superscript cast being mapped to a pseudo-type ∞, which is larger than all other types. $\gg$ is the multiset ordering induced by the supertype ordering $\sqsupseteq$, with the mentioned extension for ∞.

So let us assume as induction hypothesis that $\mathcal{S} \models \psi$ for all $\psi \in H$ with $\phi \succ \psi$, except for the case that ψ is a type-predicate literal with a superscript cast on the top-level.

The case that ϕ is normalized is covered in Lemma 4. Otherwise, ϕ contains

(a) a redundant cast $(B)u$, or
(b) a subterm u that is a function or cast application without superscript cast, and that is not at the top-level of a type predicate atom, or
(c) a subterm u that is a function or cast application u^B with a superscript cast to B where a more specific type $C \sqsubseteq B$ is known, i.e. $u \sqsubseteq C \in H$.

In case (a), saturation w.r.t. cast-del guarantees the existence of a formula $\phi' \in H$ that results from deleting the superfluous cast (B) from u. Since ϕ' has one operator less than ϕ, $sct(\phi')$ results from $sct(\phi)$ by deleting one element.[2] Therefore $\phi \succ \phi'$, and the induction hypothesis tells us that $\mathcal{S} \models \phi'$. We also know that $\mathrm{val}_{\mathcal{S}}((B)u) = \mathrm{val}_{\mathcal{S}}(u)$. Let ψ be such that $\phi = \psi[z/(B)u]$ and $\phi' = \psi[z/u]$. The substitution lemma gives us $\mathcal{S} \models \phi$ iff $\mathcal{S} \models \psi[z/(B)u]$ iff $\mathcal{S}, \{z \leftarrow \mathrm{val}_{\mathcal{S}}((B)u)\} \models \psi$ iff $\mathcal{S}, \{z \leftarrow \mathrm{val}_{\mathcal{S}}(u)\} \models \psi$ iff $\mathcal{S} \models \psi[z/u]$ iff $\mathcal{S} \models \phi'$.

In case (b), due to the type completeness of H, there is some literal $u \sqsubseteq C \in H$. This literal must be smaller than ϕ: if ϕ is a type predicate literal, u must be below the top level, so $sct(u \sqsubseteq C)$ is only a subset of that for ϕ, otherwise, $sct(u \sqsubseteq C)$ lacks at least the top level superscript cast of u. Thus, we can apply the induction hypothesis, to get $\mathcal{S} \models u \sqsubseteq C$, and therefore $\mathrm{val}_{\mathcal{S}}((C)u) = \mathrm{val}_{\mathcal{S}}(u)$. Saturation w.r.t. cast-add tells us that there is also $\phi' \in H$ in where ϕ' results from replacing u by u^C in t. Since the term u without superscript cast is considered to be

---

[2] In the case of deleting a redundant cast $(B)u \sqsubseteq A$ at the top level of a type predicate atom, the removed element is actually the superscript cast of u and not that of $(B)u$.

larger than u^C by our ordering, $\phi \succ \phi'$ and the ind. hyp. gives us $\mathcal{S} \models \phi'$. Like for (a), the result follows by an application of the substitution lemma.

Finally, in case (c), it is saturation w.r.t. cast-strengthen which tells us that if $\phi \in H$, then also $\phi' \in H$, where ϕ is the result of replacing u^B by u^C in ϕ. Now, $u \sqsubseteq C \in H$, and this literal is smaller than ϕ under $\succ$, as in case (b). Hence, $\mathcal{S}$ satisfies $u \sqsubseteq C$, and since $B \sqsubseteq C$, also $u \sqsubseteq B$. Thus, $\mathrm{val}_{\mathcal{S}}((B)u) = \mathrm{val}_{\mathcal{S}}(u) = \mathrm{val}_{\mathcal{S}}((C)u)$. Like for (b), we can apply the ind. hyp. to get $\mathcal{S} \models \phi'$, and the substitution lemma gives us $\mathcal{S} \models \phi$. □

This is easily extended to also allow type predicate literals *with* top-level superscript casts:

Lemma 6. *Let H be a saturated open branch and $\mathcal{S}$ the same structure as before.* $\mathcal{S} \models \phi$ *for any literal $\phi \in H$.*

Proof. Most literals ϕ are covered by Lemma 5. For the special case of type predicate literals $(\neg)t^A \sqsubseteq B$ for terms with superscript casts, Lemma 1 guarantees that $t \sqsubseteq A$ will also be present, so $\kappa_H(t) \sqsubseteq A$, and therefore $\mathrm{val}_{\mathcal{S}}(t^A) = \mathrm{val}_{\mathcal{S}}(t)$. Due to saturation w.r.t. cast-type, we must have $(\neg)t \sqsubseteq B \in H$, and by Lemma 5, $\mathcal{S} \models (\neg)t \sqsubseteq B$. Since t and t^A are evaluated the same we also get, using the substitution lemma, $\mathcal{S} \models (\neg)t^A \sqsubseteq B$. □

Now, we can finally show that $\mathcal{S}$ is actually a model for all of H.

Lemma 7 (Model Lemma). *Let H be a saturated open branch and let $\mathcal{S} = (\mathcal{D}, \mathcal{I})$ be the structure defined previously. $\mathcal{S}$ is a model for H, i.e. $\mathcal{S} \models \phi$ for all $\phi \in H$.*

Proof. We prove this as usual by an induction on the number of $\wedge, \vee, \forall, \exists$ occurring in ϕ.

Let $\phi \in H$, and assume that $\mathcal{S} \models \psi$ for all ψ smaller than ϕ.

If ϕ is a literal, the result is provided by Lemma 6.

The cases for conjunction, disjunction, and existential quantifiers are completely standard. For universal quantifiers, we convince ourselves that instantiation with normalized ground terms is enough:

Let $\phi = \forall x.\phi_1$ with $x : A$. Then $\phi_1[x/t] \in H$ for all normalized ground terms $t \in T_B \cap \mathrm{Norm}_H$ of static type $B \sqsubseteq A$. For all such t, we may apply the ind. hyp. , since ϕ_1 lacks the occurrence of $\forall$, so $\mathcal{S} \models \phi[x/t]$. Due to the substitution lemma, $\mathcal{S}, \{x \leftarrow \mathrm{val}_{\mathcal{S}}(t)\} \models \phi$, and because of Lemma 3, $\mathcal{S}, \{x \leftarrow [t]\} \models \phi$. Now $\{[t] \mid t \in \mathrm{Norm}_H, \sigma(t) \sqsubseteq A\} = \mathcal{D}_A$ in our structure, so $\mathcal{S} \models \phi$. □

Proof Transformation to Original Calculus. Now that we have shown completeness of the calculus with superscript casts, it only remains to show how to emulate a proof in that calculus using the original rule set of Fig. 1.

The idea is quite simple: every occurrence of $f^A(\dots)$ can be replaced by $(A)f(\dots)$, and every $(B)^A t$ by $(A)(B)t$. The modified cast-add, cast-strengthen, and cast-type rules can then be emulated by the original ones as follows:

$$\frac{\phi[z/t], t \sqsubseteq A}{\phi[z/t^A]} \text{ cast-add} \quad \rightsquigarrow \quad \frac{\phi[z/t], t \sqsubseteq A}{\phi[z/(A)t]} \text{ cast-add} \quad ,$$

$$\frac{(\neg)t^A \sqsubseteq B, t \sqsubseteq A}{(\neg)t \sqsubseteq B} \text{ cast-type} \quad \rightsquigarrow \quad \frac{(\neg)(A)t \sqsubseteq B, t \sqsubseteq A}{(\neg)t \sqsubseteq B} \text{ cast-type} \quad,$$

and for $C \sqsubseteq A$:

$$\frac{\phi[z/t^A], t \sqsubseteq C}{\phi[z/t^C]} \text{ cast-strengthen} \quad \rightsquigarrow \quad \begin{array}{c} \dfrac{\phi[z/(A)t], t \sqsubseteq C}{\phi[z/(A)(C)t]} \text{ cast-add} \\ \overline{\phi[z/(C)t]} \text{ cast-del} \end{array} \quad.$$

5 Conclusion and Future Work

We have defined a logic with static subtyping, type coercions and type predicates, given a tableau calculus for this logic, and shown its completeness.

Future work includes the accommodation of free variables. This should be possible without difficulties along the lines of [3,2], using order-sorted unification for branch closure. More interestingly, we believe that our cast normalization rules cast-add, cast-del, etc. can be formulated as *destructive* simplification rules. In that case the completeness proof would probably have to be based on saturation modulo redundancy in the style of [1].

The presented calculus is not very efficient as an automated theorem proving calculus, even with destructive cast normalization rules: One should at least get rid of the type-static rule. One can also think of a criterion that limits the insertion of casts to those cases where they can actually be of help in closing some branch. Maybe cast normalization can be restricted to cases where two literals are already unifiable except for some casts.

Acknowledgments

The author would like to thank Bernhard Beckert and Richard Bubel for the interesting discussions that led to these results, and the anonymous referees for their useful comments.

References

1. Leo Bachmair and Harald Ganzinger. Resolution theorem proving. In Alan Robinson and Andrei Voronkov, editors, *Handbook of Automated Reasoning*, volume I, chapter 2, pages 19–99. Elsevier Science, 2001.
2. Christoph Weidenbach. First-order tableaux with sorts. *Journal of the Interest Group in Pure and Applied Logics, IGPL*, 3(6):887–906, 1995.
3. Wolfgang Wernecke and Peter H. Schmitt. Tableau calculus for order-sorted logic. In U. Hedstück, C.-R. Rollinger, and K. H. Bläsius, editors, *Sorts and Types in Artificial Intelligence, Proc. of the Workshop, Ehringerfeld*, volume 418 of *Lecture Notes in AI*, pages 49–60. Springer Verlag, 1990.

A Tableau Calculus with Automaton-Labelled Formulae for Regular Grammar Logics

Rajeev Goré[1] and Linh Anh Nguyen[2]

[1] The Australian National University and NICTA,
Canberra ACT 0200, Australia
Rajeev.Gore@anu.edu.au
[2] Institute of Informatics, University of Warsaw,
ul. Banacha 2, 02-097 Warsaw, Poland
nguyen@mimuw.edu.pl

Abstract. We present a sound and complete tableau calculus for the class of regular grammar logics. Our tableau rules use a special feature called automaton-labelled formulae, which are similar to formulae of automaton propositional dynamic logic. Our calculus is cut-free and has the analytic superformula property so it gives a decision procedure. We show that the known EXPTIME upper bound for regular grammar logics can be obtained using our tableau calculus. We also give an effective Craig interpolation lemma for regular grammar logics using our calculus.

1   Introduction

Multimodal logics (and their description logic cousins) are useful in many areas of computer science: for example, multimodal logics are used in knowledge representation by interpreting $[i]\varphi$ as "agent i knows/believes that φ is true" [7,15,1]. Grammar logics are normal multimodal logics characterised by "inclusion" axioms like $[t_1] \ldots [t_h]\varphi \supset [s_1] \ldots [s_k]\varphi$, where $[t_i]$ and $[s_j]$ are modalities indexed by members t_i and s_j from some fixed set of indices. Thus $[1][2]\varphi \rightarrow [1]\varphi$ captures "if agent one knows that agent two knows φ, then agent one knows φ".

Inclusion axioms correspond in a strict sense to grammar rules of the form $t_1 t_2 \ldots t_h \rightarrow s_1 s_2 \ldots s_k$ when the index set is treated as a set of atomic words and juxtaposition is treated as word composition. Various refinements ask whether the corresponding grammar is left or right linear, or whether the language generated by the corresponding grammar is regular, context-free etc.

Grammar logics were introduced by Fariñas del Cerro and Penttonen in [8] and have been studied widely [3,4,20,11,5]. Baldoni et al. [3] gave a prefixed tableau calculus for grammar logics and used it to show that the general satisfiability problem of right linear grammar logics is decidable and the general satisfiability problem of context-free grammar logics is undecidable. But the techniques of Baldoni et al. cannot be easily extended to regular grammar logics.

While trying to understand why the decidability proof by Baldoni et al. [3,2] cannot be naturally extended to left linear grammars, Demri [4] observed

B. Beckert (Ed): TABLEAUX 2005, LNAI 3702, pp. 138–152, 2005.

that although right linear grammars generate the same class of languages as left linear grammars, this correspondence is not useful at the level of regular grammar logics. By using a transformation into the satisfiability problem for propositional dynamic logic (PDL), Demri was able to prove that the general satisfiability problem of regular grammar logics is EXPTIME-complete and that the general satisfiability problem of linear grammar logics is undecidable. In [5], Demri and de Nivelle gave a translation of the satisfiability problem for grammar logics with converse into the two-variable guarded fragment GF^2 of first-order logic, and showed that the general satisfiability problem for regular grammar logics with converse is in EXPTIME. The relationship between grammar logics and description logics was considered, among others, in [11,20].

Thus, various methods have been required to obtain complexity results and decision procedures for regular grammar logics. We show that it is possible to give a (non-prefixed) tableau calculus which is a decision procedure for the whole class of regular grammar logics, and which also gives an estimate of the complexity of these logics. Efficient tableaux for propositional multimodal (description) logics are highly competitive with translation methods, so it is not at all obvious that the translation into GF^2 from [5] is the best method for deciding these logics.

The naive way to encode inclusion axioms in a non-prefixed tableau calculus is to add a rule like ([t]) shown below at left. But such rules cannot lead to a *general* decision procedure because there are well-known examples like transitivity $[t]\varphi \supset [t][t]\varphi$, whose analogous rule is shown below at right, which immediately cause an infinite branch by adding $[t][t]\varphi$, and then $[t][t][t]\varphi$, and so on:

$$([t]) \quad \frac{X; [t]\varphi}{X; [t]\varphi; [s_1][s_2]\ldots[s_k]\varphi} \qquad\qquad (4t) \quad \frac{X; [t]\varphi}{X; [t]\varphi; [t][t]\varphi}$$

Our calculus uses a special feature called automaton-labelled formulae, which are similar to formulae of APDL [10]. Informally, whenever a formula $[t]\varphi$ is true at a tableau node w, we add an automaton labelled formula that tracks the modal transitions from w. If a sequence of transitions leads to a tableau node u, and this sequence corresponds to a word $s_1s_2\ldots s_k$ recognised by the automaton labelled formula, then we add the formula φ to u. This captures the effect of the rule ([t]) above left in a tractable manner since the influence of $[t]\varphi$ being true at w can be computed directly from the content of the automaton labelled formulae in node u. Our tableau calculus is sound, complete, cut-free and has the analytic superformula property, so it is a decision procedure. As usual for tableau calculi, it allows efficient implementation and good complexity estimation.

In Section 2, we define regular grammar logics and automaton-labelled formulae. In Section 3, we present our tableau calculus for regular grammar logics, and prove it sound. In Section 4, we prove it complete. In Section 5, we prove that the general satisfiability problem of regular grammar logics is in EXPTIME by using our tableau rules in a systematic way. In Section 6, we use our calculus to prove effective Craig interpolation for regular grammar logics. Further work and concluding remarks are in Section 7. The Appendix contains an example.

2   Preliminaries

2.1   Definitions for Multimodal Logics

Our modal language is built from two disjoint sets: $\mathcal{MOD}$ is a finite set of modal indices and $\mathcal{PROP}$ is a set of primitive propositions. We use p and q for elements of $\mathcal{PROP}$ and use t and s for elements of $\mathcal{MOD}$. Formulae of our primitive language are recursively defined using the BNF grammar below:

$$\varphi ::= p \mid \neg\varphi \mid \varphi \wedge \varphi \mid \varphi \vee \varphi \mid \varphi \supset \varphi \mid [t]\varphi \mid \langle t\rangle\varphi$$

A *Kripke frame* is a tuple $\langle W, \tau, \{R_t \mid t \in \mathcal{MOD}\}\rangle$, where W is a nonempty set of possible worlds, $\tau \in W$ is the current world, and each R_t is a binary relation on W, called the accessibility relation for $[t]$ and $\langle t\rangle$. If $R_t(w, u)$ holds then we say that the world u is accessible from the world w via R_t.

A *Kripke model* is a tuple $\langle W, \tau, \{R_t \mid t \in \mathcal{MOD}\}, h\rangle$, where $\langle W, \tau, \{R_t \mid t \in \mathcal{MOD}\}\rangle$ is a Kripke frame and h is a function mapping worlds to sets of primitive propositions. For $w \in W$, the set of primitive propositions "true" at w is $h(w)$.

A *model graph* is a tuple $\langle W, \tau, \{R_t \mid t \in \mathcal{MOD}\}, H\rangle$, where $\langle W, \tau, \{R_t \mid t \in \mathcal{MOD}\}\rangle$ is a Kripke frame and H is a function mapping worlds to formula sets. We sometimes treat model graphs as models with H restricted to $\mathcal{PROP}$.

Given a Kripke model $M = \langle W, \tau, \{R_t \mid t \in \mathcal{MOD}\}, h\rangle$ and a world $w \in W$, the *satisfaction relation* $\models$ is defined as usual for the classical connectives with two extra clauses for the modalities as below:

$$M, w \models [t]\varphi \quad \text{iff} \quad \forall v \in W.R_t(w, v) \text{ implies } M, v \models \varphi$$
$$M, w \models \langle t\rangle\varphi \quad \text{iff} \quad \exists v \in W.R_t(w, v) \text{ and } M, v \models \varphi.$$

We say that φ *is satisfied at* w *in* M if $M, w \models \varphi$. We say that φ *is satisfied in* M and call M a *model of* φ if $M, \tau \models \varphi$.

If we consider only Kripke models, with no restrictions on R_t, we obtain a normal multimodal logic with a standard Hilbert-style axiomatisation K_n.

Note: We now assume that formulae are in *negation normal form*, where $\supset$ is translated away and $\neg$ occurs only directly before primitive propositions. Every formula φ has a logically equivalent formula φ' which is in negation normal form.

2.2   Regular Grammar Logics

Recall that a *finite automaton* A is a tuple $\langle \Sigma, Q, I, \delta, F\rangle$, where Σ is the alphabet (for our case, $\Sigma = \mathcal{MOD}$), Q is a finite set of states, $I \subseteq Q$ is the set of initial states, $\delta \subseteq Q \times \Sigma \times Q$ is the transition relation, and $F \subseteq Q$ is the set of accepting states. A *run* of A on a word $s_1 \ldots s_k$ is a finite sequence of states $q_0, q_1, \ldots, q_k$ such that $q_0 \in I$ and $\delta(q_{i-1}, s_i, q_i)$ holds for every $1 \leq i \leq k$. It is an *accepting run* if $q_k \in F$. We say that A *accepts* word w if there exists an accepting run of A on w. The set of all words accepted/recognised by A is denoted by $\mathcal{L}(A)$.

Given two binary relations R_1 and R_2 over W, their relational composition $R_1 \circ R_2 = \{(x, y) \mid \exists y \in W.R_1(x, y) \,\&\, R_2(y, z)\}$ is also a binary relation over W.

A *grammar logic* is a multimodal logic extending K_n with "inclusion axioms" of the form $[t_1]\ldots[t_h]\varphi \supset [s_1]\ldots[s_k]\varphi$, where $\{t_1,\ldots t_h, s_1,\ldots s_k\} \subseteq \mathcal{MOD}$. Each inclusion axiom corresponds to the restriction $R_{s_1}\circ\ldots\circ R_{s_k} \subseteq R_{t_1}\circ\ldots\circ R_{t_h}$ on accessibility relations where the corresponding side stands for the identity relation if $k = 0$ or $h = 0$. For a grammar logic L, the *L-frame restrictions* are the set of all such corresponding restrictions. A Kripke model is an *L-model* if its frame satisfies all L-frame restrictions. A formula φ is *L-satisfiable* if there exists an L-model satisfying it. A formula φ is *L-valid* if it is satisfied in all L-models.

An inclusion axiom $[t_1]\ldots[t_h]\varphi \supset [s_1]\ldots[s_k]\varphi$ can also be seen as the grammar rule $t_1\ldots t_h \to s_1\ldots s_k$ where the corresponding side stands for the empty word if $k = 0$ or $h = 0$. Thus the inclusion axioms of a grammar logic L capture a grammar $\mathcal{G}(L)$. Here we do not distinguish terminal symbols and nonterminal symbols. $\mathcal{G}(L)$ is *context-free* if its rules are of the form $t \to s_1\ldots s_k$, and is *regular* if it is context-free and for every $t \in \mathcal{MOD}$ there exists a finite automaton A_t that recognises the words derivable from t using $\mathcal{G}(L)$.

A *regular grammar logic* L is a grammar logic whose inclusion axioms correspond to grammar rules that collectively capture a regular grammar $\mathcal{G}(L)$. A regular language is traditionally specified either by a regular expression or by a left/right linear grammar or by a finite automaton. The first two forms can be transformed in PTIME to an equivalent finite automaton that is at most polynomially larger. But there is no syntactic way to specify the class of regular (context-free) grammars, and checking whether a context-free grammar generates a regular language is undecidable (see, e.g., [14]). Hence, we cannot compute these automata if we are given an arbitrary regular grammar logic. We therefore assume that for each $t \in \mathcal{MOD}$ we are given an automaton A_t recognising the words derivable from t using $\mathcal{G}(L)$. These are the *automata specifying L*.

Lemma 1. *Let L be a regular grammar logic and let $\{A_t \mid t \in \mathcal{MOD}\}$ be the automata specifying L. Then the following conditions are equivalent:*

(i) *the word $s_1\ldots s_k$ is accepted by A_t*
(ii) *the formula $[t]\varphi \supset [s_1]\ldots[s_k]\varphi$ is L-valid*
(iii) *the inclusion $R_{s_1}\circ\ldots\circ R_{s_k} \subseteq R_t$ is a consequence of the L-frame restrictions.*

Proof. The equivalence $(ii) \Leftrightarrow (iii)$ is well-known from correspondence theory [19]. The implication $(i) \Rightarrow (ii)$ follows by induction on the length of the derivation of $s_1\ldots s_k$ from t by the grammar $\mathcal{G}(L)$, using substitution, the K-axiom $[t](\varphi \supset \psi) \supset ([t]\varphi \supset [t]\psi)$ and the modal necessitation rule $\varphi/[t]\varphi$. The implication $(iii) \Rightarrow (i)$ follows by induction on the length of the derivation of $R_{s_1} \circ\ldots\circ R_{s_k} \subseteq R_t$ from the L-frame restrictions. See also [3,4] for details.

Example 1. Let $\mathcal{MOD} = \{1,\ldots,m\}$ for a fixed m. Consider the grammar logic with the inclusion axioms $[i]\varphi \supset [j][i]\varphi$ for any $i,j \in \mathcal{MOD}$ and $[i]\varphi \supset [j]\varphi$ if $i > j$. This is a regular grammar logic because the set of words derivable from i using the corresponding grammar is $\{1,\ldots,m\}^*.\{1,\ldots,i\}$. This set is recognised by the automaton $A_i = \langle \mathcal{MOD}, \{p,q\}, \{p\}, \delta_i, \{q\}\rangle$ with $\delta_i = \{(p,j,p) \mid 1 \leq j \leq m\} \cup \{(p,j,q) \mid 1 \leq j \leq i\}$. Note that the corresponding grammar is not "linear" in that at most one symbol in the right hand side of a rule can be nonterminal.

2.3   Automaton-Labelled Formulae

If A is a finite automaton, Q is a subset of the states of A, and φ is a formula in the primitive language then $(A, Q) : \varphi$ is an *automaton-labelled formula*.

Fix a regular grammar logic L and let $\{A_t = \langle \mathcal{MOD}, Q_t, I_t, \delta_t, F_t \rangle \mid t \in \mathcal{MOD}\}$ be the automata specifying L. Let $\delta_t(Q, s) = \{q' \mid \exists q \in Q.(q, s, q') \in \delta_t\}$ be the states which can be reached from Q via an s-transition using A_t. The intuitions of automaton labelled formulae are as follows:

Tagging: A formula of the form $[t]\varphi$ in a world u is represented by $(A_t, I_t) : \varphi$.

Tracking: If $(A_t, Q) : \varphi$ occurs at u and $R(u, v)$ holds then we add the formula $(A_t, \delta_t(Q, s)) : \varphi$ to v. In particular, if $(A_t, I_t) : \varphi$ appears in world u and $R_s(u, v)$ holds then we add $(A_t, \delta_t(I_t, s)) : \varphi$ to the world v.

Acceptance: If $(A_t, Q) : \varphi$ occurs at u and Q contains an accepting state of A_t, then we add φ to u.

The formal semantics of automaton-labelled formulae are defined as follows. Let ε be the empty word and define $\widetilde{\delta}_t(Q, \varepsilon) = Q$ and $\widetilde{\delta}_t(Q, s_1 \ldots s_k) = \widetilde{\delta}_t(\delta_t(Q, s_1), s_2 \ldots s_k)$. If M is a Kripke model, w is a world of M, and $A_t = \langle \mathcal{MOD}, Q_t, I_t, \delta_t, F_t \rangle$ is an automaton, then $M, w \models (A_t, Q) : \varphi$ iff there exist worlds $w_0, \ldots, w_k = w$ (of M) and indices $s_1, \ldots, s_k \in \mathcal{MOD}$ such that $M, w_0 \models [t]\varphi$, $R_{s_i}(w_{i-1}, w_i)$ holds for $1 \leq i \leq k$, and $\widetilde{\delta}_t(I_t, s_1 \ldots s_k) = Q$. Pictorially: $M, w \models (A_t, Q) : \varphi$ iff

$$w_0 \xrightarrow{\;s_1\;} w_1 \xrightarrow{\;s_2\;} \cdots \qquad\qquad w_{k-1} \xrightarrow{\quad s_k \quad} w_k = w$$

$$[t]\varphi \qquad \widetilde{\delta}(I_t, s_1) \qquad \cdots \qquad \widetilde{\delta}(I_t, s_1 \ldots s_{k-1}) \qquad \widetilde{\delta}(I_t, s_1 \ldots s_k) = Q$$

We can see the soundness of these definitions by the following inter-derivable sequence of validities of multimodal tense logic which use the residuation properties of $\langle s \rangle^{-1}$ and $[s]$ shown at right where $\langle s \rangle^{-1}$ is the converse of $\langle s \rangle$:

$$\frac{\dfrac{[t]\varphi \supset [s_1][s_2] \ldots [s_k]\varphi}{\langle s_1 \rangle^{-1}[t]\varphi \supset [s_2] \ldots [s_k]\varphi}}{\overline{\langle s_k \rangle^{-1} \ldots \langle s_1 \rangle^{-1}[t]\varphi \supset \varphi}} \qquad\qquad \frac{\varphi \supset [s]\psi}{\langle s \rangle^{-1}\varphi \supset \psi}$$

That is, if we want to ensure that $[t]\varphi \supset [s_1][s_2] \ldots [s_k]\varphi$ is valid, then it suffices to ensure that $\langle s_k \rangle^{-1} \ldots \langle s_1 \rangle^{-1}[t]\varphi \supset \varphi$ is valid instead. But converse modalities are not part of our official language so we use the occurrence of $[t]\varphi$ at w_0 to start an automaton A_t which tracks the following constraint: φ must be true at any world w reachable from w_0 by the path/word $s_1 \ldots s_k$.

Our automaton-labelled formulae are similar to formulae of automaton propositional dynamic logic (APDL) [10]. A formula involving automata in

APDL is of the form $[A]\varphi$, where A is a finite automata with *one* initial state and *one* accepting state. An automaton labelled formula like our $(A_t, Q) : \varphi$ with $Q = \{q_1, q_2, \ldots, q_k\}$ can be simulated by the APDL formula $[B_1]\varphi \vee \ldots \vee [B_k]\varphi$ where each B_i is the automaton A_t restricted to start at the initial state q_i. Thus our formulation uses a more compact representation in which APDL formulae that differ only in their initial state are grouped together. Moreover, we do not treat different "states" of an automaton as different automata. Our compact representation not only saves memory but also increases efficiency of deduction.

From now on, by a *formula* we mean either a formula in the primitive language (as defined in Section 2.1) or an automaton-labelled formula.

2.4   Definitions for Tableau Calculi

As in our previous works on tableau calculi [9,17], our tableau formulation trace their roots to Hintikka via [18]. A *tableau rule* σ consists of a numerator N above the line and a (finite) list of denominators $D_1, D_2, \ldots, D_k$ (below the line) separated by vertical bars. The numerator is a finite formula set, and so is each denominator. As we shall see later, each rule is read downwards as "if the numerator is L-satisfiable, then so is one of the denominators". The numerator of each tableau rule contains one or more distinguished formulae called the *principal formulae*. A *tableau calculus* CL for a logic L is a finite set of tableau rules.

A CL-tableau for X is a tree with root X whose nodes carry finite formula sets obtained from their parent nodes by instantiating a tableau rule with the proviso that if a child s carries a set Z and Z has already appeared on the branch from the root to s then s is an *end node*.

Let Δ be a set of tableau rules. We say that Y is *obtainable from X by applications of rules from Δ* if there exists a tableau for X which uses only rules from Δ and has a node that carries Y. A branch in a tableau is *closed* if its end node carries only $\bot$. A tableau is *closed* if every one of its branches is closed. A tableau is *open* if it is not closed. A finite formula set X in the primitive language is said to be CL-*consistent* if every CL-tableau for X is open. If there is a closed CL-tableau for X then we say that X is CL-*inconsistent*.

A tableau calculus CL is *sound* if for all finite formula sets X in the primitive language, X is L-satisfiable implies X is CL-consistent. It is *complete* if for all finite formula sets X in the primitive language, X is CL-consistent implies X is L-satisfiable. Let σ be a rule of CL. We say that σ is sound w.r.t. L if for every instance σ' of σ, if the numerator of σ' is L-satisfiable then so is one of the denominators of σ'. Any CL containing only rules sound w.r.t. L is sound.

3   A Tableau Calculus for Regular Grammar Logics

Fix a regular grammar logic L and let $\{A_t = \langle \mathcal{MOD}, Q_t, I_t, \delta_t, F_t \rangle \mid t \in \mathcal{MOD}\}$ be the automata specifying L. Recall that formulae are in negation normal form. We use X for a formula set, and semicolon to separate elements of a formula set. We have deliberately used "s.t." for "such that" in the set notation of the

$$(\bot) \; \frac{X; p; \neg p}{\bot} \qquad\qquad (\wedge) \; \frac{X; \varphi \wedge \psi}{X; \varphi; \psi} \qquad\qquad (\vee) \; \frac{X; \varphi \vee \psi}{X; \varphi \mid X; \psi}$$

$$(\text{label}) \; \frac{X; [t]\varphi}{X; (A_t, I_t): \varphi} \qquad\qquad (\text{add}) \; \frac{X; (A_t, Q): \varphi}{X; (A_t, Q): \varphi; \varphi} \; \text{if } Q \cap F_t \neq \emptyset$$

$$(\text{trans}) \; \frac{X; \langle t \rangle \varphi}{\{(A_s, \delta_s(Q, t)): \psi \; s.t. \; (A_s, Q): \psi \in X\}; \varphi}$$

Fig. 1. Tableau Rules

denominator of the (trans)-rule because we use colon in automaton-labelled formulae and the alternative | indicates a branching rule! The tableau calculus $\mathcal{C}L$ is given in Figure 1. The first five rules are *static* rules, and the last rule is a *transitional* rule. An example is in the appendix.

A tableau calculus $\mathcal{C}L$ has the *analytic superformula* property iff to every finite set X we can assign a finite set $X_{\mathcal{C}L}^*$ which contains all formulae that may appear in any tableau for X. We write $Sf(\varphi)$ for the set of all subformulae of φ, and $Sf(X)$ for the set $\bigcup_{\varphi \in X} Sf(\varphi) \cup \{\bot\}$. Our calculus has the analytic superformula property, with $X_{\mathcal{C}L}^* = Sf(X) \cup \{(A_t, Q): \varphi \; s.t. \; [t]\varphi \in Sf(X) \text{ and } Q \subseteq Q_t\}$.

Lemma 2. *The tableau calculus $\mathcal{C}L$ is sound.*

Proof. We show that $\mathcal{C}L$ contains only rules sound w.r.t. L as follows. Suppose that the numerator of the considered rule is satisfied at a world w in a model $M = \langle W, \tau, \{R_t \mid t \in \mathcal{MOD}\}, h \rangle$. We have to show that at least one of the denominators of the rule is also satisfiable. For the static rules, we show that some denominator is satisfied at w itself. For the transitional rule (trans), we show that its denominator is satisfied at some world reachable from w via R_t.

$(\bot), (\wedge), (\vee)$: These cases are obvious.

(label): If $M, w \models X; [t]\varphi$ then $M, w \models (A_t, I_t): \varphi$ by definition.

(add): Suppose that $M, w \models X; (A_t, Q): \varphi$ and $Q \cap F_t \neq \emptyset$. By definition, there exist worlds $w_0, \ldots, w_{k-1}, w_k = w$ and indices $s_1, \ldots, s_k \in \mathcal{MOD}$ such that $M, w_0 \models [t]\varphi$, and $R_{s_i}(w_{i-1}, w_i)$ holds for $1 \leq i \leq k$, and $\widetilde{\delta}_t(I_t, s_1 \ldots s_k) = Q$. Since $Q \cap F_t \neq \emptyset$, the word $s_1 \ldots s_k$ is accepted by A_t. By Lemma 1, it follows that $M, w_0 \models [s_1] \ldots [s_k]\varphi$. Since $w = w_k$ and $R_{s_i}(w_{i-1}, w_i)$ holds for $1 \leq i \leq k$, we must have $M, w \models \varphi$.

(trans): Suppose that $M, w \models X; \langle t \rangle \varphi$. Then there exists some u such that $R_t(w, u)$ holds and $M, u \models \varphi$. For each $(A_s, Q): \psi \in X$, we have $M, w \models (A_s, Q): \psi$, and by the semantics of automaton-labelled formulae, it follows that $M, u \models (A_s, \delta_s(Q, t)): \psi$. Hence, the denominator is satisfied at u.

4   Completeness

We prove completeness of our calculus via model graphs following [18,9,16,17] by giving an algorithm that accepts a finite $\mathcal{C}L$-consistent formula set X in the primitive language and constructs an L-model graph (defined in Section 4.2) for X that satisfies every one of its formulae at the appropriate world.

4.1   Saturation

In the rules $(\wedge)$, $(\vee)$, (label) the principal formula does not occur in the denominators. For any of these rules δ, let δ' denote the rule obtained from δ by adding the principal formula to each of the denominators. Let $\mathcal{SCL}$ denote the set of static rules of $\mathcal{CL}$ with $(\wedge)$, $(\vee)$, (label) replaced by $(\wedge')$, $(\vee')$, (label$'$). For every rule of $\mathcal{SCL}$, except $(\bot)$, the numerator is included in each of the denominators.

For a finite $\mathcal{CL}$-consistent formula set X, a formula set Y is called a $\mathcal{CL}$-saturation of X if Y is a maximal $\mathcal{CL}$-consistent set obtainable from X by applications of the rules of $\mathcal{SCL}$. A set X is closed w.r.t. a tableau rule if applying that rule to X gives back X as one of the denominators.

Lemma 3. *Let X be a finite $\mathcal{CL}$-consistent formula set and Y a $\mathcal{CL}$-saturation of X. Then $X \subseteq Y \subseteq X^*_{\mathcal{CL}}$ and Y is closed w.r.t. the rules of $\mathcal{SCL}$. Furthermore, there is an effective procedure that, given a finite $\mathcal{CL}$-consistent formula set X, constructs some $\mathcal{CL}$-saturation of X.*

Proof. It is clear that $X \subseteq Y \subseteq X^*_{\mathcal{CL}}$. Observe that if a rule of $\mathcal{SCL}$ is applicable to Y, then one of the corresponding instances of the denominators is $\mathcal{CL}$-consistent. Since Y is a $\mathcal{CL}$-saturation, Y is closed w.r.t. the rules of $\mathcal{SCL}$.

We construct a $\mathcal{CL}$-saturation of X as follows: let $Y = X$; while there is some rule δ of $\mathcal{SCL}$ applicable to Y such that one of its corresponding denominator instance Z is $\mathcal{CL}$-consistent and strictly contains Y, set $Y = Z$. At each iteration, $Y \subset Z \subseteq X^*_{\mathcal{CL}}$. Hence the above process always terminates. It is clear that the resulting set Y is a $\mathcal{CL}$-saturation of X.

4.2   Proving Completeness via Model Graphs

A model graph is an *L-model graph* if its frame is an L-frame. An L-model graph $\langle W, \tau, \{R_t \,|\, t \in \mathcal{MOD}\}, H \rangle$ is *saturated* if every $w \in W$ satisfies:

And: if $\varphi \wedge \psi \in H(w)$ then $\{\varphi, \psi\} \subseteq H(w)$;
Or: if $\varphi \vee \psi \in H(w)$ then $\varphi \in H(w)$ or $\psi \in H(w)$;
Box: if $[t]\varphi \in H(w)$ and $R_t(w, u)$ holds then $\varphi \in H(u)$;
Dia: if $\langle t \rangle \varphi \in H(w)$ then there exists a $u \in W$ such that $R_t(w, u)$ and $\varphi \in H(u)$.

A saturated model graph is *consistent* if no world contains $\bot$, and no world contains $\{p, \neg p\}$. Our model graphs merely denote a data structure, while Rautenberg's model graphs are required to be saturated and consistent.

Lemma 4. *If $M = \langle W, \tau, \{R_t \,|\, t \in \mathcal{MOD}\}, H \rangle$ is a consistent saturated L-model graph, then M satisfies all formulae of $H(\tau)$ which are in the primitive language.*

Proof. By proving $\varphi \in H(w)$ implies $M, w \models \varphi$ by induction on the length of φ.

Given a finite $\mathcal{CL}$-consistent set X in the primitive language, we construct a consistent saturated L-model graph $M = \langle W, \tau, \{R_t \,|\, t \in \mathcal{MOD}\}, H \rangle$ such that $X \subseteq H(\tau)$, thereby giving an L-model for X.

4.3   Constructing Model Graphs

In the following algorithm, the worlds of the constructed model graph are marked either as *unresolved* or as *resolved*.

Algorithm 1
Input: a finite CL-consistent set X of formulae in the primitive language.
Output: an L-model graph $M = \langle W, \tau, \{R_t \mid t \in \mathcal{MOD}\}, H \rangle$ satisfying X.

1. Let $W = \{\tau\}$, $H(\tau)$ be a CL-saturation of X, and $R'_t = \emptyset$ for all $t \in \mathcal{MOD}$. Mark τ as unresolved.
2. While there are unresolved worlds, take one, say w, and do the following:
   (a) For every formula $\langle t \rangle \varphi$ in $H(w)$:
      i. Let $Y = \{(A_s, \delta_s(Q, t)) : \psi \text{ s.t. } (A_s, Q) : \psi \in H(w)\} \cup \{\varphi\}$ be the result of applying rule (trans) to $H(w)$, and let Z be a CL-saturation of Y.
      ii. If $\exists u \in W$ on the path from the root to w with $H(u) = Z$, then add the pair (w, u) to R'_t. Otherwise, add a new world w_φ with content Z to W, mark it as unresolved, and add the pair (w, w_φ) to R'_t.
   (b) Mark w as resolved.
3. Let R_t be the least extension of R'_t for $t \in \mathcal{MOD}$ such that $\langle W, \tau, \{R_t \mid t \in \mathcal{MOD}\} \rangle$ is an L-frame.

This algorithm always terminates: eventually, for every w, either w contains no $\langle t \rangle$-formulae, or there exists an ancestor with $H(u) = Z$ at Step 2(a)ii because all CL-saturated sets are drawn from the finite and fixed set X^*_{CL}.

Lemma 5. *Suppose $R_t(w, u)$ holds via Step 3. Then there exist $w_0, \dots, w_k$ in M with $w_0 = w$, $w_k = u$, and indices $s_1, \dots, s_k \in \mathcal{MOD}$ such that $R'_{s_i}(w_{i-1}, w_i)$ holds for $1 \le i \le k$, and $R_{s_1} \circ \dots \circ R_{s_k} \subseteq R_t$ follows from the L-frame restrictions.*

Proof. By induction on number of inferences in deriving $R_t(w, u)$ when extending R'_s to R_s for $s \in \mathcal{MOD}$, with L-frame restrictions of the form $R_{t_1} \circ \dots \circ R_{t_h} \subseteq R_s$. ☐

4.4   Completeness Proof

Lemma 6. *Let X be a finite CL-consistent set of formulae in the primitive language and $M = \langle W, \tau, \{R_t \mid t \in \mathcal{MOD}\}, H \rangle$ be the model graph for X constructed by Algorithm 1. Then M is a consistent saturated L-model graph satisfying X.*

Proof. It is clear that M is an L-model graph and for any $w \in W$, the set $H(w)$ is CL-consistent. We want to show that M is a saturated model graph. It suffices to show that, for every $w, u \in W$, if $[t]\varphi \in H(w)$ and $R_t(w, u)$ holds then $\varphi \in H(u)$. Suppose that $[t]\varphi \in H(w)$ and $R_t(w, u)$ holds. By Lemma 5, there exist worlds $w_0, \dots, w_k$ with $w_0 = w$, $w_k = u$ and indices $s_1, \dots, s_k \in \mathcal{MOD}$ such that $R'_{s_i}(w_{i-1}, w_i)$ holds for $1 \le i \le k$ and $R_{s_1} \circ \dots \circ R_{s_k} \subseteq R_t$ is a consequence of the L-frame restrictions. Since $H(w)$ is a CL-saturation, we have that $(A_t, I_t) : \varphi \in H(w)$. By Step 2a of Algorithm 1, $(A_t, \widetilde{\delta}_t(I_t, s_1 \dots s_i)) : \varphi \in H(w_i)$ for $1 \le i \le k$. Thus $(A_t, \widetilde{\delta}_t(I_t, s_1 \dots s_k)) : \varphi \in H(u)$. Since $R_{s_1} \circ \dots \circ R_{s_k} \subseteq R_t$ is a consequence of the L-frame restrictions, by Lemma 1, the word $s_1 \dots s_k$ is accepted by A_t. Hence $\widetilde{\delta}_t(I_t, s_1 \dots s_k) \cap F_t \ne \emptyset$. It follows that $\varphi \in H(u)$, since $H(u)$ is a CL-saturation.

The following theorem follows from Lemmas 2 and 6.

Theorem 1. *The tableau calculus CL is sound and complete.*

5 Complexity

The satisfiability problem of a logic L is to check the L-satisfiability of an input formula φ. The general satisfiability problem of a class C of logics is to check L-satisfiability of an input formula φ in an input logic $L \in C$.

Demri [4] proved that the general satisfiability problem of regular grammar logics is EXPTIME-complete by a transformation into satisfiability for PDL. We now obtain the upper bound EXPTIME using our tableaux calculus.

We need a rule $(\cup)$ to coalesce $(A_t, Q){:}\varphi$ and $(A_t, Q'){:}\varphi$ into $(A_t, Q \cup Q'){:}\varphi$

$$(\cup)\ \frac{X\ ;\ (A_t, Q){:}\varphi\ ;\ (A_t, Q'){:}\varphi}{X\ ;\ (A_t, Q \cup Q'){:}\varphi}$$

Observe that $X; (A_t, Q){:}\varphi; (A_t, Q'){:}\varphi$ is CL-consistent iff $X; (A_t, Q \cup Q'){:}\varphi$ is CL-consistent. This follows from the facts that $\delta_t(Q, s) \cup \delta_t(Q', s) = \delta_t(Q \cup Q', s)$ and $((Q \cap F_t \neq \emptyset) \vee (Q' \cap F_t \neq \emptyset)) \equiv ((Q \cup Q') \cap F_t \neq \emptyset)$. Thus, rule $(\cup)$ can be added to CL as a *static* rule, and used whenever possible without affecting soundness and completeness. Let CLu be CL plus $(\cup)$.

Allowing $(\cup)$ requires a change in the semantics of automaton-labelled formulae to: if M is a Kripke model, w is a world of M, and $A_t = \langle MOD, Q_t, I_t, \delta_t, F_t \rangle$ is an automaton, then $M, w \models (A_t, Q) : \varphi$ iff for every $q \in Q$ there exist worlds $w_0, \ldots, w_{k-1}, w_k = w$ (of M) and indices $s_1, \ldots, s_k \in MOD$ such that $M, w_0 \models [t]\varphi$, and $R_{s_i}(w_{i-1}, w_i)$ holds for $1 \leq i \leq k$, and $q \in \tilde{\delta}_t(I_t, s_1 \ldots s_k) \subseteq Q$.

Let L be a regular logic and X a finite formula set in the primitive language. Let n be the sum of the sizes of the formulae in X and the sizes of the automata specifying L. To check whether X is L-satisfiable we can search for a closed CLu-tableau for X, or equivalently, examine an and-or tree for X constructed by using Algorithm 1 to apply our CLu-tableau rules in a systematic way. In such a tree, and-branching is caused by all possible applications of rule (trans), while or-branching is caused by an application of rule $(\vee)$. The other CLu-tableau rules, including $(\cup)$, are applied locally for each node whenever possible.

There are at most $O(n)$ unlabelled subformulae of X, and at most $2^{O(n)}$ different labels. By using the rule $(\cup)$ whenever possible, each subformula of X occurs in a node with at most two labels, so a node contains at most $2n$ i.e. $O(n)$ labelled formulae. Hence there are at most $(2^{O(n)})^{O(n)} = 2^{O(n^2)}$ different nodes. Without the rule $(\cup)$, there are at most $2^{2^{O(n)}}$ different nodes, which breaks EXPTIME worst-case complexity, so the $(\cup)$ rule is absolutely essential. But it is not necessary for the satisfiability problem of a fixed logic.

Algorithm 1 terminates in general because it checks for repeated ancestors: this check is built into the definition of an end-node, and also in Step 2(a)ii. Thus the same node can appear on multiple branches. In the worst case, Algorithm 1 therefore requires $2^{2^{O(n^2)}}$ time. We therefore refine it into Algorithm 2 below:

Algorithm 2
Input: a finite set X of formulae in the primitive language.
Output: a finite graph $G = (V, E)$

1. Let $G =< V, E >=< \{X\}, \emptyset >$, and mark X as unresolved.
2. While V contains unresolved nodes, take one, say n, and do:
   (a) If $(\cup)$ is applicable to n then apply it to obtain denominator d_1
   (b) Else if any static rule of CL is applicable to n then apply it to obtain denominator(s) d_1 (and possibly d_2)
   (c) Else, for every formula $\langle t \rangle \varphi_i$ in n, let $d_i = \{(A_s, \delta_s(Q, t)) : \psi_i \text{ s.t. } (A_s, Q) : \psi_i \in n\} \cup \{\varphi_i\}$ be the denominator obtained by applying (trans) to n
   (d) Mark n as resolved (n is an or/and node if the applied rule is/isn't $(\vee)$)
   (e) For every denominator $d = d_1, \cdots, d_k$:
       i. If some proxy $c \in V$ has $c = d$, then add the edge (n, c) to E
       ii. Else add d to V, add (n, d) to E, and mark d as unresolved.

Algorithm 2 builds an and-or graph G monotonically by "caching" previously seen nodes (but not their open/closed status). The graph G contains a node d for *every* applicable static rule denominator, not just their CL-saturation as in Algorithm 1. Each node appears only once because repetitions are represented by "cross-tree" edges to their first occurrence, so G has at most $2^{O(n^2)}$ nodes.

We now make passes of the and-or graph G, marking nodes as *false* in a monotonic way. In the first pass we mark the node containing $\perp$, if it exists, since *false* captures inconsistency. In each subsequent pass we mark any unmarked or-node with two *false*-marked children, and mark any unmarked and-node with at least one *false*-marked child. We stop making passes when some pass marks no node. Otherwise, we must terminate after $2^{O(n^2)}$ passes since the root must then be marked with *false*. Note that once a node is marked with *false* this mark is never erased. Finally, mark all non-*false* nodes with *true* giving graph G_f.

Lemma 7. *If node $n \in G_f$ is marked* false *then n is CLu-inconsistent.*

Proof. By induction on the number of passes needed to mark n with *false*.

Lemma 8. *If a node $n \in G_f$ is marked* true *then it is CLu-consistent.*

Proof. An easy proof is to take the sub-graph G_n generated by n; replace each sequence of *true*-marked static rule denominators by one *true*-marked node containing their union, which represents their CLu-saturation; and turn the resulting sub-graph into an L-frame by appropriately extending it as in Step 3 of Algorithm 1. For each node x, putting $p \in h(x)$ iff $p \in x$ gives an L-model for n since: all eventualities in a *true*-marked node are fulfilled by its children, and these are guaranteed to be marked *true*; and each *true*-marked or-node has at least one *true*-marked child. By the completeness of CLu, every CLu-tableau for n must be open. A slightly trickier proof converts G_n into an and-or *tree* by mimicking the rule applications from G_n but unwinding edges to non-ancestor-proxies by making a copy of the proxy. This reproduces all CLu-tableaux for n constructible by Algorithm 1 (*sic*) and each one is open by its construction.

Algorithm 2 and the creation of G_f runs in time $(2^{O(n^2)})^2 = 2^{2 \cdot O(n^2)}$ and so the general satisfiability problem of regular grammar logics is in EXPTIME.

$(I\bot)\; Z \xrightarrow{\top} Z'; p; \neg p$

$(I\bot)\; Z; p \xrightarrow{\bot} Z'; \neg p$

$(I\wedge)\; \dfrac{Z \xrightarrow{\zeta} Z'; \varphi \wedge \psi}{Z \xrightarrow{\zeta} Z'; \varphi; \psi}$

$(I\vee)\; \dfrac{Z \xrightarrow{\zeta \wedge \xi} Z'; \varphi \vee \psi}{Z \xrightarrow{\zeta} Z'; \varphi \mid Z \xrightarrow{\xi} Z'; \psi}$

$(I\text{label})\; \dfrac{Z \xrightarrow{\zeta} Z'; [t]\varphi}{Z \xrightarrow{\zeta} Z'; (A_t, I_t){:}\varphi}$

$(I\text{add})\; \dfrac{Z \xrightarrow{\zeta} Z'; (A_t, Q){:}\varphi}{Z \xrightarrow{\zeta} Z'; (A_t, Q){:}\varphi; \varphi} \; \text{if } Q \cap F_t \neq \emptyset$

$(I\text{trans})\; \dfrac{Z \xrightarrow{[t]\zeta} Z'; \langle t\rangle\varphi}{\{(A_s, \delta_s(Q, t)){:}\psi \text{ s.t. } (A_s, Q){:}\psi \in Z\} \xrightarrow{\zeta} \{(A_s, \delta_s(Q, t)){:}\psi \text{ s.t. } (A_s, Q){:}\psi \in Z'\}; \varphi}$

Fig. 2. Rules of the Calculus for Interpolation

6 Effective Interpolation

We say that ζ is an *interpolation formula in L for the formula* $\varphi \supset \psi$ if all primitive propositions of ζ are common to φ and ψ, and $\varphi \supset \zeta$ and $\zeta \supset \psi$ are both *L*-valid. The Craig interpolation lemma for *L* states that if $\varphi \supset \psi$ is *L*-valid, then there exists an interpolation formula in *L* for $\varphi \supset \psi$. This lemma is *effective* if the proof of the lemma actually constructs the interpolation formula.

Assume our language contains $\top$ with the usual semantics. We prove effective Craig interpolation for all regular grammar logics using the method of [16].

Our tableau calculi are refutation calculi, so we use an indirect formulation of interpolation. Given two sets X and Y of formulae, and using $\overline{\zeta}$ to denote the negation normal form of $\neg\zeta$, we say that ζ is an interpolation formula w.r.t. $\mathcal{C}L$ for the pair $\langle X, Y\rangle$, and also that $X \xrightarrow{\zeta} Y$ *is* $\mathcal{C}L$-*valid*, if: all primitive propositions of ζ are common to X and Y, the formula ζ does not contain automaton-labelled formulae, and the sets $X; \overline{\zeta}$ and $\zeta; Y$ are both $\mathcal{C}L$-inconsistent. Since $\mathcal{C}L$ is sound and complete, it follows that if $\varphi \xrightarrow{\zeta} \overline{\psi}$ is $\mathcal{C}L$-valid, then $\varphi \supset \zeta$ and $\zeta \supset \psi$ are both *L*-valid, and hence that ζ is an interpolation formula in *L* for $\varphi \supset \psi$.

We now show that for any finite formula sets X and Y, if $X; Y$ is $\mathcal{C}L$-inconsistent, then there exists an interpolation formula w.r.t. $\mathcal{C}L$ for the pair $\langle X, Y\rangle$. It follows that the Craig interpolation lemma holds for *L*.

Observe that $Y \xrightarrow{\overline{\zeta}} X$ is $\mathcal{C}L$-valid iff $X \xrightarrow{\zeta} Y$ is $\mathcal{C}L$-valid. We call $Y \xrightarrow{\overline{\zeta}} X$ the *reverse form* of $X \xrightarrow{\zeta} Y$.

The rule $(I\delta)$ below left is an *interpolation rule* if the inference step below right is an instance of the tableau rule with name (δ):

$(I\delta)\; \dfrac{N \xrightarrow{\varphi} N'}{D_1 \xrightarrow{\varphi_1} D_1' \mid \ldots \mid D_k \xrightarrow{\varphi_k} D_k'}$

$\dfrac{N; N'}{D_1; D_1' \mid \ldots \mid D_k; D_k'}$

Provided that (δ) is a $\mathcal{C}L$-tableau rule, the interpolation rule $(I\delta)$ is $\mathcal{C}L$-*sound* if $\mathcal{C}L$-validity of all $D_1 \xrightarrow{\varphi_1} D_1', \ldots, D_k \xrightarrow{\varphi_k} D_k'$ implies $\mathcal{C}L$-validity of $N \xrightarrow{\varphi} N'$.

Figure 2 contains the interpolation rules obtained from the tableau rules for regular grammar logics. Each tableau rule of CL except $(I\bot)$ has one corresponding interpolation rule. Rule $(\bot)$ has an interpolation rule for each of its two principal formulae but these rules have no denominator because it is not necessary. Rule $(\cup)$ has no interpolation rule since it is just an optimisation rule.

Lemma 9. *The above interpolation rules are CL-sound.*

Proof. We consider $(I\text{trans})$ only, the others are similar. Let $X = \{(A_s, \delta_s(Q,t)):$ ψ $s.t.\,(A_s, Q)\colon \psi \in Z\}$ and $Y = \{(A_s, \delta_s(Q,t))\colon \psi$ $s.t.\,(A_s, Q)\colon \psi \in Z'\}$. Suppose that $X \overset{\zeta}{\to} Y; \varphi$ is CL-valid. Thus, both $X; \zeta$ and $\zeta; Y; \varphi$ are CL-inconsistent, and have closed CL-tableaux. We show that $Z \xrightarrow{[t]\zeta} Z'; \langle t\rangle\varphi$ is CL-valid by giving closed CL-tableaux for both $Z; \langle t\rangle\overline{\zeta}$ and $[t]\zeta; Z'; \langle t\rangle\varphi$:

$$\cfrac{\cfrac{\cfrac{\cfrac{[t]\zeta; Z'; \langle t\rangle\varphi}{(A_t, I_t)\colon \zeta; Z'; \langle t\rangle\varphi}\text{ (label)}}{(A_t, \delta_t(I_t,t))\colon \zeta; Y; \varphi}\text{ (trans)}}{(A_t, \delta_t(I_t,t))\colon \zeta; \zeta; Y; \varphi}\text{ (add)}}{\cfrac{\zeta; Y; \varphi}{\bot}\text{ (assumption)}}\text{ (wk)}$$

$$\cfrac{\cfrac{Z; \langle t\rangle\overline{\zeta}}{X; \overline{\zeta}}\text{ (trans)}}{\bot}\text{ (assumption)}$$

Applying (add) above right is justified because $\delta_t(I_t, t) \cap F_t \neq \emptyset$ since A_t accepts word t. Also, the rule (wk) of weakening is obviously admissible.

These rules build the numerator's interpolant from those of the denominators. Using Lemma 9, and the technique of [16, Lemmas 13 and 14] we obtain:

Theorem 2. *Regular grammar logics enjoy effective Craig interpolation.*

7   Further Work and Conclusions

Our main contribution is a tableau calculus that forms a decision procedure for the whole class of regular grammar logics. Our automaton-labelled formulae are similar to formulae of APDL [10], but with a more compact representation using sets of states instead of single states. We have shown that automaton-labelled formulae work well with the traditional techniques of proving soundness and completeness. Our calculus gives a simple estimate of the upper complexity bound of regular grammar logics, and can be used to obtain effective Craig interpolation for these logics. We have since found that Craig interpolation for regular grammar logics follows from [13, Corollary B4.1] and [12].

The prefixed tableaux of Baldoni *et al.* give a decision procedure only for right linear logics. A prefixed calculus that simulates our calculus would be less efficient because it would repeatedly search the current branch for computation, not just for loops as in our case. Moreover, it is well-known that loop checking can be done efficiently using, e.g., a hash table. Finally, the transformation of

Demri and de Nivelle into GF^2 is based on states, but not sets of states, which reduces efficiency. Also their resulting formula sets are much larger because they keep a copy of the formulae defining an automaton A_t for each formula $[t]\varphi$, whereas we can keep only t and Q for (A_t, Q) in $(A_t, Q){:}\,\varphi$. Similar observations have been stated for formulae of APDL.

By propagating *false* "on the fly", we believe we can prove global caching sound for checking satisfiability in multimodal **K** with global assumptions i.e. "checking $\mathcal{ALC}$-satisfiability of a concept w.r.t. a TBox with general axioms" [6].

Acknowledgements. We are grateful to Pietro Abate, Stéphane Demri, Marcus Kracht and an anonymous reviewer for their helpful comments and pointers.

References

1. F. Baader and U. Sattler. An overview of tableau algorithms for description logics. *Studia Logica*, 69:5–40, 2001.
2. M. Baldoni. *Normal Multimodal Logics: Automatic Deduction and Logic Programming Extension*. PhD thesis, Dip. di Inf., Univ. degli Studi di Torino, Italy, 1998.
3. M. Baldoni, L. Giordano, and A. Martelli. A tableau for multimodal logics and some (un)decidability results. In *TABLEAUX'1998, LNCS 1397:44-59*, 1998.
4. S. Demri. The complexity of regularity in grammar logics and related modal logics. *Journal of Logic and Computation*, 11(6):933–960, 2001 (see also the long version).
5. S. Demri and H. de Nivelle. Deciding regular grammar logics with converse through first-order logic. *Journal of Logic, Language and Information*, 2005. To appear.
6. F Donini and F Massacci. EXPTIME tableaux for $\mathcal{ALC}$. *Artificial Intelligence*, 124:87–138, 2000.
7. R Fagin, J Y Halpern, Y Moses, and M Y Vardi. *Reasoning About Knowledge*. MIT Press, 1995.
8. L. Fariñas del Cerro and M. Penttonen. Grammar logics. *Logique et Analyse*, 121-122:123–134, 1988.
9. R. Goré. Tableau methods for modal and temporal logics. In D'Agostino et al, editor, *Handbook of Tableau Methods*, pages 297–396. Kluwer, 1999.
10. D. Harel, D. Kozen, and J. Tiuryn. *Dynamic Logic*. MIT Press, 2000.
11. I. Horrocks and U. Sattler. Decidability of SHIQ with complex role inclusion axioms. *Artificial Intelligence*, 160(1-2):79–104, 2004.
12. M. Kracht. Reducing modal consequence relations. *JLC*, 11(6):879–907, 2001.
13. M. Marx and Y. Venema. *Multi-dimensional Modal Logic*. Kluwer, 1997.
14. A. Mateescu and A. Salomaa. Formal languages: an introduction and a synopsis. In *Handbook of Formal Languages - Volume 1*, pages 1–40. Springer, 1997.
15. J.-J.Ch. Meyer and W. van der Hoek. *Epistemic Logic for Computer Science and Artificial Intelligence*. Cambridge University Press, 1995.
16. L.A. Nguyen. Analytic tableau systems and interpolation for the modal logics KB, KDB, K5, KD5. *Studia Logica*, 69(1):41–57, 2001.
17. L.A. Nguyen. Analytic tableau systems for propositional bimodal logics of knowledge and belief. In *TABLEAUX 2002, LNAI 2381:206-220*. Springer, 2002.
18. W. Rautenberg. Modal tableau calculi and interpolation. *JPL*, 12:403–423, 1983.
19. J. van Benthem. Correspondence theory. In D. Gabbay and F. Guenthner, editors, *Handbook of Philosophical Logic, Vol II*, pages 167–247. Reidel, Dordrecht, 1984.
20. M. Wessel. Obstacles on the way to qualitative spatial reasoning with description logics: Some undecidability results. In *Description Logics 2001*.

Appendix. An Example

Example 2. Let $\mathcal{MOD} = \{0, 1, 2\}$. Consider the grammar logic L with the inclusion axioms $[0]\varphi \supset \varphi$ and $[i]\varphi \supset [j][k]\varphi$ if $i = (j + k) \bmod 3$. This is a regular grammar logic because the corresponding grammar is regular. We have $A_i = \langle \mathcal{MOD}, \mathcal{MOD}, \{0\}, \delta, \{i\} \rangle$ for $i \in \mathcal{MOD}$, where $\delta = \{(j, k, l) \mid j, k, l \in \{0, 1, 2\}$ and $l = (j + k) \bmod 3\}$.

We give a closed $\mathcal{CL}$-tableau for $X = \{\langle 0 \rangle p, [0](\neg p \vee \langle 1 \rangle q), [1](\neg q \vee \langle 2 \rangle r), [0]\neg r\}$, in which principal formulae of nodes are underlined. The arrows stand for rule applications and are annotated with the rule name. The labels R_i for $i \in \{0, 1, 2\}$ to the right of the arrows marked with (trans)-rule applications stand for the label on the associated edges in the underlying model being explored by the tableau.

$$\langle 0 \rangle p;\ [0](\neg p \vee \langle 1 \rangle q);\ [1](\neg q \vee \langle 2 \rangle r);\ [0]\neg r$$

$\downarrow$ 3 × (label)

$$\underline{\langle 0 \rangle p};\ (A_0, \{0\}) : (\neg p \vee \langle 1 \rangle q);\ (A_1, \{0\}) : (\neg q \vee \langle 2 \rangle r);\ (A_0, \{0\}) : \neg r$$

$\downarrow$ (trans) R_0

$$p;\ \underline{(A_0, \{0\}) : (\neg p \vee \langle 1 \rangle q)};\ (A_1, \{0\}) : (\neg q \vee \langle 2 \rangle r);\ (A_0, \{0\}) : \neg r$$

$\downarrow$ (add)

$$p;\ (A_0, \{0\}) : (\neg p \vee \langle 1 \rangle q);\ \underline{\neg p \vee \langle 1 \rangle q};\ (A_1, \{0\}) : (\neg q \vee \langle 2 \rangle r);\ (A_0, \{0\}) : \neg r$$

(∨) ⟍ (∨) ↓

$$p;\ \langle 1 \rangle q;\ (A_0, \{0\}) : (\neg p \vee \langle 1 \rangle q);\ (A_1, \{0\}) : (\neg q \vee \langle 2 \rangle r);\ (A_0, \{0\}) : \neg r$$

$p;\ \neg p;\ \dots$

(⊥) ↓

$\bot$

$\downarrow$ (trans) R_1

$$q;\ (A_0, \{1\}) : (\neg p \vee \langle 1 \rangle q);\ \underline{(A_1, \{1\}) : (\neg q \vee \langle 2 \rangle r)};\ (A_0, \{1\}) : \neg r$$

$\downarrow$ (add)

$$q;\ (A_0, \{1\}) : (\neg p \vee \langle 1 \rangle q);\ (A_1, \{1\}) : (\neg q \vee \langle 2 \rangle r);\ \underline{\neg q \vee \langle 2 \rangle r};\ (A_0, \{1\}) : \neg r$$

(∨) ⟍ (∨) ↓

$$q;\ \langle 2 \rangle r;\ (A_0, \{1\}) : (\neg p \vee \langle 1 \rangle q);\ (A_1, \{1\}) : (\neg q \vee \langle 2 \rangle r);\ (A_0, \{1\}) : \neg r$$

$q;\ \neg q;\ \dots$

(⊥) ↓

$\bot$

$\downarrow$ (trans) R_2

$$r;\ (A_0, \{0\}) : (\neg p \vee \langle 1 \rangle q);\ (A_1, \{0\}) : (\neg q \vee \langle 2 \rangle r);\ \underline{(A_0, \{0\}) : \neg r}$$

$\downarrow$ (add)

$$\underline{r};\ (A_0, \{0\}) : (\neg p \vee \langle 1 \rangle q);\ (A_1, \{0\}) : (\neg q \vee \langle 2 \rangle r);\ (A_0, \{0\}) : \neg r;\ \underline{\neg r}$$

$\downarrow$ (⊥)

$\bot$

Comparing Instance Generation Methods
for Automated Reasoning*

Swen Jacobs and Uwe Waldmann

Max-Planck-Institut für Informatik,
Saarbrücken, Germany

Abstract. The clause linking technique of Lee and Plaisted proves the unsatisfiability of a set of first-order clauses by generating a sufficiently large set of instances of these clauses that can be shown to be propositionally unsatisfiable. In recent years, this approach has been refined in several directions, leading to both tableau-based methods, such as the *Disconnection Tableau Calculus*, and saturation-based methods, such as *Primal Partial Instantiation* and *Resolution-based Instance Generation*. We investigate the relationship between these calculi and answer the question to what extent refutation or consistency proofs in one calculus can be simulated in another one.

1 Introduction

In recent years, there has been a renewed interest in instantiation-based theorem proving for first-order logic. Much of the recent work in this field is based on the research of Plaisted and Lee [LP92]. They showed that the interleaving of production of instances with recombination of clauses, as done in resolution, leads to duplication of work in subsequent inference steps. As a means of avoiding this duplication, they proposed the *clause linking* approach. In clause linking, *links* between complementary unifiable literals are used to generate instances of the given clauses, based on the unifier of the linked literals. Unlike in the resolution calculus, the generated instances are not recombined, but added to the set of instances as they are. As a consequence, in order to check satisfiability of the generated set of clauses, an additional SAT solving procedure is needed, which is usually in the spirit of the Davis-Putnam-Logemann-Loveland (DPLL) procedure [DLL62].

Today, there are several methods which are based to some extent on clause linking and/or DPLL. They may be distinguished by the means they use to detect unsatisfiability. There are calculi which arrange instances in a tree or a tableau, integrating an implicit satisfiability check. This approach is used by the disconnection tableau calculus [Bil96, LS01, Ste02], as well as by Baumgartner's FDPLL [Bau00] and the model evolution calculus [BT03]. Other procedures

---

* This work was partly supported by the German Research Council (DFG) as part of the Transregional Collaborative Research Center "Automatic Verification and Analysis of Complex Systems" (SFB/TR 14 AVACS). See www.avacs.org for more information.

B. Beckert (Ed): TABLEAUX 2005, LNAI 3702, pp. 153–168, 2005.

separate instance generation and satisfiability test. These usually use a mono-tonically growing set of clause instances and call an external SAT solver on this set. Representatives of this approach are Plaisted's (ordered semantic) hyper linking [LP92, Pla94], the partial instantiation methods [HRCR02] of Hooker et al., as well as resolution-based instance generation [GK03] by Ganzinger and Korovin. As they use interchangeable SAT solvers, the main difference between these methods lies in the guidance of instance generation.

The fact that tableau-based and saturation-based instance generation meth-ods are somehow related can be considered as folklore knowledge; it has been mentioned repeatedly in the literature (e.g., [BT03, HRCR02]). The precise rela-tionship is, however, rather unclear. In this work, we will compare four instance generation methods that stay relatively close to the original clause linking ap-proach, and can be seen as direct refinements of it:

The *Disconnection Calculus* (DCC) integrates the instance generation of the clause linking approach into a clausal tableau. The linking rule of DCC only allows inferences based on links which are on the current branch, strongly re-stricting the generation of clauses in the tableau. In this tableau structure, un-satisfiability of the given set of clauses can be decided by branch closure. Given a fair inference strategy, the calculus is refutationally complete.

Resolution-based Instance Generation (Inst-Gen) consists of a single resolu-tion-like inference rule, representing the clause linking approach of generating instances. After a given set has been saturated under this inference rule, satisfi-ability of the original clause set is equivalent to propositional satisfiability and can be checked by any propositional decision procedure.

SInst-Gen is a refinement of Inst-Gen by semantic selection, based on a propositional model for the given set of clauses. Inconsistencies, which arise when extending this model to a model for the first-order clauses, are used to guide the inferences of the extension.

The *Primal Partial Instatiation* method (PPI) is treated as a special case of SInst-Gen.

After introducing the calculi, we will consider simulation of derivations from one method in the other, and we will show to what extent refutation or consis-tency proofs in one calculus can be simulated in the other one.

2   Introducing the Calculi

This section gives a short introduction to the methods we will compare. For a comprehensive description, we refer to Letz and Stenz [LS01, Ste02] for DCC, Ganzinger and Korovin [GK03] for SInst-Gen, and Hooker et al. [HRCR02] for PPI.

2.1   Logical Prerequisites

We use the usual symbols, notation and terminology of first-order logic with standard definitions. We consider all formulas to be in clausal normal form. This allows us to consider a formula as a set of clauses, thought to be connected by conjunction. In all of our methods, every clause of a formula is implicitly $\forall$-quantified, while the methods themselves always work on quantifier-free formulas. Furthermore, all clauses are implicitly considered to be variable-disjoint.

If F is some formula and σ a substitution, then $F\sigma$ is an *instance* or *instantiation* of F. It is a *ground instance*, if it is variable-free. Otherwise it is a *partial instantiation*. It is a *proper instance*[1], if at least one variable is replaced by a non-variable term. A *variable renaming* on a formula F is an injective substitution mapping variables to variables. Two formulas (clauses, literals) K and L are *variants* of each other, if there is a variable renaming σ such that $L\sigma = K$. If F' is an instantiation of a formula F, then F is a *generalization* of F'. Given a set of formulas S, we say that $F \in S$ is a *most specific generalization* of F' with respect to S if F generalizes F' and there is no other formula $G \in S$ such that G generalizes F', F generalizes G, and F is not a variant of G.

For any literal L, $\overline{L}$ denotes its complement. By $\bot$ we denote both a distinguished constant and the substitution mapping all variables to this constant. We say that two literals L, L' are $\bot$-*complementary* if $L\bot = \overline{L'}\bot$. A set of clauses S is called *propositionally unsatisfiable* if and only if $S\bot$ is unsatisfiable.

2.2   The Disconnection Tableau Calculus

The *Disconnection Tableau Calculus* calculus has been developed by Billon [Bil96] and Letz and Stenz [LS01, Ste02]. In order to define development of a disconnection tableau, we need the notions of links, paths and tableaux:

A *literal occurrence* is a pair $\langle C, L \rangle$, where C is a clause and $L \in C$ a literal. If $\langle C, L \rangle$ and $\langle D, \overline{K} \rangle$ are two literal occurrences such that there is a most general unifier (mgu) σ of L and K, then the set $l = \{\langle C, L \rangle, \langle D, \overline{K} \rangle\}$ is called a *link* (between C and D). $C\sigma$ and $D\sigma$ are *linking instances* of C and D with respect to l.

A *path* P through a set of clauses (or occurrences of clauses) S is a set of literal occurrences such that P contains exactly one literal occurrence $\langle C, L \rangle$ for every $C \in S$. A path P is $\bot$-*complementary* if it contains literal occurrences $\langle C, L \rangle$ and $\langle D, \overline{K} \rangle$ such that $L\bot = K\bot$, otherwise it is *open*.

A *disconnection tableau* (tableau, for short) is a (possibly infinite) downward tree with literal labels at all nodes except the root. Given a set of clauses S, a *tableau for* S is a tableau in which, for every tableau node N, the set of literals $C = L_1, ..., L_m$ at the immediate successor nodes $N_1, ..., N_m$ of N is an instance of a clause in S. Every N_i is associated with the clause C and the literal occurrence $\langle C, L_i \rangle$.

Construction of a tableau starts from an *initial path* P_S through the set S of input clauses. The initial path may be chosen arbitrarily, but remains fixed through the construction of the tableau.

A *branch* of a tableau T is any maximal sequence $B = N_1, N_2, ...$ of nodes in T such that N_1 is an immediate successor of the root node and any N_{i+1} is an immediate successor of N_i. With every branch B we associate a path P_B containing

---

[1] This definition is due to Ganzinger and Korovin [GK03] and may deviate from definitions in other areas.

the literal occurrences associated with the nodes in B. The union $P_S \cup P_B$ of the initial path and the path of a branch B is called a *tableau path* of B. Note that, in contrast to the initial path and the path of a branch, the tableau path may contain two literal occurrences for (two different occurrences of) the same clause.

To develop a tableau from the initial path and the empty tableau consisting of only the root node, we define the *linking rule*: Given an initial path P_S and a tableau branch B with literal occurrences $\langle C, L \rangle$ and $\langle D, \overline{K} \rangle$ in $P_S \cup P_B$, such that $l = \{ \langle C, L \rangle, \langle D, \overline{K} \rangle \}$ is a link with mgu σ, the branch B is expanded with a linking instance with respect to l of one of the two clauses, say with $C\sigma$, and then, below the node labeled with $L\sigma$, the branch is expanded with the other linking instance with respect to l, $D\sigma$.

As all clauses are variable-disjoint, *variant-freeness* is required in order to restrict proof development to inferences which introduce "new" instances: A disconnection tableau T is *variant-free*, if no node N with clause C in T has an ancestor node N' with clause D in T such that C and D are variants of each other. In practice, variant-freeness is assured by two restrictions: First, a link that has already been used on the current branch may not be used again. Secondly, when a linking step is performed, variants of clauses which are already on the branch are not added to the tableau. This can result in linking steps where only one of the linking instances is added. By definition, variant-freeness does not extend to the initial path, i.e., variants of input clauses can be added to the tableau, and may be needed for completeness. This may lead to tableau paths with two literal occurrences for the same clause, in which case the occurrence from the initial path is redundant, as we have shown in Jacobs [Jac04].

Next, we define when tableau construction will terminate: A tableau branch B is *closed*, if P_B is $\bot$-complementary; if not, it is called *open*. A tableau is *closed* if it has no open branches. Similarly to the notion of variant-freeness, closure does not extend to the initial path, i.e., literals on the initial path may not be used to close a tableau branch. An exception can be made for unit clauses on the initial path, as adding those to the tableau would result in only one branch which would directly be closed. A branch B in a (possibly infinite) tableau T is called *saturated*, if B is open and there is no link on B which produces at least one linking instance which is not a variant of any clause on B. A tableau is *saturated* if either all its branches are closed or it has a saturated branch. A saturated branch represents a model for the set of input clauses of the tableau.

With these definitions, we have a sound and functional calculus. Starting from the initial path, the linking rule develops our tableau, restricted by variant-freeness. The disconnection calculus terminates if we can close all branches of the tableau, thus proving unsatisfiability of the input clauses, or if at least one branch can be saturated in finite time, thereby proving that the set of input clauses is satisfiable. If the choice of linking steps is fair, i.e., if all infinite branches are saturated, termination is guaranteed for every unsatisfiable input set.[2]

---

[2] For examples of open, closed and saturated tableaux, see Fig. 1, 2 and 3, respectively.

2.3   Resolution-Based Instance Generation

The *Inst-Gen* calculus is due to Ganzinger and Korovin [GK03]. It uses the following inference rule:

$$\frac{C \vee L \qquad D \vee \overline{K}}{(C \vee L)\sigma \quad (D \vee \overline{K})\sigma},$$

where σ is the mgu of K and L and a proper instantiator of either $\overline{K}$ or L.

For a set of clauses saturated under Inst-Gen, the satisfiability test can be reduced to the propositional case. As saturation may take infinitely long, however, satisfiability testing cannot be postponed until saturation is reached.

There is a formal notion of redundancy for Inst-Gen, which is, however, out of the scope of this work. The only clauses we will consider as *redundant* are variants of clauses which are already present. An inference is redundant if it only produces such variants.

SInst-Gen is an extension of Inst-Gen, which uses *semantic selection* in order to restrict the search space: Let S be a set of clauses such that $S\perp$ is satisfiable. Let $I_\perp$ be a model of $S\perp$. We define the *satisfiers* of a clause to be the set $sat_\perp(C) = \{L \in C \mid I_\perp \models L\perp\}$. Now consider *selection functions* on clauses (modulo renaming), which select for every clause in S one of its satisfiers. Instance generation, based on a selection function sel, is defined as follows:

$$\frac{C \vee L \qquad D \vee \overline{K}}{(C \vee L)\sigma \quad (D \vee \overline{K})\sigma},$$

where σ is the mgu of K and L and both $\overline{K}$ and L are selected by sel.[3]

A selection function can be considered to represent a model for the grounded set of clauses $S\perp$. When trying to extend it to a model of S, complementary unifiable literals represent inconsistencies in the extension. Every inference of SInst-Gen resolves such an inconsistency.

In order to allow hyper-inferences, more than one satisfier can be selected. If for every selected literal in a clause, a complementary unifiable selected literal in another clause is found, then a hyper-inference produces all instances that the individual inferences between these literals would produce.

In order to ensure that unsatisfiable input sets are saturated within finite time, the choice of inferences must be *fair*. An informal idea of fairness is that any inference which is available infinitely often must either be taken or become redundant by application of other inferences at some time.

Propositional satisfiability of the generated set of clauses S' is tested after every inference step by searching for a model of $S'\perp$. If this does not exist, unsatisfiability of the input set S has been proved. If it does, another inference follows, until either no model can be found or no inferences are possible. In the latter case, the model of $S'\perp$ can be extended to a model of S without conflicts,

---

[3] The second condition of the Inst-Gen rule is implied by the use of semantic selection.

i.e., satisfiability of S has been shown. If the inferences are chosen in a fair manner, SInst-Gen is refutationally complete.

The primal partial instantiation (PPI) method [HRCR02][4] is equivalent to the basic case of SInst-Gen without redundancy elimination and hyper-inferences, except for the saturation strategy: PPI uses a counter which specifies the maximal term-depth of unified literals in an inference step. The counter is not reset to 0 after selection of literals is changed, and this may lead to a behaviour that is not fair with respect to the definition by Ganzinger and Korovin [Jac04]. As PPI is complete nonetheless, this may be an indication that the fairness condition of SInst-Gen is stricter than actually necessary.

3   Comparing Refutation Proofs

To compare different reasoning methods, several patterns of comparison can be used. Assuming that the objects of investigation are abstract calculi, rather than concrete implementations, one can analyze for which inputs proofs are found, how long the proofs are, or which formulas are derived during the proof. Which of these choices are appropriate for a comparison of DCC and SInst-Gen?

Both calculi are refutationally complete, and so they find (refutation) proofs for exactly the same input sets. (They may differ, however, in their ability to prove the consistency of satisfiable input sets. We will discuss this case later.)

DCC checks propositional satisfiability internally, whereas SInst-Gen uses an external program to solve this (NP-complete) subproblem. Since the number of inference steps of the external SAT solver is unknown, a meaningful comparison of the length of proofs is impossible.

The only remaining choice is to compare the internal structure of the proofs, or more precisely, the sets of clauses that are generated during the proof.

Definition 1. *A proof of method A simulates a given proof of method B if the instances generated by the simulating A-proof are a subset of the instances generated in the original B-proof.*

Thus, if method A can simulate all proofs of method B, B can be seen as a special case (or refinement) of A. With respect to this definition, Inst-Gen is the most general of the instance generation methods we have introduced, as it can simulate DCC, PPI and SInst-Gen proofs. That is, all of these calculi are refinements of Inst-Gen. A method that can simulate any refutation proof of another method is more general, but usually not better in the sense that it finds more proofs with limited resources (time, space). On the contrary, a strictly more general method will usually have a larger search space.

The definition of simulation is also motivated by the following two lemmas:

Lemma 2. *Let S be an unsatisfiable set of clauses, let T be a closed disconnection tableau for S. Then the set S' of all instances on the tableau is propositionally unsatisfiable.*

---

[4] The calculus originally presented by Hooker et al. is unsound, but can be corrected (see Jacobs [Jac04]).

Proof. Suppose S' was not propositionally unsatisfiable. Then there must be an open path through $S'\bot$. Using the literal occurrences in this path, we can identify a branch in T which selects the same literals. This branch cannot be closed, contradicting the assumption that T is closed.

The lemma implies that an SInst-Gen proof will terminate as soon as all instances from a closed disconnection tableau have been generated. Thus, we know that, if generation of these instances is possible, the requirements for our notion of simulation will be fulfilled by SInst-Gen. The next lemma does the same for the other direction of simulation:

Lemma 3. *Let S be an unsatisfiable set of clauses. If an SInst-Gen proof terminates after generating the set of instances $S' \supseteq S$, then there exists a closed tableau containing only instances from S'.*

Proof. We may simply add instances from S' to each branch of the tableau until it is closed. As there are no open paths through S', this means that every branch containing all clauses from S' must be closed.

Note, however, that neither Lemma 2 nor Lemma 3 guarantee that the required clauses can actually be generated using the construction rules of the simulating calculus. Moreover, in the disconnection calculus, even if the instances can be generated somewhere in the tableau, this does not necessarily mean that they can be generated where they are needed.

Thus, we need to compare instance generation and the guidance of proofs in both methods in order to see if simulation is possible. Instance generation itself, represented by the inference rules, is identical in both methods: the main premise is a pair of selected literals L_1, L_2 in clauses C_1, C_2 such that $L_1\sigma = \overline{L_2}\sigma$ for some mgu σ. If those are present, the instances $C_1\sigma$ and $C_2\sigma$ are generated. In both methods, one literal per clause is selected and variants of existing clauses will not be added again. Open branches correspond essentially to selection functions; if a branch is closed in DCC, then in SInst-Gen a selection function for the set of clauses is not allowed to select the same literals. The other direction, however, does not hold, because of the special role of the initial path in DCC: It is chosen and fixed at the beginning of the proof, and literals on it must not be used to close branches. There is no equivalent notion in SInst-Gen. However, adding variants of clauses from the initial path to the tableau effectively allows a different selection on and closure by input clauses, albeit only if there is a link which allows generation of the variant. The fact that literals on the initial path do not close branches may also lead to the generation of non-proper instances of input clauses. This cannot happen in SInst-Gen, as it requires two $\bot$-complementary literals to be selected. The main difference between the two approaches, however, is that instances generated in DCC will only be available on the current branch, while in SInst-Gen all instances are available at all times, i.e., regardless of the current selection.

3.1   From SInst-Gen to DCC

Theorem 4. *There exist refuting SInst-Gen proofs that cannot be simulated by any DCC proof.*[5]

Proof. The following is an SInst-Gen proof for an unsatisfiable set of clauses. We claim that it cannot be simulated by any disconnection tableau, i.e., DCC cannot finish the proof with the same set of instances as SInst-Gen, or a subset thereof. The reason for this is that the needed instances cannot be generated on all branches without generating additional instances.

The proof which we consider starts with the set of input clauses

$$\neg P(x,y) \vee \neg P(y,z) \vee P(x,z), \quad \underline{\neg P(x,y)} \vee P(fx,c),$$
$$\underline{P(a,b)}, \quad \underline{P(b,c)}, \quad \underline{\neg P(fa,c)},$$

where a selection function is given by the underlined literals. SInst-Gen goes on to produce the following instances in the given order:

$$\neg P(a,b) \vee \underline{\neg P(b,z)} \vee P(a,z), \quad \neg P(a,b) \vee \neg P(b,c) \vee \underline{P(a,c)},$$
$$\neg P(a,c) \vee P(fa,c)$$

One can easily see that addition of the last clause makes the set propositionally unsatisfiable. SInst-Gen terminates, indicating unsatisfiability. The DCC derivation in Figure 1 tries to reproduce all steps of this SInst-Gen proof.

Within the frame we have the input clauses. The initial path of the tableau selects the same literals as the initial selection function of the SInst-Gen proof does. Links between literals are marked by dashed lines with a circle numbering the link. On the tableau, the circle above the generated clause shows the number of the link which was used. One can confirm that links number 1, 5 and 6 are equivalent to the inference steps in SInst-Gen, as they produce the same instances. Branches which are closed are marked by a ∗. We see that there is one open branch. Our definition of simulation would allow us to close this branch by generating again any of the instances from the given proof. The available links on this branch are those with numbers 2 to 4 from the initial path, as well as two additional links to $P(a,z)$ which are not displayed in the figure. However, none of those links generate one of the needed instances. Therefore, simulation has failed with this strategy.

It might still be possible that there is a tableau simulating the given SInst-Gen proof which generates instances in a different order or uses a different initial path. We have shown that even in this case simulation of the given derivation is not possible. The proof is rather lengthy and can be found in Jacobs [Jac04].

As the SInst-Gen proof given above obeys the term-depth restriction of Hooker's PPI method, it shows also that there are PPI proofs that cannot be simulated by any DCC proof:

---

[5] It is not known whether this result still holds if one considers some of the extensions of DCC from Stenz [Ste02].

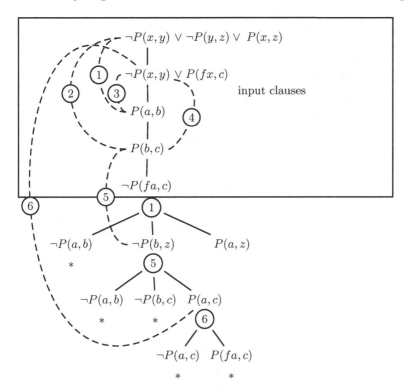

Fig. 1. DCC needs additional instances

Corollary 5. *There exist refuting PPI proofs that cannot be simulated by any DCC proof.*

3.2   From DCC to SInst-Gen

The fact that SInst-Gen proofs cannot always be simulated by DCC proofs is a consequence of the tree structure of DCC proofs. One might expect that in the other direction there is no such obstacle, but surprisingly, this is not the case.

Theorem 6. *There exist refuting DCC proofs that cannot be simulated by any SInst-Gen proof.*

Proof. Figure 2 shows a closed disconnection tableau for an unsatisfiable set of input clauses. We claim that this tableau cannot be simulated by SInst-Gen.

In order to verify this claim, let us consider all possible SInst-Gen proofs for the given set of clauses:

$$P(a, x, y, z) \vee Q(a, b, z), \quad \neg P(x, y, z, b) \vee R(x), \quad P(a, x, y, b) \vee S(y),$$
$$\neg Q(x, y, z) \vee \neg P(x, b, z, y), \quad \underline{\neg S(b)}, \quad \underline{\neg R(a)}$$

An underlined literal means that no other selection is possible in that clause. If there is no underlined literal, we consider all possible selections. There is

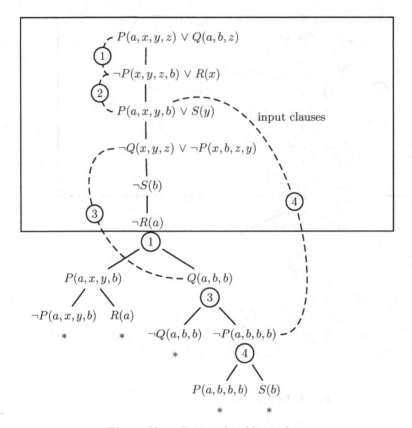

Fig. 2. SInst-Gen needs additional instances

an inconsistency between $P(a, x, y, z)$ in the first and $\neg P(x, y, z, b)$ in the second clause, which is equivalent to link number 1 in the tableau (i.e., produces the same instances). Also, there are inconsistencies between $\neg P(x, y, z, b)$ in the second and $P(a, x, y, b)$ in the third clause, and between $R(x)$ in the second clause and the unit clause $\neg R(a)$, which both produce only one of these instances, $\neg P(a, x, y, b) \vee R(a)$. There are four other possible inferences on this set of clauses, but all of them produce instances which are not on the given tableau.

Let us first consider the simulating proof which only produces $\neg P(a, x, y, b) \vee R(a)$ in the first step. Except the one mentioned first above, none of the possible inferences between the input clauses is usable for a simulation. Thus, we only need to consider inconsistencies referring to the newly generated clause. Moreover, as $R(a)$ is complementary to the given unit clause $\neg R(a)$, we only need to consider inconsistencies between $\neg P(a, x, y, b)$ and literals of the given clauses. Only one new inconsistency is introduced by the new clause, related to $P(a, x, y, z)$ in the first input clause. Furthermore, both of the inferences which are admissible for a simulating proof produce the same instance, which is $P(a, x, y, b) \vee Q(a, b, b)$.

Thus, after either taking the first inference step mentioned above, or one of the other two possible inference steps and one of the two admissible subsequent inferences, we arrive at the following set of instances:

$$P(a, x, y, z) \lor Q(a, b, z), \quad \neg P(x, y, z, b) \lor R(x), \quad P(a, x, y, b) \lor \underline{S(y)},$$
$$\neg Q(x, y, z) \lor \neg P(x, b, z, y), \quad \underline{\neg S(b)}, \underline{\neg R(a)}, \underline{\neg P(a, x, y, b)} \lor R(a),$$
$$P(a, x, y, b) \lor \underline{Q(a, b, b)}$$

All admissible inferences between input clauses have been carried out and selection is fixed on all of the generated instances. Thus, we have only one possible inference, which is between $Q(a, b, b)$ in the last and $\neg Q(x, y, z)$ in the fourth clause. This step is equivalent to linking step number 3 in the tableau, generating the instance $\neg Q(a, b, b) \lor \neg P(a, b, b, b)$.

Now, we have to select $\neg P(a, b, b, b)$ in the last instance, which only gives us one new inference, connected to $P(a, x, y, z)$ in the first clause. Note that the inference equivalent to linking step number 4 is not possible, as $P(a, x, y, b)$ is $\bot$-complementary to $\neg P(a, x, y, b)$, which has to be selected in the seventh clause. The new inference generates $P(a, b, b, b) \lor Q(a, b, b)$, which is not on the given tableau. At this point, there is no inference which does not violate our simulation property, which means that simulation has failed.

This result also holds if hyper-inferences are allowed in SInst-Gen, as one can easily verify that the possible hyper-inferences either produce instances which violate simulation, or only produce the same instances as the standard inferences.

3.3   Weak Simulation

We have shown that simulation of refutational SInst-Gen (or PPI) proofs by DCC, or vice versa, fails in general. We can get positive simulation results, however, if the definitions of the calculi are slightly changed and if the definition of simulation is modified in the following way:

Definition 7. *A proof by method A simulates a given proof by method B weakly, if every instance generated by the simulating A-proof is a generalization of an instance generated in the original B-proof.*

Recently, Letz and Stenz [LS04] introduced a more general version of the linking rule, which uses instead of standard unification a special form of a unifier: Given two literals L and L', a substitution σ is called a $\bot$-unifier of L and L' if $L\sigma\bot = L'\sigma\bot$. σ is called a most general $\bot$-unifier of L and L' if it is more general than every $\bot$-unifier of L and L'. E.g., $\bot$-unification of $P(a, x, y)$ and $P(x', y', y')$ results in $P(a, x, y)$ and $P(a, y', y')$.

It has already been stated [GK03] that the degree of instantiation of an SInst-Gen inference can be chosen flexibly, as long as at least one variable is instantiated properly. As for properly instantiated variables there is no difference between standard and $\bot$-unification, we can safely assume that we can use $\bot$-unification also for SInst-Gen.

Theorem 8. *Let S be an unsatisfiable set of clauses, S' a set of instances of clauses from S such that $(S \cup S')\bot$ is unsatisfiable. If M is a finite subset of $S \cup S'$ such that $M\bot$ is unsatisfiable, then SInst-Gen with $\bot$-unification can prove unsatisfiability of S by only generating generalizations of clauses from M.*

Proof. As the set of all generalizations of clauses in M is finite (up to renaming), it is not necessary to consider infinite derivations using such clauses. So the only way how the construction of an SInst-Gen proof using generalizations of clauses in M can fail is that at some point of the proof, SInst-Gen has generated a set of clauses M_1 such that $M_1\bot$ is satisfiable and every possible inference on M_1 results in generation of an instance which is not a generalization of any clause in M. As $M_1\bot$ is satisfiable, we can choose a selection function sel on M_1. Every clause $C\sigma \in M$ has at least one most specific generalization C with respect to M_1. Suppose we choose one of these most specific generalizations for every $C\sigma \in M$ and select the literal $L\sigma \in C\sigma$ if L is selected by sel in C. As $M\bot$ is unsatisfiable, we must have selected at least one pair of $\bot$-complementary literals, say $L_1\sigma_1 \in C_1\sigma_1$ and $L_2\sigma_2 \in C_2\sigma_2$.

Thus, in M_1, sel selects $L_1 \in C_1$ and $L_2 \in C_2$. As $L_1\sigma_1$ and $L_2\sigma_2$ are $\bot$-complementary, we can state that L_1 and L_2 are complementary unifiable, say by τ. As they are not $\bot$-complementary, there is an SInst-Gen inference with $\bot$-unification between them. The substitution used in this inference is a most general $\bot$-unifier, therefore the clauses produced by this inference will also be generalizations of clauses from M. Note that this would not hold for SInst-Gen without $\bot$-unification. The inference generates at least one proper instance with respect to the premises, say $C_1\tau$ is a proper instance of C_1. There cannot be a variant of $C_1\tau$ in M_1, as C_1 was chosen to be a most specific generalization of $C_1\sigma$. Thus, we have produced a new generalization of a clause from M, contradicting our assumption that no such inference is possible. $\qquad\square$

Corollary 9. *For every refuting DCC proof (with or without $\bot$-unification) there exists a weakly simulating SInst-Gen proof (with $\bot$-unification).*

Proof. For a DCC proof that shows the unsatisfiability of a set of clauses S, let S' be the finite set of all clauses on the DCC tableau. Since $(S \cup S')\bot$ is unsatisfiable, we can apply the previous theorem. $\qquad\square$

Theorem 10. *Let S be an unsatisfiable set of clauses, S' a set of instances of clauses from S such that $(S \cup S')\bot$ is unsatisfiable. If M is a finite subset of $S \cup S'$ such that $M\bot$ is unsatisfiable, then DCC with $\bot$-unification can prove unsatisfiability of S by only generating generalizations of clauses from M.*

Proof. Again, the fact that the set of all generalizations of clauses in M is finite ensures that unfair derivations need not be considered. Suppose a DCC tableau for S has an open branch B that cannot be extended without generating instances that are not generalizations of clauses in M. Every clause in $C\sigma \in M$ has at least one most specific generalization C with respect to the clauses on the tableau path $P_S \cup P_B$. We select in each $C\sigma \in M$ the literal corresponding to the literal of C on the tableau path. As $M\bot$ is unsatisfiable, $\bot$-complementary literals $L\sigma \in C\sigma$ and $K\tau \in D\tau$ are selected. Thus the literals L and K of the most specific generalizations C and D are on the tableau path. The DCC inference with $\bot$-unification from C and D uses a most general $\bot$-unifier, so the instances produced by this inference are again generalizations of $C\sigma$ and $D\tau$; moreover, at least one of them is a proper instance of a premise. This instance cannot be a variant of a clause on the tableau path, since C and D were chosen as most specific generalizations of $C\sigma$ and $D\tau$. Therefore, the tableau can be extended with a generalization of a clause in M, contradicting our assumption.

Corollary 11. *For every refuting SInst-Gen or PPI proof (with or without $\bot$-unification) there exists a weakly simulating DCC proof (with $\bot$-unification).*

4   Comparing Consistency Proofs

4.1   From SInst-Gen to DCC

The case of consistency proofs differs in several aspects from the case of refuting proofs. First, it is clear that no theorem proving method is guaranteed to terminate for satisfiable sets of clauses. Second, in order to declare a set of clauses unsatisfiable, SInst-Gen must check *all* propositional interpretations for the given set of clauses and show that none of them is a model. In contrast to this, termination on a satisfiable set of clauses only depends on *one* interpretation which can be extended to a first-order model of the input clauses. Essentially, the same holds for DCC. Simulation can therefore be based on the final set of clauses and selection function. In the following, we will show that this fact enables us to simulate any SInst-Gen proof which terminates on satisfiable input by DCC.

Theorem 12. *Let S be a satisfiable set of input clauses, $S \cup S'$ a finite set of clauses saturated under SInst-Gen proof with selection function* sel. *Then the given consistency proof can be simulated by DCC.*

Proof. We prove our claim by induction on the number of proof steps of the simulating proof, where a proof step consists of both the selection of literals and the generation of instances. Simulation is not based on the steps of the given proof, but only on the final set of clauses $S \cup S'$ and the selection function sel. We will show that every step of the simulating proof produces only instances from $S \cup S'$, while the tableau path we follow is always equivalent to sel for the clauses which are on this path.

First, we choose the initial path P_S of our simulating proof to select the same literals as sel on S. Then, every link on P_S can only produce instances from $S \cup S'$, as otherwise there would be an inconsistency of sel. Thus, we may carry out an arbitrary link from those which are available on P_S, resulting in a set of clauses which is a subset of $S \cup S'$.

Now, suppose an arbitrary number of steps has been carried out by our simulating proof, always following a tableau path which selects the same literals as sel for clauses on the path and only generating instances from $S \cup S'$. We can extend the path from the last step such that it selects the same literals as sel on the new instances. As sel does not select $\perp$-complementary literals, the path must also be open. By the same argument as above, we may again carry out any possible linking step, producing only instances from $S \cup S'$.

In this way, we carry out all possible linking steps. As we choose our tableau path such that only instances from $S \cup S'$ are produced, the process must terminate after a finite number of steps. As termination of the process means that we have saturated the current path, we can state that we have reached our goal to simulate the given proof.

It may happen that a linking step adds two instances, but sel is such that we cannot select a tableau path which is equivalent to the selection function *and* considers both of these clauses. This happens if both of the selected literals are not linked literals of the linking step that produces them. In this case, our tableau path can only consider one of the produced clauses. However, as we want to saturate the branch we are following, missing clauses and links are not a problem, but a benefit in this case.

4.2   From DCC to SInst-Gen

We have shown that every SInst-Gen consistency proof can be simulated by DCC. The reverse, however, does not hold: Figure 3 shows a saturated disconnection tableau for a satisfiable set of input clauses. Saturation of the tableau is achieved by generating a single non-proper instance of an input clause.

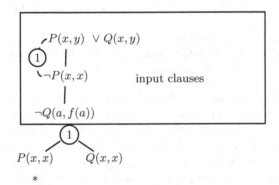

Fig. 3. Saturated disconnection tableau

An SInst-Gen proof for the given set of clauses would have to select $Q(x, y)$ in the first input clause, as $\perp$-complementary literals must not be selected. Thus, we have an inconsistency between this literal and $\neg Q(a, f(a))$, which produces an instance we do not have on the tableau. Without producing this instance however, satisfiability of the input set cannot be detected by SInst-Gen.

If ⊥-unification is used for both DCC and SInst-Gen, examples like this one are obsolete, as only proper instances or variants of input clauses will be added to the tableau. However, even in this case simulation is in general not possible, as we have demonstrated by a more complicated counterexample [Jac04].

As for refutation proofs, this result still holds when hyper-inferences are allowed in the simulating SInst-Gen proof.

5   Conclusions

We have compared the four instance generation methods DCC, Inst-Gen, SInst-Gen, and PPI. Inst-Gen, which does not make any attempt to restrict the search space, is obviously the most general of these calculi: any DCC, SInst-Gen, or PPI proof can be simulated by an Inst-Gen proof.

PPI is essentially a special case of SInst-Gen, except for its term-depth-based saturation strategy, which ensures completeness of the calculus but is not subsumed by the fairness criterion of SInst-Gen. It would be interesting to search for a more relaxed notion of fairness which allows all possible strategies of both PPI and SInst-Gen.

For DCC and SInst-Gen, we have demonstrated that, for refutation proofs, simulation between (the basic versions of) the methods is in general not possible. This implies in particular that neither of the methods can be considered as a special case of the other. In case of consistency proofs, we have shown that SInst-Gen proofs that terminate on satisfiable input can always be simulated by DCC, while this does not hold for the other direction. All of these results still hold when SInst-Gen is allowed to use hyper-inferences.

We face a very different situation when we consider weak simulation, so that we can not only use clauses from the given proof but also generalizations thereof. We have shown that DCC and SInst-Gen with ⊥-unification can weakly simulate each other; in fact we conjecture that DCC and SInst-Gen with ⊥-unification can weakly simulate any instance-based calculus.

For DCC and SInst-Gen there are various refinements which are out of the scope of this work. It is not clear how our results would translate to the refined calculi, e.g. DCC with lemma generation and subsumption or SInst-Gen with redundancy elimination. Additionally, for both methods extensions to equality reasoning are available. A comparison of the different approaches to these refinements and extensions might give more useful insights on the relation between DCC and SInst-Gen.

References

[Bau00]   Peter Baumgartner. FDPLL – A First-Order Davis-Putnam-Logeman-Loveland Procedure. In *CADE-17, LNAI 1831*, pages 200–219. Springer, 2000.

[Bil96]   Jean-Paul Billon. The disconnection method: a confluent integration of unification in the analytic framework. In *Tableaux 1996, LNAI 1071*, pages 110–126. Springer, 1996.

[BT03]   Peter Baumgartner and Cesare Tinelli. The model evolution calculus. In
           CADE-19, LNAI 2741, pages 350–364. Springer, 2003.
[DLL62]   Martin Davis, George Logemann, and Donald Loveland. A machine pro-
           gram for theorem proving. *Communications ACM*, 5:201–215, 1962.
[GK03]   Harald Ganzinger and Konstantin Korovin.   New directions in
           instantiation-based theorem proving. In *LICS-03*, pages 55–64, Ottawa,
           Canada, 2003. IEEE.
[HRCR02]   John N. Hooker, Gabriela Rago, Vijay Chandru, and Anjul Rivastava.
           Partial instantiation methods for inference in first-order logic. *Journal of
           Automated Reasoning 28*, pages 371–396, 2002.
[Jac04]   Swen Jacobs.   Instance generation methods for automated reason-
           ing. Diploma Thesis, Universität des Saarlandes, 2004. Available at
           http://www.mpi-sb.mpg.de/~sjacobs/publications.
[LP92]   Shie-Jue Lee and David A. Plaisted. Eliminating duplication with the
           hyper-linking strategy. *Journal of Automated Reasoning*, 9:25–42, 1992.
[LS01]   Reinhold Letz and Gernot Stenz. Proof and model generation with discon-
           nection tableaux. In *LPAR 2001, LNAI 2250*, pages 142–156. Springer,
           2001.
[LS04]   Reinhold Letz and Gernot Stenz. Generalised handling of variables in
           disconnection tableaux. In *IJCAR 2004, LNCS 3097*, pages 289–306.
           Springer, 2004.
[Pla94]   David A. Plaisted. Ordered semantic hyper-linking. Technical Report
           MPI-I-94-235, Max-Planck-Institut für Informatik, Saarbrücken, Germany,
           1994.
[Ste02]   Gernot Stenz. *The Disconnection Calculus*. PhD thesis, TU München,
           2002.

An Order-Sorted Quantified Modal Logic for Meta-ontology

Ken Kaneiwa[1] and Riichiro Mizoguchi[2]

[1] National Institute of Informatics, Japan
kaneiwa@nii.ac.jp
[2] Osaka University, Japan
miz@ei.sanken.osaka-u.ac.jp

Abstract. The notions of meta-ontology enhance the ability to process knowledge in information systems; in particular, *ontological property classification* deals with the kinds of properties in taxonomic knowledge based on a philosophical analysis. The goal of this paper is to devise a reasoning mechanism to check the *ontological and logical consistency* of knowledge bases, which is important for reasoning services on taxonomic knowledge. We first consider an ontological property classification that is extended to capture individual existence and time and situation dependencies. To incorporate the notion into logical reasoning, we formalize an order-sorted modal logic that involves rigidity, sortality, and three kinds of modal operators (temporal/situational/any world). The sorted expressions and modalities establish axioms with respect to properties, implying the truth of properties in different kinds of possible worlds and in varying domains in Kripke semantics. We provide a prefixed tableau calculus to test the satisfiability of such sorted modal formulas, which validates the ontological axioms of properties.

1 Introduction

Formal ontology deals with the kinds of entities in the real world, such as properties, events, processes, objects, and parts [17]. In this field of research, Guarino and Welty [11] have defined meaningful property classifications as a meta-ontology where properties of individuals are rigorously classified into sortal/non-sortal, rigid/anti-rigid/non-rigid, etc., by a philosophical analysis. The notions of meta-ontology describe the general features of knowledge, which can be applied to enhance knowledge processing in information systems.

On the other hand, order-sorted logic has been recognized as a useful tool for providing logical reasoning systems on taxonomic knowledge [3,18,5,15,12]. Kaneiwa and Mizoguchi [13] noticed that the ontological property classification [19] fits the formalization of order-sorted logic, and they refined the sorted logic by means of the ontological notion of rigidity and sortality. By using Kripke semantics, *rigid* properties are true in any possible world and *sortal* properties consist of individuals whose parts do not have the same properties. However, they did not cover *individual existence* and *temporal and situational*

B. Beckert (Ed): TABLEAUX 2005, LNAI 3702, pp. 169–184, 2005.

aspects of properties for realistic reasoning services on taxonomic knowledge (only certain temporal aspects were axiomatized by modal and tense operators in [10]).

The first aim of this paper is to present an ontological property classification extended to include the notions of individual existence and time/situation/time-situation dependencies, which are based on the following:

- Entities of properties cannot exist forever in ontological analysis, i.e., every (physical) object will cease to exist at some time.
- One property (e.g., baby) holds depending only on time and is situationally stable, while another (e.g., weapon) holds depending on its use situation and is temporally unstable. For example, a knife can be a weapon in a situation, but it is usually employed as a tool for eating.

These ideas lead to *varying domains, times*, and *situations* in possible worlds, which inspire us to define rigidity with individual existence and to further classify anti-rigid properties. In order to model them, we distinguish times and situations from other possible worlds and include varying domains in Kripke semantics.

In order to establish the extensions, we make use of the techniques of quantified modal and temporal logics. Although the logics are usually formalized in *constant domains*, several quantified modal logics address philosophical problems such as *varying domains, non-rigid terms*, and local terms. Garson [8] discussed different systems for variants of quantified modal logics. Fitting and Mendelsohn [4] treated rigidity of terms and constant/varying domains by means of a tableau calculus and a predicate abstraction. Meyer and Cerrito [14] proposed a prefixed tableau calculus for all the variants of quantified modal logics with respect to cumulative/varying domains, rigid/non-rigid terms, and local/non-local terms.

Our second aim is to propose an order-sorted modal logic for capturing the extended property classification. This logic requires a combination of order-sorted logic, quantified modal logic, and temporal logic due to their respective features:

1. Order-sorted logic has the advantage that sorted terms and formulas adequately represent properties based on the ontological property classification.
2. Meyer and Cerrito's quantified modal logic provides us with a prefixed tableau calculus for supporting varying domains and non-rigid terms, which can be extended to order-sorted terms/formulas and multi-modal operators.
3. Temporal logic contains temporal representation; however, the standard reasoning systems are propositional [16,9] or the first-order versions [7,6,2] adopt constant domains since they are not easily extended to the first-order temporal logic with varying domains, as discussed in [8].

The proposed logic provides a logical reasoning system for checking the ontological and logical consistency of knowledge bases with respect to properties. Unary predicates and sorts categorized on the basis of rigidity, sortality, and dependencies can be used to represent properties that are appropriately interpreted in modalities and varying domains. We redesign a prefixed tableau calculus for testing the satisfiability of sorted formulas comprising three kinds of modal operators (temporal/situational/any world). This calculus is a variant of Meyer and Cerrito's prefixed tableau calculus that is extended by denoting the kinds of possible worlds in prefixed formulas and by adjusting and supplementing the tableau rules for sorted expressions and multi-modalities supporting individual existence.

2 Property Classification in Semantics

We consider the meaning of properties on the basis of ontological analysis. We begin by characterizing the rigidity of properties in Kripke semantics where a set W of possible worlds w_i is introduced and properties are interpreted differently in each world. Let U be the set of individuals (i.e., the universe), and let $I = \{I_w \mid w \in W\}$ be the set of interpretation functions I_w for all possible worlds $w \in W$. Sortal properties are called *sorts*. We specify that every sort s is interpreted by $I_w(s) \subseteq U$ for each world w, and a subsort relation $s_i < s_j$ is interpreted by $I_w(s_i) \subseteq I_w(s_j)$.

Unlike anti-rigid sorts, substantial sorts (called *types*), constants, and functions are rigid and yield the following semantic constraints. Let τ be a type, c be a constant, and f be a function. For all possible worlds $w_i, w_j \in W$, $I_{w_i}(\tau) = I_{w_j}(\tau)$, $I_{w_i}(c) = I_{w_j}(c)$, and $I_{w_i}(f) = I_{w_j}(f)$ hold. In addition, for each world $w \in W$, every sort s and its sort predicate p_s (as the unary predicate denoted by a sort) are identical in the interpretation and are defined by $I_w(s) = I_w(p_s)$. Standard order-sorted logic does not include the intensional semantics that reflects the rigidity of sorts and sort predicates.

The semantics can be further sophisticated in terms of dependencies of time and situation. As special possible worlds, we exploit *time* and *situation* in order to capture distinctions among anti-rigid sorts (as non-substantial properties). We introduce the set W_{Tim} of times tm_i and the set W_{Sit} of situations st_i where $W_{\mathrm{Tim}} \cup W_{\mathrm{Sit}} \subseteq W$. They do not violate rigidity in the interpretation if types, constants, and functions preserve their rigidity in any time and situation. We show dependencies on time and situation that classify anti-rigid sorts as follows:

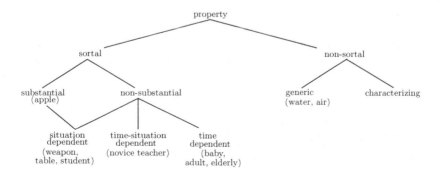

Fig. 1. Ontological property classification

time dependent: baby, child, youth, adult, elderly
situation dependent: weapon, table, student
time-situation dependent: novice teacher

In Fig.1, these are added to the property classification. The time dependency implies that the truth of a property depends only on time or the meaning of a property is decided essentially by time. For example, the property *baby* is time dependent, so that its entities have the denoting property in a particular time or period.

The situation dependency indicates that the truth of a property is dependent on situation but not on time. Moreover, the situation dependency obtained from extending types (such as *weapon*, *table*, but not *student*) involves a complex idea as mentioned below. We can regard the property *weapon* as a substantial sort (type); however, it is anti-rigid as situation-dependent if it is used as a role expressed by the sort predicate p_{weapon}. E.g., the properties *weapon* and *table* have two kinds of entities: (i) guns and dining tables that innately possess the property of *weapon* and *table* and (ii) knives and boxes that play the roles of *weapon* and *table*. In the latter case, they are not really the aforementioned artifacts and are just referred to as *weapon* and *table*. Thus, knives play the role of a *weapon* only when they are used to attack or kill someone. In the language of order-sorted logic, the former case is an instantiation of a sort (e.g., c_{weapon}), and the latter case is an entity characterized by a sort predicate (e.g., $p_{weapon}(c)$). This consideration disproves the fact that sorts and their sort predicates are interpreted identically in semantics.

Specification 1 *Let τ be a type. If the type predicate p_τ is situation dependent, then $I_w(\tau) \subsetneq I_w(p_\tau)$.*

For example, $gun1 \in I_w(weapon) \cap I_w(p_{weapon})$, whereas $knife1 \notin I_w(weapon)$ and $knife1 \in I_w(p_{weapon})$.

The time-situation dependency is defined such that the truth of a property depends on both time and situation. For example, the property *novice_teacher* is time-situation dependent. Since each novice teacher will become a veteran teacher after a number of years, the property holds only at a particular time under the situation. In other words, the time-situation dependency implies time dependency under a situation, while the situation dependency bears no relationship to time.

We define those dependencies semantically in possible worlds. The basic notion of interpreting time dependency is that for every time-dependent property p and for every individual $d \in U$, if $d \in I_{tm}(p)$ with $tm \in W_{\text{Tim}}$, then another time $tm_j \in W_{\text{Tim}}$ exists such that $d \notin I_{tm_j}(p)$. This is based on the assumption that the same entities (individuals) exist in all possible worlds (called constant domains). However, this assumption does not appear to be realistic since there may be different entities in each possible world. Let U_w be the set of individuals existing in a possible world w. This enables us to consider the case where U_{w_1} and U_{w_2} do not coincide for some possible words $w_1, w_2 \in W$. Consider the following example: every entity of the property *person* ceases to exist at some time, because no person can live forever. Therefore, we redefine the rigidity of sorts, constants, and functions by supporting individual existence:

Specification 2 (Rigidity) *For all possible worlds $w_i, w_j \in W$, $I_{w_i}(c) = I_{w_j}(c)$ and $I_{w_i}(f) = I_{w_j}(f)$. Let $w \in W$, let $d \in U_w$, and let $R' \subseteq W \times W$ be an accessibility relation. For every type τ, if $d \in I_w(\tau)$ and $(w, w') \in R'$, then $d \in U_{w'}$ implies $d \in I_{w'}(\tau)$. For every anti-rigid sort σ, if $d \in I_w(\sigma)$, then there exists $w_j \in W$ with $(w, w_j) \in R'$ such that $d \notin I_{w_j}(\sigma)$ with $d \in U_{w_j}$.*

By considering the existence of individuals in each world, we specify time/ situation/time-situation dependencies where $R_{\text{Tim}} \subseteq W \times W_{\text{Tim}}$ and $R_{\text{Sit}} \subseteq W \times W_{\text{Sit}}$ are accessibility relations from worlds to times and situations, respectively.

Specification 3 (Time Dependency) *Let p be a time-dependent predicate and let $w \in W$.*

1. *(temporally unstable) for all $(w, tm) \in R_{\text{Tim}}$ and for all $d \in U_{tm}$, if $d \in I_{tm}(p)$, then there exists $tm_j \in W_{\text{Tim}}$ with $(tm, tm_j) \in R_{\text{Tim}}$ such that $d \notin I_{tm_j}(p)$ with $d \in U_{tm_j}$.*
2. *(situationally stable under time) if $d \in I_{tm}(p)$ with $(w, tm) \in R_{\text{Tim}}$, then for all situations $st \in W_{\text{Sit}}$ with $(tm, st) \in R_{\text{Sit}}$, $d \in U_{st}$ implies $d \in I_{st}(p)$.*

The *temporally unstable* implies that for every time tm accessible from a world w, if an individual d has the property p at tm, we can find a time tm_j accessible from tm where it does not have the property. The *situationally stable under time* defines the fact that for every time tm accessible from a world w, if an individual d has the property p at the time tm, then it has this property in any situation st accessible from the time tm as long as the individual exists. Similar to it, situation dependency can be defined as follows:

Specification 4 (Situation Dependency) *Let p be a situation-dependent predicate and let $w \in W$.*

1. *(situationally unstable) for all $(w, st) \in R_{\text{Sit}}$ and for all $d \in U_{st}$, if $d \in I_{st}(p)$, then there exists $st_j \in W_{\text{Sit}}$ with $(st, st_j) \in R_{\text{Sit}}$ such that $d \notin I_{st_j}(p)$ with $d \in U_{st_j}$.*
2. *(temporally stable under situation) if $d \in I_{st}(p)$ with $(w, st) \in R_{\text{Sit}}$, then for all times $tm \in W_{\text{Tim}}$ with $(st, tm) \in R_{\text{Tim}}$, $d \in U_{tm}$ implies $d \in I_{tm}(p)$.*

Moreover, we define time-situation dependency as follows:

Specification 5 (Time-Situation Dependency) *Let p be a time-situation dependent predicate and let $w \in W$.*

1. *(situationally unstable) the same as in the above.*
2. *(temporally unstable under situation) if $d \in I_{st}(p)$ with $(w, st) \in R_{\text{Sit}}$, then there are some $tm_i, tm_j \in W_{\text{Tim}}$ with $(st, tm_i), (st, tm_j) \in R_{\text{Tim}}$ such that $d \in I_{tm_i}(p)$ and $d \notin I_{tm_j}(p)$ with $d \in U_{tm_i} \cap U_{tm_j}$.*

Besides the situational unstability, the *temporally unstable under situation* implies that for every situation st accessible from a world w, if an individual d has the property p in the situation st, then there are times tm_i, tm_j accessible from st such that it has the property p at tm_i, but not at tm_j.

By interpreting such dependencies in possible worlds, the semantics determines ontological differences among anti-rigid properties related to partial rigidity. That is, even if a property is anti-rigid, it may be rigid over a particular kind of possible worlds (e.g., time and situation), as a subset of W. For example, the property *baby* is time dependent, i.e., it is temporally unstable and situationally stable under time. When Bob is a baby, the time dependency derives the fact that Bob is a baby in any situation within the time. Formally, if $bob \in I_{tm}(baby)$, then for any situation st accessible from tm, $bob \in U_{st}$ implies $bob \in I_{st}(baby)$ (if Bob exists in st, he is a baby). It can be viewed as rigid over situations.

3   Order-Sorted Modal Logic

We define the syntax and semantics of an order-sorted modal logic. The alphabet of a sorted first-order modal language $\mathcal{L}$ with rigidity and sort predicates comprises the following symbols: a countable set T of type symbols (including the greatest type $\top$), a countable set S_A of anti-rigid sort symbols ($T \cap S_A = \emptyset$), a countable set C of constant symbols, a countable set F_n of n-ary function symbols, a countable set P_n of n-ary predicate symbols (including the existential predicate symbol E; the set $P_{T \cup S_A}$ of sort predicate symbols $\{p_s \mid s \in T \cup S_A\}$; and a countable set P_{non} of non-sortal predicate symbols), the connectives $\wedge, \vee, \rightarrow, \neg$, the modal operators $\square_i, \diamond_i, \blacksquare, \blacklozenge$, and the auxiliary symbols $(,)$.

We generally refer to type symbols τ or anti-rigid sort symbols σ as *sort symbols* s. $T \cup S_A$ is the set of sort symbols. V_s denotes an infinite set of variables x_s of sort s. We abbreviate variables $x_\top$ of sort $\top$ as x. The set of variables of all sorts is denoted by $V = \bigcup_{s \in T \cup S_A} V_s$. The unary predicates $p_s \in P_1$ indexed by the sorts s (called *sort predicates*) are introduced for all sorts $s \in T \cup S_A$. In particular, the predicate p_τ indexed by a type τ is called a *type predicate*, and the predicate p_σ indexed by an anti-rigid sort σ is called an *anti-rigid sort predicate*. Hereafter, we assume that the language $\mathcal{L}$ contains all the sort predicates in $P_{T \cup S_A}$. Types can be situation dependent (no type has time/time-situation dependencies), while anti-rigid sorts can be either time, situation, or time-situation dependent (e.g., the type *weapon* is situation dependent, and the anti-rigid sort *adult* is time dependent). Each sort predicate p_s has the same dependency as its sort s.

Definition 1. *A signature of a sorted first-order modal language $\mathcal{L}$ with rigidity and sort predicates (called sorted signature) is a tuple $\Sigma = (T, S_A, \Omega, \leq)$ such that (i) $(T \cup S_A, \leq)$ is a partially ordered set of sorts where $T \cup S_A$ is the union of the set of type symbols and the set of anti-rigid sort symbols in $\mathcal{L}$ and each ordered pair $s_i \leq s_j$ is a subsort relation (implying that s_i is a subsort of s_j) fulfilling the following:*

- *every subsort of types is a sort ($s \leq \tau$) and every subsort of anti-rigid sorts is an anti-rigid sort ($\sigma \leq \sigma'$);*
- *every subsort of time dependent sorts is time or time-situation dependent; every subsort of situation dependent sorts is situation or time-situation dependent; and every subsort of time-situation dependent sorts is time-situation dependent,*

(ii) if $c \in C$, then there is a unique constant declaration $c\colon \to \tau \in \Omega$, (iii) if $f \in F_n$ ($n > 0$), then there is a unique function declaration $f\colon \tau_1 \times \cdots \times \tau_n \to \tau \in \Omega$, and (iv) if $p \in P_n$, then there is a unique predicate declaration $p\colon s_1 \times \cdots \times s_n \in \Omega$ (in particular, if $p_s \in P_{T \cup S_A}$, then there is a unique sort predicate declaration $p_s\colon \tau \in \Omega$ where $s \leq \tau$, and if $p \in P_{non}$, then $p\colon undef \in \Omega$).

A partially ordered set $(T \cup S_A, \leq)$ constructs a sort-hierarchy by suitably ordering different kinds of sorts. A subsort of anti-rigid sorts cannot be a type, a subsort of situation/time-situation dependent sorts cannot be time dependent, and a subsort of time/time-situation dependent sorts cannot be situation dependent. These conditions are guaranteed by the fact that each sort inherits (temporal and situational) unstability and anti-rigidity from its supersorts. For example, the sort *novice_teacher* must be situationally unstable (as time-situation dependent) if the supersort *teacher* is situation dependent.

In sorted signatures, the sorts of constants, functions, and predicates have to be declared by adhering to the rigidity in Specification 2, i.e., since every constant and function is rigid, their sorts have to be rigid. The sort declarations of constants c and functions f are denoted by the forms $c\colon \to \tau$ and $f\colon \tau_1 \times \cdots \times \tau_n \to \tau$ where types τ_i, τ are used to declare the sorts. On the other hand, the sort declarations of predicates are denoted by the form $p\colon s_1 \times \cdots \times s_n$ where types and anti-rigid sorts s_i can be used to declare the sorts.

Although the declarations of sort predicates are defined by the greatest sort $\top$ (i.e., $p_s\colon \top$) in [1], we reconsider it in this paper. For each anti-rigid sort σ, there is a basic type τ to be an entity of σ, i.e., every entity of the sort σ must be an entity of the type τ. For example, the anti-rigid sorts *student* and *husband* respectively have the basic types *person* and *male*, defined as the firstness of being able to play the roles. Hence, the declaration of each sort predicate p_s is defined by a type τ such that $s \leq \tau$ (i.e., $p_s\colon \tau$) if it is anti-rigid. Unlike anti-rigid predicates, the declaration of a type predicate $p_{\tau'}$ is simply defined by a necessary condition for the predicate, that is a supersort of the target type (i.e., $p_{\tau'}\colon \tau$ where $\tau' \leq \tau$). For example, the type *person* may have a necessary condition *animal*.

In contrary, each non-sortal property has no such a general type. For instance, a necessary condition of the property *red* appears to be *thing*. However, when considering it as the necessary condition of *red_light*, it is difficult to determine of whether *light* is a thing. Moreover, the property *water* may have the general property *substance*, but it is a non-sortal property (not a type). To avoid such a problem, we express every non-sortal property by a unary predicate (in P_{non}) without a particular sort declaration (denoted instead by $p\colon undef$).

Following the sorted signature, we introduce the three kinds of terms: *typed term*, *anti-rigid sorted term*, and *sorted term* in a sorted first-order modal language $\mathcal{L}_\Sigma$.

Definition 2. *Let $\Sigma = (T, S_A, \Omega, \leq)$ be a sorted signature. The set $\mathcal{T}_\tau^-$ of terms of type τ (called typed terms) is the smallest set such that (i) for every $x_\tau \in V_\tau$, $x_\tau \in \mathcal{T}_\tau^-$, (ii) for every $c \in C$ with $c: \to \tau \in \Omega$, $c_\tau \in \mathcal{T}_\tau^-$, (iii) if $t_1 \in \mathcal{T}_{\tau_1}^-, \ldots, t_n \in \mathcal{T}_{\tau_n}^-$, $f \in F_n$, and $f: \tau_1 \times \cdots \times \tau_n \to \tau \in \Omega$, then $f_{\tau^*,\tau}(t_1, \ldots, t_n) \in \mathcal{T}_\tau^-$ with $\tau^* = \tau_1, \ldots, \tau_n$, and (iv) if $t \in \mathcal{T}_{\tau'}^-$ and $\tau' \leq \tau$, then $t \in \mathcal{T}_\tau^-$. The set $\mathcal{T}_\sigma^-$ of terms of anti-rigid sort σ (called anti-rigid sorted terms) is the smallest set such that (i) for every $x_\sigma \in V_\sigma$, $x_\sigma \in \mathcal{T}_\sigma^-$ and (ii) if $t \in \mathcal{T}_{\sigma'}^-$ and $\sigma' \leq \sigma$, then $t \in \mathcal{T}_\sigma^-$. The set $\mathcal{T}_s$ of terms of sort s (called sorted terms) is the smallest set such that (i) $\mathcal{T}_s^- \subseteq \mathcal{T}_s$ and (ii) if $t \in \mathcal{T}_{s'}$ and $s' \leq s$, then $t \in \mathcal{T}_s$.*

Due to the rigidity of types and anti-rigid sorts, any anti-rigid sorted term (in $\mathcal{T}_\sigma^-$) must be a variable term whereas typed terms (in $\mathcal{T}_\tau^-$) can contain constants and functions. In other words, every anti-rigid sorted term is not rigid (e.g., $x_{student}$) and every typed term is rigid (e.g., c_{person}). We denote $sort(t)$ as the sort of a term t, i.e., $sort(t) = s$ if t is of the form x_s, c_s, or $f_{\tau^*,s}(t_1, \ldots, t_n)$. Next, we define sorted modal formulas in the language $\mathcal{L}_\Sigma$.

Definition 3. *The set $\mathcal{F}$ of formulas is the smallest set such that (i) if $t_1 \in \mathcal{T}_{s_1}, \ldots, t_n \in \mathcal{T}_{s_n}$, $p \in P_n$, and $p: s_1 \times \cdots \times s_n \in \Omega$, then $p(t_1, \ldots, t_n)$ is a formula, (ii) if $t \in \mathcal{T}_\tau$, $p \in P_{T \cup S_A}$, and $p_s: \tau \in \Omega$, then $p_s(t)$ is a formula, (iii) if $t \in \mathcal{T}_\top$, then $E(t)$ and $p(t)$ are formulas where $p \in P_{non}$, and (iv) if F, F_1, and F_2 are formulas, then $\neg F$, $(\forall x_s)F$, $(\exists x_s)F$, $\Box_i F$, $\Diamond_i F$, $\blacksquare F$, $\blacklozenge F$, $F_1 \wedge F_2$, and $F_1 \vee F_2$ are formulas where $i \in \{\mathbf{Tim}, \mathbf{Sit}\}$.*

The modal formulas are constructed by the modal operators $\blacksquare, \blacklozenge$ (any world), $\Box_{\mathbf{Tim}}, \Diamond_{\mathbf{Tim}}$ (temporal), and $\Box_{\mathbf{Sit}}, \Diamond_{\mathbf{Sit}}$ (situational). To axiomatize rigidity and dependencies with individual existence, the modality $\blacksquare F$ and $\Box_i F$ asserts that F holds for any accessible world whenever individuals exist. For example, the sorted modal formula

$$\Box_{\mathbf{Tim}} p_{male}(bob_{person})$$

implies that for any time accessible from a world, Bob is a male person *as long as he exists*. The existential predicate formula $E(t)$ merely asserts that a term t exists. The formula $\neg F_1 \vee F_2$ abbreviates to $F_1 \to F_2$.

We define the semantics for a sorted first-order modal language $\mathcal{L}_\Sigma$. A sorted Σ-structure M is a tuple (W, w_0, R, R', U, I) such that (i) W is a superset of $\bigcup_{1 \leq i \leq n} W_i$ where W_i is a non-empty set of worlds and $W_i \cap W_j = \emptyset$ ($i \neq j$); (ii) $R = (R_1, \ldots, R_n)$ where R_i is a subset of $W \times W_i$; (iii) R' is a superset of $R_1 \cup \cdots \cup R_n$; (iv) U is a superset of $\bigcup_{w \in W} U_w$ where U_w is the set of individuals in world w[1]; and (v) $I = \{I_w \mid w \in W\}$ is the set of interpretation functions I_w for all worlds $w \in W$ with the following conditions:

---

[1] Each world can have a different domain (possibly $U_{w_1} \neq U_{w_2}$).

1. if $s \in T \cup S_A$, then $I_w(s) \subseteq U_w$ (in particular, $I_w(\top) = U_w$). In addition, $I(s)$ is a superset of $\bigcup_{w \in W} I_w(s)$ such that $U_w \cap I(s) \subseteq I_w(s)^2$,
2. if $s_i \leq s_j$ with $s_i, s_j \in T \cup S_A$, then $I_w(s_i) \subseteq I_w(s_j)$,
3. if $c \in C$ and $c: \ \to \tau \in \Omega$, then $I_w(c) \in I(\tau)$,
4. if $f \in F_n$ and $f: \tau_1 \times \cdots \times \tau_n \to \tau \in \Omega$, then $I_w(f): I(\tau_1) \times \cdots \times I(\tau_n) \to I(\tau)$,
5. if $p \in P_n$ and $p: s_1 \times \cdots \times s_n \in \Omega$, then $I_w(p) \subseteq I_w(s_1) \times \cdots \times I_w(s_n)$ (in particular, if $p_s \in P_{T \cup S_A}$ and $p_s: \tau \in \Omega$, then $I_w(p_s) \subseteq I_w(\tau)$),
6. if $p \in P_{non}$ and $p: undef \in \Omega$, then $I_w(p) \subseteq U_w \cup \Delta_w$ where Δ_w is an uncountably infinite set.

The semantic difference between types τ and anti-rigid sorts σ (i.e., rigidity) is characterized in the following definition using Specification 2. By restricting sorted Σ-structures in terms of rigidity and time and situation dependencies (introducing accessibility relations from worlds to times and situations), we obtain a class of sorted Σ-structures as follows:

Definition 4. *A sorted Σ-structure with rigidity and time/situation dependencies (called sorted Σ^+-structure) is a sorted Σ-structure $M = (W, w_0, R, R', U, I)$ such that*
(rigidity)
1. *$R' \supseteq R_{\text{Tim}} \cup R_{\text{Sit}}$ is reflexive and transitive,*
2. *Specification 2 (in Section 2),*
3. *for every generic predicate[3] $p \in P_{non}$, if $d \in I_w(p)$ and $(w, w') \in R'$, then $d \in U_{w'} \cup \Delta_{w'}$ implies $d \in I_{w'}(p)$,*
(time and situation dependencies)
4. *W is a superset of $W_{\text{Tim}} \cup W_{\text{Sit}}$ where W_{Tim} is the set of times and W_{Sit} is the set of situations ($W_{\text{Tim}} \cap W_{\text{Sit}} = \emptyset$),*
5. *$R = (R_{\text{Tim}}, R_{\text{Sit}})$ where $R_{\text{Tim}} \subseteq W \times W_{\text{Tim}}$ is reflexive and transitive over $W_{\text{Tim}} \times W_{\text{Tim}}$ and $R_{\text{Sit}} \subseteq W \times W_{\text{Sit}}$ is reflexive and transitive over $W_{\text{Sit}} \times W_{\text{Sit}}$,*
6. *Specifications 3-5 (in Section 2).*

Further, there exist the correspondences between sorts and their sort predicates (based on Specification 1). If a type τ is situation dependent, then it and its type predicate are *extensible* with $I_w(\tau) \subsetneq I_w(p_\tau)$ (in addition, $I_w(p_\tau) \subseteq I_w(p_{\tau'})$ if $\tau \leq \tau'$ and τ, τ' are extensible). Any other sort and its sort predicate are *inextensible* with $I_w(s) = I_w(p_s)$. For every extensible type predicate p_τ, we assume that there exists an anti-rigid sort σ as the role of type τ with $\sigma \leq \tau$ such that $I_w(\sigma) = I_w(p_\tau) \backslash I_w(\tau)$. E.g., the anti-rigid sort *temporary_weapon* is the role of the type *weapon*.

To define satisfiability of formulas, we employ the existence of terms in each world. Let $M = (W, w_0, R, R', U, I)$ be a sorted Σ^+-structure, let $w \in W$, and let $[\![t]\!]_w$ be the denotation of a term t in w. The set Nex_w of formulas with terms nonexisting in w is the smallest set such that (i) $p(t_1, \ldots, t_n) \in Nex_w$ *iff* for some

---

[2] If an individual in $I(s)$ exists in a world w, then it must belong to the interpretation $I_w(s)$ in w. That is, $I(s)$ may be constructed by $\bigcup_{w \in W} I_w(s)$ and individuals nonexisting in any world.

[3] A non-sortal predicate is called generic if it is rigid.

ground term $t \in \{t_1, \ldots, t_n\}$, $[\![t]\!]_w \notin U_w$; (ii) $\neg F$, $(\forall x_s)F$, $(\exists x_s)F \in Nex_w$ iff $F \in Nex_w$; (iii) $\Box_i F$, $\Diamond_i F$, $\blacksquare F$, $\blacklozenge F \notin Nex_w$; (iv) $F_1 \wedge F_2 \in Nex_w$ iff $F_1 \in Nex_w$ or $F_2 \in Nex_w$; and (v) $F_1 \vee F_2 \in Nex_w$ iff $F_1 \in Nex_w$ and $F_2 \in Nex_w$. This set is important for the interpretation of modality. In Definition 5, the modal formula $\blacksquare F$ is satisfied in a world w if for any world w' accessible from w, F is satisfied in w' ($w' \models F$) or some ground terms in F do not exist in w' ($F \in Nex_{w'}$).

Definition 5. *The Σ^+-satisfiability relation $w \models F$ is defined inductively as follows:*

1. $w \models p(t_1, \ldots, t_n)$ *iff* $([\![t_1]\!]_w, \ldots, [\![t_n]\!]_w) \in I_w(p)$.
2. $w \models E(t)$ *iff there exists $d \in U_w$ such that $[\![t]\!]_w = d$.*
3. $w \models (\forall x_s)F$ *iff for all $d \in I_w(s)$, $w \models F[x_s/\bar{d}]$.*
4. $w \models (\exists x_s)F$ *iff for some $d \in I_w(s)$, $w \models F[x_s/\bar{d}]$.*
5. $w \models \Box_i F$ *(resp. $\blacksquare F$) iff for all $w' \in W_i$ with $(w, w') \in R_i$ (resp. R'), $w' \models F$ or $F \in Nex_{w'}$.*
6. $w \models \Diamond_i F$ *(resp. $\blacklozenge F$) iff for some $w' \in W_i$ with $(w, w') \in R_i$ (resp. R'), $w' \models F$ and $F \notin Nex_{w'}$.*

The formulas $\neg F$, $F_1 \wedge F_2$, and $F_1 \vee F_2$ are satisfied in the usual manner of first-order logic. Let F be a formula. It is Σ^+-true in M if $w_0 \models F$ (M is a Σ^+-model of F). If F has a Σ^+-model, it is Σ^+-satisfiable, otherwise, it is Σ^+-unsatisfiable. F is Σ^+-valid if every sorted Σ^+-structure is a Σ^+-model of F.

Proposition 1. *Let p be an inextensible type predicate $p_{\tau'}$ with $p_{\tau'} : \tau \in \Omega$ or generic non-sortal predicate (in this case, $\tau = \top$), and p_σ be an anti-rigid sort predicate with $p_\sigma : \tau \in \Omega$. The following axioms are Σ^+-valid.*

1. **Rigid predicate axiom:**
   $(\forall x_\tau)(p(x_\tau) \rightarrow \blacksquare p(x_\tau))$
2. **Anti-rigid predicate axiom:**
   $(\forall x_\tau)(p_\sigma(x_\tau) \rightarrow \blacklozenge(\neg p_\sigma(x_\tau)))$
3. **Time dependency axiom:**
   $\Box_{\mathbf{Tim}}(\forall x_\tau)(p_\sigma(x_\tau) \rightarrow \Diamond_{\mathbf{Tim}}(\neg p_\sigma(x_\tau)))$
   $\Box_{\mathbf{Tim}}(\forall x_\tau)(p_\sigma(x_\tau) \rightarrow \Box_{\mathbf{Sit}} p_\sigma(x_\tau))$
4. **Situation dependency axiom:**
   $\Box_{\mathbf{Sit}}(\forall x_\tau)(p_\sigma(x_\tau) \rightarrow \Diamond_{\mathbf{Sit}}(\neg p_\sigma(x_\tau)))$
   $\Box_{\mathbf{Sit}}(\forall x_\tau)(p_\sigma(x_\tau) \rightarrow \Box_{\mathbf{Tim}} p_\sigma(x_\tau))$
5. **Time-situation dependency axiom:**
   $\Box_{\mathbf{Sit}}(\forall x_\tau)(p_\sigma(x_\tau) \rightarrow \Diamond_{\mathbf{Sit}}(\neg p_\sigma(x_\tau)))$
   $\Box_{\mathbf{Sit}}(\forall x_\tau)(p_\sigma(x_\tau) \rightarrow (\Diamond_{\mathbf{Tim}} p_\sigma(x_\tau) \wedge \Diamond_{\mathbf{Tim}}(\neg p_\sigma(x_\tau))))$

These axioms reflect the intension of rigidity and time and situation dependencies. We denote the set of axioms by $\mathcal{A}_{\mathcal{L}_\Sigma}$.

Example 1. Let p_{apple} be an inextensible type predicate and $p_{nov\_teacher}$ be an anti-rigid sort predicate (time-situation dependent) where $p_{apple} : fruit$ and

$p_{nov\_teacher} : person$ in Ω. Then, the two sorted modal formulas $p_{apple}(c_{fruit}) \rightarrow$
$\blacksquare p_{apple}(c_{fruit})$ and

$$\Box \mathbf{Sit}(p_{nov\_teacher}(john_{person}) \rightarrow (\Diamond \mathbf{Tim} p_{nov\_teacher}(john_{person}) \wedge$$
$$\Diamond \mathbf{Tim} \neg p_{nov\_teacher}(john_{person})))$$

are Σ^+-valid. The subformula $\blacksquare p_{apple}(c_{fruit})$ expresses rigidity with *individual existence*. This implies "c_{fruit} is an apple in any world as long as it exists," (for any accessible world $w \in W$, if $[\![c_{fruit}]\!]_w \in U_w$ (it exists), then $[\![c_{fruit}]\!]_w \in I_w(p_{apple})$), but this does not imply "c_{fruit} is an apple forever" or "c_{fruit} exists forever." Moreover, the subformula $\Diamond \mathbf{Tim} \neg p_{nov\_teacher}(john_{person})$ indicates "there is a time where $john_{person}$ exists but is not a novice teacher."

4   Tableau Calculus

We present a prefixed tableau calculus for the order-sorted modal logic. Let A be a closed formula in negation normal form (i.e., negation occurs only in front of an atomic formula) and S be a finite set of closed formulas in negation normal form. We define S^n as the set, A^n as the formula, and t^n as the term obtained from S, A, and t by annotating each non-annotated constant and function with level n ($\in \mathbb{N}$), such that (i) if $S = \{A_1, \ldots, A_m\}$ then $S^n = \{A_1^n, \ldots, A_m^n\}$, (ii) if $A_i = p(t_1, \ldots, t_m)$ then $A_i^n = p(t_1^n, \ldots, t_m^n)$, (iii) if $A_i \neq p(t_1, \ldots, t_m)$ then $A_i^n = A_i$, (iv) if $t = x_s$ then $t^n = x_s$, (v) if $t = c_\tau$ then $t^n = c_\tau^n$, and (vi) if $t = f_{\tau^*, \tau}(t_1, \ldots, t_m)$ then $t^n = f_{\tau^*, \tau}^n(t_1^n, \ldots, t_m^n)$. The annotated term t^n syntactically implies that it exists in the world corresponding to level n. Each node in a tableau is labeled with a prefixed formula set $(i, n) \colon S$ where $i \in \{W, Tim, Sit\}$ and $n \in \mathbb{N}$. The initial tableau for S is the single node $(W, 1) \colon (S^+)^1$ where S^+ is the smallest superset of S obtained by adding $\blacksquare F$ for all axioms F in $\mathcal{A}_{\mathcal{L}_\Sigma}$. The initial tableau $(W, 1) \colon (S^+)^1$ plays the key role in deciding Σ^+-satisfiability for S since it includes the formulas $\blacksquare F$ for all axioms F in $\mathcal{A}_{\mathcal{L}_\Sigma}$. The axioms characterize the meta-features of properties, and the attached operator $\blacksquare F$ validates the axioms in any world by applications of π_j-/$\pi_{i \rightarrow j}$-/π_W-rules.

   We introduce a set of tableau rules as follows: A ground term t is with level n if the annotated term t^n occurs in an ancestor. Let $i \in \{W, Tim, Sit\}$, $j \in \{Tim, Sit\}$, let t be any ground term with level n, and let comma be the union of sets (i.e., $S_1, S_2 = S_1 \cup S_2$, $A, S = \{A\} \cup S$, and $A, B = \{A\} \cup \{B\}$).

Conjunction and disjunction rules
$\dfrac{(i, n) \colon A \wedge B, S}{(i, n) \colon A^n, B^n, S} \; (\alpha) \qquad \dfrac{(i, n) \colon A \vee B, S}{(i, n) \colon A^n, S \quad (i, n) \colon B^n, S} \; (\beta)$

In α-rule and β-rule, the decomposed formulas A and B are annotated with level n (such as A^n and B^n) since they may be atomic formulas. For example, if $p(t) \wedge F$ is decomposed to $p(t)$ and F by α-rule, then we obtain the annotated atomic formula $p(t^n)$.

Existential predicate rules

$$\frac{(i,n)\colon \neg E(t), S}{(i,n)\colon \bot, \neg E(t), S} \ (E) \qquad \frac{(i,n)\colon S}{(i,n)\colon E(a_n^n), p_s(a_n^n), S} \ (I)$$

In I-rule, a_n is the dummy constant for level n such that $sort(a_n) \le s$ for all sorts $s \in T \cup S_A$ (a_n^n is the annotated term of a_n with level n). By an application of I-rule, a_n is introduced as a ground term with level n. The dummy constant for each level is used to guarantee the non-empty domain of each world.

In the modal operation rules, $*S$ denotes $\{*F \mid F \in S\}$ for $* \in \{\blacksquare, \Box_i\}$ (possibly $*S = \emptyset$) and $\langle *_1, *_2 \rangle S$ denotes $*_1 S \cup *_2 S$ (i.e., $\langle \Box_i, \blacksquare \rangle S = \Box_i S \cup \blacksquare S$). Let $\mathcal{T}_0$ be the set of ground terms. The translation $\mathcal{E}\colon \mathcal{F} \to \mathcal{F}$ is defined as follows: (i) $\mathcal{E}(p(t_1, \ldots, t_n)) = \emptyset$ if $\{t_1, \ldots, t_n\} \nsubseteq \mathcal{T}_0$, otherwise $\mathcal{E}(p(t_1, \ldots, t_n))$ $= \bigwedge_{t \in \{t_1, \ldots, t_n\} \cap \mathcal{T}_0} E(t)$; (ii) $\mathcal{E}(*F) = \mathcal{E}(F)$ for every $* \in \{\neg, \forall x_s, \exists x_s\}$; (iii) $\mathcal{E}(*F)$ $= \emptyset$ for every $* \in \{\Box_i, \Diamond_i, \blacksquare, \blacklozenge\}$; (iv) $\mathcal{E}(F_1 \wedge F_2) = \mathcal{E}(F_1) \wedge \mathcal{E}(F_2)$; and (v) $\mathcal{E}(F_1 \vee F_2) = \mathcal{E}(F_1) \vee \mathcal{E}(F_2)$. Moreover, we define $S \vee \neg \mathcal{E}(S) = \{F \vee \neg \mathcal{E}(F) \mid F \in S\}$.

Modal operator rules

$$\frac{(j,n)\colon \Box_j A, S}{(j,n)\colon A \vee \neg\mathcal{E}(A), \Box_j A, S} \ (\nu_j) \qquad \frac{(j,n)\colon \Diamond_j A, \langle \Box_j, \blacksquare \rangle S, S'}{(j,n+1)\colon A \wedge \mathcal{E}(A), S \vee \neg\mathcal{E}(S), \langle \Box_j, \blacksquare \rangle S} \ (\pi_j)$$

$$\frac{(i,n)\colon \Diamond_j A, \langle \Box_j, \blacksquare \rangle S, S'}{(j,n+1)\colon A \wedge \mathcal{E}(A), S \vee \neg\mathcal{E}(S), \blacksquare S} \ (\pi_{i \mapsto j}) \qquad \frac{(i,n)\colon \blacklozenge A, \blacksquare S, S'}{(W,n+1)\colon A \wedge \mathcal{E}(A), S \vee \neg\mathcal{E}(S), \blacksquare S} \ (\pi_W)$$

$$\frac{(i,n)\colon \blacksquare A, S}{(i,n)\colon A \vee \neg\mathcal{E}(A), \Box_{\mathbf{Tim}} A, \Box_{\mathbf{Sit}} A, \blacksquare A, S} \ (\blacksquare\Box) \qquad \frac{(i,n)\colon \Diamond_j A, S}{(i,n)\colon \blacklozenge A, \Diamond_j A, S} \ (\Diamond\blacklozenge)$$

In $\pi_{i \mapsto j}$-rule, $i \ne j$, in π_j-/$\pi_{i \mapsto j}$-rules, S' is a set of closed formulas without the forms $\blacksquare F$ and $\Box_j F$, and in π_W-rule, S' is a set of closed formulas without the form $\blacksquare F$.

Sorted quantifier rules

$$\frac{(i,n)\colon \forall x_\tau A, S}{(i,n)\colon A[x_\tau/t]^n, \forall x_\tau A, S} \ (\gamma_\tau) \qquad \frac{(i,n)\colon p_{s'}(t^n), \forall x_s A, S}{(i,n)\colon p_{s'}(t^n), A[x_s/t]^n, \forall x_s A, S} \ (\gamma_s)$$

$$\frac{(i,n)\colon \exists x_\tau A, S}{(i,n)\colon E(c_\tau^n), A[x_\tau/c_\tau]^n, \exists x_\tau A, S} \ (\delta_\tau) \qquad \frac{(i,n)\colon \exists x_\sigma A, S}{(i,n)\colon p_\sigma(c_\tau^n), A[x_\sigma/c_\tau]^n, \exists x_\sigma A, S} \ (\delta_\sigma)$$

In γ_τ-rule, $sort(t) \le \tau$, in γ_s-rule, $s' \le s$ and if s is extensible, then $p_{s'}$ is an anti-rigid sort predicate, in δ_τ-rule, c_τ is a constant not in $\{\exists x_\tau A\} \cup S$, and in δ_σ-rule, c_τ is a constant not in $\{\exists x_\sigma A\} \cup S$ where $p_\sigma\colon \tau \in \Omega$.

Sort predicate rules

$$\frac{(i,n)\colon S}{(i,n)\colon p_\tau(t^n), S} \ (p_\tau) \qquad \frac{(i,n)\colon p_s(t^n), S}{(i,n)\colon p_{s'}(t^n), p_s(t^n), S} \ (<)$$

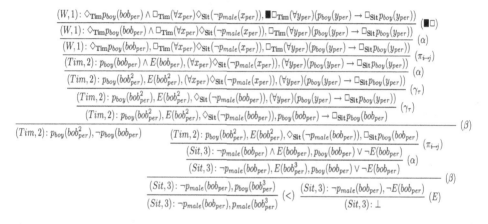

Fig. 2. A proof of satisfiability

Fig. 3. A proof of satisfiability

In p_τ-rule, $sort(t) \leq \tau$, and in $<$-rule, $s < s'$.

A tableau rule is called *static* if it does not change the level n (i.e., $(i,n)\colon S$ is expanded to $(i,n)\colon S'$ by an application of the rule), it is called *dynamic* otherwise (e.g., π_j-/$\pi_{i\to j}$-/π_W-rules are dynamic). The set of closed nodes in a tableau for $(i,n)\colon S$ is defined as follows: (i) if a node contains two complementary literals ($\neg A$ and A^n) or the clash symbol $\bot$, then it is closed and (ii) if all the children of a node are closed, then it is closed. A tableau is closed if the root is closed.

Theorem 1 (Completeness). *There exists a closed tableau for S if and only if S is Σ^+-unsatisfiable.*

Let us prove that the following sorted modal formula is Σ^+-valid by using the calculus.

$$F = p_{apple}(c_{fruit}) \to \blacksquare p_{apple}(c_{fruit})$$

(if c_{fruit} is an apple, then it is an apple in any world as long as it exists) where $T = \{apple, fruit, \top\}$, $S_A = \emptyset$, $\leq\ = \{(apple, fruit)\}$, $C = \{c\}$, $P = \{p_{apple}, p_{fruit}, p_\top\}$, $\Omega = \{c\colon\ \to fruit, p_{apple}\colon fruit, p_{fruit}\colon \top, p_\top\colon \top\}$, and $apple$ and $fruit$ are inextensible in Σ.

In order to determine the validity of this formula, it is sufficient to check the satisfiability of its negation $\neg F$, i.e., F is Σ^+-valid if and only if $\neg F$ is Σ^+-unsatisfiable. To test the satisfiability of any closed formula, we need to transform the formula into an equivalent one in negation normal form (i.e., negation occurs only in front of an atomic formula). $FV(F)$ denotes the set of free variables occurring in a formula F. Let F_1 and F_2 be formulas where $FV(F_1) \subseteq FV(F_2)$ and $FV(F_1) \cup FV(F_2) = \{x_{s_1}^1, \ldots, x_{s_n}^n\}$. $F_1 \simeq F_2$ is a semantic equivalence if for every sorted Σ^+-structure $M = (W, w_0, R, R', U, I)$ and for every $w \in W$, $w \models F_1[x_{s_1}^1/\bar{d}_1, \ldots, x_{s_n}^n/\bar{d}_n]$ if and only if $w \models F_2[x_{s_1}^1/\bar{d}_1, \ldots, x_{s_n}^n/\bar{d}_n]$.

By the semantic equivalences, the formula $\neg F$ is transformed into an equivalent one in negation normal form as follows:

$$\neg(p_{apple}(c_{fruit}) \rightarrow \blacksquare p_{apple}(c_{fruit})) \simeq p_{apple}(c_{fruit}) \wedge \neg \blacksquare p_{apple}(c_{fruit})$$
$$\simeq p_{apple}(c_{fruit}) \wedge \blacklozenge(\neg p_{apple}(c_{fruit}))$$

Fig.2 illustrates a proof of testing the satisfiability of the formula $\neg F$ where every tableau for $S = \{\neg F\}$ is closed. This derives that the formula $\neg F$ is Σ^+-unsatisfiable, and hence F is Σ^+-valid.

Furthermore, consider testing the validity of the following sorted modal formula:

$$F' = \lozenge_{\textbf{Tim}} p_{boy}(bob_{person}) \rightarrow \lozenge_{\textbf{Tim}}(\exists x_{person})(\square_{\textbf{Sit}} p_{male}(x_{person}))$$

(if Bob is a boy at a time, a person exists at a time who is male in any situation within the time) where $T = \{person, male, animal, \top\}$, $S_A = \{boy\}$, $\leq = \{(boy, person), (boy, male), (boy, animal), (person, animal), (male, animal)\}$, $C = \{bob\}$, $P = \{p_{person}, p_{male}, p_{animal}, p_{boy}, p_\top\}$, $\Omega = \{bob\colon \rightarrow person$, $p_{person}\colon animal, p_{male}\colon animal, p_{animal}\colon \top, p_{boy}\colon person, p_\top\colon \top\}$, and boy is time dependent in Σ. The formula $\neg F'$ is transformed into an equivalent one in negation normal form as follows:

$$\neg(\lozenge_{\textbf{Tim}} p_{boy}(bob_{person}) \rightarrow \lozenge_{\textbf{Tim}}(\exists x_{person})(\square_{\textbf{Sit}} p_{male}(x_{person})))$$
$$\simeq \lozenge_{\textbf{Tim}} p_{boy}(bob_{person}) \wedge \neg \lozenge_{\textbf{Tim}}(\exists x_{person})(\square_{\textbf{Sit}} p_{male}(x_{person}))$$
$$\simeq \lozenge_{\textbf{Tim}} p_{boy}(bob_{person}) \wedge \square_{\textbf{Tim}} \neg(\exists x_{person})(\square_{\textbf{Sit}} p_{male}(x_{person}))$$
$$\simeq \lozenge_{\textbf{Tim}} p_{boy}(bob_{person}) \wedge \square_{\textbf{Tim}}(\forall x_{person})(\neg \square_{\textbf{Sit}} p_{male}(x_{person}))$$
$$\simeq \lozenge_{\textbf{Tim}} p_{boy}(bob_{person}) \wedge \square_{\textbf{Tim}}(\forall x_{person}) \lozenge_{\textbf{Sit}}(\neg p_{male}(x_{person}))$$

In Fig.3, we show a proof of testing the satisfiability of the formula $\neg F'$. Since every tableau for $S' = \{\neg F'\}$ is closed, F' is Σ^+-valid.

5   Conclusion

The main results of this paper are (i) a refinement of the ontological property classification by means of individual existence and time and situation dependencies and (ii) an integration of sort predicates and sorted terms (in order-sorted logic), modalities and varying domains (in quantified modal logic), and temporal operators (in first-order temporal logic) in order to model the ontological distinctions among properties.

We formalized the syntax, semantics, and inference system (with axioms) for an order-sorted modal logic such that they are well-suited to deal with the ontological notions. The formal semantics of properties is practically and theoretically useful in deciding the *ontological* and philosophical suitability of property descriptions in information systems and for guaranteeing *logical* consistency in reasoning about properties. We presented a prefixed tableau calculus by extending Meyer and Cerrito's calculus. To handle the multi-modal operators with individual existence, our calculus derives existential predicate formulas and processes formulas prefixed by a pair (i, n) of the kind of worlds i and a natural number n (as a world).

References

1. C. Beierle, U. Hedtsück, U. Pletat, P.H. Schmitt, and J. Siekmann. An order-sorted logic for knowledge representation systems. *Artificial Intelligence*, 55:149–191, 1992.
2. S. Cerrito, M. C. Mayer, and S. Praud. First order linear temporal logic over finite time structures. In *Proceedings of the 6th International Conference on Logic for Programming Artificial Intelligence and Reasoning (LPAR'99)*, pages 62–76, 1999.
3. A. G. Cohn. Taxonomic reasoning with many sorted logics. *Artificial Intelligence Review*, 3:89–128, 1989.
4. M. Fitting and R. L. Mendelsohn. *First-Order Modal Logic*. 1998.
5. A. M. Frisch. The substitutional framework for sorted deduction: fundamental results on hybrid reasoning. *Artificial Intelligence*, 49:161–198, 1991.
6. D. M. Gabbay, A. Kurucz, F. Wolter, and M. Zakharyaschev. *Many-Dimensional Modal Logics: Theory and Applications*. Elsevier, 2003.
7. D. Gabelaia, R. Kontchakov, A. Kurucz, F. Wolter, and M. Zakharyaschev. On the computational complexity of spatio-temporal logics. In *Proc. of FLAIRS*, pages 460–464, 2003.
8. J. W. Garson. Quantification in modal logic. In D. Gabbay and F. Guenthner, editors, *Handbook of Philosophical Logic, Vol.II*, pages 249–307. 1984.
9. R. Goré. Tableau methods for modal and temporal logics. In M. D'Agostino, D. Gabbay, R. Hähnle, and J. Posegga, editors, *Handbook of Tableau Methods*. Kluwer, 1999.
10. N. Guarino, M. Carrara, and P. Giaretta. An ontology of meta-level categories. In *Proc. of the 4th Int. Conf. on the Principles of Knowledge Representation and Reasoning*, pages 270–280, 1994.
11. N. Guarino and C. Welty. Ontological analysis of taxonomic relationships. In *Proceedings of ER-2000: The Conference on Conceptual Modeling*, 2000.
12. K. Kaneiwa. Order-sorted logic programming with predicate hierarchy. *Artificial Intelligence*, 158(2):155–188, 2004.
13. K. Kaneiwa and R. Mizoguchi. Ontological knowledge base reasoning with sort-hierarchy and rigidity. In *Proc. of the 9th Int. Conf. on the Principles of Knowledge Representation and Reasoning*, pages 278–288, 2004.

14. M. C. Mayer and S. Cerrito. Ground and free-variable tableaux for variants of quantified modal logics. *Studia Logica*, 69(1):97–131, 2001.
15. M. Schmidt-Schauss. *Computational Aspects of an Order-Sorted Logic with Term Declarations*. Springer-Verlag, 1989.
16. P. H. Schmitt and J. Goubault-Larrecq. A tableau system for linear-TIME temporal logic. In *Proc. of the 3rd International Conference on Tools and Algorithms for Construction and Analysis of Systems (TACAS'97)*, pages 130–144, 1997.
17. B. Smith. Basic concepts of formal ontology. In *Formal Ontology in Information Systems*. 1998.
18. C. Walther. A mechanical solution of Schubert's steamroller by many-sorted resolution. *Artificial Intelligence*, 26(2):217–224, 1985.
19. C. Welty and N. Guarino. Supporting ontological analysis of taxonomic relationships. *Data and Knowledge Engineering*, 39(1):51–74, 2001.

A Redundancy Analysis of Sequent Proofs

Tatjana Lutovac[1] and James Harland[2]

[1] Department of Applied Mathematics, Faculty of Electrical Engineering,
University of Belgrade, P.O. Box 35-54, 11120 Belgrade, Serbia
tlutovac@eunet.yu
[2] School of Computer Science and Information Technology, RMIT University,
GPO Box 2476V, Melbourne, 3001, Australia
jah@cs.rmit.edu.au

Abstract. Proof search often involves a choice between alternatives which may result in redundant information once the search is complete. This behavior can manifest itself in proof search for sequent systems by the presence of redundant formulae or subformulae in a sequent for which a proof has been found. In this paper we investigate the detection and elimination of redundant parts of a provable sequent by using labels and Boolean constraints to keep track of usage information. We illustrate our ideas in propositional linear logic, but we believe the general approach is applicable to a variety of sequent systems, including other resource-sensitive logics.

1 Introduction

A good reasoning system should include the tools for not only generating proofs, but also for analysing and manipulating them. A central aspect of proof search is the identification of and control over various forms of redundancies in the search space. Proof search often involves managing information which later turns out to be redundant. In this paper, we consider the problem of detection of unnecessary parts of a sequent proof and elimination of redundant formulae that does not alter the search strategy applied. We propose a mechanism for distinguishing between the necessary and unnecessary formulae in a proof. The mechanism is incorporated into the sequent rules and is independent of the search strategy used. In particular, we use labelling and Boolean constraints to track formula occurrences during backward chaining search. We illustrate our ideas about detection and elimination of unused formulae and redundant parts of a proof through examples in propositional linear logic (LL).

There have been various approaches for detecting whether or not a formula occurrence is actually used in a derivation [9,14]. Also, the technique of labelling has been employed in a number of systems being developed for the resource management in linear logic proof search ([16,5,6,12,13]). None of the existing algorithms for efficient implementation of proof search (in linear logic, at least) can distinguish between unused formulae that can be unconditionally eliminated from the proof and unused formulae whose elimination invalidates the proof, as far as we are aware. Our labelled system makes such a distinction.

B. Beckert (Ed): TABLEAUX 2005, LNAI 3702, pp. 185–200, 2005.

Furthermore it allows selection in a sense that a redundant (sub)formula can be either eliminated or replaced by an arbitrary formula. This induces a class of equivalent formulae (in terms of provability) and a class of equivalent proofs modulo the redundant parts. This allows, for example, more flexible reuse of previously successful searches and is potentially useful for implementations.

Example 1. Consider the proof below. We denote by $\mathcal{P}$ and $\mathcal{G}$ respectively the antecedent and succedent of the sequent to be proved. As t and q are *unused*

$$
\cfrac{
  \cfrac{
    \cfrac{s \vdash s}{!s \vdash s}\,!L \quad
    \cfrac{
      \cfrac{
        \cfrac{
          \cfrac{r \vdash r}{}Ax \quad \cfrac{\cfrac{p \vdash p}{}Ax \quad \cfrac{\vdash \top,t}{}\top}{p \vdash p \otimes \top, t}\otimes R
        }{r \wp p \vdash r, t, \top \otimes p}\wp L
      }{r \wp p \vdash ?r, t, \top \otimes p}?R
      \quad
      }{r \wp p \vdash ?r, t, ((\top \otimes p) \oplus q)}\oplus R
    }{r \wp p \vdash ?r \wp t, ((\top \otimes p) \oplus q)}\wp R
  }{!s, r \wp p \vdash ?r \wp t, ((\top \otimes p) \oplus q) \otimes s}\otimes R
}{!s, r \wp p \vdash (?r \wp t) \wp (((\top \otimes p) \oplus q) \otimes s)}\wp R
$$

subformulae, both can be omitted or replaced with another formula. Hence we may think of the underlined parts of the formula $\mathcal{G} : \underline{(?r \wp t)}\ \wp\ (\underline{((\top \otimes p) \oplus q) \otimes s})$ as necessary parts of $\mathcal{G}$, in that the search process establishes that $(?r \wp ((\top \otimes p) \otimes s))$ succeeds, and hence deducing the success of $\mathcal{G}$.

Furthermore, we can refine this process by omitting the (redundant) constant $\top$ as well as the connective ?, resulting in the formula $\mathcal{G}' = r\wp(p \otimes s)$. In this way an analyzer could find a formula $\mathcal{G}'$ such that both $\mathcal{P} \vdash \mathcal{G}'$ and $\mathcal{G}' \vdash \mathcal{G}$ are provable. This may be thought of as calculating an *interpolant formula* $\mathcal{G}'$ from $\mathcal{P}$ and $\mathcal{G}$. Note that the transformation from $\mathcal{G}$ to $\mathcal{G}'$ does not alter the search strategy used, in that the order of application of the rules is not changed as shown below.

$$
\cfrac{
  \cfrac{s \vdash s}{!s \vdash s}\,!L \quad
  \cfrac{
    \cfrac{
      \cfrac{r \vdash r}{}Ax \quad \cfrac{p \vdash p}{}Ax
    }{r \wp p \vdash r, p}\wp L
  }{!s, r \wp p \vdash r, p \otimes s}\otimes R
}{!s, r \wp p \vdash r \wp (p \otimes s)}\wp R
$$

This formula $\mathcal{G}'$ can be thought of as a representative of a family of formulae whose derivations, for the given formula $\mathcal{P}$, will require no effort to establish. For the above example, the obligatory part (i.e., the minimal information which must be present in $\mathcal{G}$) of $\mathcal{G}$ is $r\wp(p\otimes s)$ while a general template for successful formulae based on $\mathcal{G}$ could be $(\ [?]\ r\ \wp[F]\ )\ \wp\ (\ (([\top]\otimes\ \boldsymbol{p})\oplus[Q])\otimes\ \boldsymbol{s}\ )$ where F and Q are arbitrary formulae, and [] denotes parts of the original formula $\mathcal{G}$ that can be omitted. This knowledge allows later computations to make use of earlier work. So a proof-search strategy can retain the results of a previous successful search and to apply and combine them to a new situation. The knowledge about redundant and eliminable formulae can be potentially useful when composing programs (and hence proofs), for debugging, and for teaching purposes. For a given search strategy, we may thus consider this work as an initial requirements analysis of the properties of proofs.

Example 2.   Consider the (linear logic) proof Π below:

$$
\Pi :\quad
\cfrac{
  \cfrac{t \vdash t}{}Ax \quad
  \cfrac{
    \cfrac{
      \cfrac{
        \cfrac{
          \cfrac{r \vdash r}{}Ax
        }{r \vdash ?p, r}?wR
      }{r \vdash ?q, ?p, r}?wR
    }{r \vdash ?q\wp?p, r}\wp R
  }{r \vdash (?q\wp?p) \oplus s, r}\oplus R
}{r, t \vdash t \otimes ((?q\wp?p) \oplus s), r}\otimes R
$$

$$
\mathcal{D}_1 :\quad \cfrac{\quad ? \quad}{r, t \vdash t, r}
\qquad
\mathcal{D}_2 :\quad
\cfrac{
  \cfrac{t \vdash t}{}Ax \quad \cfrac{\quad ? \quad}{r \vdash s, r}
}{r, t \vdash t \otimes s, r}\otimes R
$$

The (sub)formulae p, q and s are *unused*, but only s can be freely deleted from the proof while formulae p and q cannot be simultaneously eliminated. Note that p, q and s are subformulae of the active formula $(?q\wp?p) \oplus s$ of the multiplicative rule $\otimes R$. Elimination of the whole formula $(?q\wp?p) \oplus s$ will disable proof branching i.e. distribution of formulae across the multiplicative branches of the proof (see derivation $\mathcal{D}_1$ above). Elimination of the subformula $?q\wp?p$ will also lead to the unprovable sequent (derivation $\mathcal{D}_2$ on the right-hand side above). So, we have that p, q and s are unused and that p and q cannot be simultaneously eliminated from the proof. For each unused atom we have three possibilities: to omit it from the proof; to leave the atom unchanged or to replace it with an arbitrary formula. So proof Π can be thought of as a template for $(3^2 - 1) \cdot 3$ proofs (i.e. some variations of the given proof) which can be generated by alterations of p, q and s. All that proofs do not alter the search strategy used, in that the order of application of the rules is not changed.

This paper is organized as follows. In Section 2 we explain our approach to redundancy detection. In Section 3 we introduce a labelled sequent calculus with constraints for partial redundancy elimination. Section 4 gives a detailed description of the algorithm. In Section 5 we give the formal results. In Section 6 we briefly point out some possible applications of our system. In Section 7 we present our conclusions.

2   Our Approach to Automated Detection of Redundant Formulae

We present an algorithm for *partial* elimination of redundant formulae (Algorithm *PRE*) from a given proof. Our intention is not to find all different proofs of a given sequent but to generate all the concrete simplifications which are instances of a generated proof. By partial elimination of redundant formulae we mean:

- elimination which is independent of the search strategy used;
- elimination which does not alter the search strategy applied (i.e. rule instances may be deleted);
- no additional proof search i.e. redundant formulae remaining in the resulting proof cannot be eliminated without additional proof search (i.e. rule instances may not be added);
- preserving the multiplicative branching structure of the proof.

In the rest of this section we will explain the intuition behind this points. Let us firstly recall the basic notions related to categorization of formulae in an inference rule [3,8]. The *active* formulae of an inference are the formulae which are present in the premise(s), but not in the conclusion. The *principal* formula of an inference is the formula which is present in the conclusion, but not in the premise(s). Intuitively, the inference converts the active formulae into the principal formula.

Let us now, explain our view of redundant formulae in a given proof i.e. let us try to categorize formulae which can be safely eliminated from the proof. Intuitively, we say a (sub)formula F which occurs in a proof Π of a propositional sequent $\Gamma \vdash \Delta$ is *passive* iff none of its subformulae have appeared in *axiom* leaves or as the principal formula(e) of leaves that correspond to initial rules ($\top R$, $\mathbf{1}R$, $\perp L$, $\mathbf{0}L$). Thus, for example, for the proof Π of Example 2 the set of all passive (sub)formulae is the set $\{ s, ?q, ?p, ?q\wp?p, (?q\wp?p) \oplus s \}$. Note that formula $(?q\wp?p) \oplus s$ is passive although it is the principal formula of the $\oplus R$ rule. As Example 2 indicates, special care is needed at multiplicative branches. The formula $(?q\wp?p) \oplus s$ is passive and the active formula of a multiplicative rule, so it can not be safely deleted i.e. it cannot be considered as redundant (or unused) in the proof Π.

Note that elimination of passive formulae which are the active formulae of a multiplicative rule may not always result in an invalid proof. As it is illustrated in the next example elimination of such formulae may cause some other 'problems' (such as loss of information concerning the usage of some computational resources (formulae)).

Example 3.

In the left-hand proof the only passive formula is $?s$. The proof can be simplified as shown on the right-side. Both proof simplifications are valid. The simplification denoted ii) implies additional proof search due to some rearrangements of the rules. In the simplification denoted i) (in which the passive formula $?s$ is eliminated) the formulae p, q, and $p \multimap q$ become passive although they are not so in the left-hand proof. In particular, we lost the following subproof:

$$\frac{\overline{p \vdash p} \; Ax \qquad \overline{q \vdash q} \; Ax}{p, \, p \multimap q \vdash q} \multimap L$$

Let us now consider a linear logic (sub)proof of the following structure. Let us assume that formula ϕ can be eliminated from the subproof Π_1. What meaning-

$$\frac{\Pi_1 \qquad \Pi_2}{\Gamma_1 \vdash \phi, \Delta_1 \qquad \Gamma_2 \vdash \psi, \Delta_2} \otimes R \\ \overline{\Gamma_1, \Gamma_2 \vdash \phi \otimes \psi, \Delta_1, \Delta_2}$$

ful information (for example about the usage of computational resources (formulae) can be extracted from the shown multiplicative branch node? As $\Gamma_1 \vdash \Delta_1$ is provable, Γ_1 and Δ_1 are responsible for the success in the left branch. Success of the right branch means that resources Γ_2 provides a context within which the requirements ψ, Δ_2 are met. Elimination of the formula ϕ leads to the sequent $\Gamma_1, \Gamma_2 \vdash \psi, \Delta_1, \Delta_2$. Without additional proof search, such elimination will result in a valid proof iff either subproof Π_1 or Π_2 can accept additional formulae (distributed from the opposite, omitted branch).

$$
\dfrac{\overset{\Pi_1}{\Gamma_1 \vdash \phi, \Delta_1} \quad \overset{\Pi_2}{\Gamma_2 \vdash \psi, \Delta_2}}{\Gamma_1, \Gamma_2 \vdash \phi \otimes \psi, \Delta_1, \Delta_2} \otimes R
\quad \mapsto \quad
\dfrac{\overset{\Pi_1}{\Gamma_1 \vdash \Delta_1} \quad \overset{\Pi_2}{\Gamma_2 \vdash \psi, \Delta_2}}{\Gamma_1, \Gamma_2 \vdash \psi, \Delta_1, \Delta_2}
\quad \mapsto \quad
\begin{cases}
\dfrac{\overset{\Pi_1}{}}{\Gamma_1, \Gamma_2 \vdash \psi, \Delta_1, \Delta_2} \\[2mm]
\dfrac{\overset{\Pi_2}{}}{\Gamma_1, \Gamma_2 \vdash \psi, \Delta_1, \Delta_2}
\end{cases}
$$

In this way such formulae (distributed from the omitted branch) become passive and the original information about their usage is lost.

Note that for the case of additive branch nodes deleting any active formula of a binary additive rule and, consequently, omitting the corresponding branch do not imply adding of additional formulae to the branch to be left over:

$$
\dfrac{\Gamma \vdash A_1, \Delta \quad \Gamma \vdash A_2, \Delta}{\Gamma \vdash A_1 \& A_2, \Delta} \& R
\quad \mapsto \quad \Gamma \vdash A_i, \Delta
$$

We allow free elimination of any active formula(e) of additive rules.

3   A Labelled Sequent Calculus for Partial Elimination of Redundant Formulae

In this section we define a sequent calculus with labels and constraints that allow us to store the necessary information about the usage of formulae in a proof.

At first, the sequent to be proved has to be 'rectified' so that distinct occurrences of a same atom or constant have distinct indices. For example the sequent $p \otimes q, !p \vdash p \otimes q, r \wp s$ will be rectified as $p \otimes q, !p_1 \vdash p_2 \otimes q_1, r \wp s$.
Our sequent has the form: $A_{1,[v_1]}, A_{2,[v_2]}, \ldots A_{k,[v_k]} \vdash B_{1,[w_1]}, B_{2,[w_2]}, \ldots B_{n,[w_n]} - \mathcal{C}$
where:
- $\mathcal{C}$ is a set of constraints being generated so far on the branch of a proof tree by application of the rules $\oplus R$, $\& L$, $\multimap R$, the contraction rules $c?R$ and $c!L$, as well as by application of the multiplicative binary rules $\otimes R$, $\wp L$ and $\multimap L$.
- Labels $[v_1], \ldots, [w_1], \ldots [w_n]$ are called *contraction-labels*. We use *contraction*-labels to trace formulae (and their subformulae) duplicated by the contraction rule. $v_1, \ldots v_k, w_1, \ldots w_n$ are words (of arbitrary length) on an infinite alphabet $\Sigma = \{x, y, z, x_1, y_1, z_1, \ldots\}$. Let us adopt the any subformula of a formula $F_{[w]}$ has the same ($[w]$) contraction-label.
We begin with the empty *contraction*-label (denoted as $[\,]$) on each (sub)formula of the sequent to be proved and with the empty set $\mathcal{C}$. For each application of the contraction rule, the conclusion's contraction-label is 'extended' with fresh, distinct letter in order to distinguish and trace different copies of a same formula:

$$
\dfrac{\Gamma \vdash ?F_{i,[wx_1]}, ?F_{i,[wx_2]}, \Delta \ - \mathcal{C} \cup \mathcal{C}_1}{\Gamma \vdash ?F_{i,[w]}, \Delta \ - \mathcal{C}} \, c?R
$$

All other rules propagate up with contraction-labels unchanged. Example 4 illustrates our method for tracking formula occurrences connected with the application of contraction rules.
The labelled sequent calculus, denoted LL^{PRE}, for partial redundancy elimination in propositional linear logic is defined as shown in Figure 1. The labels are

omitted when they are unchanged. Contraction-labels and constraints have no effect on proof-search. It is straightforward to show the following result:

Proposition 1. *A sequent* $\Gamma \vdash \Delta$ *is provable in propositional LL iff* $\Gamma_{[]} \vdash \Delta_{[]} - \emptyset$ *is provable in the labelled system* LL^{PRE}.

In particular, on the successful completion of a proof our algorithm (Algorithm *PRE*, Section 4) takes as input the labelled proof tree. More precisely, our approach in Algorithm PRE is to interpret the set of accumulated constraints via the set of Boolean constraints and to find an assignment for Boolean variables. Intuitively, atoms that correspond to Boolean variables being annotated value 0, i.e. formulae made up of such atoms can be safely eliminated (deleted) from the proof.

$$\frac{-\mathcal{C}\cup\{p=1\}}{p\vdash p \quad -\mathcal{C}}\,Ax \qquad \frac{\Gamma\vdash\phi_{[w]},\Delta - \mathcal{C}}{\Gamma,(\neg\phi)_{[w]}\vdash\Delta - \mathcal{C}}\,\neg L \qquad \frac{\Gamma,\phi_{[w]}\vdash\Delta - \mathcal{C}}{\Gamma\vdash(\neg\phi)_{[w]},\Delta - \mathcal{C}}\,\neg R \qquad \frac{-\mathcal{C}\cup\{\bot=1\}}{\bot\vdash \quad -\mathcal{C}}\,\bot L$$

$$\frac{\Gamma\vdash\Delta - \mathcal{C}}{\Gamma\vdash\bot,\Delta - \mathcal{C}}\,\bot R \qquad \frac{-\mathcal{C}\cup\{\top=1\}}{\Gamma\vdash\top,\Delta - \mathcal{C}}\,\top R \qquad \frac{-\mathcal{C}\cup\{0=1\}}{\Gamma,0\vdash\Delta - \mathcal{C}}\,0L \qquad \frac{\Gamma\vdash\Delta - \mathcal{C}}{\Gamma,1\vdash\Delta - \mathcal{C}}\,1L$$

$$\frac{-\mathcal{C}\cup\{1=1\}}{\vdash 1 - \mathcal{C}}\,1R \qquad \frac{\Gamma,\phi_{[w]}\vdash\Delta - \mathcal{C}}{\Gamma,(!\phi)_{[w]}\vdash\Delta - \mathcal{C}}\,!L \qquad \frac{!\Gamma,\phi_{[w]}\vdash?\Delta - \mathcal{C}}{!\Gamma,(?\phi)_{[w]}\vdash?\Delta - \mathcal{C}}\,?L \qquad \frac{\Gamma\vdash\Delta - \mathcal{C}}{\Gamma,!\phi\vdash\Delta - \mathcal{C}}\,w!L$$

$$\frac{!\Gamma\vdash\phi_{[w]},?\Delta - \mathcal{C}}{!\Gamma\vdash(!\phi)_{[w]},?\Delta - \mathcal{C}}\,!R \qquad \frac{\Gamma\vdash\phi_{[w]},\Delta - \mathcal{C}}{\Gamma\vdash(?\phi)_{[w]},\Delta - \mathcal{C}}\,?R \qquad \frac{\Gamma\vdash\Delta - \mathcal{C}}{\Gamma\vdash?\phi,\Delta - \mathcal{C}}\,w?R$$

$$\frac{\Gamma,\phi_{[w]},\psi_{[w]}\vdash\Delta - \mathcal{C}}{\Gamma,(\phi\otimes\psi)_{[w]}\vdash\Delta - \mathcal{C}}\,\otimes L \qquad \frac{\Gamma\vdash\phi_{[w]},\Delta - \mathcal{C}\cup\{\phi_{[w]}>0\} \quad \Gamma'\vdash\psi_{[w]},\Delta' - \mathcal{C}\cup\{\psi_{[w]}>0\}}{\Gamma,\Gamma'\vdash(\phi\otimes\psi)_{[w]},\Delta,\Delta' - \mathcal{C}}\,\otimes R$$

$$\frac{\Gamma,\psi_{[w]}\vdash\Delta - \mathcal{C}\cup\{\phi_{[w]}\leq\psi_{[w]}\}}{\Gamma,(\phi\&\psi)_{[w]}\vdash\Delta - \mathcal{C}}\,\&L \qquad \frac{\Gamma\vdash\phi_{[w]},\Delta - \mathcal{C} \quad \Gamma\vdash\psi_{[w]},\Delta - \mathcal{C}}{\Gamma\vdash(\phi\&\psi)_{[w]},\Delta - \mathcal{C}}\,\&R$$

$$\frac{\Gamma,\phi_{[w]}\vdash\Delta - \mathcal{C} \quad \Gamma,\psi_{[w]}\vdash\Delta - \mathcal{C}}{\Gamma,(\phi\oplus\psi)_{[w]}\vdash\Delta - \mathcal{C}}\,\oplus L \qquad \frac{\Gamma\vdash\psi_{[w]},\Delta - \mathcal{C}\cup\{\phi_{[w]}\leq\psi_{[w]}\}}{\Gamma\vdash(\phi\oplus\psi)_{[w]},\Delta - \mathcal{C}}\,\oplus R$$

$$\frac{\Gamma,\phi_{[w]}\vdash\Delta - \mathcal{C}\cup\{\phi_{[w]}>0\} \quad \Gamma',\psi_{[w]}\vdash\Delta' - \mathcal{C}\cup\{\psi_{[w]}>0\}}{\Gamma,\Gamma',(\phi\wp\psi)_{[w]}\vdash\Delta,\Delta' - \mathcal{C}}\,\wp L$$

$$\frac{\Gamma\vdash\phi_{[w]},\Delta - \mathcal{C}\cup\{\phi_{[w]}>0\} \quad \Gamma',\psi_{[w]}\vdash\Delta' - \mathcal{C}\cup\{\psi_{[w]}>0\}}{\Gamma,\Gamma',(\phi\multimap\psi)_{[w]}\vdash\Delta,\Delta' - \mathcal{C}}\,\multimap L$$

$$\frac{\Gamma,\phi_{[w]}\vdash\psi_{[w]},\Delta - \mathcal{C}\cup\{\phi_{[w]}\leq\psi_{[w]}\}}{\Gamma\vdash(\phi\multimap\psi)_{[w]},\Delta - \mathcal{C}}\,\multimap R \qquad \frac{\Gamma\vdash\phi_{[w]},\psi_{[w]},\Delta - \mathcal{C}}{\Gamma\vdash(\phi\wp\psi)_{[w]},\Delta - \mathcal{C}}\,\wp R$$

$$\frac{\Gamma,!F_{i,[wx_1]},!F_{i,[wx_2]}\vdash\Delta - \mathcal{C}\cup\mathcal{C}_1}{\Gamma,!F_{i,[w]}\vdash\Delta - \mathcal{C}}\,c!L \qquad \frac{\Gamma\vdash?F_{i,[wx_1]},?F_{i,[wx_2]},\Delta - \mathcal{C}\cup\mathcal{C}_1}{\Gamma\vdash?F_{i,[w]},\Delta - \mathcal{C}}\,c?R$$

Where $\mathcal{C}_1 = \{\,\langle\natural F_{i,[wx_1]}\rangle = \langle\natural F_{i,[wx_2]}\rangle \vee \langle\natural F_{i,[wx_1]}\rangle = (0,\ldots 0) \vee \langle\natural F_{i,[wx_2]}\rangle = (0,\ldots,0),$
$\langle\natural F_{i,[w]}\rangle = \langle\natural F_{i,[wx_1]}\rangle + \langle\natural F_{i,[wx_2]}\rangle\,\}, \qquad (\natural\in\{?,!\})$

Fig. 1. LL^{PRE} sequent calculi

Let us now explain constraints of the labelled system LL^{PRE} in more detail.

3.1 Constraints

The constraints accumulated during a proof construction in the labelled system LL^{PRE} place some restrictions on the elimination of the corresponding formulae.

Constraints are of the following form: ($\natural \in \{?, !\}$)

I. $\psi_{i,[w]} \leq \phi_{i,[w]}$,   **II.** $\phi_{i,[w]} > 0$,   **III.** $\phi_{i,[w]} = 1$,
IV. $\langle\natural\phi_{i,[w]}\rangle = \langle\natural\phi_{i,[wx_1]}\rangle + \langle\natural\phi_{i,[wx_2]}\rangle$, $\langle\natural\phi_{i,[wx_1]}\rangle = \langle\natural\phi_{i,[wx_2]}\rangle$, $\langle\natural\phi_{i,[wx_1]}\rangle = (0, \ldots 0)$

Let us explain each type of constraint in more detail.

I. Constraints connected with the application of the rules $\oplus R$, $\&L$, $\multimap R$.
• Every application of the $\oplus R$ or $\&L$ rule 'sets' a constraint of the form
$discharged - formula^1 \leq active - formula$. The intuition underlying such
constraint is that elimination of a formula being discarded by the $\oplus R$ or $\&L$ rule
is a necessary condition for elimination of the corresponding active formula. Let
us analyse the $\oplus R$ rule and explain our motivations for the constraint. Consider
the left-hand proof below and note the passive formulae: $?a$, b and $?a \oplus b$.

$$\cfrac{\cfrac{\cfrac{\overline{q_1 \vdash q}\;Ax}{q_1 \vdash q, ?a}\;w?R}{q_1 \vdash q, ?a \oplus b}\;\oplus R \quad \overline{p_1 \vdash p}\;Ax}{p_1, q_1 \vdash p \otimes q, ?a \oplus b}\;\otimes R \qquad\longmapsto\qquad \cfrac{\cfrac{?}{q_1 \vdash q, b} \quad \overline{p_1 \vdash p}\;Ax}{p_1, q_1 \vdash p \otimes q, b}\;\otimes R$$

Simplification (on the right-hand side above) shows that for the $\oplus R$ rule elimination of the active formula independently of elimination of the formula being
discarded by this rule may endanger the existence of the proof. It is not hard
to find an example where elimination of the active formula of the $\oplus R$ rule does
not endanger existence of the proof, but implies additional proof search.
• Every application of the $\multimap R$ rule 'sets' a constraint of the form
$active - formula_{from\ the\ antecedent} \leq active - formula_{from\ the\ succedent}$. The intuition underlying such constraint is that elimination of the active formula from
the antecedent is a necessary condition for elimination of the active formula from
the succedent. For example, consider the left-hand proof and note the passive
formula $?s$. It cannot be consider as redundant formula as its elimination results
in a derivation which is not a valid proof (as shown on the right side):

$$\cfrac{\cfrac{\overline{p \vdash p_1}\;Ax}{p \vdash ?s, p_1}\;w?R}{\vdash p \multimap ?s, p_1}\;\multimap R \qquad\longmapsto\qquad \cfrac{?}{\vdash p, p_1}$$

**II. Constraints connected with the application of multiplicative binary
rules.**   Every application of the $\otimes R$, $\multimap L$ or $\wp L$ rule 'sets' a constraint of
the form $active\text{-}formula > 0$. The intuition underlying such constraint is not
to allow an active formula of the multiplicative rule to be whole eliminated.
We have seen in Examples 2 and 3 that elimination of an active formula of
a multiplicative rule may endanger the existence of the proof or may imply
additional proof search.

III. Constraints connected with the leaves of the proof tree. At the leaves
for the proof tree, the constraints assign the value 1 to atoms which appear at
Axioms and value 1 to principal formula of non-axiom leaf.

---

[1] A formula present in the conclusion but not in the premise.

IV. Constraints connected with the application of the contraction rule.

Let us, firstly, adopt the following notation: let $\langle \natural F_{i,[w]} \rangle$ denotes the n-tuples of all atoms occurring, from the left-hand to right-hand side, in a formula $\natural F_{i,[w]}$ (where $\natural \in \{?, !\}$). For example $\langle ?(p_1 \wp (s \otimes ?p))_{[w]} \rangle \; = \; (\; p_{1,[w]}, \; s_{[w]}, \; p_{[w]} \;)$.

For each application of the contraction rule the conclusion's set $\mathcal{C}$ is extended with the new constraints as shown below for the $c?R$ rule:

$$\frac{\Gamma \; \vdash \; ?F_{i,[wx_1]}, \; ?F_{i,[wx_2]}, \; \Delta \quad - \mathcal{C} \cup \mathcal{C}_1}{\Gamma \; \vdash \; ?F_{i,[w]}, \; \Delta \quad - \mathcal{C}} \; c?R$$

where $\mathcal{C}_1 \; = \; \{ \; \langle ?F_{i,[w]} \rangle \; = \; \langle ?F_{i,[wx_1]} \rangle \; + \; \langle ?F_{i,[wx_2]} \rangle$,

$\langle ?F_{i,[wx_1]} \rangle \; = \; \langle ?F_{i,[wx_2]} \rangle \; \vee \; \langle ?F_{i,[wx_1]} \rangle \; = \; (0, \ldots 0) \; \vee \; \langle ?F_{i,[wx_2]} \rangle \; = \; (0, \ldots 0) \; \}$

Although the behaviour of formulae $?F_{i,[wx_1]}$ and $?F_{i,[wx_2]}$ is independent, elimination of their redundant subformulae must be coordinated. From the elimination point of view, the possible choices are:
- either the same subformulae may be simultaneously eliminated from $?F_{i,[wx_1]}$ and $?F_{i,[wx_2]}$ (this choice is captured by the constraint $\langle ?F_{i,[wx_1]} \rangle = \langle ?F_{i,[wx_2]} \rangle$);
- or one of the formulae $?F_{i,[wx_1]}$ and $?F_{i,[wx_2]}$ may be eliminated in total, leaving the remaining formula without any constraints for elimination. This choice is 'implemented' by the constraint $\langle ?F_{i,[wx_1]} \rangle = (0, \ldots 0) \; \vee \; \langle ?F_{i,[wx_2]} \rangle = (0, \ldots 0)$.
In either case formulae which will be eliminated in the premisses (and above them in the proof tree) must be coordinated with the formulae which will be eliminated in the conclusion of the contraction rule and below it in the proof tree. This is implemented by the constraint $\langle ?F_{i,[w]} \rangle \; = \; \langle ?F_{i,[wx_1]} \rangle \; + \; \langle ?F_{i,[wx_2]} \rangle$.

We denote by $\mathcal{C}_{final}$ the union of sets $\mathcal{C}$ being occurred at the leaves of the proof tree.

Thus, going back to Example 2 (from Section 2), for the proof Π we have the labelled proof tree $\Pi^{LL^{PRE}}$ (generated in the system LL^{PRE}) as shown below and $\mathcal{C}_{final} = \{ (?q \wp ?p) \oplus s > 0, \; s \leq ?q \wp ?p, \; r = 1, \; r_1 = 1, \; t = 1, \; t_1 = 1, t_1 > 0 \}$.

$$\frac{\displaystyle \frac{\displaystyle \frac{\displaystyle \frac{\displaystyle \frac{\displaystyle - \; \{ (?q \wp ?p) \oplus s > 0, \; s \; \leq \; ?q \wp ?p, \; r = 1 \; r_1 = 1 \}}{r \vdash r_1 \quad - \; \{ (?q \wp ?p) \oplus s > 0, \; s \; \leq \; ?q \wp ?p \}} \; Ax}{r \vdash ?p, r_1 \quad - \; \{ (?q \wp ?p) \oplus s > 0, \; s \; \leq \; ?q \wp ?p \}} \; w?R}{r \vdash ?q, ?p, r_1 \quad - \; \{ (?q \wp ?p) \oplus s > 0, \; s \; \leq \; ?q \wp ?p \}} \; w?R}{r \vdash ?q \wp ?p, r_1 \quad - \; \{ (?q \wp ?p) \oplus s > 0, \; s \; \leq \; ?q \wp ?p \}} \; \wp R}{\displaystyle \frac{- \; \{ t = 1, \; t_1 = 1, t_1 > 0 \}}{t \vdash t_1 \quad - \; \{ t_1 > 0 \}} \; Ax \qquad \frac{r \vdash (?q \wp ?p) \oplus s, r_1 \quad - \quad \{ (?q \wp ?p) \oplus s > 0 \}}{} \; \oplus R}{r_{[]}, t_{[]} \; \vdash \; (t_1 \otimes ((?q \wp ?p) \oplus s))_{[]}, r_{1,[]} \quad - \emptyset} \; \otimes R$$

4   Algorithm for Partial Elimination of Redundant Formulae

In this section we describe the algorithm for partial elimination of redundant formulae (Algorithm PRE). Let π be a proof generated in the system LL^{PRE}.

Algorithm PRE (input: proof π)

1. *Generate Boolean expressions and constraints on Boolean expressions;*
2. *Calculate possible assignments for Boolean variables;*
3. *If there is an assignment with at least one Boolean variable being assigned the value 0 then: Delete atoms being assigned 0 i.e. delete formulae made up of such atoms and the corresponding inferences*
   Else EXIT: 'Simplification of proof π is not possible'

We are now going to explain steps 1. to 3. in more detail.

Step 1: Generation of Boolean expressions and Boolean constraints.
A specific Boolean expression is assigned to each atom from the proof π i.e. to each formula occurring in the set $\mathcal{C}_{final}$. We denote by $F_{[w]}[e]$ a formula $F_{[w]}$ 'connected' to Boolean expression e. Consider the following mapping i.e. a bijection from the set of all propositional letters and constants into the set of Boolean variables: $p_{i,[v]} \mapsto p_{i,[v]}$, $const_{j,[w]} \mapsto cconst_{j,[w]}$, $const \in \{\top, \bot, \mathbf{1}, \mathbf{0}\}$.

For a given proof π generated in the system LL^{PRE}, any formula $F_{i,[w]}$ being appeared in $\mathcal{C}_{final}$ or in a leaf of a proof tree is associated with the Boolean sum of Boolean variables that correspond (under the above mapping) to the participating atoms or constants. For example consider the following formulae and the associated Boolean expressions:

$(?(p\otimes?q)\wp p_1)_{[w]}[p_{[w]}+q_{[w]}+p_{1,[w]}]$, $p_{5,[w]}[p_{5,[w]}]$, $?(\top\oplus?p_i)_{[w]}[c\top_{[w]}+p_{i,[w]}]$.
The set $\mathcal{C}_{final}$ 'transfers' the constraints to Boolean expressions of the corresponding formulae. For example, in Example 2 (we have seen that $\mathcal{C}_{final} = \{(?q\wp?p) \oplus s > 0,\ s \leq ?q\wp?p,\ r = 1,\ r_1 = 1,\ t = 1,\ t_1 = 1, t_1 > 0\}$) we will get the following Boolean constraints on Boolean variables: $q + p + s > 0$, $s \leq q + p$, $r = 1$, $r_1 = 1$, $t = 1$, $t_1 = 1$, $t_1 > 0$.

Step 2: Calculation of possible assignments for Boolean variables. Values of Boolean variables are calculated in according to Boolean constraints being 'inherited' from the set $\mathcal{C}_{final}$. The values of Boolean variables are intended to specify that the corresponding atom, constant i.e. formula can be safely eliminated from proof π. Intuitively, formulae associated with Boolean expression being assigned to 0 can be safely deleted from the proof. Any assignment with *at least one variable being assigned the value 0* (if it exists) reflects a possible simplification of the original proof.

Step 3: Elimination of redundant formulae. For a selected assignment of Boolean variables in which at least one variable has been assigned the value 0 (if any such assignment): (1) eliminate (i.e. delete) every appearance of atoms 'annotated' to 0, i.e. every appearance of (sub)formulae made up of such atoms; (2) delete any rule inference with the same premise and conclusion (for example by deleting the premise of the inference) except for the $\&R$ rule (i.e. $\oplus L$ rule). For the $\&R$ rule (i.e. $\oplus L$ rule) distinguish cases when just one or both active formulae are deleted. Let us adopt the following notation: let $\widetilde{\Gamma}$, $\widetilde{F}$ and $\widetilde{\pi}$

respectively denote the result of eliminating (i.e. deleting) atoms 'annotated' to 0 from multiset Γ, from formula F and form proof π. In the case when just one active formula is deleted, delete the inference and the whole branch that corresponds to the deleted active formula, as illustrated on the left-hand side for the $\&R$ rule.

$$
\cfrac{\overset{\pi_1}{\tilde{\Gamma} \vdash \tilde{\phi}_{[w]}, \tilde{\Delta}} \quad \overset{\pi_2}{\tilde{\Gamma} \vdash \tilde{\Delta}}}{\tilde{\Gamma} \vdash \tilde{\phi}_{[w]}, \tilde{\Delta}} \&R \quad \longmapsto \quad \overset{\pi_1}{\tilde{\Gamma} \vdash \tilde{\phi}_{[w]}, \tilde{\Delta}} \qquad\qquad \cfrac{\overset{\pi_1}{\tilde{\Gamma} \vdash \tilde{\Delta}} \quad \overset{\pi_2}{\tilde{\Gamma} \vdash \tilde{\Delta}}}{\tilde{\Gamma} \vdash \tilde{\Delta}} \; \sharp \quad \longmapsto \quad \overset{\pi_i}{\tilde{\Gamma} \vdash \tilde{\Delta},}
$$

$$\sharp \in \{\&R, \oplus L\}, \quad i \in \{1,2\}$$

In the case when both active formulae are deleted, then select which branch to delete, as shown on the right-hand side above.

Finishing Example 2   In according to Boolean constraints (being 'inherited' from the set $\mathcal{C}_{final}$) there are five assignments for the (unassigned) Boolean variables: $(p, q, s) \in \{ (0,1,0), (1,0,0), (1,1,0), (1,1,0), (1,0,1) \}$. Hence, there are five possible simplifications of proof Π. Below we give some of them.

$$
(p,q,s) = (0,1,0) \longmapsto \cfrac{\cfrac{}{t \vdash t_1} Ax \quad \cfrac{\cfrac{}{r \vdash r_1} Ax}{r \vdash ?q, r_1} ?wR}{r,t \vdash t_1 \otimes ?q, r_1} \otimes R
$$

$$
(p,q,s) = (0,1,1) \longmapsto \cfrac{\cfrac{}{t \vdash t_1} Ax \quad \cfrac{\cfrac{\cfrac{}{r \vdash r_1} Ax}{r \vdash ?q, r_1} ?wR}{r \vdash ?q \oplus s, r_1} \oplus R}{r,t \vdash t_1 \otimes (?q \oplus s), r_1} \otimes R
$$

Example 4. illustrates our method for tracking formula occurrences connected with the application of contraction rules.

Example 4.   Consider the following LL^{PRE} proof:

$$
\cfrac{-\{q_1 = 1, q = 1, q > 0\} \; \cfrac{\cfrac{\cfrac{-\mathcal{C}_2 \cup \{a_1 = 1, a = 1, a > 0, a_1 > 0\}}{a_1 \vdash a - \mathcal{C}_2 \cup \{a > 0, a_1 > 0,\}} Ax}{d_1 \wp a_1 \vdash (?(d \oplus s))_{[x_1]}, a - \mathcal{C}_2 \cup \{a > 0\}} \wp L \quad \Pi_1}{d_1 \wp a_1, b_1 \vdash (?(d \oplus s))_{[x_1]}, (?(d \oplus s))_{[x_2]}, a \otimes b - \mathcal{C}_1 \cup \{?(d \oplus s)_{[]} > 0\}} \otimes R}{\cfrac{q_1 \vdash q - \{q > 0\}}{} Ax \quad \cfrac{d_1 \wp a_1, b_1 \vdash (?(d \oplus s))_{[]}, a \otimes b - \{?(d \oplus s)_{[]} > 0\}}{} c?R}{q_1, d_1 \wp a_1, b_1 \vdash (?(d \oplus s) \otimes q)_{[]}, (a \otimes b)_{[]} - \emptyset} \otimes R
$$

where the subproofs Π_1 and Π_2 are respectively as follows:

$$
\cfrac{\cfrac{-\mathcal{C}_2 \cup \{b_1 = 1, b = 1, b > 0\}}{b_1 \vdash b - \mathcal{C}_2 \cup \{b > 0\}} Ax}{b_1 \vdash ?(d \oplus s)_{[x_2]}, b - \mathcal{C}_2 \cup \{b > 0\}} w?R
$$

$$
\cfrac{\cfrac{\cfrac{-\mathcal{C}_2 \cup \{d_1 = 1, d_{[x_1]} = 1, s_{[x_1]} \le d_{[x_1]}, d_1 > 0, a > 0\}}{d_1 \vdash d_{[x_1]} - \mathcal{C}_2 \cup \{s_{[x_1]} \le d_{[x_1]}, d_1 > 0, a > 0\}} Ax}{d_1 \vdash (d \oplus s)_{[x_1]} - \mathcal{C}_2 \cup \{d_1 > 0, a > 0\}} \oplus R}{d_1 \vdash (?(d \oplus s))_{[x_1]} - \mathcal{C}_2 \cup \{d_1 > 0, a > 0\}} ?R
$$

and the set $\mathcal{C}_2 = \mathcal{C}_1 \cup \{?(d \oplus s)_{[]} > 0\}$, $\mathcal{C}_1 = \{ (d_{[]}, s_{[]}) = (d_{[x_1]}, s_{[x_1]}) + (d_{[x_2]}, s_{[x_2]})$, $(d_{[x_1]}, s_{[x_1]}) = (d_{[x_2]}, s_{[x_2]}) \vee (d_{[x_1]}, s_{[x_1]}) = (0,0) \vee (d_{[x_2]}, s_{[x_2]}) = (0,0) \}$. We have the following solutions: $q_1 = q = 1$, $a_1 = a = 1$, $b_1 = b = 1$, $d_1 = d_{[x_1]} = 1$, $(s_{[x_1]}, d_{[x_2]}, s_{[x_2]}, s, d) \in \{(0,1,0,0,1), (0,0,0,0,1), (1,0,0,1,1)\}$ and hence three possible simplifications. For example the simplifications which correspond to the first two assignments are as follows:

$$\cfrac{\cfrac{\cfrac{\cfrac{\cfrac{d_1 \vdash d}{d_1 \vdash ?d}\;{}^{Ax}}{d_1 \wp a_1 \vdash ?d, a}\;{}^{\wp L}\quad \cfrac{b_1 \vdash b}{b_1 \vdash ?d, b}\;{}^{Ax}_{w?R}}{d_1 \wp a_1, b_1 \vdash ?d, ?d, a \otimes b}\;{}^{\otimes R}}{\cfrac{d_1 \wp a_1, b_1 \vdash ?d, a \otimes b}{}}\;{}^{?cR}}{q_1, d_1 \wp a_1, b_1 \vdash ?d \otimes q, a \otimes b}}\;{}^{\otimes R}$$

$$\cfrac{q_1 \vdash q\;{}^{Ax}\quad \cfrac{\cfrac{\cfrac{d_1 \vdash d}{d_1 \vdash ?d}\;{}^{Ax}}{d_1 \wp a_1 \vdash ?d, a}\;{}^{\wp L}\quad \cfrac{b_1 \vdash b}{}\;{}^{Ax}}{d_1 \wp a_1, b_1 \vdash ?d, a \otimes b}\;{}^{\otimes R}}{q_1, d_1 \wp a_1, b_1 \vdash ?d \otimes q, a \otimes b}\;{}^{\otimes R}$$

5   Formal Results

In this section we give the formal results which establish soundness and completeness of Algorithm *PRE*. We begin with the definitions of generated-subformulae, eliminable, passive and redundant formulae in a propositional proof π generated in the labelled system LL^{PRE}.

Definition 1. *(generated-subformulae) Let $F_{[w]}$ be a (sub)formula which occurs in proof π. We say a formula $\phi_{[wu]}$ is a* generated-subformula *of formula $F_{[w]}$ iff they satisfy either of the following conditions:*
1) $u \equiv w$ (i.e. u is empty word) and $\phi_{[w]} \equiv F_{[w]}$
2) $\phi_{[\,]}$ is a subformula of $F_{[\,]}$ and $\phi_{[wu]}$ is the active formula of some rule occurrence in π.

For example, for the proof of Example 4 the generated-subformulae of the formula $(?(d \oplus s) \otimes q)_{[\,]}$ are: $q_{[\,]}$, $?(d \oplus s)_{[\,]}$, $?(d \oplus s)_{[x_1]}$, $?(d \oplus s)_{[x_2]}$, $d_{[x_1]}$, while generated-subformulae of the formula $(a \otimes b)_{[\,]}$ are: $a_{[\,]}$, $b_{[\,]}$.

Definition 2. *(eliminable, passive, redundant formula in the system LL^{PRE}) Let $F_{[w]}$ be a (sub)formula which occurs in a propositional proof π.*
- We say $F_{[w]}$ is eliminable *in π iff $F_{[w]}$ is not the active formula of any multiplicative rule occurrence in π and all generated-subformule of $F_{[w]}$ are eliminable.*
- We say $F_{[w]}$ is passive *in π iff none of its generated-subformulae have appeared in axiom leaves or as the principal formula(e) of leaves that correspond to initial rules ($\top R$, $\mathbf{1}R$, $\perp L$, $\mathbf{0}L$).*
- We say $F_{[w]}$ is redundant *in π iff it satisfies the following conditions:*
1) $F_{[w]}$ either appears at the root sequent of π or is the active formula of the contraction rule occurrence in π;
2) $F_{[w]}$ is not the active formula of the $\oplus R$, $\&L$ or $-\circ R^2$ rule occurrence in π;
3) $F_{[w]}$ is passive and eliminable.
Otherwise, we say $F_{[w]}$ is not redundant.

Let us recall that our approach is that a formula may be connected as redundant iff its elimination does not endanger the existence of the proof, does not require additional proof search and whose elimination implies just elimination of the corresponding subformulae. For example, consider the proof shown on the left side and note redundant formula $?a \oplus b$. The (sub)formula b is redundant while $?a$ is not redundant.

$$\cfrac{\cfrac{\cfrac{c \vdash c}{c \vdash ?a, c}\;{}^{Ax}_{w?R}}{c \vdash ?a \oplus b, c}\;{}^{\oplus R}}{}$$

---

[2] Active formula which belong to the succedent.

Let π be a proof (in the labelled system LL^{PRE}) and $\mathcal{I}$ be an assignment of Boolean variables (generated in the step 2. of the Algorithm PRE) such that at least one atom has been assigned the value 0. Let $\mathcal{R}$ be the set of all (sub)formulae (from π) made up of atoms being assigned the value 0 under the assignment $\mathcal{I}$.

The following results hold:

Lemma 1. *Let $F_{[w]}$ (i.e. $F_{[w]}[0]$) be a formula from $\mathcal{R}$.*
1) If $F_{[w]}$ is the active formula (respectively active formula from the succedent) of the $\oplus R$ or $\&L$ (respectively of the $\multimap R$) rule occurrence in π then there is a formula $P_{[w]}$ which satisfies the conditions:
1.a) $P_{[w]} \in \mathcal{R}$,
1.b) $P_{[w]} \oplus F_{[w]} \in \mathcal{R}$ or $F_{[w]} \oplus P_{[w]} \in \mathcal{R}$ or $P_{[w]} \multimap F_{[w]} \in \mathcal{R}$ or $P_{[w]} \& F_{[w]} \in \mathcal{R}$ or $F_{[w]} \& P_{[w]} \in \mathcal{R}$.
2) $F_{[w]}$ is redundant formula in π or $\exists Q_{[v_1]} \in \mathcal{R}$ ($Q_{[v_1]}$ is redundant in π and $F_{[w]}$ is a subformula of $Q_{[v_1]}$).

Proof. 1) Let us just prove the case when $F_{[w]}$ is the active formula of the $\oplus R$ rule occurrence in π. Thus, in the proof π there is an inference of the form as shown on the left. The constraint $P_{[w]} \leq F_{[w]}$ means that the same inequality

$$\frac{\Gamma \vdash F_{[w]}, \Delta \quad - \quad \mathcal{C} \cup \{P_{[w]} \leq F_{[w]}\}}{\Gamma \vdash (F \oplus P)_{[w]}, \Delta \quad - \quad \mathcal{C}} \; \oplus R$$

holds for the corresponding Boolean expressions. So as the formula $F_{[w]} \in \mathcal{R}$ (i.e. its corresponding Boolean expression) has been assigned the value 0 the same value must be assigned to the formula $P_{[w]}$ and hence to the formula $(F \oplus P)_{[w]}$ i.e. it must be $P_{[w]} \in \mathcal{R}$, and $(F \oplus P)_{[w]} \in \mathcal{R}$.

2) As $F_{[w]} \in \mathcal{R}$ the set $\mathcal{C}_{final}$ does not contain constraints of the form: $F_{[w]} > 0$, $F_{1,[wv_n]} > 0$, $\dots F_{n,[wv_n]} > 0$, $F_{[w]} = 1$, $F_{1,[wv_1]} = 1, \dots \dots F_{n,[wv_n]} = 1$ where formulae $F_{1,[wv_1]}, F_{2,[wv_2]}, \dots F_{n,[wv_n]}$ are all generated-subformulae of formula $F_{[w]}$. Thus, formula $F_{[w]}$ must be passive and eliminable formula in π.
Let us assume that $F_{[w]} \in \mathcal{R}$ is not redundant formula. Thus formula $F_{[w]}$ is either (1) the active formula of the $\oplus R$, $\&L$ or $\multimap R$ rule or (2) does not appear at the root sequent of π and is not the active formula of any contraction rule occurrence in π. In the case of (1) the assertion holds as a consequence of the first part of this Lemma. In the case of (2) $F_{[w]}$ must be a subformula of an active formula of (at least one) contraction rule occurrence in π. Note the contraction rule occurrence which is closest to the root sequent of π and the active formula $\sharp Q_{1,[w]}$ ($\sharp \in \{?, !\}$) which 'contains' $F_{[w]}$ as a subformula. Since $F_{[w]} \in \mathcal{R}$ i.e. $F_{[w]}$ is assigned the value 0, we further distinguish two subcases:
(2.1) $Q_{1,[w]} \in \mathcal{R}$ In this subcase we have that $Q_{1,[w]}$ is redundant, so the above assertion holds (with $Q_{[v_1]} = \sharp Q_{1,[w]}$).
(2.2) $Q_{1,[w]} \notin \mathcal{R}$ In this subcase the root sequent must contain the (sub)formula $F_{[\,]}$ which has been assigned the value 0. It is obvious that $F_{[\,]}$ is redundant and $F_{[\,]} \in \mathcal{R}$, so the above assertion holds (with $Q_{[v_1]} = F_{[\,]}$). $\qquad \square$

Proposition 2 (Soundness of the Algorithm *PRE*). *Denoting the proof tree that results from applying step 3. of the Algorithm* PRE *to* π *by* π', *we have:* π' *is a valid LL proof, with the same multiplicative structure as* π, *with the same inference rules and the same order of inference rules modulo deleting some of them.*

Proof. By induction on the height of the proof π. □

Proposition 3 (Completeness of the Algorithm *PRE*). *Let* π *be a propositional* LL^{PRE} *proof of the sequent* $\Gamma_{[]} \vdash \Delta_{[]} - \emptyset$. *Let* $F^1_{[w_1]}, \ldots F^n_{[w_n]}$ *be redundant (sub)formulae in* π. *Then there is assignment* $\mathcal{I}$ *of Boolean variables in which all participating atoms and constants in formulae* $F^1_{[w_1]}, \ldots F^n_{[w_n]}$ *are assigned the value 0.*

Proof. As $F^i_{[w_i]}$ is eliminable and passive, the set $\mathcal{C}_{final}$ does not contain the constraints that do not allow to any participating (subformula i.e.) atom of formula $F^i_{[w_i]}$ to be assigned the value 0. Thus, there is an assignment $\mathcal{I}$ with all atoms and constants from $F^1_{[w_1]}, \ldots F^n_{[w_n]}$ being annotated to 0. □

6   Possible Applications

Our labelling system is independent of the search strategy used and independent of any implementation technique. It can be combined with any existing context distribution mechanism (for example, with the existing lazy splitting mechanism for distribution of multiplicative resources [16]) in order to both accelerate proof search and to reduce the search space. For example, consider the following situation with the multiplicative rule where proof is successfully completed in the left branch. If A is redundant in the subproof Π_1 and if the Π_1 can accept addi-

$$\frac{\begin{array}{c}\Pi_1\\ \Gamma_1 \vdash A, \Delta_1 \end{array} \quad \Gamma_2 \vdash B, \Delta_2}{\Gamma_1, \Gamma_2 \vdash A \otimes B, \Delta_1, \Delta_2} \otimes R$$

tional formulae (it is obvious that such proof must contain an occurrence of the $\top$ inference) then it is straightforward to conclude provability of the sequents: $\Gamma_1 \vdash \Delta_1$, and $\Gamma_1, \Gamma_2 \vdash B, \Delta_1, \Delta_2$. If we are interested only to find a proof, then after completion of the left branch of the $\otimes R$ rule the search can be terminated with success without examination of the right branch.

Another possibility for reduction of proof search space we have at additive branch nodes as shown below. If A is unused in the subproof Π_1 (and hence

$$\frac{\begin{array}{c}\Pi_1\\ \Gamma_1 \vdash A, \Delta_1 \quad \Gamma_1 \vdash B, \Delta_1\end{array}}{\Gamma_1 \vdash A\&B, \Delta_1} \&R$$
$$\vdots$$
$$\Gamma \vdash A\&B, \Delta$$

$\Gamma_1 \vdash \Delta_1$ is provable), an analyzer could conclude (without any examination) provability for the sequents $\Gamma \vdash A, \Delta$ and $\Gamma \vdash \Delta$. Also, if B is *similar enough* to A, it could be concluded (on the basis of possible replacement of A with formula B and hence without examination of the right branch) that sequent $\Gamma_1 \vdash B, \Delta_1$ is provable too. This is similar to the approach of Cervesato, Hodas and Pfenning [1] for optimising search in such cases.

An interesting point could be to analyze, in the case of failure of the original search, how the labels and constraints of the system LL^{PRE} and Algorithm PRE

can be used and/or extended to extract the appropriate information about the failure. For example, some resources may remain unconsumed, preventing the proof being successfully completed. When faced with such a situation, a simple answer *no* is not particularly helpful. However, as there are intuitively more ways for a linear sequent to fail than a classical one (as the lack of weakening and contraction limits the ability to remove "unwanted" resources), there is more potential for a variety of methods to extract information from failed proof attempts. For example, consider the left-hand failed proof attempt given below.

$$
\frac{
\dfrac{
\dfrac{
\dfrac{!\mathcal{P}, r \vdash r \quad\quad !\mathcal{P}, p, q \vdash p}{!\mathcal{P}, p, q, r \vdash p \otimes r}
}{!\mathcal{P}, p, q, r \vdash s}
}{
\dfrac{!\mathcal{P} \vdash p^{\perp} \wp q^{\perp} \wp r^{\perp} \wp s}{!(p^{\perp} \wp q^{\perp} \wp r^{\perp} \wp s \multimap t), !(p \otimes r \multimap s) \vdash t}} \; R^{\perp} * 3, \wp R * 3
}{!\mathcal{P}}
$$

$$
\frac{\dfrac{?}{s, p \vdash p} \quad \dfrac{?}{r \vdash q, r}}{s, p, r \vdash p \otimes q, r} \; \otimes R
$$

Let us 'extend' the LL^{PRE} system with the rule $\dfrac{- \mathcal{C} \cup \{p = 1, p_1 = 1\}}{\Gamma, p_1 \vdash p, \Delta \quad - \mathcal{C}}$ which would be applicable in the case of failure for a (non-terminal) leaf node. Thus, for the above left-hand failed proof attempt, by using the same constraint and labelling technique and Algorithm PRE we will get that q is the only redundant formula. So, its elimination will lead to a valid proof. Consider now the above right-hand failed proof attempt. Formulae s and q are 'reasons' for the failure. According to our approach s is the only redundant formula, while q is passive but not redundant. This indicates that s can be simply eliminated (i.e. just omitted) while elimination of formula q 'requires' more subtlety in order to get a valid proof.

7   Conclusions and Future Work

We have shown how some form of labelled deduction can be used to extract information about the necessary and unnecessary formulae in a proof. As we have explained our approach assumes (among others) preserving the multiplicative branching structure of the original proof. Naturally, the next step is to extend the approach to extract information about redundant multiplicative rules whose elimination will not require additional proof search and/or endanger existence of the proof. It is clear that an active formula of a multiplicative binary rule can be considered as redundant if the corresponding context formulae are redundant. Thus, seems that by reformulating the rule as

$$
\frac{\Gamma_1 \vdash \phi_{[w]}, \Delta_1 \quad -\mathcal{D}, \mathcal{C} \cup \{\phi_{[w]} \geq \Gamma_1 + \Delta_1\} \quad\quad \Gamma_2 \vdash \psi_{[w]}, \Delta_2 \quad -\mathcal{D}, \mathcal{C} \cup \{\psi_{[w]} \geq \Gamma_2 + \Delta_2\}}{\Gamma_1, \Gamma_2 \vdash (\phi \otimes \psi)_{[w]}, \Delta_1, \Delta_2 \quad -\mathcal{D}, \mathcal{C}} \; \otimes R
$$

i.e. by attaching the constraints $\phi \geq \Gamma_1 + \Delta_1$ and $\psi \geq \Gamma_2 + \Delta_2$ to $\mathcal{C}$ we are able to use the same constraint technique as in Algorithm PRE to detect redundant multiplicative rules.

It should be noted that the multiplicative binary rules are quite intricate in that they seem to always expose new implementation problems. In addition to the problem of non-determinism in the distribution of linear formulae between the multiplicative branches, these rules require extra treatment and an extra amount of subtlety in the detection of redundant formulae.

Whilst our running example is propositional linear logic, we do not believe that our techniques are limited to these fragment; indeed, it seems straightforward to extend these to an arbitrary set of sequent rules. For example, in terms of refined structure of sequent rules, proposed in [8], one possibility is the following generalization for the unary right rules where $C_1 = C \cup \{ (Principal\text{-}part - Active\text{-}part)^3 \leq Active\text{-}part \}$:

$$\frac{\Gamma \vdash Active\text{-}part \; ; \; Quasi\text{-}active\text{-}part \; ; \; Context \; ; \qquad\quad - \quad C_1}{\Gamma \vdash Principal\text{-}part \; ; \; Quasi\text{-}active\text{-}part \; ; \; Context \; ; \; Extra\text{-}part \quad - C}$$

Our work is intended as a contribution to a library of automatic support tools for managing redundancies in sequent calculi proof search. Our work is more aimed at producing proof assistants, rather than proof checkers (such as LEGO) or proof generators (i.e. theorem provers). The techniques we have proposed can be implemented and utilized by means of an automated proof assistant such as Twelf [11], possibly in conjunction with constraint logic programming techniques [10]. A natural direction for future work is complexity analysis of the Algorithm *PRE*. Our technique is limited to sequent proofs and thereby differs from dead-code elimination in functional languages. Developing more general techniques for program slicing and dead-code elimination in advanced logic programming languages are items of future work.

References

1. Cervesato,I., Hodas, J., Pfenning, F.: Efficient Resource Management for Linear Logic Proof Search, in R. Dyckhoff, H. Herre, and P. Schroeder-Heister (eds.), Proceedings of the 5th International Workshop on Extensions of Logic Programming Leipzig, March, 1996. LNAI **1050** (1996) 67–82
2. Galmiche, D., Méry, D.: Resource Graphs and Countermodels in Resource Logics, Proceedings of the IJCAR 2004 Workshop on disproving: Non-Theorems, Non-Validity, Non-Provability, Cork, Ireland (2004) 59–75
3. Galmiche, D., Perrier, G.: On Proof Normalisation in Linear Logic, Theoretical Computer Science **135** (1994) 67–110
4. Harland, J.: An Algebraic Approach to Proof Search in Sequent Calculi, short paper presented at the International Joint Conference on Automated Reasoning, Siena, July (2001)
5. Harland, J., Pym, D.: Resource-distribution via Boolean constraints, ACM Transactions on Computational Logic **4:1** (2003) 56–90
6. Hodas, J., Lopez, P., Polakow, J., Stoilova L., Pimentel, E.: A Tag-Frame System of Resource Management for Proof Search in Linear Logic Programming, in J. Bradfield (ed.), Proceedings of the Annual Conference of the European Association for Computer Science Logic (CSL 2002), Edinburgh, September (2002) 167–182
7. Lopez, P., et al.: Isolating Resource Consumption in Linear Logic Proof Search (extended abstract), Electronic Notes in Theoretical Computer Science **70(2)** (2002)

---

[3] This set contains each (sub)formula from the principal part that does not appear in the active part.

 8. Lutovac, T., Harland J.: Issues in the Analysis of Proof-search Strategies in Sequential Presentations of Logics, Proceedings of the IJCAR'04 Workshop on Strategies in Automated Deduction, Electronic Notes in Theoretical Computer Science **125(2)** (2005) 115–147
 9. Massacci, F.: Efficient Approximate Deduction and an Application to Computer Security, PhD thesis, Università degeli Studi di Roma 'La Sapienza' (1998)
10. Marriot, K., Stuckey, P.: Programming with Constraints, MIT Press (1998)
11. Pfenning, F., Schürmann, C., Twelf — a meta-logical framework for deductive systems, H. Ganzinger (ed.), Proceedings of the 16th International Conference on Automated Deduction (CADE-16), Trento, Italy, July 1999. LNAI **1632** (1999) 202–206
12. Polakov, J.: Linear Logic Programming with an Ordered Context, in Proceedings of the Second International ACM SIGPLAN Conference on Principles and Practice of Declarative Programming (PPDP'00) Montreal, (2000) 68–79
13. Polakov, J.: Linearity Constraints as Bounded Intervals in Linear Logic Programming, in D, Galmiche, D. Pym and P. O'Hearn (eds.), Proceedings of the LICS'04 Workshop on Logic for Resources, Process and Programs (LRPP) (2004) 173–182
14. Schmitt, S., Kreitz, C.: Deleting Redundancy in Proof Reconstruction, in: H. de Swart, (ed.), International Conference TABLEAUX-98, LNAI **1397** (1998) 262–277
15. Wallen, L.: Automated Proof Search in Non-classical Logic, MIT Press (1990)
16. Winikoff, M., Harland, J.: Implementing the Linear Logic Programming Language Lygon, Proceedings of the International Logic Programming Symposium, Portland, December (1995) 66–80

A Tableau Algorithm for Description Logics with Concrete Domains and GCIs

Carsten Lutz and Maja Miličić*

Institute of Theoretical Computer Science,
TU Dresden, Germany
{lutz, milicic}@tcs.inf.tu-dresden.de

Abstract. In description logics (DLs), *concrete domains* are used for defining concepts based on concrete qualities of their instances such as the weight, age, duration, and spatial extension. So-called *general concept inclusions (GCIs)* play an important role for capturing background knowledge. It is well-known that, when combining concrete domains with GCIs, reasoning easily becomes undecidable. In this paper, we identify a general property of concrete domains that is sufficient for proving decidability of DLs with both concrete domains and GCIs. We exhibit some useful concrete domains, most notably a spatial one based on the RCC-8 relations, which have this property. Then, we present a tableau algorithm for reasoning in DLs equipped with concrete domains and GCIs.

1 Introduction

Description Logics (DLs) are an important family of logic-based knowledge representation formalisms [4]. In DL, one of the main research goals is to provide a toolbox of logics such that, given an application, one may select a DL with adequate expressivity. Here, adequate means that, on the one hand, all relevant concepts from the application domain can be captured. On the other hand, no unessential means of expressivity should be included to prevent a (potential) increase in computational complexity. For several relevant applications of DLs such as the semantic web and reasoning about ER and UML diagrams, there is a need for DLs that include, among others, the expressive means *concrete domains* and *general concept inclusions (GCIs)* [3,8,15]. The purpose of concrete domains is to enable the definition of concepts with reference to concrete qualities of their instances such as the weight, age, duration, and spatial extension. GCIs play an important role in modern DLs as they allow to represent background knowledge of application domains by stating that the extension of a concept is included in the extension of another concept.

Unfortunately, combining concrete domains with GCIs easily leads to undecidabilty. For example, it has been shown in [18] that the basic DL $\mathcal{ALC}$ extended

---

* supported by DFG under grant GRK 334/3.

B. Beckert (Ed): TABLEAUX 2005, LNAI 3702, pp. 201–216, 2005.

with GCIs and a rather inexpressive concrete domain based on the natural numbers and providing for equality and incrementation predicates is undecidable, see also the survey paper [16]. In view of this discouraging result, it is a natural question whether there are *any* useful concrete domains that can be combined with GCIs in a decidable DL. A positive answer to this question has been given in [17] and [14], where two such well-behaved concrete domains are identified: a temporal one based on the Allen relations for interval-based temporal reasoning, and a numerical one based on the rationals and equipped with various unary and binary predicates such as "$\leq$", "$>_5$", and "$\neq$". Using an automata-based approach, it has been shown in [17,14] that reasoning in the DLs $\mathcal{ALC}$ and $\mathcal{SHIQ}$ extended with these concrete domains and GCIs is decidable and ExpTime-complete.

The purpose of this paper it to advance the knowledge about decidable DLs with both concrete domains and GCIs. Our contribution is two-fold: first, instead of focussing on particular concrete domains as in previous work, we identify a *general* property of concrete domains, called ω-admissibility, that is sufficient for proving decidability of DLs equipped with concrete domains and GCIs. For defining ω-admissibility, we concentrate on a particular kind of concrete domains: *constraint systems*. Roughly, a constraint system is a concrete domain that only has binary predicates, which are interpreted as jointly exhaustive and pairwise disjoint (JEPD) relations. We exhibit two example constraint systems that are ω-admissible: a temporal one based on the rational line and the Allen relations [1], and a spatial one based on the real plane and the RCC8 relations [6,20]. The proof of ω-admissibility turns out to be relatively straightforward in the Allen case, but is somewhat cumbersome for RCC8. We believe that there are many other useful constraint systems that can be proved ω-admissible.

Second, for the first time we develop a *tableau algorithm* for DLs with both concrete domains and GCIs. This algorithm is used to establish a general decidability result for $\mathcal{ALC}$ equipped with GCIs and any ω-admissible concrete domain. In particular, we obtain decidability of $\mathcal{ALC}$ with GCIs and the Allen relations as first established in [17], and, as a new result, get decidability of $\mathcal{ALC}$ with GCIs and the RCC8 relations as a concrete domain. As state-of-the-art DL reasoners such as FaCT and RACER are based on tableau algorithms similar to the one described in this paper [11,10], we view our algorithm as a first step towards an efficient implementation of description logics with (ω-admissible) concrete domains and GCIs.

This paper is organized as follows: in Section 2, we introduce constraint systems and identify some properties of constraint systems that will be useful for defining ω-admissibility. In Section 3, we introduce the description logic $\mathcal{ALC}(\mathcal{C})$ that incorporates constraint systems and GCIs. The tableau algorithm for deciding satisfiability in $\mathcal{ALC}(\mathcal{C})$ is developed in Section 4. In Section 5, we briefly discuss the implementability of our algorithm. This paper is accompanied by a technical report containing full proofs [19].

2   Constraint Systems

We introduce a general notion of *constraint system* that is intended to capture standard constraint systems based on a set of jointly-exhaustive and pairwise-disjoint (JEPD) binary relations.

Let Var be a countably infinite set of variables and Rel a finite set of binary relation symbols. A Rel-*constraint* is an expression $(v\ r\ v')$ with $v, v' \in$ Var and $r \in$ Rel. A Rel-*network* is a (finite or infinite) set of Rel-constraints. Let N be a Rel-network. We use V_N to denote the variables used in N and say that N is *complete* if, for all $v, v' \in V_N$, there is exactly one constraint $(v\ r\ v') \in N$.

To assign a semantics to networks in an abstract way, we use complete networks as models: N is a *model of a network* N' if N is complete and there is a mapping $\tau : V_{N'} \to V_N$ such that $(v\ r\ v') \in N'$ implies $(\tau(v)\ r\ \tau(v')) \in N$. In this context, the nodes in N, although from the set Var, are to be understood as values rather than variables (see below for examples).

A *constraint system* $C = \langle \text{Rel}, \mathfrak{M} \rangle$ consists of a finite set of binary relation symbols Rel and a set $\mathfrak{M}$ of complete Rel-networks (the *models* of C). A Rel-network N is *satisfiable* in C if $\mathfrak{M}$ contains a model of N.

We give two examples of constraint systems: a constraint system for temporal reasoning based on the Allen relations in the rational line, and a constraint system for spatial reasoning based on the RCC8 relations in the real plane. Both constraint systems have been extensively studied in the literature.

In artificial intelligence, constraint systems based on Allen's interval relations are a popular tool for the representation of temporal knowledge [1]. Let

$$\text{Allen} = \{\text{b, a, m, mi, o, oi, d, di, s, si, f, fi,} =\}$$

denote the thirteen Allen relations. Examples of these relations are given in Figure 1. As the flow of time, we use the rational numbers with the usual ordering. Let $\text{Int}_\mathbb{Q}$ denote the set of all closed intervals $[q_1, q_2]$ over $\mathbb{Q}$ with $q_1 < q_2$,

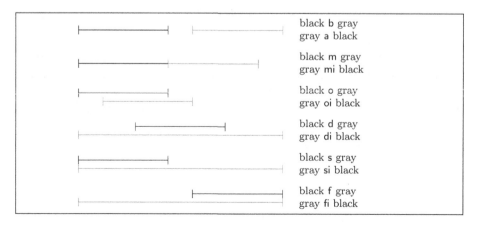

Fig. 1. The thirteen Allen relations, equality omitted

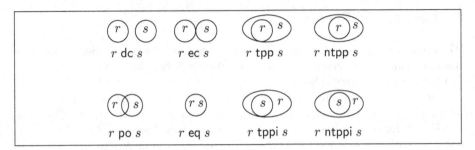

Fig. 2. The eight RCC8 relations

i.e., point-intervals are not admitted. The extension $r^{\mathbb{Q}}$ of each Allen relation r is a subset of $\mathsf{Int}_{\mathbb{Q}} \times \mathsf{Int}_{\mathbb{Q}}$. It is defined in terms of the relationships between endpoints in the obvious way, c.f. Figure 1. We define the constraint system $\mathsf{Allen}_{\mathbb{Q}} = \langle \mathsf{Allen}, \mathfrak{M}_{\mathbb{Q}} \rangle$ by setting $\mathfrak{M}_{\mathbb{Q}} := \{N_{\mathbb{Q}}\}$, where $N_{\mathbb{Q}}$ is defined by fixing a variable $v_i \in \mathsf{Var}$ for every $i \in \mathsf{Int}_{\mathbb{Q}}$ and setting

$$N_{\mathbb{Q}} := \{(v_i \; r \; v_j) \mid r \in \mathsf{Allen}, \; i,j \in \mathsf{Int}_{\mathbb{Q}} \text{ and } (i,j) \in r^{\mathbb{Q}}\}.$$

Whether we use the rationals or the reals for defining this constraint system has no impact on the satisfiability of (finite and infinite) constraint networks.

The RCC8 relations describe the possible relation between two regions in a topological space [20]. In this paper, we use the standard topology of the real plane, one of the most natural topologies for spatial reasoning. Let

$$\mathsf{RCC8} = \{\mathsf{eq}, \mathsf{dc}, \mathsf{ec}, \mathsf{po}, \mathsf{tpp}, \mathsf{ntpp}, \mathsf{tppi}, \mathsf{ntppi}\}$$

denote the RCC8 relations. Examples of these relations are given in Figure 2. Recall that a topological space is a pair $\mathfrak{T} = (U, \mathbb{I})$, where U is a set and $\mathbb{I}$ is an *interior operator* on U, i.e., for all $s, t \subseteq U$, we have

$$\mathbb{I}(U) = U \quad \mathbb{I}(s) \subseteq s \quad \mathbb{I}(s) \cap \mathbb{I}(t) = \mathbb{I}(s \cap t) \quad \mathbb{II}(s) = \mathbb{I}(s).$$

As usual, the closure operator $\mathbb{C}$ is defined as $\mathbb{C}(s) = \overline{\mathbb{I}(\overline{s})}$, where $\overline{t} = U \setminus t$, for $t \subseteq U$. As the *regions* of a topological space $\mathfrak{T} = (U, \mathbb{I})$, we use the set of non-empty, regular closed subsets of U, where a subset $s \subseteq U$ is called *regular closed* if $\mathbb{CI}(s) = s$. Given a topological space $\mathfrak{T}$ and a set of regions $U_{\mathfrak{T}}$, we define the extension of the RCC8 relations as the following subsets of $U_{\mathfrak{T}} \times U_{\mathfrak{T}}$:

$$
\begin{aligned}
(s,t) \in \mathsf{dc}^{\mathfrak{T}} &\text{ iff } s \cap t = \emptyset \\
(s,t) \in \mathsf{ec}^{\mathfrak{T}} &\text{ iff } \mathbb{I}(s) \cap \mathbb{I}(t) = \emptyset \wedge s \cap t \neq \emptyset \\
(s,t) \in \mathsf{po}^{\mathfrak{T}} &\text{ iff } \mathbb{I}(s) \cap \mathbb{I}(t) \neq \emptyset \wedge s \setminus t \neq \emptyset \wedge t \setminus s \neq \emptyset \\
(s,t) \in \mathsf{eq}^{\mathfrak{T}} &\text{ iff } s = t \\
(s,t) \in \mathsf{tpp}^{\mathfrak{T}} &\text{ iff } s \cap \overline{t} = \emptyset \wedge s \cap \overline{\mathbb{I}(t)} \neq \emptyset \\
(s,t) \in \mathsf{ntpp}^{\mathfrak{T}} &\text{ iff } s \cap \overline{\mathbb{I}(t)} = \emptyset \\
(s,t) \in \mathsf{tppi}^{\mathfrak{T}} &\text{ iff } (t,s) \in \mathsf{tpp}^{\mathfrak{T}} \\
(s,t) \in \mathsf{ntppi}^{\mathfrak{T}} &\text{ iff } (t,s) \in \mathsf{ntpp}^{\mathfrak{T}}.
\end{aligned}
$$

Let $\mathfrak{T}_{\mathbb{R}^2}$ be the standard topology on $\mathbb{R}^2$ induced by the Euclidean metric, and let $\mathcal{RS}_{\mathbb{R}^2}$ be the set of all non-empty regular-closed subsets of $\mathfrak{T}_{\mathbb{R}^2}$. Intuitively, regular closedness is required to eliminate sub-dimensional regions such as 0-dimensional points and 1-dimensional spikes. We define the constraint system $\text{RCC8}_{\mathbb{R}^2} = \langle \text{RCC8}, \mathfrak{M}_{\mathbb{R}^2} \rangle$ by setting $\mathfrak{M}_{\mathbb{R}^2} := \{N_{\mathbb{R}^2}\}$, where $N_{\mathbb{R}^2}$ is defined by fixing a variable $v_s \in \mathsf{Var}$ for every $s \in \mathcal{RS}_{\mathbb{R}^2}$ and setting

$$N_{\mathbb{R}^2} := \{(v_s \; r \; v_t) \mid r \in \text{RCC8}, \; s, t \in \mathcal{RS}_{\mathbb{R}^2} \text{ and } (s, t) \in r^{\mathfrak{T}_{\mathbb{R}^2}} \}.$$

2.1 Properties of Constraint Systems

We will use constraint systems as a concrete domain for description logics. To obtain sound and complete reasoning procedures for DLs with such concrete domains, we require constraint system to have certain properties.

Definition 1 (Patchwork Property, Compactness). *Let $\mathcal{C} = \langle \mathsf{Rel}, \mathfrak{M} \rangle$ be a constraint system. If N is a Rel-network and $V \subseteq V_N$, we write $N|_V$ to denote the network $\{(v \, r \, v') \in N \mid v, v' \in V\} \subseteq N$. We say that*

- *$\mathcal{C}$ has the* patchwork property *if the following holds: for all finite, complete, and satisfiable Rel-networks N, M that agree on their (possibly empty) intersection (i.e. $N|_{V_N \cap V_M} = M|_{V_N \cap V_M}$), $N \cup M$ is satisfiable;*
- *$\mathcal{C}$ has the* compactness property *if the following holds: a Rel-network N with V_N infinite is satisfiable in $\mathcal{C}$ if and only if, for every finite $V \subseteq V_N$, the network $N|_V$ is satisfiable in $\mathcal{C}$.*

Intuitively, the patchwork property ensures that satisfiable networks (satisfying some additional conditions) can be "patched" together to a joint network that is also satisfiable. Compactness ensures that this even works when patching together an infinite number of satisfiable networks. In [5], where constraint systems are combined with linear temporal logic, Balbiani and Condotta formulate a property closely related to ours. This property requires that partial models of networks can be extended to complete models. For our purposes, such a property could be used alternatively to the patchwork property and compactness (in fact, it implies both of them).

In the technical report [19], we prove the following:

Theorem 1. $\text{RCC8}_{\mathbb{R}^2}$ *and* $\mathsf{Allen}_{\mathbb{Q}}$ *satisfy the patchwork property and the compactness property.*

The proof of compactness works by devising a satisfiability-preserving translation of constraint networks to sets of first-order formulas, and then appealing to compactness of the latter. In the case of $\mathsf{Allen}_{\mathbb{Q}}$, we need compactness of first-order logic on structures $\langle \mathbb{Q}, < \rangle$, while arbitrary structures are sufficient for $\text{RCC8}_{\mathbb{R}^2}$. The proof of the patchwork property is relatively straightforward in the case of $\mathsf{Allen}_{\mathbb{Q}}$: given two finite, satisfiable, and complete networks N and M that agree on the overlapping part, we show how models of N and M can be manipulated into a model of $N \cup M$. Finally, the proof of the patchwork property of $\text{RCC8}_{\mathbb{R}^2}$ requires quite some machinery. We consider RCC8-networks interpreted on topologies that are induced by so-called fork frames, and then use the standard translation of RCC8-networks into the model logic S4 and repeated careful applications of a theorem from [9] to establish the patchwork property.

3   Syntax and Semantics

We introduce the description logic $\mathcal{ALC}(\mathcal{C})$ that allows to define concepts with reference to the constraint system $\mathcal{C}$. Different incarnations of $\mathcal{ALC}(\mathcal{C})$ are obtained by instantiating it with different constraint systems.

Let $\mathcal{C} = (\text{Rel}, \mathfrak{M})$ be a constraint system, and let $\mathsf{N_C}$, $\mathsf{N_R}$, and $\mathsf{N_{cF}}$ be mutually disjoint and countably infinite sets of *concept names*, *role names*, and *concrete features*. We assume that $\mathsf{N_R}$ has a countably infinite subset $\mathsf{N_{aF}}$ of *abstract features*. A *path* is a sequence $R_1 \cdots R_k g$ consisting of roles $R_1, \ldots, R_k \in \mathsf{N_R}$ and a concrete feature $g \in \mathsf{N_{cF}}$. A path $R_1 \cdots R_k g$ with $\{R_1, \ldots, R_k\} \subseteq \mathsf{N_{aF}}$ is called *feature path*. The set of $\mathcal{ALC}(\mathcal{C})$-concepts is built according to the following syntax rule

$$C ::= A \mid \neg C \mid C \sqcap D \mid C \sqcup D \mid \exists R.C \mid \forall R.C \mid \exists U_1, U_2.r \mid \forall U_1, U_2.r$$

where A ranges over $\mathsf{N_C}$, R ranges over $\mathsf{N_R}$, r ranges over Rel, and U_1, U_2 are both feature paths or $U_1 = Rg_1$ and $U_2 = g_2$ with $R \in \mathsf{N_R}$ and $g_1, g_2 \in \mathsf{N_{cF}}$ or vice versa. Throughout this paper, we use $\top$ as abbreviation for an arbitrary propositional tautology and $C \to D$ for $\neg C \sqcup D$.

A *general concept inclusion axiom (GCI)* is an expression of the form $C \sqsubseteq D$, where C and D are concepts. A finite set of GCIs is called *TBox*. The TBox formalism introduced here is often called *general TBox* since it subsumes several other, much weaker variants [7,13]. We use $C \doteq D$ to abbreviate $C \sqsubseteq D$ and $D \sqsubseteq C$. For example, the following TBox describes some properties of cities using the concrete domain $\text{RCC}_{\mathbb{R}^2}$:

$$\text{City} \sqsubseteq \forall\text{waters}.(\text{River} \sqcup \text{Lake} \sqcup \text{Ocean}) \sqcap \forall\text{trade-partner}.\text{City}$$
$$\text{RegionalTrader} \doteq \text{City} \sqcap \exists(\text{trade-partner loc}), (\text{province loc}).\text{ntpp}$$
$$\text{HarborCity} \doteq \exists(\text{waters loc}), \text{loc}.\text{po} \sqcap \exists(\text{port loc}), \text{loc}.\text{ntpp}$$
$$\sqcap \exists(\text{waters loc}), (\text{port loc}).\text{ec}$$

Here, trade−partner is a role, province, waters, and port are abstract features, and loc is a concrete feature. The second GCI says that RegionalTraders trade with at least one city located in the same province. The third GCI says that HarborCitys overlap some water and contain a port externally connected to this water.

The semantics of $\mathcal{ALC}(\mathcal{C})$ is defined in terms of interpretations as usual. To deal with the *constraint constructors* $\exists U_1, U_2.r$ and $\forall U_1, U_2.r$, interpretations comprise a model of $\mathcal{C}$ as an additional component: an *interpretation* $\mathcal{I}$ is a tuple $(\Delta_{\mathcal{I}}, \cdot^{\mathcal{I}}, M_{\mathcal{I}})$, where $\Delta_{\mathcal{I}}$ is a set called the *domain*, $\cdot^{\mathcal{I}}$ is the *interpretation function*, and $M_{\mathcal{I}} \in \mathfrak{M}$. The interpretation function maps

- each concept name C to a subset $C^{\mathcal{I}}$ of $\Delta_{\mathcal{I}}$,
- each role name R to a subset $R^{\mathcal{I}}$ of $\Delta_{\mathcal{I}} \times \Delta_{\mathcal{I}}$,
- each abstract feature f to a partial function $f^{\mathcal{I}}$ from $\Delta_{\mathcal{I}}$ to $\Delta_{\mathcal{I}}$, and
- each concrete feature g to a partial function $g^{\mathcal{I}}$ from $\Delta_{\mathcal{I}}$ to the set of variables $V_{M_{\mathcal{I}}}$ of $M_{\mathcal{I}}$.

The interpretation function is extended to arbitrary concepts as follows:

$$
\begin{aligned}
\neg C^{\mathcal{I}} &:= \Delta^{\mathcal{I}} \setminus C^{\mathcal{I}}, \\
(C \sqcap D)^{\mathcal{I}} &:= C^{\mathcal{I}} \cap D^{\mathcal{I}}, \\
(C \sqcup D)^{\mathcal{I}} &:= C^{\mathcal{I}} \cup D^{\mathcal{I}}, \\
(\exists R.C)^{\mathcal{I}} &:= \{d \in \Delta^{\mathcal{I}} \mid \exists e \in \Delta^{\mathcal{I}} : (d,e) \in R^{\mathcal{I}} \text{ and } e \in C^{\mathcal{I}}\}, \\
(\forall R.C)^{\mathcal{I}} &:= \{d \in \Delta^{\mathcal{I}} \mid \forall e \in \Delta^{\mathcal{I}} : (d,e) \in R^{\mathcal{I}} \text{ implies } e \in C^{\mathcal{I}}\}, \\
(\exists U_1, U_2.r)^{\mathcal{I}} &:= \{d \in \Delta^{\mathcal{I}} \mid \exists x_1 \in U_1^{\mathcal{I}}(d),\ x_2 \in U_2^{\mathcal{I}}(d) : (x_1 r x_2) \in M_{\mathcal{I}}\} \\
(\forall U_1, U_2.r)^{\mathcal{I}} &:= \{d \in \Delta^{\mathcal{I}} \mid \forall x_1 \in U_1^{\mathcal{I}}(d),\ x_2 \in U_2^{\mathcal{I}}(d) : (x_1 r x_2) \in M_{\mathcal{I}}\}
\end{aligned}
$$

where, for every path $U = R_1 \cdots R_k g$ and $d \in \Delta_{\mathcal{I}}$, $U^{\mathcal{I}}(d)$ is defined as

$$
\begin{aligned}
\{x \in V_{M_{\mathcal{I}}} \mid \exists e_1, \ldots, e_{k+1} : d &= e_1, \\
(e_i, e_{i+1}) &\in R_i^{\mathcal{I}} \text{ for } 1 \leq i \leq k, \text{ and } g^{\mathcal{I}}(e_{k+1}) = x\}.
\end{aligned}
$$

An interpretation $\mathcal{I}$ is a *model* of a concept C iff $C^{\mathcal{I}} \neq \emptyset$. $\mathcal{I}$ is a *model* of a TBox $\mathcal{T}$ iff it satisfies $C^{\mathcal{I}} \subseteq D^{\mathcal{I}}$ for all GCIs $C \sqsubseteq D$ in $\mathcal{T}$.

The most important reasoning tasks for DLs are satisfiability and subsumption: a concept C is called *satisfiable with respect to a TBox* $\mathcal{T}$ iff there exists a common model of C and $\mathcal{T}$. A concept D *subsumes* a concept C with respect to $\mathcal{T}$ (written $C \sqsubseteq_{\mathcal{T}} D$) iff $C^{\mathcal{I}} \subseteq D^{\mathcal{I}}$ holds for each model $\mathcal{I}$ of $\mathcal{T}$. It is well-known that subsumption can be reduced to (un)satisfiability: $C \sqsubseteq_{\mathcal{T}} D$ iff $C \sqcap \neg D$ is unsatisfiable w.r.t. $\mathcal{T}$. Therefore, in the current paper we only consider concept satisfiability.

4 Tableau Algorithm

In this section, we present a tableau algorithm which decides satisfiability of $\mathcal{ALC(C)}$-concepts w.r.t. TBoxes. Tableau algorithms are the most popular decision procedures for description logics since, despite not always yielding tight upper complexity bounds, they are amenable to various optimizations and can often be efficiently implemented. In general, tableau algorithms for DLs decide satisfiability of a concept by trying to construct a model for it. The underlying data structure is a tree which, in case of a successful run of the algorithm, represents a *single* tree model of the input concept and TBox in a straightforward way: the nodes of the tree are the domain elements and the edges denote the extension of roles. Note that this is in contrast to many modal and first-order tableaux, where models of the input formula correspond to *branches* of the tree generated by the tableau algorithm.

In particular, we assume a certain normal form for concepts and TBoxes: negation is only allowed in front of concept names, and the length of paths is restricted.

A concept is said to be in *negation normal form (NNF)* if negation occurs only in front of concept names. We now show that NNF can be assumed without loss of generality: for every $\mathcal{ALC}(\mathcal{C})$-concept, an eqi-satisfiable one in NNF can be computed in linear time. Note that usual NNF transformations are even equivalence-preserving, which cannot be achieved in our case. We assume that the constraint system $\mathcal{C}$ has an equality predicate "=", i.e., $= \in \mathsf{Rel}$ such that, for all $M \in \mathfrak{M}$ and $v \in V_M$, we have $(v = v) \in M$.

Lemma 1 (NNF Conversion). *Exhaustive application of the following rewrite rules translates $\mathcal{ALC}(\mathcal{C})$-concepts to eqi-satisfiable ones in NNF. The number of rule applications is linear in the length of the original concept.*

$$\neg\neg C \rightsquigarrow C \qquad \neg(C \sqcap D) \rightsquigarrow \neg C \sqcup \neg D \qquad \neg(C \sqcup D) \rightsquigarrow \neg C \sqcap \neg D$$

$$\neg(\exists R.C) \rightsquigarrow (\forall R.\neg C) \qquad\qquad \neg(\forall R.C) \rightsquigarrow (\exists R.\neg C)$$

$$\neg(\forall U_1, U_2.r) \rightsquigarrow \bigsqcup_{r' \in \mathsf{Rel}, r' \neq r} \exists U_1, U_2.r'$$

$$\neg(\exists U_1, U_2.r) \rightsquigarrow \bigsqcup_{r' \in \mathsf{Rel}, r' \neq r} \forall U_1, U_2.r' \qquad \text{where } U_1, U_2 \text{ are feature paths}$$

$$\neg(\exists Rg_1, g_2.r) \rightsquigarrow (\forall Rg^*, g_2. =) \sqcap \bigsqcup_{r' \in \mathsf{Rel}, r' \neq r} \forall R.(\forall g_1, g^*.r')$$

$$\text{where } R \in \mathsf{N_R} \setminus \mathsf{N_{aF}} \text{ and } g^* \text{ is a fresh concrete feature}$$

By $\mathrm{nnf}(C)$, we denote the result of converting C into NNF using the above rules.

In the last transformation, the fresh concrete feature g^* is used to propagate the value of g_2 to all R successors. This transformation is the reason for the fact that our NNF translation is not equivalence-preserving. Intuitively, giving an equivalence preserving-translation would require to allow the formation of complex $\mathcal{C}$-relations from atomic ones by means of union.

We now introduce path normal form for $\mathcal{ALC}(\mathcal{C})$-concepts and TBoxes. Path normal form was first considered in [17,14].

Definition 2 (Path Normal Form). *An $\mathcal{ALC}(\mathcal{C})$-concept C is in path normal form (PNF) iff it is in NNF and, for all subconcepts $\exists U_1, U_2.r$ and $\forall U_1, U_2.r$ of C, we have either*

1. *$U_1 = g_1$ and $U_2 = g_2$ for some $g_1, g_2 \in \mathsf{N_{cF}}$ or*
2. *$U_1 = Rg_1$ and $U_2 = g_2$ for some $R \in \mathsf{N_R}$ and $g_1, g_2 \in \mathsf{N_{cF}}$ or*
3. *$U_1 = g_1$ and $U_1 = Rg_2$ for some $R \in \mathsf{N_R}$ and $g_1, g_2 \in \mathsf{N_{cF}}$.*

An $\mathcal{ALC}(\mathcal{C})$-TBox $\mathcal{T}$ is in path normal form iff all concepts in $\mathcal{T}$ are in PNF.

The following lemma shows that we can w.l.o.g. assume $\mathcal{ALC}(\mathcal{C})$-concepts and TBoxes to be in PNF.

Lemma 2. *Satisfiability of $\mathcal{ALC}(\mathcal{C})$-concepts w.r.t. TBoxes can be polynomially reduced to satisfiability of $\mathcal{ALC}(\mathcal{C})$-concepts in PNF w.r.t. TBoxes in PNF.*

Proof. Let C be an $\mathcal{ALC(C)}$-concept. For every feature path $u = f_1 \cdots f_n g$ used in C, we assume that $[g], [f_n g], \ldots, [f_1 \cdots f_n g]$ are concrete features not used in C. We inductively define a mapping λ from feature paths u in C to concepts as follows:

$$\lambda(g) = \top \qquad \lambda(fu) = (\exists f[u], [fu]. =) \sqcap \exists f.\lambda(u)$$

For every $\mathcal{ALC(C)}$-concept C, a corresponding concept $\rho(C)$ is obtained by

- first replacing all subconcepts $\forall u_1, u_2.r$, where $u_i = f_1^{(i)} \cdots f_{k_i}^{(i)} g_i$ for $i \in \{1, 2\}$, with

$$\forall f_1^{(1)}. \cdots \forall f_{k_1}^{(1)}.\forall g_1, g_1.r^{\neq} \sqcup \forall f_1^{(2)}. \cdots \forall f_{k_2}^{(2)}.\forall g_2, g_2.r^{\neq} \sqcup \exists u_1, u_2.r$$

  where $r^{\neq} \in \mathsf{Rel} \setminus \{=\}$ is arbitrary, but fixed;
- and then replacing all subconcepts $\exists u_1, u_2.r$ with $\exists [u_1], [u_2].r \sqcap \lambda(u_1) \sqcap \lambda(u_2)$.

We extend the mapping ρ to TBoxes in the obvious way: replace each GCI $C \sqsubseteq D$ with $\rho(C) \sqsubseteq \rho(D)$. To convert a concept to PNF, we may first convert to NNF and then apply the above translation ρ. It is easily verified that (un)satisfiability is preserved, and that the translation can be done in polynomial time. $\qquad \square$

In what follows, we generally assume that all concepts and TBoxes are in path normal form. Moreover, we require that constraint systems are ω-admissible:

Definition 3 (ω-admissible). *Let $\mathcal{C} = (\mathsf{Rel}, \mathfrak{M})$ be a constraint system. We say that $\mathcal{C}$ is ω-admissible iff the following holds:*

1. *satisfiability in $\mathcal{C}$ is decidable;*
2. *$\mathcal{C}$ has the patchwork property;*
3. *$\mathcal{C}$ has the compactness property.*

In Section 2, we have shown that $\mathsf{RCC8}_{\mathbb{R}^2}$ and $\mathsf{Allen_Q}$ have the patchwork and compactness property. Moreover, satisfiability in $\mathsf{RCC8}_{\mathbb{R}^2}$ and $\mathsf{Allen_Q}$ is NP-complete [21,22]. Thus, these constraint systems are ω-admissible and may be used with our tableau algorithm.

Let C_0 be a concept and $\mathcal{T}$ a TBox such that satisfiability of C_0 w.r.t. $\mathcal{T}$ is to be decided. The *concept form* $C_{\mathcal{T}}$ is defined as

$$C_{\mathcal{T}} = \bigsqcap_{C \sqsubseteq D \in \mathcal{T}} \mathsf{nnf}(C \to D).$$

We define the set of subconcepts $\mathsf{sub}(C_0, \mathcal{T}) := \mathsf{sub}(C_0) \cup \mathsf{sub}(C_{\mathcal{T}})$, with $\mathsf{sub}(C)$ denoting the set of all subconcepts of C, including C.

As already noted, our algorithm uses trees as the main data structure, and nodes of this tree represent elements of the interpretation domain. Due to the presence of concrete domains, trees have two types of nodes: abstract ones that represent individuals of the logic domain $\Delta_{\mathcal{I}}$, and concrete ones representing values of the concrete domain. Likewise, edges represent either roles or concrete features.

Definition 4 (Completion system). *Let O_a and O_c be disjoint and countably infinite sets of* abstract *and* concrete nodes. *A completion tree for C_0, $\mathcal{T}$ is a finite, labelled tree $T = (V_a, V_c, E, \mathcal{L})$ with nodes $V_a \cup V_c$, such that $V_a \subseteq O_a$, $V_c \subseteq O_c$, and all nodes from V_c are leaves. The tree is labelled as follows:*

1. *each node $a \in V_a$ is labelled with a subset $\mathcal{L}(a)$ of $\mathsf{sub}(C_0, \mathcal{T})$,*
2. *each edge $(a, b) \in E$ with $a, b \in V_a$ is labelled with a role name $\mathcal{L}(a, b)$ occurring in C_0 or $\mathcal{T}$;*
3. *each edge $(a, x) \in E$ with $a \in V_a$ and $x \in V_c$ is labelled with a concrete feature $\mathcal{L}(a, x)$ occurring in C_0 or $\mathcal{T}$.*

A node $b \in V_a$ is an R-successor of a node $a \in V_a$ if $(a, b) \in E$ and $\mathcal{L}(a, b) = R$, while an $x \in V_c$ is a g-successor of a if $(a, x) \in E$ and $\mathcal{L}(a, x) = g$. The notion u-successor for a path u is defined in the obvious way. A completion system for C_0 and $\mathcal{T}$ is a tuple $S = (T, \mathcal{N})$ where $T = (V_a, V_c, E, \mathcal{L})$ is a completion tree for C_0 and $\mathcal{T}$ and $\mathcal{N}$ is a Rel-network with $V_{\mathcal{N}} = V_c$.

To decide the satisfiability of C_0 w.r.t. $\mathcal{T}$ (both in PNF), the tableau algorithm is started with the initial completion system

$$S_{C_0} = (T_{C_0}, \emptyset), \text{ where } T_{C_0} = (\{a_0\}, \emptyset, \emptyset, \{a_0 \mapsto \{C_0\}\}).$$

The algorithm applies completion rules to the completion system until an obvious inconsistency (clash) is detected or no completion rule is applicable any more. Before we define the completion rules for $\mathcal{ALC(C)}$, we introduce an operation that is used by completion rules to add new nodes to completion trees.

Definition 5 ($\oplus$ Operation). *An abstract or concrete node is called* fresh *w.r.t. a completion tree T if it does not appear in T. Let $S = (T, \mathcal{N})$ be a completion system with $T = (V_a, V_c, E, \mathcal{L})$. We use the following operations:*

- *$S \oplus aRb$ ($a \in V_a$, $b \in O_a$ fresh in T, $R \in N_R$) yields a completion system obtained from S in the following way:*
  - *if $R \notin N_{aF}$ or $R \in N_{aF}$ and a has no R-successors, then add b to V_a, (a, b) to E and set $\mathcal{L}(a, b) = R$, $\mathcal{L}(b) = \emptyset$;*
  - *if $R \in N_{aF}$ and there is a $c \in V_a$ such that $(a, c) \in E$ and $\mathcal{L}(a, c) = R$ then rename c in T with b.*
- *$S \oplus agx$ ($a \in V_a$, $x \in O_c$ fresh in T, $g \in N_{cF}$) yields a completion system obtained from S in the following way:*
  - *if a has no g-successors, then add x to V_c, (a, x) to E and set $\mathcal{L}(a, x) = g$;*
  - *if a has a g-successor y, then rename y in T and $\mathcal{N}$ with x.*

Let $u = R_1 \cdots R_n g$ be a path. With $S \oplus aux$, where $a \in V_a$ and $x \in O_c$ is fresh in T, we denote the completion system obtained from S by taking distinct nodes $b_1, ..., b_n \in O_a$ which are fresh in T and setting

$$S' := S \oplus aR_1b_1 \oplus \cdots \oplus b_{n-1}R_nb_n \oplus b_ngx$$

To ensure termination of the tableau algorithm, we need a mechanism for detecting cyclic expansions, commonly called *blocking*. Informally, we detect nodes in the completion tree "similar" to previously created ones and "block" them, i.e., apply no more completion rules to such nodes. To define the blocking condition, we need a couple of notions. For $a \in V_a$, define:

$$\mathsf{cs}(a) := \{g \in N_{cF} \mid a \text{ has a } g\text{-successor}\}$$
$$\mathcal{N}(a) := \{(g \; r \; g') \mid \text{ there are } x, y \in V_c \text{ such that } x \text{ is a } g\text{-successor of } a,$$
$$y \text{ is a } g'\text{-successor of } a, \text{ and } (x \; r \; y) \in \mathcal{N}\}$$
$$\mathcal{N}'(a) := \{(x \; r \; y) \mid \text{there exist } g, g' \in \mathsf{cs}(a) \text{ s.t. } x \text{ is a } g\text{-successor of } a,$$
$$y \text{ is a } g'\text{-successor of } a, \text{ and } (x \; r \; y) \in \mathcal{N}\}$$

A *completion* of a Rel-network N is a satisfiable and complete Rel-network N' such that $V_N = V_{N'}$ and $N \subseteq N'$.

Definition 6 (Blocking). *Let $S = (T, \mathcal{N})$ be a completion system for a concept C_0 and a TBox $\mathcal{T}$ with $T = (V_a, V_c, E, \mathcal{L})$. Let $a, b \in V_a$. We say that $a \in V_a$ is*

- *potentially blocked by b if b is an ancestor of a in T, $\mathcal{L}(a) \subseteq \mathcal{L}(b)$, and $\mathsf{cs}(a) = \mathsf{cs}(b)$.*
- *directly blocked by b if a is potentially blocked by b, $\mathcal{N}(a)$ and $\mathcal{N}(b)$ are complete, and $\mathcal{N}(a) = \mathcal{N}(b)$.*

Finally, a is blocked *if it or one of its ancestors is directly blocked.*

We are now ready to define the completion rules, which are given in Figure 3. Among the rules, there are three non-deterministic ones: $R\sqcup$, Rnet and Rnet$'$. All rules except Rnet and Rnet$'$ are rather standard, as they are variants of the corresponding rules from existing algorithms for DLs with concrete domains, see e.g. [2]. The purpose of these additional rules is to resolve potential blocking situations into actual blocking situations (or non-blocking situations) by completing the parts of the network $\mathcal{N}$ that correspond to the "blocked" and "blocking" node. To ensure an appropriate interplay between Rnet/Rnet$'$, and the blocking condition and thus to guarantee termination, we apply these rules with highest precedence.

Note that the blocking mechanism obtained in this way is *dynamic* in the sense that blocking situations can be broken again after they have been established. Also note that the conditions $\mathcal{L}(a) \subseteq \mathcal{L}(b)$ and $\mathsf{cs}(a) = \mathsf{cs}(b)$ can be viewed as a refinement of pairwise blocking as known from [12]: due to path normal form, pairwise blocking is a strictly sharper condition than the above two.

The algorithm applies completion rules until no more rules are applicable (such a completion system is called *complete*), or a clash is encountered.

Definition 7 (Clash). *Let $S = (T, \mathcal{N})$ be a completion system for a concept C and a TBox $\mathcal{T}$ with $T = (V_a, V_a, E, \mathcal{L})$. S is said to contain a clash if*

- *there is an $a \in V_a$ and an $A \in N_C$ such that $\{A, \neg A\} \subseteq \mathcal{L}(a)$, or*
- *$\mathcal{N}$ is not satisfiable in $\mathcal{C}$.*

$R\sqcap$	if $C_1 \sqcap C_2 \in \mathcal{L}(a)$, a is not blocked, and $\{C_1, C_2\} \not\subseteq \mathcal{L}(a)$, then set $\mathcal{L}(a) := \mathcal{L}(a) \cup \{C_1, C_2\}$
$R\sqcup$	if $C_1 \sqcup C_2 \in \mathcal{L}(a)$, a is not blocked, and $\{C_1, C_2\} \cap \mathcal{L}(a) = \emptyset$, then set $\mathcal{L}(a) := \mathcal{L}(a) \cup \{C\}$ for some $C \in \{C_1, C_2\}$
$R\exists$	if $\exists R.C \in \mathcal{L}(a)$, a is not blocked, and there is no R-successor of a such that $C \in \mathcal{L}(b)$, then set $S := S \oplus aRb$ for a fresh $b \in O_a$ and $\mathcal{L}(b) := \mathcal{L}(b) \cup \{C\}$
$R\forall$	if $\forall R.C \in \mathcal{L}(a)$, a is not blocked, and b is an R-successor of a such that $C \notin \mathcal{L}(b)$, then set $\mathcal{L}(b) := \mathcal{L}(b) \cup \{C\}$
$R\exists_c$	if $\exists U_1, U_2.r \in \mathcal{L}(a)$, a is not blocked, and there exist no $x_1, x_2 \in V_c$ such that x_i is a U_i-successor of a for $i = 1, 2$ and $(x_1 \ r \ x_2) \in \mathcal{N}$ then set $S := (S \oplus aU_1 x_1 \oplus aU_2 x_2)$ with $x_1, x_2 \in O_c$ fresh and $\mathcal{N} := \mathcal{N} \cup \{(x_1 \ r \ x_2)\}$
$R\forall_c$	if $\forall U_1, U_2.r \in \mathcal{L}(a)$, a is not blocked, and there are $x_1, x_2 \in V_c$ such that x_i is a U_i-successor of a for $i = 1, 2$ and $(x_1 \ r \ x_2) \notin \mathcal{N}$, then set $\mathcal{N} := \mathcal{N} \cup \{(x_1 \ r \ x_2)\}$
Rnet	if a is potentially blocked by b and $\mathcal{N}(a)$ is not complete, then non-deterministically guess a completion $\mathcal{N}'$ of $\mathcal{N}'(a)$ and set $\mathcal{N} := \mathcal{N} \cup \mathcal{N}'$
Rnet$'$	if a is potentially blocked by b and $\mathcal{N}(b)$ is not complete, then non-deterministically guess a completion $\mathcal{N}'$ of $\mathcal{N}'(b)$ and set $\mathcal{N} := \mathcal{N} \cup \mathcal{N}'$
Rgci	if $C_T \notin \mathcal{L}(a)$, then set $\mathcal{L}(a) := \mathcal{L}(a) \cup \{C_T\}$

Fig. 3. The Completion Rules

Note that the existence of clashes is decidable since we require that satisfiability in $\mathcal{C}$ is decidable. In an actual implementation of our algorithm, checking for clashes would require calling an external reasoner for satisfiability in the constraint system used. The tableau algorithm checks for clashes before each rule application. It returns "satisfiable" if there is a way to apply the non-deterministic rules such that a complete and clash-free completion system is generated. Otherwise, it returns "unsatisfiable". In actual implementations of our algorithm, non-determinism has to be replaced by backtracking and search.

Note that checking for clashes before every rule application ensures that Rnet and Rnet$'$ are well-defined: if Rnet is applied, then there indeed exists a completion $\mathcal{N}'$ of $\mathcal{N}(a)$ to be guessed: due to clash checking, the network $\mathcal{N}$ is satisfiable, and it is readily checked that this implies the existence of the required completion.

Theorem 2. *If $\mathcal{C}$ is an ω-admissible constraint system, the tableau algorithm decides satisfiability of $\mathcal{ALC}(\mathcal{C})$ concepts w.r.t. general TBoxes.*

Proof. Termination of the algorithm is ensured by the blocking condition, the Rnet and Rnet$'$ rules, and the fact that these rules are executed with highest precedence. Completeness can be proved in the standard way, by showing that if the input concept C_0 and TBox $\mathcal{T}$ have a common model $\mathcal{I}$, we can guide the (non-deterministic parts of) the tableau algorithm according to $\mathcal{I}$, such that it ends up with a clash-free completion system. Detailed proofs are given in [19].

Here we sketch the soundness proof. We have to show, that, if the tableau algorithm returns "satisfiable", then the input concept C_0 is satisfiable w.r.t. the input TBox $\mathcal{T}$. If the tableau algorithm returns "satisfiable", then there exists a complete and clash-free completion system $S = (T, \mathcal{N})$ of C_0 and $\mathcal{T}$. Let $T = (V_a, V_c, E, \mathcal{L})$, and let root $\in V_a$ denote the root of T. Our aim is to define a model $\mathcal{I}$ of C_0 and $\mathcal{T}$. We proceed in several steps.

Let blocks be a function that for every directly blocked $b \in V_a$, returns an unblocked $a \in V_a$ such that b is blocked by a in S. It can easliy seen that, by definition of blocking, such a node a always exists. A *path* in S is a (possibly empty) sequence of pairs of nodes $\frac{a_1}{b_1}, \ldots, \frac{a_n}{b_n}$, with $a_1, \ldots, a_n$ and $b_1, \ldots, b_n$ from V_a, such that, for $1 \leq i < n$, b_{i+1} is a successor of a_i in T and

$$a_{i+1} := \begin{cases} b_{i+1} & \text{if } b_{i+1} \text{ is not blocked,} \\ \text{blocks}(b_{i+1}) & \text{otherwise.} \end{cases}$$

We use Paths to denote the set of all paths in S, head(p) to denote the first pair of a path p and tail(p) to denote the last pair of p (if p is nonempty). We now define the "abstract part" of the the model $\mathcal{I}$ we are constructing:

$$\Delta_{\mathcal{I}} := \{p \in \text{Paths} \mid p \text{ non-empty and head}(p) = \frac{\text{root}}{\text{root}}\}$$

$$A^{\mathcal{I}} := \{p \in \Delta_{\mathcal{I}} \mid \text{tail}(p) = \frac{a}{b} \text{ and } A \in \mathcal{L}(a)\},$$

$$R^{\mathcal{I}} := \{(p, p \cdot \frac{a}{b}) \in \Delta_{\mathcal{I}} \times \Delta_{\mathcal{I}} \mid \text{tail}(p) = \frac{a'}{b'} \text{ and } b \text{ is } R\text{-successor of } a' \text{ in } T \}$$

for all $A \in N_C$ and $R \in N_R$. Observe that $\Delta_{\mathcal{I}}$ is non-empty since $\frac{\text{root}}{\text{root}} \in \Delta_{\mathcal{I}}$, and that $f^{\mathcal{I}}$ is functional for every $f \in N_{aF}$, which is ensured by the "$\oplus$" operation and by definition of Paths.

Intuitively, the abstract part of $\mathcal{I}$ as defined above is "patched together" from (copies of) parts of the completion tree T. For defining the concrete part of $\mathcal{I}$, we make this patching explicit: For $p, q \in$ Paths,

- p is called a *hook* if $p = \frac{\text{root}}{\text{root}}$ or tail$(p) = \frac{a}{b}$ with $a \neq b$ (and thus b blocked by a). We use Hooks to denote the set of all hooks.
- we call p a *q-companion* if q is a hook and there exists $q' \in$ Paths such that $p = qq'$ and all nodes $\frac{a}{b}$ in q' satisfy $a = b$, with the possible exception of tail(q').

Intuitively, the hooks, which are induced by blocking situations in T, are the points where we patch together parts of T. The part of T patched at a hook p with tail$(p) = \frac{a}{b}$ is comprised of (copies of) all the nodes c in T that are reachable from a, except indirectly blocked ones. Formally, the part of $\mathcal{I}$ belonging to the hook p is defined as $P(p) := \{q \in \Delta_{\mathcal{I}} \mid q \text{ is a } p\text{-companion}\}$. For $p, q \in$ Hooks, q is called a *successor* of p if q is a p-companion and $p \neq q$. Observe that, for each hook p, $P(p)$ includes all successor hooks of p. Intuitively, this means that the parts patched together to obtain the abstract part of $\mathcal{I}$ are overlapping at the hooks.

For space limitations, we only sketch how the concrete part of $\mathcal{I}$ is defined. The full construction with proofs can be found in [19]. Since the completion system S is clash-free, its constraint network $\mathcal{N}$ is satisfiable. Therefore, there exists a completion $\mathcal{N}^c$ of $\mathcal{N}$. For every $p \in \mathsf{Hooks}$, we define a constraint network $N(p)$ that defines the constraints that have to be satisfied by the concrete part of $\mathcal{I}$ corresponding to $P(p)$. More precisely, $N(p)$ is defined as (a copy of) the part of $\mathcal{N}^c$ that corresponds to the part of T patched at p.

Then the network $\mathbf{N} = \bigcup_{p \in \mathsf{Hooks}} N(p)$ describes the constraints that have to be satisfied by the concrete part of the whole model $\mathcal{I}$. By construction, the networks $N(p)$ are finite, complete, satisfiable, and overlap at the hooks. Due to the blocking condition, their overlapping parts are identical. Thus, we can use the patchwork and compactness property of $\mathcal{C}$ to show that $\mathbf{N}$ is satisfiable in $\mathcal{C}$. Then a model $M_{\mathcal{I}} \in \mathfrak{M}$ of $\mathbf{N}$ becomes the last argument of our interpretation $\mathcal{I}$, and we can define extensions of concrete features in $\mathcal{I}$. To show that $\mathcal{I}$ is indeed a model of C_0 and T, we can prove by structural induction that for all $p \in \Delta_{\mathcal{I}}$ with $\mathsf{tail}(p) = \frac{a}{b}$ and for all $C \in \mathsf{sub}(C_0, T)$ the following holds: if $C \in \mathcal{L}(a)$ then $p \in C^{\mathcal{I}}$. Since $C_0 \in \mathcal{L}(\mathsf{root})$ we have that $\frac{\mathsf{root}}{\mathsf{root}} \in C_0^{\mathcal{I}}$. Finally, $C_T \in \mathcal{L}(a)$ for all unblocked $a \in \mathsf{V_a}$ implies that $p \in C_T^{\mathcal{I}}$ for all $p \in \Delta_{\mathcal{I}}$, and thus $\mathcal{I}$ models T.

5   Conclusion

We have proved decidability of $\mathcal{ALC}$ with ω-admissible constraint systems and GCIs. We conjecture that, by mixing the techniques from the current paper with those from [17,14], it is possible to prove ExpTime-completeness of satisfiability in $\mathcal{ALC}(\mathcal{C})$ provided that satisfiability in $\mathcal{C}$ can be decided in ExpTime. Various language extensions, both on the logical and concrete side, should also be possible in a straightforward way.

We also exhibited the first tableau algorithm for DLs with concrete domains and GCIs in which the concrete domain constructors are not limited to concrete features. We view this algorithm as a first step towards an implementation, although there is clearly room for improvements: the rules Rnet and Rnet$'$ add considerable non-determinism, clash checking involves the whole network $\mathcal{N}$ rather than only a local part of it, and blocking can be further refined.

We believe that, in general, getting rid of the additional non-determinism introduced by Rnet and Rnet$'$ is difficult. One possible way out may be to permit only a single concrete feature: then Rnet and Rnet$'$ become deterministic (in fact they can be omitted), and "potentially blocking" coincides with "directly blocking". We believe that having only one concrete feature is actually rather natural: for the Allen/RCC8 concrete domains, the concrete feature could be hasTime and hasLocation, respectively.

However, a complication is posed by the fact that path normal form introduces additional concrete features. Simply requiring, as an additional restriction, that only concepts and TBoxes in PNF are allowed is rather severe: it can be seen that, then, satisfiability in $\mathcal{ALC}(\mathcal{C})$ instantiated with the RCC8 and Allen constraint systems can be decided by adding some simple clash conditions. In particular, there is no need to use an external reasoner for the constraint system at all. Therefore, it is more interesting to find a tableau algorithm for $\mathcal{ALC}(\mathcal{C})$ with only one concrete feature that does not rely on PNF, but still avoids the non-determinism and global satisfiability check of $\mathcal{N}$.

References

1. J. Allen. Maintaining knowledge about temporal intervals. *Communications of the ACM*, 26(11), 1983.
2. F. Baader and P. Hanschke. A scheme for integrating concrete domains into concept languages. In *Proc. of IJCAI-91*, pages 452–457, Morgan Kaufman. 1991.
3. F. Baader, I. Horrocks, and U. Sattler. Description logics as ontology languages for the semantic web. In *Festschrift in honor of Jörg Siekmann*, LNAI. Springer-Verlag, 2003.
4. F. Baader, D. L. McGuiness, D. Nardi, and P. Patel-Schneider. *The Description Logic Handbook: Theory, implementation and applications*. Cambridge University Press, 2003.
5. P. Balbiani and J.-F. Condotta. Computational complexity of propositional linear temporal logics based on qualitative spatial or temporal reasoning. In *Proc. of FroCoS 2002*, number 2309 in LNAI, pages 162–176. Springer, 2002.
6. B. Bennett. Modal logics for qualitative spatial reasoning. *Journal of the IGPL*, 4(1), 1997.
7. D. Calvanese. Reasoning with inclusion axioms in description logics: Algorithms and complexity. In *Proc. of ECAI-96*, pages 303–307, 1996.
8. D. Calvanese, M. Lenzerini, and D. Nardi. Description logics for conceptual data modeling. In *Logics for Databases and Information Systems*, pages 229–263. Kluwer, 1998.
9. D. M. Gabbay, A. Kurucz, F. Wolter, and M. Zakharyaschev. *Many-Dimensional Modal Logics: Theory and Applications*. Elsevier, 2003.
10. V. Haarslev and R. Möller. RACER system description. In *Proc. of IJCAR'01*, number 2083 in LNAI, pages 701–705. Springer-Verlag, 2001.
11. I. Horrocks. Using an expressive description logic: Fact or fiction? In *Proc. of KR98*, pages 636–647, 1998.
12. I. Horrocks, U. Sattler, and S. Tobies. Practical reasoning for expressive description logics. In *Proc. of LPAR'99*, number 1705 in LNAI, pages 161–180. Springer, 1999.
13. C. Lutz. Complexity of terminological reasoning revisited. In *Proc. of LPAR'99*, number 1705 in LNAI, pages 181–200. Springer, 1999.
14. C. Lutz. Adding numbers to the $\mathcal{SHIQ}$ description logic—First results. In *Proc. of KR2002*, pages 191–202. Morgan Kaufman, 2002.
15. C. Lutz. Reasoning about entity relationship diagrams with complex attribute dependencies. In *Proc. of DL2002*, number 53 in CEUR-WS (http://ceur-ws.org/), pages 185–194, 2002.
16. C. Lutz. Description logics with concrete domains—a survey. In *Advances in Modal Logics Volume 4*, pages 265–296. King's College Publications, 2003.
17. C. Lutz. Combining interval-based temporal reasoning with general tboxes. *Artificial Intelligence*, 152(2):235–274, 2004.
18. C. Lutz. NExpTime-complete description logics with concrete domains. *ACM Transactions on Computational Logic*, 5(4):669–705, 2004.

19. C. Lutz and M. Miličić.   A tableau algorithm for DLs with concrete do-
    mains and GCIs. LTCS-Report 05-07, TU Dresden, 2005. See http://lat.inf.tu-
    dresden.de/research/reports.html.
20. D. A. Randell, Z. Cui, and A. G. Cohn. A spatial logic based on regions and
    connection. In *Proc. of KR'92*, pages 165–176. Morgan Kaufman, 1992.
21. J. Renz and B. Nebel. On the complexity of qualitative spatial reasoning: A max-
    imal tractable fragment of the region connection calculus. *Artificial Intelligence*,
    108(1–2):69–123, 1999.
22. M. Vilain, H. Kautz, and P. van Beek. Constraint propagation algorithms for
    temporal reasoning: a revised report. In *Readings in qualitative reasoning about
    physical systems*, pages 373–381. Morgan Kaufmann, 1990.

The Space Efficiency of OSHL

Swaha Miller and David A. Plaisted

University of North Carolina, Chapel Hill, NC 27599, USA

Abstract. Ordered semantic hyper-linking (OSHL) is a first-order theorem prover that tries to take advantage of the speed of propositional theorem proving techniques. It instantiates first-order clauses to ground clauses, and applies propositional techniques to these ground clauses. OSHL-U extends OSHL with rules for unit clauses to speed up the instantiation strategy. OSHL-U obtains many of the same proofs as Otter does. This shows that many first-order theorems can be obtained without true unification, so techniques used to speed up propositional provers may be applicable to them. OSHL-U, in finding proofs, also generates and stores significantly fewer clauses than resolution prover Otter on many TPTP problems. On some TPTP groups, OSHL-U finds more proofs than Otter, despite a slower inference rate.

1   Introduction

OSHL-U is a non-resolution theorem prover for first-order logic that extends the Ordered Semantic Hyperlinking (OSHL) [1] strategy. OSHL can be characterized as a tableau method with semantic restrictions, where each branch B has an interpretation $I_0[B]$ associated with it, where I_0 is an interpretation supplied initially and $I_0[B]$ is derived from I_0 in such a way that $I_0[B]$ is the minimal interpretation that satisfies B. The branch B can be expanded only with clauses that $I_0[B]$ falsifies. A tableaux characterization of OSHL was given in [2]. DCTP [3] is another first-order prover that attempts to apply propositional techniques. It can also be seen as a tableau method. Current propositional provers perform at tremendous speed and OSHL tries to import the speed of propositional theorem proving into first-order theorem proving. OSHL-U employs syntactic strategies to improve on the instance generation strategy of OSHL. However, features such as unification, semantics, special methods for equality, and term rewriting, are not implemented on it. We describe the OSHL-U rules and present results of running OSHL-U on TPTP v2.5.0 [4] problems, compare its space efficiency to that of a resolution-based strategy and discuss the advantages and disadvantages of OSHL-U.

OSHL instantiates first-order clauses to ground clauses (fully instantiated clauses) and applies propositional methods to these ground clauses. Ground instances are generated with guidance from models that are supplied by the user and refined as the algorithm progresses. OSHL-U extends OSHL with unit rules to improve on the instance generation strategy. Addition of the unit rules, that were described in [5], helped to get more proofs on the TPTP problems than

B. Beckert (Ed): TABLEAUX 2005, LNAI 3702, pp. 217–230, 2005.
© Springer-Verlag Berlin Heidelberg 2005

with OSHL; an earlier implementation of the prover obtained 238 proofs without the unit rules and 900 proofs with the unit rules, using trivial semantics. In this paper, we attempt to give a simpler notation for the OSHL-U rules. We run OSHL-U as a purely syntactic prover with trivial semantic models. Unlike most current syntactic provers for FOL, OSHL-U is not based on resolution [6] or on model elimination [7]. OSHL and OSHL-U also differ from other propositional approaches to first-order theorem proving such as FDPLL [8] and DCTP [3] in that OSHL and OSHL-U work completely at the propositional level. FDPLL does not work completely at the ground level, and uses the idea that a special case of a literal overrides a more general literal. DCTP employs true unification to generate instances. SATCHMO [9], though it has similarities to OSHL-U, does not use orderings or models in the way OSHL-U does and does not seem to have a counterpart for all of the unit rules. The capability of OSHL-U for using sophisticated semantic guidance, that these other provers lack, also makes it an interesting system for study.

2   Background

Early theorem proving strategies (such as that of [10]) were based on the idea of instantiating a set of first-order clauses to obtain a set of propositional clauses, and then applying a propositional decision procedure to test satisfiability. Some recent provers such as SATCHMO [9] and Baumgartner's first-order DPLL method [8] continue in this tradition. However, since Robinson's groundbreaking paper on resolution [6] and Loveland's work on model elimination [7], the focus of the field has largely shifted to these and other similar approaches. Despite their successes, a shortcoming of such strategies, in a fully automated mode, is their weakness on non-Horn problems.

As shown in [11], methods such as DPLL are much faster than resolution on non-Horn propositional problems, suggesting that similar methods might be efficient for non-Horn first-order problems. The OSHL strategy uses propositional techniques and does not involve unification between non-ground literals. A deficiency of OSHL is the blind enumeration of instances; this is constrained somewhat by interleaving instantiation and model searching, and by the use of semantics, but it is still a problem. OSHL-U attempts to overcome this deficiency by making instance generation more intelligent when possible. This is done by relaxing the OSHL constraint that the instance generated must be a minimal instance contradicting a specified interpretation. OSHL-U permits the instance to be non-minimal and to contradict another interpretation, in exchange for avoiding blind enumeration, when possible. The result is a significant improvement in performance, even with trivial semantics as was shown in [5]. OSHL-U uses a combination of strategies including case analysis and ground UR resolution. Only when all these strategies fail to find an instance does OSHL-U resort to the enumeration strategy of OSHL.

OSHL-U proves more than half the number of TPTP problems that resolution prover Otter [12] proves in 30 seconds, and on several TPTP groups, especially those having many non-Horn clauses, has equivalent or superior performance;

these results were shown in [5]. The fact that so many proofs can be obtained by pure propositional techniques suggests that propositional techniques have the potential to speed up a significant number of first-order proofs and that a propositional prover may be superior to resolution on certain kinds of problems. Of course, other provers such as SPASS [13], GANDALF [14], SETHEO [15] may perform better than Otter and OSHL-U on these problems. These provers have been thoroughly optimized and have efficient data structures, which give them a considerable advantage over OSHL-U, which lacks such extensive optimization. Some of these provers use case analysis (splitting), which is really a form of propositional reasoning imported into FOL. Otter is suitable for comparison because it uses a simpler more uniform strategy selection that helps to guarantee that it has not been especially tuned to a particular problem set. Some other provers break the problems into 50 or more categories [14,16,17,18] so it is difficult to know whether the performance of the prover is due to the underlying method or the way in which the categories are selected. Some modern resolution theorem provers are based on the theory of resolution and superposition with redundancy rather than Otter's use of resolution, but it does not appear that such provers would give significantly different results, and even if they did, the comparison of OSHL-U with Otter is significant in itself. We compare the Otter and OSHL-U strategies independent of implementation using their search spaces (number of clauses generated) and storage spaces (number of clauses kept). However, Otter's efficient data structures still give it an execution speed advantage over OSHL-U.

OSHL and prior propositional provers have performed well on near-propositional problems and on problems involving definition expansion [19]. However, many non-Horn problems can not be solved simply by definition expansion. OSHL-U, not being oriented to definition expansion, demonstrates improved performance on such problems. OSHL with user specified semantics [20] solved some situation calculus planning problems that were difficult for resolution, but this required the user to input natural semantics. OSHL-U has not been tested with sophisticated semantics, as was the prover of [20], although OSHL-U has the potential for such semantic guidance. OSHL-U has no special rules for term-rewriting or the equality axioms; its performance might be significantly enhanced by special rules for equality axioms, better data structures, and semantic guidance.

3   OSHL-U Rules

We give a simplified notation for OSHL-U rules. Let S be the set of input clauses. The rules operate on an *ascending sequence* which is a sequence $C_1 C_2 \ldots C_n$ of ground clauses. Initially the ascending sequence is empty. The proof search stops when an ascending sequence containing the empty clause is derived; this indicates that S is unsatisfiable. There are *basic rules* and U *rules*. The basic rules by themselves are complete and work essentially at the ground level. The U rules, though not necessary, perform operations involving unit clauses and frequently help to get proofs faster.

The ascending sequence consists of a (possibly empty) sequence of *basic clauses* (created by basic rules) followed by a (possibly empty) sequence of U *clauses* (created by U rules). U clauses need not be unit clauses. Each clause in the ascending sequence has a selected literal called an *eligible literal*. L_i denotes the eligible literal of clause C_i in the sequence. E is the set $\{L_1, L_2, \ldots, L_n\}$ of eligible literals and $\overline{E}$ is the set of its complements. Several rules add a clause C to the end of an ascending sequence; this is always subject to the restriction $C \cap E = \{\}$ where E consists of the eligible literals before C is added.

There is a total syntactic ordering $<_{lit}$ on ground atoms that is extended to literals by $L <_{lit} M$ iff $at(L) <_{lit} at(M)$ where $at(\neg L) = L$ and if L is an atom, $at(L) = L$. This ordering restricts which literals may be selected from clauses. For basic clauses C_i, L_i is the $<_{lit}$ maximum literal in C_i. For U clauses C_i, if C_i contains a literal L that is not complementary to existing eligible literals, then some such L must be selected. Otherwise, a literal must be selected that is complementary to L_j for the maximum possible j. There is an ordering $<_{el}$ on the eligible literals. Suppose L_i and L_j are the selected literals from clauses C_i and C_j in the ascending sequence. Let i' be minimal such that L_i or $\neg L_i$ is the selected literal from clause $C_{i'}$, and similarly for j'. Then $L_i <_{el} L_j$ iff $i' < j'$.

3.1 Basic Rules

Extension. Let $C\Theta$ be a ground instance of an input clause C such that $C\Theta \cap E = \{\}$. Add $C\Theta$ to the end of the ascending sequence and select a literal from it.

Resolution. If the selected literals L and M of the last two clauses C and D of the ascending sequence are complementary, remove these clauses and add their resolvent $(C - \{L\}) \cup (D - \{M\})$ to the ascending sequence. At least one of C and D must be a basic clause.

Clause Deletion. If $L_n < L_{n-1}$, delete C_{n-1} from the ascending sequence. The effect of clause deletion is to delete clauses whose selected literals are "out of order". Resolution removes successive clauses whose selected literals have the same atom. If clause deletion and resolution are done eagerly, then after they are both finished, it will be true that for all i and j, $L_i <_{el} L_j$ iff $i < j$.

3.2 U Rules

UR Resolution. If C is a non-unit clause then a sequence of resolutions between C and unit clauses producing a unit clause is called a unit resultant(UR) resolution. The UR resolution rule in OSHL-U is a special case of UR resolution. Find $C \in S$ which resolves with eligible literals to give unit clause $\{L\}$ such that L is a ground literal and $L \notin E$. Let $C\Theta$ be an instance of C such that $C\Theta \subset \overline{E} \cup \{L\}$. Add $C\Theta$ to the end of the ascending sequence and select the literal L from it.

For example, given the sequence $[\{s(a), \mathbf{p(b)}\}, \{t(a), \mathbf{q(b)}\}]$ and the clause $\{not\ p(X), not\ q(X), r(X)\}$, where the highlighted literals are the eligible literals, the sequence $[\{s(a), \mathbf{p(b)}\}, \{t(a), \mathbf{q(b)}\}, \{not\ p(b), not\ q(b), \mathbf{r(b)}\}]$ is created.

Unit Filtering. Let D be obtainable from $C \in S$ by zero or more unit resolutions with unit clauses in S. Let $D\Theta$ be an instance of D such that $D\Theta \subset \overline{E}$. Add $D\Theta$ to the end of the ascending sequence. Note that after this rule is done, either a resolution or a clause deletion (or their U counterparts) will be possible, so the ascending sequence will get shorter. If $D = C$, this is called filtering. We implemented filtering, a restricted form of this rule which permits resolution with only propositional unit clauses.

For example, given the sequence $[\{s(a), \mathbf{p(b)}\}, \{t(a), \mathbf{q(b)}\}]$ and the clause $\{not\ p(X), not\ q(X)\}$, where the highlighted literals are the eligible literals, the sequence $[\{s(a), \mathbf{p(b)}\}, \{t(a), \mathbf{q(b)}\}, \{not\ p(b), \mathbf{not\ q(b)}\}]$ is created.

Case Analysis. Let C be an input clause and L be a literal of C containing all the variables of C. Let $L\Theta$ be an instance of L such that $L\Theta \in \overline{E}$ and $C\Theta \cap E = \{\}$. Add $C\Theta$ to the end of the ascending sequence.

For example, given the sequence $[\{s(a), \mathbf{p(b)}\}, \{t(a), \mathbf{q(b)}\}]$ and the clause $\{not\ q(X), r(X), s(X)\}$, where the highlighted literals are the eligible literals, the sequence $[\{s(a), \mathbf{p(b)}\}, \{t(a), \mathbf{q(b)}\}, \{not\ q(b), r(b), \mathbf{s(b)}\}]$ is created by the case analysis rule.

U Resolution. This is like resolution but applies to two U clauses.

U Clause Deletion. This is like basic clause deletion but it applies to U clauses. It is also possible to apply clause deletion if C_n is a U clause and C_{n-1} is a basic clause.

3.3   Implementation

The implementation has additional restrictions; for example, the instance $C\Theta$ used for extension must be a minimal clause contradicting a minimal interpretation of E where clauses and interpretations are ordered in a specified way. Also, after an extension operation, resolution and clause deletion are done as often as possible. Then the U rules are applied to find a clause C such that all literals of C are complements of selected literals from basic clauses. If such a clause C can be found, then all the U clauses are removed from the sequence, C is added to the end, and resolutions and clause deletions are done as appropriate. If such a clause C cannot be derived within a reasonable time, then all the U clauses are removed from the sequence and another extension operation is performed. When the U rules are being used, U resolution and U clause deletion are performed whenever possible; then unit filtering, then UR resolution and case analysis.

Further, the implementation is complicated by the details of OSHL and the necessity to be able to use semantics, topics that are not covered here. For example, the construction of the ground instances requires the application of disunification.

3.4 An Example

The following example illustrates the operation of OSHL-U. The empty clause is derived by application of OSHL-U rules. The theorem being proved is that the set, a, equals its union with itself ($a = a \cup a$). Following are the clauses that are given as input – 9 set theory axioms and the conjecture clause, which is the negation of the theorem. The predicates *member*, *subset*, *equal_sets* and *union* have been replaced by the symbols $\in$, $\subset$, $=$, and $\cup$, respectively, for better readibility.

% membership_in_subsets, axiom 1.
$\{\text{not}(X \in Y), \text{not}(Y \subset Z), X \in Z\}$.
% subsets_axiom1, axiom 2.
$\{Y \subset Z, g(Y, Z) \in Y\}$.
% subsets_axiom2, axiom 3.
$\{\text{not}(g(Y, Z) \in Z), Y \subset Z\}$.
% set_equal_sets_are_subsets1, axiom 4.
$\{\text{not}(Y = Z), Y \subset Z\}$.
% set_equal_sets_are_subsets2, axiom 5.
$\{\text{not}(Z = Y), Y \subset Z\}$.
% subsets_are_set_equal_sets, axiom 6.
$\{\text{not}(X \subset Y), \text{not}(Y \subset X), Y = X\}$.
% union_definition_1, axiom 7.
$\{\text{not}(A \in B \cup C), A \in B, A \in C\}$.
% union_definition_2, axiom 8.
$\{\text{not}(A \in B), A \in B \cup C\}$.
% union_definition_3, axiom 9.
$\{\text{not}(A \in B), A \in C \cup B\}$.
% conjecture.
$\{\text{not}(a = a \cup a)\}$.

The successive ascending sequences resulting from the application of the OSHL-U rules are given below. The eligible literal in a clause is highlighted. The initial interpretation is all-positive, that is every predicate is interpreted to "true". Initially, the ascending sequence is empty. The conjecture clause is added to the sequence by Extension. Then case analysis is applied on an instance of axiom 6, obtained by using the substitution $\{X \to a \cup a, Y \to a\}$. Each of the literals, $\text{not}(a \cup a \subset a)$, $\text{not}(a \subset a \cup a)$, is removed by applying a sequence of UR Resolution and Unit Filter; the remaining literal is removed by Resolution with the conjecture clause, generating the empty clause. Note that the two cases being removed with U rules correspond to proving the subgoals, $a \cup a \subset a$ and $a \subset a \cup a$.

$S_0 : [\,]$
$S_1 : [\{\textbf{not}(\mathbf{a = a \cup a})\}]$ (Extension, conjecture)
$S_2 : [\{\textbf{not}(\mathbf{a = a \cup a})\}, \{\textbf{not}(\mathbf{a \cup a \subset a})$, $\text{not}(a \subset a \cup a), a = a \cup a\}$.] (Case Analysis, 6)
$S_3 : [\{\textbf{not}(\mathbf{a = a \cup a})\}, \{\textbf{not}(\mathbf{a \cup a \subset a})$, $\text{not}(a \subset a \cup a), a = a \cup a\}, \{\textbf{not}(\mathbf{g(a \cup a, a) \in a})$, $a \cup a \subset a\}]$ (UR Resolution, 3)

S_4 : [{**not**($a = a \cup a$)}, {**not**($a \cup a \subset a$), not($a \subset a \cup a$), $a = a \cup a$}, {**not**(**g**(**a** $\cup$ **a**, **a**) $\in$ **a**), $a \cup a \subset a$}, {$a \cup a \subset a$, **g**($a \cup a$, **a**) $\in a \cup a$}.] (UR Resolution,2)
S_5 : [{**not**($a = a \cup a$)}, {**not**($a \cup a \subset a$), not($a \subset a \cup a$), $a = a \cup a$}, {**not**(**g**(**a** $\cup$ **a**, **a**) $\in$ **a**), $a \cup a \subset a$}, {$a \cup a \subset a$, **g**($a \cup$ **a**, **a**) $\in a \cup$ **a**}, {**not**(**g**($a \cup$ **a**, **a**) $\in a \cup a$), $g(a \cup a, a) \in a$}.] (Unit Filter, 7)
S_6 : [{**not**($a = a \cup a$)}, {**not**($a \cup a \subset a$), not($a \subset a \cup a$), $a = a \cup a$}, {**not**(**g**(**a** $\cup$ **a**, **a**) $\in$ **a**), $a \cup a \subset a$}, {$a \cup a \subset a$, **g**($a \cup$ **a**, **a**) $\in a$}] (U Resolution)
S_7 : [{**not**($a = a \cup a$)}, {**not**($a \cup a \subset a$), not($a \subset a \cup a$), $a = a \cup a$}, {**a** $\cup$ **a** $\subset a$}.] (U Resolution)
S_8 : [{**not**($a = a \cup a$)}, {**not**($a \subset a \cup a$), $a = a \cup a$}.] (U Resolution)
S_9 : [{**not**($a = a \cup a$)}, {**not**($a \subset a \cup a$), $a = a \cup a$}, {$a \subset a \cup a$, **g**($a, a \cup$ **a**) $\in a$}] (UR Resolution, 2)
S_{10} : [{**not**($a = a \cup a$)}, {**not**($a \subset a \cup a$), $a = a \cup a$}, {$a \subset a \cup a$, **g**($a, a \cup$ **a**) $\in a$}, {**not**(**g**($a, a \cup a$) $\in a \cup a$), $a \subset a \cup a$}] (UR Resolution, 3)
S_{11} : [{**not**($a = a \cup a$)}, {**not**($a \subset a \cup a$), $a = a \cup a$}, {$a \subset a \cup a$, **g**($a, a \cup$ **a**) $\in a$}, {**not**(**g**($a, a \cup a$) $\in a \cup a$), $a \subset a \cup a$}, {not(**g**($a, a \cup a$) $\in a$), **g**($a, a \cup a$) $\in a \cup a$}] (Unit Filter, 7)
S_{12} : [{**not**($a = a \cup a$)}, {**not**($a \subset a \cup a$), $a = a \cup a$}, {$a \subset a \cup a$, **g**($a, a \cup$ **a**) $\in a$}, {$a \subset a \cup a$, **not**(**g**($a, a \cup a$) $\in a$)}] (U Resolution)
S_{13} : [{**not**($a = a \cup a$)}, {**not**($a \subset a \cup a$), $a = a \cup a$}, {**a** $\subset a \cup$ **a**}] (U Resolution)
S_{14} : [{**not**($a = a \cup a$)}, {**a** $= a \cup$ **a**}] (U Resolution)
S_{15} : [{}] (Resolution)

4   Results

We ran OSHL-U on problems from TPTP v2.5.0 with trivial semantics that interpret all predicates to true (all-positive) or all predicates to false (all-negative). An all-positive semantics corresponds to back chaining, and an all-negative semantics corresponds to forward chaining. OSHL-U obtains proofs of mostly the same problems with either of the trivial semantics. The OSHL-U results presented are obtained with all-negative semantics on all but 32 problems on which proofs are obtained with all-positive semantics. OSHL-U is not designed to detect satisfiability, so we did not include the known satisfiable problems in our tests. The number of problems tested on was 4417. We ran Otter v3.3 in the autonomous mode on the same problems. On each prover, the maximum CPU time allowed for each problem was 30 seconds. We used Otter because of its simple control structure. Other provers sometimes break the TPTP into many categories and apply a different strategy to each category. This makes it difficult to know when the performance of the prover is due to the inherent strategy and when it is due to the method of breaking problems into categories. Otter, in the autonomous mode, has simpler uniform control structure and does not break the input set into categories. We compare the number of clauses generated by each prover on the same problem. This lets us compare the prover strategies independent of how well the provers are implemented. We also compared storage spaces used by the provers on the same problem by counting the largest number of clauses stored at any one time by each prover while executing on the problem. We compared the provers on problems which both provers proved within the 30 seconds limit; there were 827 such problems. We compared the clauses generated and clauses stored on the Horn and non-Horn categories of these problems. We also categorized the problems by their TPTP difficulty ratings – rating 0 or rating greater than 0 – and compared on these categories. TPTP ratings range from 0 to 1. A lower rating indicates an easier problem, based on the results of all the provers that have been tested on the TPTP, many of them being resolution-based. The tests were performed on a Pentium3 processor.

Fig. 1 shows the number of clauses generated by OSHL-U against that generated by Otter for every problem on which both provers obtained a proof. Note that these results would still be true even if OSHL-U had a faster inference generation rate. Table 1 gives a frequency distribution of the ratios of the number of clauses generated by Otter to that generated by OSHL-U. A ratio of 1

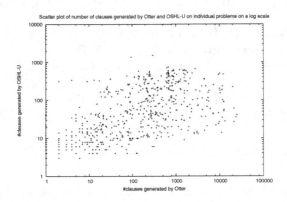

Fig. 1. Comparison of the number of clauses generated by OSHL-U against that of Otter on individual problems shown on a scatter plot

Table 1. Distribution of the ratios of the number of clauses generated by Otter to that generated by OSHL-U. Ratio > 1 means Otter generated more clauses; ratio < 1 means OSHL-U generated more clauses; ratio = 1 means Otter and OSHL-U generated the same number of clauses

Ratio of Generated Clauses by Otter to OSHL-U	Frequency of Occurence
[0,0.01)	48
[0.01,0.1)	18
[0.1,1)	253
1	11
(1,10)	350
[10,100)	117
[100,1000)	30

Table 2. Number of clauses, in thousands, generated by Otter and OSHL-U on problems for which both provers find a proof and the ratio of number of clauses for Otter to that for OSHL-U (H denotes Horn, nH non-Horn, R rating)

	All	H	nH	R=0	R>0	nH,R>0
Otter	708	90	618	357	351	348
OSHL-U	104	39	65	78	26	26
Ratio (Otter/OSHL-U)	6.8	2.3	9.5	4.6	13.5	13.5

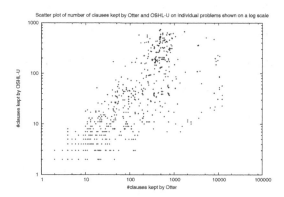

Fig. 2. Comparison of space used as the number of clauses stored by OSHL-U against that of Otter on individual problems shown on a scatter plot

means the two provers generate the same number of clauses, less than 1 means Otter generates less clauses, and greater than 1 means OSHL-U generates less clauses. Otter generates less clauses than OSHL-U on 319 problems, OSHL-U generates less clauses than Otter on 497 problems, and the provers generate the same number of clauses on 11 problems. On the 319 problems on which Otter generates less clauses, the average number of clauses generated by Otter is 39.48 and that for OSHL-U is 98.96. On the 497 problems where OSHL-U generates less clauses, the average number of clauses generated by OSHL-U is 145.36 and that for Otter is 1398.84. The best ratio for Otter is 0 on 44 of the 48 problems in the [0, 0.01) range of Table 1; the largest number of clauses that OSHL-U generates on any of these 44 problems is 191. The best ratio for OSHL-U is 794.3. The number of clauses generated on the problem for which the difference between Otter and OSHL-U is highest, that is, Otter generated fewer clauses by the largest margin is 94 on Otter and 1377 on OSHL-U. The number of clauses generated on the problem for which OSHL-U generated fewer clauses than Otter by the largest margin is 24829 on Otter and 41 on OSHL-U. This shows that OSHL-U generates fewer clauses on many problems; and that when Otter generates fewer clauses than OSHL-U, it does so by not a very large factor, except when Otter generates 0 clauses.

Table 2 shows the sum of the number of clauses generated for all the problems that both provers can prove in 30 seconds and the ratio of the sums for Otter

Table 3. Ratio of total number of clauses generated by Otter to that by OSHL-U on problems for which both provers find a proof (H denotes Horn, nH non-Horn, R rating)

Domain	All	H	nH	R=0	R>0	nH, R>0
ALG	27.2	-	27.2	27.2	-	-
CAT	3.4	2.9	6.9	1.5	7.8	-
COL	2.0	2.0	-	2.0	-	-
COM	0.9	0.6	1.5	0.9	-	-
FLD	117.4	-	117.4	117.2	117.5	117.5
GEO	44.2	0.3	13.3	53.3	53.3	53.3
GRA	0.9	-	0.9	0.9	-	-
GRP	15.3	5.1	17.7	14.1	28.9	28.9
HEN	19.7	19.7	-	19.7	-	-
HWV	3.1	-	3.1	4.1	1.0	1.0
KRS	7.2	0.5	7.5	7.2	-	-
LAT	0.0	0.0	-	0.0	-	-
LCL	2.1	2.9	0.3	2.1	-	-
MGT	2.3	0.4	2.3	1.9	3.7	3.7
MSC	10.8	-	10.8	2.9	20.7	20.7
NLP	1.9	-	1.9	1.9	-	-
NUM	5.9	1.6	6.1	5.9	-	-
PUZ	5.3	6.4	5.2	5.2	5.7	2.2
RNG	220.9	617.8	42.3	617.8	42.3	42.3
SET	6.2	1.3	6.2	4.9	14.0	10.7
SWV	4.4	0.1	4.5	4.2	7.7	7.7
SYN	2.3	1.8	6.8	2.2	29.5	29.5
TOP	1.6	-	1.6	1.6	-	-

to that for OSHL-U. Table 3 shows the ratio of the number of clauses generated for Otter to that for OSHL-U on individual domains. Also shown are the ratios for only the Horn and non-Horn problems, problems having rating 0 or greater than 0, and non-Horn problems having rating greater than 0. Ratios could not be computed when there were no problems in a particular category and are marked as - in the table. OSHL-U generates fewer clauses than Otter in many domains. The ratios indicate that OSHL-U performs best in comparison to Otter for problems which are non-Horn and which have rating greater than 0. OSHL-U seems to generate fewer clauses than Otter even for Horn clauses on several domains.

We compare storage spaces using the maximum number of clauses stored on each problem. The scatter plot in Fig. 2 shows the number of clauses stored by OSHL-U against that of Otter, and the frequency of their ratios is shown in Table 4. Tables 5 and 6 show the comparison of storage space on all domains and on individual domains, respectively. These indicate that OSHL-U stores significantly fewer clauses than Otter on some problems; this is independent of running time.

Table 7 shows the number of proofs obtained by each prover within 30 seconds. OSHL-U generates fewer clauses per second compared to Otter, which has

Table 4. Distribution of the ratios of number of clauses stored by Otter to that stored by OSHL-U. Ratio > 1 means Otter stored more clauses; ratio < 1 means OSHL-U stored more clauses; ratio = 1 means Otter and OSHL-U stored the same number of clauses.

Ratio of Stored Clauses by Otter to OSHL-U	Frequency of Occurence
[0,0.01)	0
[0,0.1)	0
[0.1,1)	69
1	31
(1,10)	605
[10,100)	100
[100,1000)	22

Table 5. Number of clauses stored, in thousands, by Otter and OSHL-U over problems for which both provers find a proof and the ratio of clauses stored by Otter to that of OSHL-U (H denotes Horn, nH non-Horn, R rating)

	All	H	nH	R=0	R>0	nH,R>0
Otter	423	81	342	230	193	192
OSHL-U	92	37	55	67	25	25
Ratio (Otter/OSHL-U)	4.6	2.2	6.2	3.4	7.7	7.7

a more sophisticated implementation. OSHL-U is implemented in OCaml and Otter in C with better data structures. Nevertheless, OSHL-U obtains more than half the number of proofs that Otter does overall and obtains more proofs than Otter on SET(set theory) and FLD(field theory) categories. This is especially interesting because OSHL-U has no special rules for equality handling. OSHL-U was not designed for Horn problems and is not necessarily better than UR resolution on this kind of problem. As shown by the number of proofs as well as the comparison of number of clauses generated and stored on the different categories of problems, the performance of OSHL-U relative to resolution seems to improve for problems that have hardness rating greater than 0 and are non-Horn.

5   Discussion

The OSHL strategy starts with a model, and progressively modifies it till the empty clause can be derived in the model. The crucial step is to generate an instance D of an input clause that is falsified by the current model and modify the model to make D true. Proofs with smaller terms are generated before those with larger terms. As mentioned in [1], OSHL had problems generating proofs with larger terms. OSHL-U extends OSHL with U rules that provide better syntactic guidance. This enabled OSHL-U to obtain more proofs than OSHL, a result already shown in [5]; the previous implementation of OSHL-U, using trivial semantics, obtained 900 proofs with the unit rules and 238 proofs without the unit rules.

Table 6. Ratio of total storage space of Otter to that of OSHL-U on problems for which both provers find a proof (H denotes Horn, nH non-Horn, R rating)

Domain	All	H	nH	R=0	R>0	nH, R>0
ALG	6.7	-	6.7	6.7	-	-
CAT	2.4	2.0	4.0	1.9	3.7	-
COL	3.5	3.5	-	3.5	-	-
COM	0.9	0.8	1.1	0.9	-	-
FLD	88.2	-	88.2	104.9	81.5	81.5
GEO	8.9	2.0	9.0	9.9	8.7	8.7
GRA	3.4	-	3.4	3.4	-	-
GRP	4.1	4.7	3.9	3.7	6.4	6.4
HEN	4.2	4.2	-	4.2	-	-
HWV	2.8	-	2.8	3.7	0.8	0.8
KRS	3.2	1.3	3.4	3.2	-	-
LAT	5.7	5.7	-	5.7	-	-
LCL	4.4	5.5	1.6	4.4	-	-
MGT	2.4	1.6	2.4	2.5	2.0	2.0
MSC	10.1	-	10.1	1.2	20.5	20.5
NLP	1.9	-	1.9	1.9	-	-
NUM	6.9	2.7	7.1	6.9	-	-
PUZ	4.5	6.0	4.3	4.4	5.2	1.7
RNG	46.1	129.6	12.7	129.6	12.7	12.7
SET	4.4	2.9	4.4	3.3	10.7	10.7
SWV	2.5	1.0	2.6	2.4	3.0	3.0
SYN	2.2	2.1	4.4	2.1	8.6	29.5
TOP	5.0	-	5.0	5.0	-	-

Table 7. Comparison of total number of proofs found by Otter and OSHL-U in 30 seconds (R denotes rating)

Prover	Total	Horn	non-Horn		
			total	R=0	R>0
Otter	1697	764	933	636	297
OSHL-U	1027	311	716	754	273

On problems for which both OSHL-U and Otter found proofs, Otter generates 7 times as many clauses as OSHL-U overall, over 9 times as many clauses on non-Horn problems, and over 13 times as many clauses on problems of TPTP rating greater than 0. The space advantage of OSHL-U over Otter appears to be greater on non-Horn problems and on problems of rating greater than 0, which are presumably harder for resolution. Our comparison of number of clauses generated is independent of implementation. However, Otter uses term-rewriting which reduces the number of clauses generated on some problems. One of our objectives is to investigate the propositional approach on problems that are difficult for resolution and find more efficient strategies for such problems. It is possible that other provers could benefit from the syntactic extensions of our propositional prover. The unit filtering rule provides a way of interfacing OSHL-U with a conventional resolution prover, because resolvents from the conventional prover can be added to the set of clauses used for filtering.

OSHL-U appears to perform best in comparison to resolution on non-Horn problems. OSHL-U even seems to generate fewer clauses for Horn problems. This can be explained by the fact that, as soon as OSHL-U has generated a propositionally unsatisfiable set of ground clauses it detects a proof, while resolution needs more time to find the proof. Also, OSHL-U can sometimes detect a proof even before generating a propositionally unsatisfiable set of ground clauses, because some of the ground clauses can be generated automatically by filtering. It is theoretically possible that if OSHL-U were run long enough, then Otter would generate significantly fewer clauses on some problems. For very hard problems, space can be a more important consideration than time because a space-efficient method can run for a very long time without exhausting memory. A prover that runs out of space will fail, no matter how fast it is.

OSHL-U generates more proofs than Otter on the TPTP groups FLD (Field theory) and SET (Set theory). This is the first time to our knowledge that a propositional prover, not performing unification on non-propositional literals, has demonstrated performance comparable to that of a resolution prover on groups of TPTP problems (FLD, SET) that are not near-propositional in structure. This helps to understand how essential unification of non-ground literals is to theorem proving, and how much theorem proving one can do without it.

OSHL is still in the stage where large speedups are taking place, such as with the U rules; so OSHL performance may improve further. On the same system, clause generation rate is much higher for Otter(10000 per second) than for OSHL-U(100 per second), one reason being Otter's efficient data structures. The use of better data structures could significantly speed up OSHL-U and it might outperform Otter in execution speed on many problems. Although other propositional style provers outperform Otter, OSHL is particularly interesting because it is one of the few propositional approaches to FOL that can use sophisticated semantic guidance and that does not use true unification.

6   Future Work

We would like to use more efficient data structures to improve the search efficiency of the implementation. Currently, OSHL-U generates far fewer clauses per second compared to Otter. If it were to generate clauses at a comparable rate to Otter, OSHL-U would be faster than Otter in execution speed on many problems. OSHL-U could also serve as a good platform for studying the use of semantics to guide proof search. The OSHL algorithm was designed to be used only in conjunction with semantic guidance. This capability of using semantic models in its search has not yet been utilized by OSHL-U, which currently uses trivial semantics.

Acknowledgements

This work was partially supported by the National Science Foundation under grant number CCR-9972118.

References

1. Plaisted, D.A., Zhu, Y.: Ordered semantic hyper linking. Journal of Automated Reasoning **25** (2000) 167–217
2. Yahya, A., Plaisted, D.A.: Ordered semantic hyper tableaux. Journal of Automated Reasoning **29** (2002) 17–57
3. Letz, R., Stenz, G.: DCTP: A Disconnection Calculus Theorem Prover. In: Lecture Notes in Artificial Intelligence. Volume 2083., Springer Verlag (2001) 381–385
4. Sutcliffe, G., Suttner, C.: The TPTP Problem Library: CNF Release v1.2.1. Journal of Automated Reasoning **21(2)** (1998) 177–203
5. Das, S., Plaisted, D.A.: An improved propositional approach to first-order theorem proving. In: Workshop on Model Computation - Principles, Algorithms, Applications at The 19th International Conference on Automated Deduction. (2003)
6. Robinson, J.A.: A machine-oriented logic based on the resolution principle. Journal of the ACM **12** (1965) 23–41
7. Loveland, D.W.: Mechanical theorem-proving by model elimination. Journal of the ACM **15** (1968) 236–251
8. Baumgartner, P.: FDPLL – A First-Order Davis-Putnam-Logeman-Loveland Procedure. In McAllester, D., ed.: CADE-17 – The 17th International Conference on Automated Deduction. Volume 1831., Springer (2000) 200–219
9. Manthey, R., Bry, F.: SATCHMO: A Theorem Prover Implemented in Prolog. In: Proceedings of the Ninth International Conference on Automated Deduction. Lecture Notes in Computer Science 310, Springer-Verlag (1988) 415–434
10. Gilmore, P.C.: A proof method for quantification theory. IBM Journal of Research and Development **4** (1960) 28–35
11. Plaisted, D.A., Lee, S.J.: Eliminating duplication with the hyper-linking strategy. Journal of Automated Reasoning **9** (1992) 25–42
12. McCune, W.W.: Otter 3.0 reference manual and guide. Technical Report ANL-94/6, Argonne National Laboratory, Argonne, Illinois (1994)
13. Weidenbach, C.: SPASS version 0.49. Journal of Automated Reasoning **18** (1997) 247–252
14. Tammet, T.: Gandalf. Journal of Automated Reasoning **18** (1997) 199–204
15. Letz, R., Schumann, J., Bayerl, S., Bibel, W.: SETHEO:A high-performance theorem prover. Journal of Automated Reasoning **8** (1992) 183–212
16. Hillenbrand, T., Jaeger, A., Löchner, B.: Waldmeister - Improvements in Performance and Ease of Use. In Ganzinger, H., ed.: Proceedings of the 16th International Conference on Automated Deduction. Number 1632 in Lecture Notes in Artificial Intelligence, Springer-Verlag (1999) 232–236
17. Sutcliffe, G., Seyfang, D.: Smart Selective Competition Parallelism ATP. In Kumar, A., Russell, I., eds.: Proceedings of the 12th Florida Artificial Intelligence Research Symposium, AAAI Press (1999) 341–345
18. Stenz, G., Wolf, A.: Strategy Selection by Genetic Programming. In Kumar, A., Russell, I., eds.: Proceedings of the 12th Florida Artificial Intelligence Research Symposium, AAAI Press (1999) 346–350
19. Plaisted, D.A., Zhu, Y.: Replacement rules with definition detection. In Caferra, R., Salzer, G., eds.: Automated Deduction in Classical and Non-Classical Logics. Lecture Notes in Artificial Intelligence 1761 (1999) 80–94 Invited paper.
20. Zhu, Y., Plaisted, D.A.: FOLPLAN: A Semantically Guided First-Order Planner. 10th International FLAIRS Conference (1997)

Efficient Query Processing with Compiled Knowledge Bases

Neil V. Murray[1] and Erik Rosenthal[2]

[1] Department of Computer Science, State University of New York, Albany, NY 12222, USA
nvm@cs.albany.edu
[2] Department of Mathematics, University of New Haven, West Haven, CT 06516, USA
erosenthal@newhaven.edu

Abstract. The goal of knowledge compilation is to enable fast queries. Prior approaches had the goal of small (i.e., polynomial in the size of the initial knowledge bases) compiled knowledge bases. Typically, query-response time is linear, so that the efficiency of querying the compiled knowledge base depends on its size. In this paper, a target for knowledge compilation called the *ri-trie* is introduced; it has the property that even if they are large they nevertheless admit fast queries. Specifically, a query can be processed in time *linear in the size of the query* regardless of the size of the compiled knowledge base.

1 Introduction

The last decade has seen a virtual explosion of applications of propositional logic. One is *knowledge representation*, and one approach to it is *knowledge compilation*. Knowledge bases can be represented as propositional theories, often as sets of clauses, and the propositional theory can then be *compiled*; i.e., preprocessed to a form that admits fast response to queries. While knowledge compilation is intractable, it is done once, in an off-line phase, with the goal of making frequent on-line queries efficient. Heretofore, that goal has not been achieved for arbitrary propositional theories.

A typical query of a propositional theory has the form, is a clause logically entailed by the theory? This question is equivalent to asking, is the conjunction of the theory and the negation of the clause unsatisfiable? Propositional logic is of course intractable (unless $\mathcal{NP} = \mathcal{P}$), so the primary goal of most research is to find relatively efficient deduction techniques. A number of languages — for example, *Horn sets*, *ordered binary decision diagrams*, sets of *prime implicates/implicants*, *decomposable negation normal form*, *factored negation normal form*, and *pairwise-linked formulas* — have been proposed as targets for knowledge compilation. (See, for example, [1,3,8,10,18,21,20,29,43,49].

Knowledge compilation was introduced by Kautz and Selman [24]. They were aware of one issue that is not discussed by all authors: The ability to answer queries in time polynomial (indeed, often linear) in the size of the compiled theory is not very fast if the compiled theory is exponential in the size of the underlying propositional theory. Most investigators who have considered this issue focused on minimizing the size of the compiled theory, possibly by restricting or approximating the original theory.

B. Beckert (Ed): TABLEAUX 2005, LNAI 3702, pp. 231–244, 2005.

Another approach is considered in this paper: admitting large compiled theories—stored off-line[1] — on which queries can be answered in time *linear in the size of the query*.

A data structure that has this property is called a *reduced implicate trie* or, more simply, an *ri-trie*, and is introduced in Section 3.2. These tries can be thought of as compact *implicate tries*, which are introduced in Section 3.1. Note that the target languages studied by the authors in [21,34] are related but nevertheless distinct from this work; they enable response times linear only in the size of the compiled theory, which (unfortunately) can be exponentially large.

The Tri operator and RIT operator, which are the building blocks of *ri*-tries, are introduced and the appropriate theorems are proved in Section 4. A direct implementation of the RIT operator would appear to have a significant inefficiency. However, the algorithm developed in Section 5 avoids this inefficiency.

2   Preliminaries

For the sake of completeness, define an *atom* to be a propositional variable, a *literal* to be an atom or the negation of an atom, and a *clause* to be a disjunction of literals. Clauses are often referred to as sets of literals. Most authors restrict attention to *conjunctive normal form* (CNF) — a conjunction of clauses — but no such restriction is required in this paper.

Consequences expressed as minimal clauses that are implied by a formula are its *prime implicates*; (and minimal conjunctions of literals that imply a formula are its *prime implicants*). Implicates are useful in certain approaches to non-monotonic reasoning [27,40,46], where all consequences of a formula — for example, the support set for a proposed common-sense conclusion — are required. The implicants are useful in situations where satisfying models are desired, as in error analysis during hardware verification. Many algorithms have been proposed to compute the prime implicates (or implicants) of a propositional boolean formula [5,14,22,23,26,39,42,48,51].

An *implicate* of a logical formula is a clause that is entailed by the formula; i.e., a clause that contains a prime implicate. Thus, if $\mathcal{F}$ is a formula and C is a clause, then C is an implicate of $\mathcal{F}$ if (and only if) C is satisfied by every interpretation that satisfies $\mathcal{F}$. Still another way of looking at implicates is to note that asking whether a given clause is entailed by a formula is equivalent to asking whether the clause is an implicate of the formula. Throughout the paper, this question is what is meant by query.

3   A Data Structure That Enables Fast Query Processsing

The goal of knowledge compilation is to enable fast queries. Prior approaches had the goal of a small (i.e., polynomial in the size of the initial knowledge base) compiled knowledge base. Typically, query-response time is linear, so that the efficiency of

---

[1] The term *off-line* is used in two ways: first, for off-line memory, such as hard drives, as opposed to on-line storage, such as RAM, and secondly, for "batch preparation" of a knowledge base for on-line usage.

querying the compiled knowledge base depends on its size. The approach considered in this paper is to admit target languages that may be large as long as they enable fast queries. The idea is for the query to be processed in time *linear in the size of the query*. Thus, if the compiled knowledge base is exponentially larger than the initial knowledge base, the query must be processed in time logarithmic in the size of the compiled knowledge base. One data structure that admits such fast queries is called a *ri-trie* (for *reduced implicate trie*).

3.1 Implicate Tries

The trie is a well-known data structure introduced by Morrison in 1968 [30]; it is a tree in which each branch represents the sequence of symbols labeling the nodes[2] on that branch, in descending order. A prefix of such a sequence may be represented along the same branch by defining a special *end symbol* and assigning an extra child labeled by this symbol to the node corresponding to the last symbol of the prefix. For convenience, it is assumed here that the node itself is simply marked with the end symbol, and leaf nodes are also so marked. One common application for tries is a dictionary. The advantage is that each word in the dictionary is present precisely as a (partial) branch in the trie. Checking a string for membership in the dictionary merely requires tracing a corresponding branch in the trie. This will either fail or be done in time linear in the size of the string.

Tries have also been used to represent logical formulas, including sets of prime implicates [46]. The nodes along each branch represent the literals of a clause, and the conjunction of all such clauses is a CNF equivalent of the formula represented by the trie. But observe that this CNF formula introduces significant redundancy. In fact, the trie can be interpreted directly as an NNF formula, recursively defined as follows: A trie consisting of a single node represents the constant labeling that node. Otherwise, the trie represents the disjunction of the label of the root with the conjunction of the formulas represented by the tries rooted at its children.

When clause sets are stored as tries, space advantages can be gained by ordering the literals and treating the clauses as ordered sets. An n-literal clause will be represented by one of the $n!$ possible sequences. If the clause set is a set of implicates, then one possibility is to store only prime implicates — clauses that are not subsumed by others — because all subsumed clauses are also implicates and thus implicitly in the set. The space savings can be considerable, but there will in general be exponentially many prime implicates. Furthermore, to determine whether clause C is in the set, the trie must be examined for any subset of C; the literal ordering helps, but the cost is still proportional to the size of the trie.

Suppose instead that *all* implicates are stored; the resulting trie is called an *implicate trie*. To define it formally, let $p_1, p_2, ..., p_n$ be the variables that appear in the input knowledge base $\mathcal{D}$, and let q_i be the literal p_i or $\neg p_i$. Literals are ordered as follows: $q_i \prec q_j$ iff $i < j$. (This can be extended to a total order by defining $\neg p_i \prec p_i, 1 \leq i \leq n$.

---

[2] Many variations have been proposed in which arcs rather than nodes are labeled, and the labels are sometimes strings rather than single symbols.

But neither queries nor branches in the trie will contain such complementary pairs.) The implicate trie for $\mathcal{D}$ is a tree defined as follows: If $\mathcal{D}$ is a tautology (contradiction), the tree consists only of a root labeled 1 (0). Otherwise, it is a tree whose root is labeled 0 and has, for any implicate $C = \{q_{i_1}, q_{i_2}, \ldots, q_{i_m}\}$, a child labeled q_{i_1}, which is the root of a subtree containing a branch with labels corresponding to $C - \{q_{i_1}\}$. The clause C can then be checked for membership in time linear in the size of C, simply by traversing the corresponding branch.

Note that the node on this branch labeled q_{i_m} will be marked with the end symbol. Furthermore, given any node labeled by q_j and marked with the end symbol, if $j < n$, it will have as children nodes labeled q_k and $\neg q_k$, $j < k \leq n$, and these are all marked with the end symbol. This is an immediate consequence of the fact that a node marked with the end symbol represents an implicate which is a prefix (in particular, subset) of every clause obtainable by extending this implicate in all possible ways with the literals greater than q_j in the ordering.

3.2 Reduced Implicate Tries

Recall that for any logical formulas $\mathcal{F}$ and α and subformula $\mathcal{G}$ of $\mathcal{F}$, $\mathcal{F}[\alpha/\mathcal{G}]$ denotes the formula produced by substituting α for every occurrence of $\mathcal{G}$ in $\mathcal{F}$. If α is a truth functional constant 0 or 1 (*false* or *true*), and if p is a negative literal, we will slightly abuse this notation by interpreting the substitution $[0/p]$ to mean that 1 is substituted for the atom that p negates.

The following simplification rules[3] are useful (even if trivial).

SR5.	$\mathcal{F}[\mathcal{G}/\mathcal{G} \vee 0]$	$\mathcal{F}[\mathcal{G}/\mathcal{G} \wedge 1]$
SR6.	$\mathcal{F}[0/\mathcal{G} \wedge 0]$	$\mathcal{F}[1/\mathcal{G} \vee 1]$
SR8.	$\mathcal{F}[0/p \wedge \neg p]$	$\mathcal{F}[1/p \vee \neg p]$

If $C = \{q_{i_1}, q_{i_2}, \ldots, q_{i_m}\}$ is an implicate of $\mathcal{F}$, it is easy to see that the node labeled q_{i_m} will become a leaf if these rules are applied repeatedly to the subtree of the implicate trie of $\mathcal{F}$ rooted at q_{i_m}. Moreover, the product of applying these rules to the entire implicate trie until no applications of them remain will be a trie in which no internal nodes are marked with the end symbol and all leaf nodes are, rendering that symbol merely a convenient indicator for leaves. The result ot this process is called a *reduced implicate trie* or simply an *ri-trie*.

Consider an example. Suppose that the knowledge base $\mathcal{D}$ contains the variables p, q, r, s, in that order, and suppose that $\mathcal{D}$ consists of the following clauses: $\{p, q, \neg s\}$, $\{p, q, r\}$, $\{p, r, s\}$, and $\{p, q\}$. Initialize the ri-trie as a single node labeled 0 and then build it one clause at a time. After the first is added to the tree, its two supersets must also be added. The resulting ri-trie is on the left in the diagram below. Adding the second clause implies that the node labeled by r is also a leaf. Then all extensions of this branch are entailed by $\{p, q, r\}$ (and thus by $\mathcal{D}$), and the corresponding child is dropped resulting in the ri-trie in the center.

---

[3] The labels of these rules come from [21].

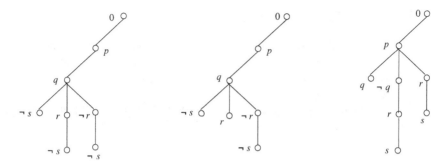

Adding the last two clauses produces the ri-trie on the right.

The complete implicate trie in the example is shown below. It has eight branches, but there are eleven end markers representing its eleven implicates (nine in the subtree rooted at q, and one each at the two rightmost occurrences of s.)

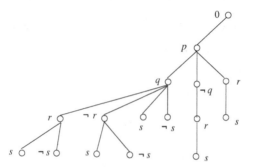

Observations (ri-tries).

1. The NNF equivalent of the ri-trie is

$$p \lor (q \land (\neg q \lor r \lor s) \land (r \lor s)).$$

2. In general, the length of a branch is at most n, the number of variables that appear in the original logical formula $\mathcal{D}$.

3. In order to have the determination of entailment of a given clause be linear in the size of the clause, enough branches must be included in the tree so that the test for entailment can be accomplished by traversing a single branch.

4. When a clause is added to the trie, the literal of highest index becomes the leaf, and *that branch need never be extended*; i.e., if a clause is being tested for entailment, and if a prefix of the clause is a branch in the trie, that the clause is entailed.

5. The ri-trie will be stored off-line. Even if it is very large, each branch is small — no longer than the number of variables — and can be traversed very quickly, even with relatively slow off-line storage.

6. If the query requires sorting, the search time will be $n \log n$, where n is the size of the query. But the sorting can be done in main memory, and the search time in the (off-line) ri-trie is still linear in the size of the query.

4 A Procedure for Computing ri-Tries

In this section a procedure for computing an ri-trie as a logical formula is presented. The *Tri operator* is described in Section 4.1; it can be thought of as a single step in the process that creates an ri-trie. The *RIT operator*, described in Section 4.2, produces the ri-trie with recursive applications of the Tri operator.

It will be convenient to assume that any constant that arises in a logical formula is simplified away with repeated applications of rules **SR5** and **SR6** (unless the formula is constant).

4.1 The Tri Operator

The Tri operator restructures a formula by substituting truth constants for a variable. Lemmas 1 and 2 and the corollary that follows provide insight into the implicates of the components of $\mathrm{Tri}(\mathcal{F}, p)$. Let $\mathrm{Imp}(\mathcal{F})$ denote the set of all implicates of $\mathcal{F}$.

The *tri-expansion*[4] of any formula $\mathcal{F}$ with respect to any atom p is defined to be

$$\mathrm{Tri}(\mathcal{F}, p) = (p \vee \mathcal{F}[0/p]) \wedge (\neg p \vee \mathcal{F}[1/p]) \wedge (\mathcal{F}[0/p] \vee \mathcal{F}[1/p]).$$

Lemma 1. Suppose that the clause C is an implicate of the logical formula $\mathcal{F}$, and that the variable p occurs in $\mathcal{F}$ but not in C. Then $C \in \mathrm{Imp}(\mathcal{F}[0/p]) \cap \mathrm{Imp}(\mathcal{F}[1/p])$.

Proof. Let I be an interpretation that satisfies $\mathcal{F}[1/p]$; we must show that I satisfies C. Extend I to $\tilde{I}$ by setting $\tilde{I}(p) = 1$. Clearly, $\tilde{I}$ satisfies $\mathcal{F}$[5], so $\tilde{I}$ satisfies C. But then, since p does not occur in C, I satisfies C. The proof for $\mathcal{F}[0/p]$ is identical, except that $\tilde{I}(p)$ must be set to 0. $\square$

Lemma 2. Let $\mathcal{F}$ and $\mathcal{G}$ be logical formulas. Then $\mathrm{Imp}(\mathcal{F}) \cap \mathrm{Imp}(\mathcal{G}) = \mathrm{Imp}(\mathcal{F} \vee \mathcal{G})$.

Proof. Suppose first that C is an implicate of both $\mathcal{F}$ and $\mathcal{G}$. We must show that C is an implicate of $\mathcal{F} \vee \mathcal{G}$, so let I be an interpretation that satisfies $\mathcal{F} \vee \mathcal{G}$. Then I satisfies $\mathcal{F}$ or I satisfies $\mathcal{G}$, say $\mathcal{F}$. Then, since C is an implicate of $\mathcal{F}$, I satisfies C.

Suppose now that $C \in \mathrm{Imp}(\mathcal{F} \vee \mathcal{G})$. We must show that $C \in \mathrm{Imp}(\mathcal{F})$ and that $C \in \mathrm{Imp}(\mathcal{G})$. To see that $C \in \mathrm{Imp}(\mathcal{F})$, let I be any satisfying interpretation of $\mathcal{F}$. Then I satisfies $\mathcal{F} \vee \mathcal{G}$, so I satisfies C. The proof that $C \in \mathrm{Imp}(\mathcal{G})$ is entirely similar. $\square$

Corollary. Let C be a clause not containing p or $\neg p$, and let $\mathcal{F}$ be any logical formula. Then C is an implicate of $\mathcal{F}$ iff C is an implicate of $\mathcal{F}[0/p] \vee \mathcal{F}[1/p]$. $\square$

4.2 The RIT Operator

The ri-trie of a formula can be obtained by applying the Tri operator successively on the variables. Let $\mathcal{F}$ be a logical formula, and let the variables of $\mathcal{F}$ be $V = \{p_1, p_2, ..., p_n\}$. Then the RIT operator is defined by

---

[4] This is tri as in three, not as in trie; the pun is probably intended.

[5] It is possible that there are variables other than p that occur in $\mathcal{F}$ but not in $\mathcal{F}[1/p]$. But such variables must "simplify away" when 1 is substituted for p, so I can be extended to an interpretation of $\mathcal{F}$ with any truth assignment to such variables.

$$\mathrm{RIT}(\mathcal{F}, V) \ = \ \begin{cases} \mathcal{F} & V = \emptyset \\[2ex] \begin{aligned} &p_i \ \vee \ \mathrm{RIT}(\mathcal{F}[0/p_i], V - \{p_i\}) \\ &\qquad\qquad \wedge \\ &\neg p_i \ \vee \ \mathrm{RIT}(\mathcal{F}[1/p_i], V - \{p_i\}) \\ &\qquad\qquad \wedge \\ &\mathrm{RIT}((\mathcal{F}[0/p_i] \ \vee \ \mathcal{F}[1/p_i]), V - \{p_i\}) \end{aligned} & p_i \in V \end{cases}$$

where p_i is the variable of lowest index in V.

Implicit in this definition is the use of simplification rules **SR5**, **SR6**, and **SR8**. Theorem 2 below essentially proves that the RIT operator produces ri-tries; reconsidering the above example illustrates this fact:

$$\mathcal{F} = \{p, q, \neg s\} \ \wedge \ \{p, q, r\} \ \wedge \ \{p, r, s\} \ \wedge \ \{p, q\},$$

so that $V = \{p, q, r, s\}$. Let $\mathcal{F}[0/p]$ and $\mathcal{F}[1/p]$ be denoted by $\mathcal{F}_0$ and $\mathcal{F}_1$, respectively. Since p occurs in every clause, $\mathcal{F}_0$ amounts to deleting p from each clause, and $\mathcal{F}_1 = 1$. Thus,

$$\begin{aligned} \mathrm{RIT}(\mathcal{F}, V) \ &= \ (p \ \vee \ \mathrm{RIT}(\mathcal{F}_0, \{q, r, s\})) \ \wedge \ (\neg p \ \vee \ 1) \ \wedge \ \mathrm{RIT}((\mathcal{F}_0 \ \vee \ 1), \{q, r, s\}) \\ &= \ (p \ \vee \ \mathrm{RIT}(\mathcal{F}_0, \{q, r, s\})) \end{aligned}$$

where

$$\mathcal{F}_0 \ = \ \{q, \neg s\} \ \wedge \ \{q, r\} \ \wedge \ \{r, s\} \ \wedge \ \{q\}.$$

Let $\mathcal{F}_0[0/q]$ and $\mathcal{F}_0[1/q]$ be denoted by $\mathcal{F}_{00}$ and $\mathcal{F}_{01}$, respectively. Observe that $\mathcal{F}_{00} = 0$, and $q \vee 0 = q$. Thus,

$$\mathrm{RIT}(\mathcal{F}_0, \{q, r, s\}) \ = \ q \ \wedge \ (\neg q \ \vee \ \mathrm{RIT}(\mathcal{F}_{01}, \{r, s\})) \ \wedge \ \mathrm{RIT}((0 \vee \mathcal{F}_{01}), \{r, s\}),$$

where

$$\mathcal{F}_{01} \ = \ \mathcal{F}_0[1/q] \ = \ 1 \ \wedge \ 1 \ \wedge \ (r \ \vee \ s) \ \wedge \ 1 \ = \ (r \ \vee \ s).$$

Observe now that $\mathcal{F}_{01}[0/r] = s$ and $\mathcal{F}_{01}[1/r] = 1$. Thus,

$$\mathrm{RIT}(\mathcal{F}_{01}, \{r, s\}) \ = \ (r \ \vee \ \mathrm{RIT}(s, \{s\})) \ \wedge \ (\neg r \ \vee \ 1) \ \wedge \ \mathrm{RIT}((s \vee 1), \{s\}).$$

Finally, since $\mathrm{RIT}(s, \{s\}) = s$, substituting back produces

$$\mathrm{RIT}(\mathcal{F}, V) \ = \ p \ \vee \ (q \ \wedge \ (\neg q \ \vee \ r \ \vee \ s) \ \wedge \ (r \ \vee \ s)),$$

which is exactly the formula obtained originally from the ri-trie.

If a logical formula contains only one variable p, then it must be logically equivalent to one of the following four formulas: $0, 1, p, \neg p$. The next lemma, which is trivial to prove from the definition of the RIT operator, says that in that case, $\mathrm{RIT}(\mathcal{F}, \{p\})$ is precisely the simplified logical equivalent.

Lemma 3. Suppose that the logical formula $\mathcal{F}$ contains only one variable p. Then $\mathrm{RIT}(\mathcal{F}, \{p\})$ is logically equivalent to $\mathcal{F}$ and is one of the formulas $0, 1, p, \neg p$. $\square$

For the remainder of the paper, assume the following notation with respect to a logical formula $\mathcal{F}$: Let $V = \{p_1, p_2, ...p_n\}$ be the set of variables of $\mathcal{F}$, and let $V_i = \{p_{i+1}, p_{i+2}, ..., p_n\}$. Thus, for example, $V_0 = V$, and $V_1 = \{p_2, p_3, ..., p_n\}$. Let $\mathcal{F}_t = \mathcal{F}[t/p_i], t = 0, 1$, where p_i is the variable of lowest index in $\mathcal{F}$.

Theorem 1. If $\mathcal{F}$ is any logical formula with variable set V, then $\mathrm{RIT}(\mathcal{F}, V)$ is logically equivalent to $\mathcal{F}$, and each branch of $\mathrm{RIT}(\mathcal{F}, V)$ is an implicate of $\mathcal{F}$.

Proof. We first prove logical equivalence. Proceed by induction on the number n of variables in $\mathcal{F}$. The last lemma takes care of the base case $n = 1$, so assume the theorem holds for all formulas with at most n variables, and suppose that $\mathcal{F}$ has $n + 1$ variables. Then we must show that

$$\mathcal{F} \equiv \mathrm{RIT}(\mathcal{F}, V) = \begin{array}{c} p_1 \vee \mathrm{RIT}(\mathcal{F}_0, V_1) \\ \wedge \\ \neg p_1 \vee \mathrm{RIT}(\mathcal{F}_1, V_1) \\ \wedge \\ \mathrm{RIT}((\mathcal{F}_0 \vee \mathcal{F}_1), V_1). \end{array}$$

By the induction hypothesis, $\mathcal{F}_0 \equiv \mathrm{RIT}(\mathcal{F}_0, V_1)$, $\mathcal{F}_1 \equiv \mathrm{RIT}(\mathcal{F}_1, V_1)$, and $(\mathcal{F}_0 \vee \mathcal{F}_1) \equiv \mathrm{RIT}((\mathcal{F}_0 \vee \mathcal{F}_1), V_1)$. Let I be any interpretation that satisfies $\mathcal{F}$, and suppose first that $I(p_1) = 1$. Then I satisfies p_1, $\mathcal{F}_1$, and $(\mathcal{F}_0 \vee \mathcal{F}_1)$, so I satisfies each of the three conjuncts of $\mathrm{RIT}(\mathcal{F}, V)$; i.e., I satisfies $\mathrm{RIT}(\mathcal{F}, V)$. The case when $I(p) = 0$ and the proof that any satisfying interpretation of $\mathrm{RIT}(\mathcal{F}, V)$ satisfies $\mathcal{F}$ are similar.

Every branch is an implicate of $\mathcal{F}$ since, by the distributive laws, $\mathrm{RIT}(\mathcal{F}, V)$ is logically equivalent to the conjunction of its branches.   □

Lemma 4. Let C be an implicate of $\mathcal{F}$ containing the literal p. Then $C - \{p\} \in \mathrm{Imp}(\mathcal{F}[0/p])$.

Proof. Let I be an interpretation that satisfies $\mathcal{F}[0/p]$. Extend I by defining $I(p) = 0$. Then I satisfies $\mathcal{F}$, so I satisfies C. Since I assigns 0 to p, I must satisfy a literal in C other than p; i.e., I satisfies $C - \{p\}$.   □

The theorem below says, in essence, that ri-tries have the desired property that determining whether a clause is an implicate can be done by traversing a single branch. If $\{q_1, q_2, ..., q_k\}$ is the clause, it will be an implicate iff for some $i \leq k$, there is a branch labeled $q_1, q_2, ..., q_i$. The clause $\{q_1, q_2, ..., q_i\}$ subsumes $\{q_1, q_2, ..., q_k\}$, but it is not an arbitrary subsuming clause. To account for this relationship, a *prefix* of a clause $\{q_1, q_2, ..., q_k\}$ is a clause of the of the form $\{q_1, q_2, ..., q_i\}$, where $0 \leq i \leq k$. Implicit in this definition is a fixed ordering of the variables; also, if $i = 0$, then the prefix is the empty clause.

Theorem 2. Let $\mathcal{F}$ be a logical formula with variable set V, and let C be an implicate of $\mathcal{F}$. Then there is a unique prefix of C that is a branch of $\mathrm{RIT}(\mathcal{F}, V)$.

Proof. Let $V = \{p_1, p_2, ...p_n\}$, and proceed by induction on n. Lemma 3 takes care of the base case $n = 1$.

Assume now that the theorem holds for all formulas with at most n variables, and suppose that $\mathcal{F}$ has $n + 1$. Let C be an implicate of $\mathcal{F}$, say $C = \{q_{i_1}, q_{i_2}, ... q_{i_k}\}$, where q_{i_j} is either p_{i_j} or $\neg p_{i_j}$, and $i_1 < i_2 < ... < i_j$. We must show that a prefix of C is a branch in $\mathrm{RIT}(\mathcal{F}, V)$, which is the formula

$$
\mathrm{RIT}(\mathcal{F}, V) = \begin{array}{c} p_1 \ \vee \ \mathrm{RIT}(\mathcal{F}_0, V_1) \\ \wedge \\ \neg p_1 \ \vee \ \mathrm{RIT}(\mathcal{F}_1, V_1) \\ \wedge \\ \mathrm{RIT}((\mathcal{F}_0 \ \vee \ \mathcal{F}_1), V_1). \end{array}
$$

Observe that the induction hypothesis applies to the third branch. Thus, if $i_1 > 1$, there is nothing to prove, so suppose that $i_1 = 1$. Then q_1 is either p_1 or $\neg p_1$. Consider the case $q_1 = p_1$; the proof when $q_1 = \neg p_1$ is entirely similar. By Lemma 4, $C - \{p_1\}$ is an implicate of $\mathcal{F}_0$, and by the induction hypothesis, there is a unique prefix B of $C - \{p_1\}$ that is a branch of $\mathrm{RIT}(\mathcal{F}_0, V_1)$. But then $A = \{p_1\} \cup B$ is a prefix of C that is a branch of $\mathrm{RIT}(\mathcal{F}, V)$.

To complete the proof, we must show that A is the only such prefix of C. Suppose to the contrary that D is another prefix of C that is a branch of $\mathrm{RIT}(\mathcal{F}, V)$. Then either D is a prefix of A or A is a prefix of D; say that D is a prefix of A. Let $D = \{p_1\} \cup E$. Then E is a prefix of B in $\mathrm{RIT}(\mathcal{F}_0, V_1)$, which in turn means that E is a prefix of $C - \{p_1\}$. But we know from the inductive hypothesis that $C - \{p_1\}$ has a unique prefix in $\mathrm{RIT}(\mathcal{F}_0, V_1)$, so $E = B$, so $D = A$. If A is a prefix of D, then it is immediate that E is a prefix of $C - \{p_1\}$ in $\mathrm{RIT}(\mathcal{F}_0, V_1)$, and, as before, $E = B$, and $D = A$. $\square$

The corollaries below are immediate because of the uniqueness of prefixes of implicates in $\mathrm{RIT}(\mathcal{F}, V)$.

Corollary 1. Every prime implicate of $\mathcal{F}$ is a branch in $\mathrm{RIT}(\mathcal{F}, V)$. $\qquad\square$

Corollary 2. Every subsuming implicate (including any prime implicate) of a branch in $\mathrm{RIT}(\mathcal{F}, V)$ contains the literal labeling the leaf of that branch. $\qquad\square$

5   An Algorithm for Computing ri-Tries

In this section, an algorithm that produces ri-tries is developed using pseudo-code. The algorithm relies heavily on Lemma 2, which states that $\mathrm{Imp}(\mathcal{F}_0 \vee \mathcal{F}_1) = \mathrm{Imp}(\mathcal{F}_0) \cap \mathrm{Imp}(\mathcal{F}_1)$; i.e., the branches produced by the third conjunct of the RIT operator are precisely the branches that occur in both of the first two (ignoring, of course, the root labels p_i and $\neg p_i$). The algorithm makes use of this lemma rather than directly implementing the RIT operator; in particular, the recursive call $\mathrm{RIT}((\mathcal{F}[0/p_i] \vee \mathcal{F}[1/p_i]), V - \{p_i\})$ is avoided. This is significant because that call doubles the size of the formula *along a single branch*.

No attempt was made to make the algorithm maximally efficient. For clarity, the algorithm is designed so that the first two conjuncts of the RIT operator are constructed in their entirety, and then the third conjunct is produced by parallel traversal of the first two.

The algorithm employs two functions: *rit* and *buildzero*. The nodes of the trie consist of five fields: *label*, which is the name of the literal that occurs in the node; *parent*, which is a pointer to the parent of the node; and *plus*, *minus*, and *zero*, which are pointers to the three children. The function rit recursively builds the first two conjuncts of the RIT operator and then calls buildzero, which recursively builds the third conjunct from the first two.

The reader may note that the algorithm builds a ternary tree rather than an n-ary trie. The reason is that the construction of the subtree representing the third conjunct of the RIT operator sets the label of the root to 0. This is convenient for the abstract description of the algorithm; it is straightforward but tedious to write the code without employing the zero nodes.

Observations

1. Recall that each (sub-)trie represents the disjunction of the label of its root with the conjunction of the sub-tries rooted at its children.
2. If either of the first two sub-tries are 1, then that sub-trie is empty. The third is also empty since it is the intersection of the first two.
3. If any child of a node is 0, then the node reduces to a leaf. In practice, in the algorithm, this will only occur in the first two branches.
4. If both of the first two sub-tries are leaves, then they are deleted by **SR8** and the root becomes a leaf.
5. No pseudocode is provided for the straightforward routines "makeleaf", "leaf", and "delete". The first two are called on a pointer to a trienode, the third is called on a trienode.

The ri-Trie Algorithm

```
declare( structure trienode(    lit: label,
                             parent: ↑trienode,
                               plus: ↑trienode,
                              minus: ↑trienode,
                               zero: ↑trienode);

              RItrie:trienode);

input(G);                        {The logical formula G has variables p₁, p₂, ..., pₙ}
RItrie ← rit(G, 1, 0);
```

input(G);   {The logical formula G has variables $p_1, p_2, ..., p_n$}
RItrie ← rit(G, 1, 0);

```
function rit(G: wff, polarity, varindex: integer): ↑ trienode;
    N ← new(trienode);
    if varindex=0 then n.lit ← 0                     {root of entire trie is 0}
        else if polarity = 0 then N.lit ← −p_varindex
            else N.lit ← p_varindex
    Gplus ← G[0/p_varindex];
    Gminus ← G[1/p_varindex];
```

```
        if (Gplus = 0 or Gminus = 0)
            then makeleaf(N); return(↑N);                    {Observation 3.}
        if Gplus = 1 then N.plus ← nil; N.zero ← nil         {Observation 2.}
            else N.plus ← rit(Gplus, 1, varindex+1);
                if N.plus.lit = 1 then delete(N.plus↑); N.plus ← nil;
        if Gminus = 1 then N.minus ← nil; N.zero ← nil       {Observation 2.}
            else N.minus ← rit(Gminus, 0, varindex+1);
                if N.minus.lit = 1 then delete(N.minus↑)
                if N.plus = nil then N.lit ← 1; makeleaf(N);
        if (leaf(N.plus) and leaf(N.minus))                  {Observation 4.}
            then delete(N.plus); delete(N.minus); makeleaf(N); return(↑N);
        if (N.plus ≠ nil and N.minus ≠ nil)
            then N.zero ← buildzero(N.plus, N.minus);
        N.zero↑.parent ← ↑N;
        N.zero↑.lit ← 0
        return(↑N);
    end rit;

function buildzero(N1, N2, ↑trienode): ↑trienode;
    Nzero ← new(trienode);
    Nzero.lit ← N1↑.lit;
    if leaf(N1) then Nzero.(plus, minus, zero) ← N2↑.(plus, minus, zero);
        return(↑Nzero)
    if leaf(N2) then Nzero.(plus, minus, zero) ← N1↑.(plus, minus, zero);
        return(↑Nzero);

    if (N1↑.plus = nil or N2↑.plus = nil)
        then Nzero.plus ← nil
        else Nzero.plus ← buildzero(N1↑.plus, N2↑.plus);
    if (N1↑.minus = nil or N2↑.minus = nil)
        then Nzero.minus ← nil
        else Nzero.minus ← buildzero(N1↑.minus, N2↑.minus);
    if (N1↑.zero = nil or N2↑.zero = nil)
        then Nzero.zero ← nil
        else Nzero.zero ← buildzero(N1↑.zero, N2↑.zero);
    if leaf(↑Nzero) then delete(↑Nzero); return(nil)
        else begin
            if Nzero.plus ≠ nil then Nzero.plus↑.parent ← ↑Nzero;
            if Nzero.minus ≠ nil then Nzero.minus↑.parent ← ↑Nzero;
            if Nzero.zero ≠ nil then Nzero.zero↑.parent ← ↑Nzero;
            return(↑Nzero)
        end

end buildzero.
```

A final straightforward observation is that by employing the dual of the Tri operator (say DTri),

$$\text{DTri}(\mathcal{F}, p) = (p \wedge \mathcal{F}[1/p]) \vee (\neg p \wedge \mathcal{F}[0/p]) \vee (\mathcal{F}[1/p] \wedge \mathcal{F}[0/p]).$$

tries that store implicants may be analogously built.

6 Future Work

The ri-trie has the very nice property that no matter how large the trie is, any query can be answered in time linear in the size of the query. This gives rise to a number of questions. Under what circumstance will the ri-trie be small enough[6] to be practical? How easy is it to maintain and update ri-tries? The size of an ri-trie is surely dependent on the variable ordering — are there good heuristics for setting the variable ordering? Might a "forest" of ri-tries be effective?

An ri-trie is considerably smaller than the full implicate trie because whenever a prefix of an implicate is also an implicate, the trie may be truncated at the last node of the prefix. But knowing that any subset of a clause is an implicate is enough to determine that the clause is itself an implicate. Can this fact be used to reduce the trie further? Are there other insights that might provide further size reduction?

The answers to these questions will assist with implementation, but the results of this paper are adequate to begin experiments with ri-tries.

References

1. Bryant, R. E., Symbolic Boolean manipulation with ordered binary decision diagrams, *ACM Comput. Surv.* **24, 3** (1992), 293–318.
2. Bibel, W. On matrices with connections. *J. ACM* **28**,4 (1981), 633 – 645.
3. Cadoli, M., and Donini, F. M., A survey on knowledge compilation, *AI Commun.* **10** (1997), 137–150.
4. Chin, R. T., and Dyer, C. R. Model-based recognition in robot vision. *ACM Computing Surveys* 18(1) (Mar. 1986) 67–108.
5. Coudert, O. and Madre, J. Implicit and incremental computation of primes and essential implicant primes of boolean functions. In *Proceedings of the 29th ACM/IEEE Design Automation Conference*, (1992) 36-39.
6. Coudert, O. and Madre, J. A new graph based prime computation technique. In *Logic Synthesis and Optimization* (T. Sasao, Ed.), Kluwer (1993), 33–58.
7. D'Agostino, M., Gabbay, D. M., Hähnle, R., and Posegga, J., *Handbook of Tableau Methods*, Kluwer Academic Publishers, 1999.
8. Darwiche, A., Compiling devices: A structure-based approach, Proc. *Int'l Conf. on Principles of Knowledge Representation and Reasoning (KR98)*, Morgan-Kaufmann, San Francisco (1998), 156–166.
9. Darwiche, A., Model based diagnosis using structured system descriptions, *Journal of A.I. Research*, **8**, 165-222.
10. Darwiche, A., Decomposable negation normal form, *J.ACM* **48**,4 (2001), 608–647.

---

[6] In this context, a large room full of large hard drives qualifies as small enough.

11. Darwiche, A. and Marquis, P., A knowledge compilation map, *J. of AI Research* **17** (2002), 229–264.

12. Davis, M. and Putnam, H. A computing procedure for quantification theory. *J.ACM*, **7** (1960), 201–215.

13. de Kleer, J. Focusing on probable diagnosis. *Proc. of AAAI-91, Anaheim, CA*, pages 842–848, 1991.

14. de Kleer, J. An improved incremental algorithm for computing prime implicants. *Proceedings of AAAI-92*, San Jose, CA, (1992) 780–785.

15. de Kleer, J., Mackworth, A. K., and Reiter, R. Characterizing diagnoses and systems, *Artificial Intelligence*, 32 (1987), 97–130.

16. de Kleer, J. and Williams, B. Diagnosing multiple faults, *Artificial Intelligence*, 56:197–222, 1987.

17. Fitting, M., *First-Order Logic and Automated Theorem Proving (2nd ed.)*, Springer-Verlag, New York, (1996).

18. Forbus, K.D. and de Kleer, J., *Building Problem Solvers*, MIT Press, Cambridge, Mass. (1993).

19. Gomes, C. P., Selman, B., and Kautz, H., Boosting combinatorial search through randomization, *Proc. AAAI-98* (1998), Madison, Wisconsin, 431-437.

20. Hai, L. and Jigui, S., Knowledge compilation using the extension rule, *J. Automated Reasoning*, 32(2), 93-102, 2004.

21. Hähnle, R., Murray N.V., and Rosenthal, E. Normal Forms for Knowledge Compilation. *Proceedings of the International Symposium on Methodologies for Intelligent Systems*, (ISMIS '05, Z. Ras ed.), Lecture Notes in Computer Science, Springer (to appear).

22. Jackson, P. and Pais, J., Computing prime implicants. *Proceedings of the 10th International Conference on Automated Deductions*, Kaiserslautern, Germany, July, 1990. In *Lecture Notes in Artificial Intelligence*, Springer-Verlag, Vol. 449 (1990), 543-557.

23. Jackson, P. Computing prime implicants incrementally. *Proceedings of the 11th International Conference on Automated Deduction*, Saratoga Springs, NY, June, 1992. In *Lecture Notes in Artificial Intelligence*, Springer-Verlag, Vol. 607 (1992) 253-267.

24. Kautz, H. and Selman, B., A general framework for knowledge compilation, in *Proceedings of the International Workshop on Processing Declarative Knowledge (PDK)*, Kaiserslautern, Germany (July, 1991).

25. Kautz, H. and Selman, B., Proceedings of the Workshop on Theory and Applications of Satisfiability Testing, *Electronic Notes in Discrete Mathematics* **9** (2001), Elsevier Science Publishers.

26. Kean, A. and Tsiknis, G. An incremental method for generating prime implicants/implicates. *Journal of Symbolic Computation* **9** (1990), 185-206.

27. Kean, A. and Tsiknis, G. Assumption based reasoning and clause management systems. *Computational Intelligence* **8**,1 (1992), 1–24.

28. Loveland, D.W. *Automated Theorem Proving: A Logical Basis*. North-Holland, New York, (1978).

29. Marquis, P., Knowledge compilation using theory prime implicates, Proc. *Int'l Joint Conf. on Artificial Intelligence (IJCAI)* (1995), Morgan-Kaufmann, San Mateo, Calif, 837-843.

30. Morrison, D.R. PATRICIA — practical algorithm to retrieve information coded in alphanumeric. *Journal of the ACM*, **15**,4, 514–34, 1968.

31. Murray, N.V. and Rosenthal, E. Inference with path resolution and semantic graphs. *J. ACM* **34**,2 (1987), 225–254.

32. Murray, N.V. and Rosenthal, E. Dissolution: making paths vanish. *J.ACM* **40**,3 (July 1993), 504–535.

33. Murray, N.V. and Rosenthal, E. On the relative merits of path dissolution and the method of analytic tableaux, Theoretical Computer Science 131 (1994), 1-28.

34. Murray N.V. and Rosenthal, E. Duality in Knowledge Compilation Techniques. *Proceedings of the International Symposium on Methodologies for Intelligent Systems*, (ISMIS '05, Z. Ras ed.), Lecture Notes in Computer Science, Springer (to appear).
35. Murray, N.V. and Ramesh, A. An application of non-clausal deduction in diagnosis. *Proceedings of the Eighth International Symposium on Artificial Intelligence*, Monterrey, Mexico, October 17-20, 1995, 378–385.
36. Murray, N.V., Ramesh, A. and Rosenthal, E. The semi-resolution inference rule and prime implicate computations. *Proceedings* of the *Fourth Golden West International Conference on Intelligent Systems*, San Fransisco, CA, (June 1995) 153–158.
37. Murray, N.V. and Rosenthal, E. "Tableaux, Path Dissolution, and Decomposable Negation Normal Form for Knowledge Compilation." *Proceedings* of the *International Conference TABLEAUX 2003 – Analytic Tableaux and Related Methods*, Rome, Italy, September 2003. In *Lecture Notes in Artificial Intelligence*, Springer-Verlag, Vol. 2796, 165-180.
38. Nelson, R. J. 'Simplest normal truth functions', *Journal of Symbolic Logic* **20**, 105-108 (1955).
39. Ngair, T. A new algorithm for incremental prime implicate generation. *Proc of IJCAI-93*, Chambery, France, (1993).
40. Przymusinski, T. C. An algorithm to compute circumscription. *Artificial Intelligence* **38** (1989), 49-73.
41. Ramesh, A. D-trie: A new data structure for a collection of minimal sets. Technical Report SUNYA-CS-95-04, December 28, 1995.
42. Ramesh, A., Becker, G. and Murray, N.V. CNF and DNF considered harmful for computing prime implicants/implicates. *Journal of Automated Reasoning* **18**,3 (1997), Kluwer, 337–356.
43. Ramesh, A. and Murray, N.V. An application of non-clausal deduction in diagnosis. *Expert Systems with Applications* **12**,1 (1997), 119-126.
44. Ramesh, A. and Murray, N.V. Parameterized Prime Implicant/Implicate Computations for Regular Logics. *Mathware & Soft Computing*, Special Issue on Deduction in Many-Valued Logics IV(2):155-179, 1997.
45. Reiter, R. A theory of diagnosis from first principles. *Artificial Intelligence*, **32** (1987), 57-95.
46. Reiter, R. and de Kleer, J. Foundations of assumption-based truth maintenance systems: preliminary report. *Proceedings of the 6th National Conference on Artificial Intelligence*, Seattle, WA, (July 12-17, 1987), 183-188.
47. Sieling, D., and Wegener, I., Graph driven BDDs – a new data structure for Boolean functions, *Theor. Comp. Sci.* **141** (1995), 283-310.
48. Slagle, J. R., Chang, C. L. and Lee, R. C. T. A new algorithm for generating prime implicants. *IEEE transactions on Computers* **C-19**(4) (1970), 304-310.
49. Selman, B., and Kautz, H., Knowledge compilation and theory approximation, *J.ACM* **43**,2 (1996), 193-224.
50. Selman, B., Kautz, H., and McAllester, D., Ten challenges in propositional reasoning and search, *Proc of IJCAI-97*, Aichi, Japan (1997).
51. Strzemecki, T. Polynomial-time algorithm for generation of prime implicants. *Journal of Complexity* **8** (1992), 37-63.
52. Ullman, J.D. *Principles of Database Systems*, Computer Science Press, Rockville, 1982.

Clausal Connection-Based Theorem Proving in Intuitionistic First-Order Logic

Jens Otten

Institut für Informatik, University of Potsdam,
August-Bebel-Str. 89, 14482 Potsdam-Babelsberg, Germany
jeotten@cs.uni-potsdam.de

Abstract. We present a clausal connection calculus for first-order intuitionistic logic. It extends the classical connection calculus by adding
prefixes that encode the characteristics of intuitionistic logic. Our calculus is based on a clausal matrix characterisation for intuitionistic logic,
which we prove correct and complete. The calculus was implemented by
extending the classical prover leanCoP. We present some details of the
implementation, called ileanCoP, and experimental results.

1 Introduction

Automated reasoning in intuitionistic first-order logic is an important task within
the formal approach of constructing verifiable correct software. Interactive proof
assistants, like NuPRL [5] and Coq [2], use constructive type theory to formalise
the notion of computation and would greatly benefit from a higher degree of
automation. Automated theorem proving in intuitionistic logic is considerably
more difficult than in classical logic, because additional non-permutabilities in
the intuitionistic sequent calculus [8] increase the search space. In classical logic
(disjunctive or conjunctive) clausal forms are commonly used to simplify the
problem of copying appropriate subformulae (so-called multiplicities). Once a
formula is converted into clausal form multiplicities can be restricted to clauses.
For intuitionistic logic a validity-preserving clausal form does not exist.

An elegant way of encoding the intuitionistic non-permutabilities was given
by Wallen [30] by extending Bibel's (non-clausal) *characterisation* of logical validity [3]. The development of proof *calculi* and implementations based on this
characterisation (e.g. [11,21,22,25]) were restricted to non-clausal procedures,
making it difficult to use more established clausal methods (e.g. [3,4,13,14]).

In this paper we present a clausal matrix characterisation for intuitionistic
logic. Extending the usual Skolemization technique makes it possible to adapt a
connection calculus that works on prefixed matrices in clausal form. It simplifies
the notation of multiplicities and the use of existing clausal connection-based
implementations for classical logic. Only a few changes are required to adapt the
classical prover leanCoP to deal with prefixed matrices.

The paper is organised as follows. In Section 2 the standard (non-clausal)
matrix characterisation is presented before Section 3 introduces a clausal characterisation using Skolemization and a prefixed connection calculus. The ileanCoP

B. Beckert (Ed): TABLEAUX 2005, LNAI 3702, pp. 245–261, 2005.

implementation with experimental results is described in Section 4. We conclude
with a summary and a brief outlook on further research in Section 5.

2   Preliminaries

We assume the reader to be familiar with the language of classical first-order
logic (see, e.g., [7]). We start by defining some basic concepts before briefly
describing the matrix characterisation for classical and intuitionistic logic.

2.1   Formula Trees, Types and Multiplicities

Some basic concepts are formula trees, types and multiplicities (see [4,7,30]).
Multiplicities encode the contraction rule in the sequent calculus [8].

Definition 1 (Formula Tree). *A formula tree is the graphical representation
of a formula F as a tree. Each node is labeled with a connective/quantifier or
atomic subformula of F and marked with a unique name, its* position, *denoted
by* $a_0, a_1, \ldots$. *The set of all positions is denoted by* Pos. *The tree ordering
$< \subseteq$ Pos $\times$ Pos is the (partial) ordering on the positions in the formula tree,
i.e. $a_i < a_j$ iff position a_i is below position a_j in the formula tree.*

Example 1. Figure 1 shows the formula tree for $F_1 = \forall x P x \Rightarrow P b \wedge P c$ with
Pos $= \{a_1, \ldots, a_5\}$. It is, e.g., $a_1 < a_2$ and $a_0 < a_4$. Note that each subformula of a
given formula corresponds to exactly one position in its formula tree, e.g. $\forall x P x$
corresponds to the position a_1.

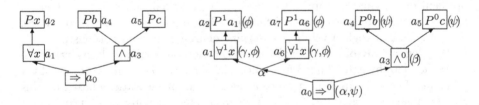

Fig. 1. Formula tree for F_1 and F_1^μ with $\mu(a_1) = 2$

Definition 2 (Types, Polarity). *The* principal *and* intuitionistic type *of a
position is determined by its label and polarity according to Table 1 (e.g. a for-
mula/position $A \wedge B$ with polarity 1 has type α). Atomic positions have no principal type and some positions have no intuitionistic type. We denote the sets of
positions of type γ, δ, ϕ, and ψ by Γ, Δ, Φ, and Ψ, respectively. The polarity
(0 or 1) of a position is determined by the label and polarity of its predecessor
in the formula tree according to table 1 (e.g. if $A \wedge B$ has polarity 1 then both
subformula A and B have polarity 1 as well). The root position has polarity 0.*

Table 1. Principal type, intuitionistic type and polarity of positions/subformulae

principal type α	$(A \wedge B)^1$	$(A \vee B)^0$	$(A{\Rightarrow}B)^0$	$(\neg A)^1$	$(\neg A)^0$
successor polarity	A^1, B^1	A^0, B^0	A^1, B^0	A^0	A^1
principal type β	$(A \wedge B)^0$	$(A \vee B)^1$	$(A{\Rightarrow}B)^1$		
successor polarity	A^0, B^0	A^1, B^1	A^0, B^1		

principal type γ	$(\forall xA)^1$	$(\exists xA)^0$	principal type δ		$(\forall xA)^0$	$(\exists xA)^1$
successor polarity	A^1	A^0	successor polarity		A^0	A^1

intuitionistic type ϕ	$(\neg A)^1$	$(A{\Rightarrow}B)^1$	$(\forall xA)^1$	P^1 (P is atomic)
intuitionistic type ψ	$(\neg A)^0$	$(A{\Rightarrow}B)^0$	$(\forall xA)^0$	P^0 (P is atomic)

Example 2. In the left formula tree for F_1 in Figure 1 (ignore the additional branch for now) each node/position is marked with its types and its polarity. The positions a_0, a_1, a_3 have principal type α, γ and β, respectively; the positions a_1, a_2 and a_0, a_4, a_5 have intuitionistic type ϕ and ψ, respectively.

Definition 3 (Multiplicity). *The* multiplicity $\mu : \Gamma \cup \Phi \to \mathbb{N}$ *assigns each position of type γ and ϕ a natural number. By F^μ we denote an* indexed formula. *In the formula tree of F^μ multiple instances of subformulae — according to the multiplicity of its positions — have been considered. The branch instances of a subformula have an implicit node/position of type α as predecessor.*

Example 3. The formula tree for F_1 with $\mu(a_1)=2$ is shown in Figure 2. Here and in the following we will replace variables in atomic formulae by their quantifier positions. Thus positions of type γ and δ appear in atomic formulae.

2.2   Matrix Characterisation for Classical Logic

To resume the characterisation for classical logic [3,4] we introduce the concepts of matrices, paths and connections. A path corresponds to a sequent and a connection to an axiom in the sequent calculus [8]. For the first-order case we also need the notation of first-order substitution and reduction ordering.

Definition 4 (Matrix,Path,Connection).

1. *In the* matrix(-representation) *of an indexed formula F^μ we place the subformulae of a formula of principal type α side by side whereas subformulae of a subformula of principal type β are placed one upon the other. Furthermore, we omit all connectives and quantifiers.*
2. *A* path *through an indexed formula F^μ is a subset of the atomic formulae of its formula tree; it is a horizontal path through the matrix of F^μ.*
3. *A* connection *is a pair of atomic formulae with the same predicate symbol but with different polarities.*

Example 4. The matrix for F_1^μ is given in Figure 2. There are two paths through F_1^μ: $\{P^1a_1, P^1a_6, P^0b\}$ and $\{P^1a_1, P^1a_6, P^0c\}$. They contain the connections $\{P^1a_1, P^0b\}$, $\{P^1a_6, P^0b\}$ and $\{P^1a_1, P^0c\}$, $\{P^1a_6, P^0c\}$, respectively. We will present a more formal definition of matrices and paths in Section 3.

$$\begin{bmatrix} & & P^0b \\ P^1a_1 & P^1a_6 & \\ & & P^0c \end{bmatrix} \qquad \dfrac{\Gamma, A \vdash}{\Gamma \vdash \neg A, \Delta} \qquad \dfrac{\Gamma, A \vdash B}{\Gamma \vdash A \Rightarrow B, \Delta} \qquad \dfrac{\Gamma \vdash A[x \backslash a]}{\Gamma \vdash \forall x A, \Delta}$$

$$\qquad\qquad\qquad\qquad \neg\text{-right} \qquad\qquad \Rightarrow\text{-right} \qquad\qquad \forall\text{-right}$$

Fig. 2. Matrix for F_1^μ and special intuitionistic sequent rules

Definition 5 (First-order Substitution, Reduction Ordering).

1. *A first-oder substitution $\sigma_Q : \Gamma \to \mathcal{T}$ assigns to each position of type γ a term $t \in \mathcal{T}$. $\mathcal{T}$ is the set of terms made up of constants, functions and elements of Γ and Δ, which are called term variables and term constants, respectively. A connection $\{P^0s, P^1t\}$ is said to be σ_Q-complementary iff $\sigma_Q(s) = \sigma_Q(t)$. σ_Q induces a relation $\sqsubset_Q \subseteq \Delta \times \Gamma$ in the following way: for all $u \in \Gamma$ $v \sqsubset_Q u$ holds for all $v \in \Delta$ occurring in $\sigma_Q(u)$.*

2. *The reduction ordering $\lhd := (< \cup \sqsubset_Q)^+$ is the transitive closure of the tree ordering $<$ and the relation $\sqsubset_Q$. A first order substitution σ_Q is said to be admissible iff the reduction ordering $\lhd$ is irreflexive.*

Theorem 1 (Characterisation for Classical Logic [4]). *A (first-order) formula F is classically valid, iff there is a multiplicity μ, an admissible first-order substitution σ_Q and a set of σ_Q-complementary connections such that every path through F^μ contains a connection from this set.*

Example 5. Let $\mu(a_1)=2$ and $\sigma_Q=\{a_1 \backslash b, a_6 \backslash c\}$, i.e. $\sigma_Q(a_1)=b$, $\sigma_Q(a_6)=c$. σ_Q is admissible, since the induced reduction ordering ($=$ tree ordering) is irreflexive. $\{\{P^1a_1, P^0b\}, \{P^1a_6, P^0c\}\}$ is a σ_Q-complementary set of connections and every path through F_1^μ contains a connection from this set. Thus F_1 is classically valid.

2.3   Matrix Characterisation for Intuitionistic Logic

A few extensions are necessary to adapt the characterisation of classical validity to intuitionistic logic. These are prefixes and an intuitionistic substitution. Prefixes encode the characteristics of the special rules (see Figure 2) in the intuitionistic sequent calculus (see [30]). Alternatively they can be seen as an encoding of the possible world semantics of intuitionistic logic (see [30]).

Definition 6 (Prefix). *Let $u_1 < u_2 < \ldots < u_n \le u$ be the positions of type ϕ or ψ that dominate the position/formula u in the formula tree. The prefix of u, denoted $pre(u)$, is a string over $\Phi \cup \Psi$ and defined as $pre(u) := u_1 u_2 \ldots u_n$.*

Example 6. For the formula F_1^μ we obtain $pre(a_2)=a_0 A_1 A_2$, $pre(a_7)= a_0 A_6 A_7$, $pre(a_4)=a_0 a_4$ and $pre(a_5)=a_0 a_5$. Positions of type ϕ are written in capitals.

Definition 7 (Intuitionistic/Combined Substitution, Admissibility).

1. *An intuitionistic substitution $\sigma_J : \Phi \to (\Phi \cup \Psi)^*$ assigns to each position of type ϕ a string over the alphabet $(\Phi \cup \Psi)$. Elements of Φ and Ψ are called prefix variables and prefix constants, respectively. σ_J induces a relation $\sqsubset_J \subseteq \Psi \times \Phi$ in the following way: for all $u \in \Phi$ $v \sqsubset_J u$ holds for all $v \in \Psi$ occurring in $\sigma_J(u)$.*

2. *A combined substitution $\sigma := (\sigma_Q, \sigma_J)$ consists of a first-order and an intuitionistic substitution. A connection $\{P^0 s, P^1 t\}$ is σ-complementary iff $\sigma_Q(s) = \sigma_Q(t)$ and $\sigma_J(pre(P^0 s)) = \sigma_J(pre(P^1 t))$. The reduction ordering $\lhd := (< \cup \sqsubset_Q \cup \sqsubset_J)^+$ is the transitive closure of $<$, $\sqsubset_Q$ and $\sqsubset_J$.*

3. *A combined substitution $\sigma = (\sigma_Q, \sigma_J)$ is said to be admissible iff the reduction ordering $\lhd$ is irreflexive and the following condition holds for all $u \in \Gamma$ and all $v \in \Delta$ occurring in $\sigma_Q(u)$: $\sigma_J(pre(u)) = \sigma_J(pre(v)) \circ q$ for some $q \in (\Phi \cup \Psi)^*$.*

Theorem 2 (Characterisation for Intuitionistic Logic [30]). *A formula F is intuitionistically valid, iff there is a multiplicity μ, an admissible combined substitution $\sigma = (\sigma_Q, \sigma_J)$, and a set of σ-complementary connections such that every path through F^μ contains a connection from this set.*

Example 7. Let $\mu(a_1) = 2$ and $\sigma = (\sigma_Q, \sigma_J)$ with $\sigma_Q = \{a_1 \backslash b, a_6 \backslash c\}$ and $\sigma_J = \{A_1 \backslash \varepsilon, A_2 \backslash a_4, A_6 \backslash \varepsilon, A_7 \backslash a_5\}$ in which ε denotes the empty string. Then σ is admissible and $\{\{P^1 a_1, P^0 b\}, \{P^1 a_6, P^0 c\}\}$ is a σ_Q-complementary set of connections and every path through F_1^μ contains a connection from this set. Therefore F_1 is intuitionistically valid.

3   Clausal Connection-Based Theorem Proving

This section presents a clausal version of the matrix characterisation for intuitionistic logic and a prefixed connection calculus based on this characterisation.

3.1   Prefixed Matrix and Skolemization

In order to develop a prefixed connection calculus more formal definitions of the concepts of a matrix and paths (as introduced in Definition 4) are required. Afterwards we will also introduce a Skolemization technique, which extends the Skolemization used for classical logic to intuitionistic logic.

A (non-clausal, i.e. nested) matrix is a set of clauses and contains α-related subformulae. A clause is a set of matrices and contains β-related subformulae. A (non-clausal) matrix is a compact representation of a (classical) formula and can be represented in the usual two dimensional graphical way. Adding prefixes will make it suitable to represent intuitionistic formulae as well. The formal definition of paths through a (matrix of a) formula is adapted accordingly.

Definition 8 (Prefixed Matrix).

1. *A prefix p is a string with $p \in (\Phi \cup \Psi)^*$. Elements of Φ and Ψ play the role of variables and constants, respectively. A prefixed (signed) formula $F^{pol}{:}p$ is a formula F marked with a polarity $pol \in \{0, 1\}$ and a prefix p.*

2. *A prefixed matrix M is a set of clauses in which a clause is a set of prefixed matrices and prefixed atomic formulae. The prefixed matrix of a prefixed formula $F^{pol}{:}p$ is inductively defined according to Table 2. In Table 2 A, B are formulae, Ps is an atomic formula, $p \in (\Phi \cup \Psi)^*$ is a prefix, $Z \in \Phi$, $a \in \Psi$*

Table 2. The prefixed matrix of a formula F^{pol}:p

Formula F^{pol}:p	Prefixed matrix of F^{pol}:p	M_A, M_B is matrix of	Formula F^{pol}:p	Prefixed matrix of F^{pol}:p	M_A, M_B is matrix of
$(P^1 s)$:p	$\{\{P^1 s{:}p{\circ}Z\}\}$		$(\neg A)^1$:p	M_A	A^0:p$\circ Z$
$(P^0 s)$:p	$\{\{P^0 s{:}p{\circ}a\}\}$		$(\neg A)^0$:p	M_A	A^1:p$\circ a$
$(A \wedge B)^1$:p	$\{\{M_A\},\{M_B\}\}$	A^1:p , B^1:p	$(A \wedge B)^0$:p	$\{\{M_A, M_B\}\}$	A^0:p , B^0:p
$(A \vee B)^0$:p	$\{\{M_A\},\{M_B\}\}$	A^0:p , B^0:p	$(A \vee B)^1$:p	$\{\{M_A, M_B\}\}$	A^1:p , B^1:p
$(A \Rightarrow B)^0$:p	$\{\{M_A\},\{M_B\}\}$	A^1:p , B^0:p	$(A \Rightarrow B)^1$:p	$\{\{M_A, M_B\}\}$	A^0:p , B^1:p
$(\forall y A)^1$:p	M_A	A_x^1:p$\circ Z$	$(\forall y A)^0$:p	M_A	A_c^0:p$\circ a$
$(\exists y A)^0$:p	M_A	A_x^0:p	$(\exists y A)^1$:p	M_A	A_c^1:p

are new (unused) elements. A_x and A_c is the formula A in which y is replaced by new (unused) elements $x \in \Gamma$ and $c \in \Delta$, respectively. The prefixed matrix of a formula F, denoted $matrix(F)$, is the prefixed matrix of F^0:ε.

Definition 9 (Path). A path *through a (prefixed) matrix* M *or clause* C *is a set of prefixed atomic formulae and defined as follows: If* $M=\{\{P^{pol}s{:}p\}\}$ *is a matrix and* Ps *an atomic formula then* $\{P^{pol}s{:}p\}$ *is the only path through* M. *Otherwise if* $M=\{C_1,..,C_n\}$ *is a matrix and* $path_i$ *is a path through the clause* C_i *then* $\bigcup_{i=1}^{n} path_i$ *is a path through* M. *If* $C=\{M_1,..,M_n\}$ *is a clause and* $path_i$ *is a path through the matrix* M_i *then* $path_i$ *(for* $1 \leq i \leq n$*) is a path through* C.

Example 8. The prefixed matrix of formula F_1 (see examples in Section 2) is $M_1=matrix(F_1)= \{\{P^1 x_1{:}a_2 Z_3 Z_4\},\{P^0 b{:}a_2 a_5, P^0 c{:}a_2 a_6\}\}$, in which submatrices of the form $\{\{M_1,..,M_n\}\}$ have been simplified to $M_1,..,M_n$. The paths through M_1 are $\{P^1 x_1{:}a_2 Z_3 Z_4, P^0 b{:}a_2 a_5\}$ and $\{P^1 x_1{:}a_2 Z_3 Z_4, P^0 c{:}a_2 a_6\}$.

An indexed formula F^μ is defined in the usual way, i.e. each subformula F' of type γ or ϕ is assigned its multiplicity $\mu(F') \in I\!N$ encoding the number of instances to be considered. Before F^μ is translated into a matrix, each F' is replaced with $(F'_1 \wedge \ldots \wedge F'_{\mu(F')})$ in which F'_i is a copy of F'. The notation of σ-complementary is slightly modified, i.e. a connection $\{P^0 s{:}p, P^1 t{:}q\}$ is σ-complementary iff $\sigma_Q(s)=\sigma_Q(t)$ and $\sigma_J(p)=\sigma_J(q)$.

For a combined substitution σ to be admissible (see Definition 7) the reduction ordering $\lhd$ induced by σ has to be irreflexive. In classical logic this restriction encodes the Eigenvariable condition in the classical sequent calculus [8]. It is usually integrated into the σ-complementary test by using the well-known Skolemization technique together with the occurs-check of term unification. We extend this concept to the intuitionistic substitution σ_J. To this end we introduce a new substitution $\sigma_<$, which is induced by the tree ordering $<$. $\sigma_<$ assigns to each constant (elements of Δ and Ψ) a Skolemterm containing all variables (elements of Γ and Φ) occurring below this constant in the formula tree. It is sufficient to restrict $\sigma_<$ to these elements, since $\sqsubset_J \subseteq \Psi \times \Phi$ and $\sqsubset_Q \subseteq \Delta \times \Gamma$.

Definition 10 ($\sigma_<$-Skolemization). *A tree ordering substitution $\sigma_< : (\Delta \cup \Psi)$ $\rightarrow \mathcal{T}$ assigns a Skolemterm to every element of $\Delta \cup \Psi$. It is induced by the tree ordering $<$ in the following way: $\sigma_< := \{ c \backslash c(x_1, ...x_n) \mid c \in \Delta \cup \Psi \text{ and for all } x \in \Gamma \cup \Phi: (x \in \{x_1, \ldots x_n\} \text{ iff } x < c)\}.$*

Example 9. The tree ordering of formula F_1 induces the substitution $\sigma_< = \{a_2 \backslash a_2(), a_5 \backslash a_5(), a_6 \backslash a_6()\}$, since no variables occur before any constant.

Note that we follow a purely proof-theoretical view on Skolemization, i.e. as a way to integrate the irreflexivity test of the reduction ordering into the condition of σ-complementary. For classical logic this close relationship was pointed out by Bibel [4]. Like for classical logic the use of Skolemization simplifies proof calculi and implementations. There is no need for an explicit irreflexivity check and subformulae/-matrices can be copied by just renaming all their variables.

Lemma 1 (Admissibility Using $\sigma_<$-Skolemization). *Let F be a formula and $<$ be its tree ordering. A combined substitution $\sigma = (\sigma_Q, \sigma_J)$ is admissible iff (1) $\sigma' = \sigma_< \cup \sigma_Q \cup \sigma_J$ is idempotent, i.e. $\sigma'(\sigma') = \sigma'$, and (2) the following holds for all $u \in \Gamma$ and all $v \in \Delta$ occurring in $\sigma_Q(u)$: $\sigma_J(pre(u)) = \sigma_J(pre(v)) \circ q$ for some $q \in (\Phi \cup \Psi)^*$.*

Proof. It is sufficient to show that $\lhd$ is reflexive iff $\sigma_< \cup \sigma_Q \cup \sigma_J$ is not idempotent. "$\Rightarrow$": Let $\lhd$ be reflexive. Then there are positions with $a_1 \lhd \ldots \lhd a_n \lhd a_1$. For each $a_i \lhd a_j$ there is $a_i < a_j$, $a_i \sqsubset_Q a_j$ or $a_i \sqsubset_J a_j$. According to Definition 10, 5 and 7 there is some substitution with $\{a_j \backslash t\}$ in $\sigma_<$, σ_Q or σ_J, respectively, in which a_i occurs in t. Then $\sigma_< \cup \sigma_Q \cup \sigma_J$ is not idempotent. "$\Leftarrow$": Let $\sigma_< \cup \sigma_Q \cup \sigma_J$ be not idempotent. Then there is $\sigma' = \{a_1 \backslash ..a_n.., a_n \backslash ..a_{n-1}.., \ldots, a_2 \backslash ..a_1..\} \subseteq \sigma_< \cup \sigma_Q \cup \sigma_J$. Each $\{a_j \backslash ..a_i..\} \in \sigma'$ is part of $\sigma_<$, σ_Q or σ_J. According to Definition 10, 5 and 7 it is $a_i < a_j$, $a_i \sqsubset_Q a_j$ or $a_i \sqsubset_J a_j$. Therefore $\lhd := (< \sqcup \sqsubset_Q \sqcup \sqsubset_J)^+$ is reflexive. □

3.2 A Clausal Matrix Characterisation

We will now define prefixed matrices in clausal form and adapt the notation of multiplicities to clausal matrices before presenting a clausal matrix characterisation for intuitionistic logic. The restriction of multiplicities to clauses makes it possible to use existing clausal calculi for which multiplicities can be increased during the proof search in an easy way. The transformation of a prefixed matrix to clausal form is done like for classical logic. Note that we apply the substitution $\sigma_<$ to the clausal matrix, i.e. to terms and prefixes of atomic formulae.

Definition 11 (Prefixed Clausal Matrix). *Let $<$ be a tree ordering and M be a prefixed matrix. The (prefixed) clausal matrix of M, denoted clausal(M), is a set of clauses in which each clause is a set of prefixed atomic formulae. It is inductively defined as follows: If M has the form $\{\{P^{pol}s{:}p\}\}$ then clausal$(M) := M$; otherwise clausal$(M) := \bigcup_{C \in M} \{\{\bigcup_{i=1}^n c_i\} \mid c_i \in clausal(M_i)$ and $C = \{M_1, \ldots, M_n\}\}$. The (prefixed) clausal matrix of a formula F (with tree ordering $<$), denoted $M_c = matrix_c(F)$, is $\sigma_<(clausal(matrix(F)))$.*

Definition 12 (Clausal Multiplicity). *The* clausal multiplicity $\mu_c : M_c \to I\!N$
*assigns each clause in M_c a natural number, specifying the number of clause
instances to be considered. $matrix_c^\mu(F)$ denotes the clausal matrix of F in which
clause instances/copies specified by μ_c have been considered. Term and prefix
variables in clause copies are renamed (Skolemfunctions are not renamed).*

Example 10. For F_1 let $\mu_c(\{P^1x_1{:}a_2()Z_3()Z_4()\}) = 2$. Then $matrix_c^\mu(F_1) =$
$\{\{P^1x_1{:}a_2()Z_3Z_4\}, \{P^1x_7{:}a_2()Z_8Z_9\}, \{P^0b{:}a_2()a_5(), P^0c{:}a_2()a_6()\}\}$.

Note that we use the same Skolemfunction (symbol) for instances of the
same subformula/clause. This is an optimisation, which has a similar effect like
the liberalised δ^+-rule for (classical) semantic tableaux [9]. Since we apply the
substitution $\sigma_<$ to the clausal matrix, term and prefix constants can now both
contain term and prefix variables. Therefore the first-order and intuitionistic
substitutions σ_Q and σ_J are extended so that they assign terms over $\Gamma \cup \Delta \cup \Phi \cup \Psi$
to term and prefix variables (elements of Γ and Φ).

Theorem 3 (Clausal Characterisation for Intuitionistic Logic). *A formula F is intuitionistically valid, iff there is a multiplicity μ_c, an admissible
combined substitution $\sigma = (\sigma_Q, \sigma_J)$, and a set of σ-complementary connections
such that every path through $matrix_c^\mu(F)$ contains a connection from this set.*

Proof. Follows directly from the following Lemma 2 and Theorem 2.   □

Lemma 2 (Equivalence of Matrix and Clausal Matrix Proofs). *There
is a clausal matrix proof for F iff there is a (non-clausal) matrix proof for F.*

*Proof. Main idea: paths through clausal matrix and (non-clausal) matrix of F
are identical and substitutions and multiplicities μ_c/μ can be transfered into each
other. Note that in the following proof "variable" refers to term and prefix variables. We assume that $\sigma_<$ is applied to the non-clausal matrix as well.*
"⇒": *Let (μ_c, σ, S) be a clausal matrix proof of F in which S is the connection
set. We will construct a matrix proof (μ, σ', S') for F. Let $M_c=matrix_c^\mu(F)$ be
the clausal matrix of F (with clause instances according to μ_c). We construct a
matrix $M=matrix(F^\mu)$ in the following way: We start with the original matrix
M. For each clause $C_i \in M_c$ we identify the vertical path through M which represents C_i (modulo Skolemfunction names); see matrices on left side of Figure
3. If this vertical path shares variables with already identified clauses in M we
copy the smallest subclause of M, so that the new path contains copies $x'_1,..,x'_n$
of all (term and prefix) variables $x_1,..,x_n$ in C_i with $\sigma(x_i){\neq}\sigma(x'_i)$. Note that
Skolemfunctions in copied subclauses of M are renamed and therefore unique.
The constructed matrix M determines μ. Let $\sigma':=\sigma$ and $S':=S$. We identify every connection $\{P^0s{:}p, P^1t{:}q\} \in S$ from M_c as $\{P^0s'{:}p', P^1t'{:}q'\}$ in M. If s', t', p'
or q' contain a Skolemfunction f_i (which is unique in M), we rename Skolemfunctions in σ' so that $\sigma'(s',p')=\sigma(t',q')$. There can be no renaming conflict for
the set S': Let $f(x_1), f(x_2),..$ be the same Skolemfunctions in M_c, which are
represented by different Skolemfunctions $f_1(x_1), f_2(x_2),..$ in M. If $\sigma(x_i){\neq}\sigma(x_j)$
then different Skolemfunctions do not matter, since $f_i(x_i)$ and $f_j(x_j)$ are never*

assigned to the same variable. The case $\sigma(x_i)=\sigma(x_j)$ does not occur according to the construction of M. If copies of the same variable are substituted by the same term, the branches in the formula tree can be folded up (see right side of Figure 3); otherwise the branches (and Skolemterms) differ. Every path through M contains a path from M_c as a subset. Therefore (μ, σ', S') is a proof for F.

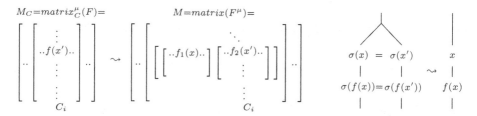

Fig. 3. Clause C_i in (non-)clausal matrix and folding up branches in formula tree

"⇐": Let (μ, σ', S') be a (non-clausal) matrix proof for F. We will construct a clausal matrix proof (μ_c, σ, S). Let $M=matrix(F^\mu)$ be the matrix of F^μ. We construct a clausal matrix $M_c=matrix_c^\mu(F)$ in the following way: We start with the clausal form of the original formula F, i.e. $M_c=matrix_c(F)$. We extend M_c by using an appropriate μ_c, so that it is the clausal form of M (modulo Skolemfunction and variable names). Let $\sigma:=\sigma'$ and $S:=S'$. We extend σ by unifying all variables $x_1, x_2, ..$ in M_c which have the same image in M. Since copied Skolemfunction names in M differ, but are identical in M_c we can simply replace them with a unique name in σ. Then all connections in S are σ-complementary. By induction we can also show that every path through M_c contains a connection from S. Therefore (μ_c, σ, S) is a clausal proof for F. □

3.3 A Prefixed Connection Calculus

The clausal *characterisation* of intuitionistic validity in Theorem 3 serves as a basis for a proof *calculus*. Calculating the first-order substitution σ_Q is done by the well-known algorithms for term unification (see, e.g., [16]). Checking that all paths contain a connection can be done by, e.g., sequent calculi [8] tableau calculi [7] or connection-driven calculi [3,4,13,14]. The clausal connection calculus is successfully used for theorem proving in classical logic (see, e.g., [12]). A basic version of the connection calculus for first-order classical logic is presented in [19]. The calculus uses a connection-driven search strategy, i.e. in each step a connection is identified along an active (sub-)path and only paths not containing the active path and this connection will be investigated afterwards. The clausal multiplicity μ_c is increased dynamically during the path checking process. We adapt this calculus to deal with intuitionistic logic (based on the clausal characterisation) by adding prefixes to the atomic formulae of each clause as specified in Definition 8 and 11.

Definition 13 (Prefixed Connection Calculus). *The axiom and the rules of the prefixed connection calculus are given in Figure 3. M is a prefixed clausal matrix, C, C_p, C_1, C_2' are clauses of M and C_p is a positive clause (i.e. contains no atomic formulae with polarity 1), $C_1 \cup \{\overline{L}\}$ contains no (term or prefix) variables, $C_2' \cup \{\overline{L'}\}$ contains at least one variable and $C_2/\overline{L}$ are copies of $C_2'/\overline{L'}$ in which all variables have been renamed. $\{L, \overline{L}\}$ and $\{L, \overline{L'}\}$ are σ-complementary connections, and Path is the active path, which is a subset of some path trough M. A formula F is valid iff there is an admissible substitution $\sigma = (\sigma_Q, \sigma_J)$ and a derivation for $matrix_c(F)$ in which all leaves are axioms.*

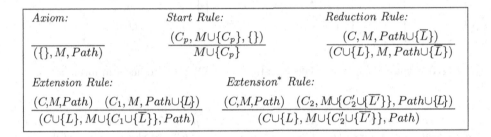

Fig. 4. The connection calculus for first-order logic

Example 11. Let $matrix_c(F_1)=\{\{P^1x_1{:}a_2()Z_3Z_4\},\{P^0b{:}a_2()a_5(),P^0c{:}a_2()a_6()\}\}$ $=M$ be the clausal matrix of F_1. The following is a derivation for M.

This derivation is a proof for F_1 with the admissible substitution $\sigma=(\sigma_Q,\sigma_J)$ and $\sigma_Q=\{x_1'\backslash b, x_2''\backslash c\}$, $\sigma_J=\{Z_3'\backslash\varepsilon, Z_4'\backslash a_2()a_5(), Z_3''\backslash\varepsilon, Z_4''\backslash a_2()a_6()\}$, i.e. F_1 is valid.

The intuitionistic substitution σ_J can be calculated using the prefix unification algorithm in [20]. It calculates a minimal set of most general unifiers for a given set of prefixes $\{s_1=s_1, .., t_n=t_n\}$ by using a small set of rewriting rules (similar to the algorithm in [16] for term unification). The following definition briefly describes this algorithm (see [20] for a detailed introduction).

Definition 14 (Prefix Unification). *The prefix unification of two prefixes s and t is done by applying the rewriting rules defined in Table 3. The rewriting rules replace a tuple (E, σ_J) by a modified tuple (E', σ_J') in which E and E' represent a prefix equation and σ_J, σ_J' are intuitionistic substitutions. Rules are applied non-deterministically. We start with the tuple $(\{s = \varepsilon|t\}, \{\})$, for technical reasons we divide the right part of t, and stop when the tuple $(\{\}, \sigma_J)$ is derived. In this case σ_J represents a most general unifier. There can be more than one* *most general unifier (see [20]). The algorithm can also be used to calculate a unifier for a set of prefix equations $\{s_1{=}t_1, .., s_n{=}t_n\}$ in a stepwise manner.*

We use these symbols: $s, t, z \in (\Phi\cup\Psi)^*$, $s^+, t^+, z^+\in(\Phi\cup\Psi)^+$, $X \in (\Phi\cup\Psi)$, V, V_1 $V' \in \Phi$ and $C, C_1, C_2 \in \Psi$. V' is a new variable which does not refer to a position in the formula tree and does not occur in the substitution σ_J computed so far.

Table 3. Rewriting rules for prefix unification

R1.	$\{\varepsilon = \varepsilon|\varepsilon\}, \sigma_J$	$\rightarrow$	$\{\}, \sigma_J$	
R2.	$\{\varepsilon = \varepsilon|t^+\}, \sigma_J$	$\rightarrow$	$\{t^+ = \varepsilon|\varepsilon\}, \sigma_J$	
R3.	$\{Xs = \varepsilon|Xt\}, \sigma_J$	$\rightarrow$	$\{s = \varepsilon|t\}, \sigma_J$	
R4.	$\{Cs = \varepsilon|Vt\}, \sigma_J$	$\rightarrow$	$\{Vt = \varepsilon|Cs\}, \sigma_J$	
R5.	$\{Vs = z|\varepsilon\}, \sigma_J$	$\rightarrow$	$\{s = \varepsilon|\varepsilon\}, \{V\backslash z\}\cup\sigma_J$	
R6.	$\{Vs = \varepsilon|C_1t\}, \sigma_J$	$\rightarrow$	$\{s = \varepsilon|C_1t\}, \{V\backslash\varepsilon\}\cup\sigma_J$	
R7.	$\{Vs = z|C_1C_2t\}, \sigma_J$	$\rightarrow$	$\{s = \varepsilon|C_2t\}, \{V\backslash zC_1\}\cup\sigma_J$	
R8.	$\{Vs^+ = \varepsilon|V_1t\}, \sigma_J$	$\rightarrow$	$\{V_1t = V|s^+\}, \sigma_J$	$(V{\neq}V_1)$
R9.	$\{Vs^+ = z^+|V_1t\}, \sigma_J$	$\rightarrow$	$\{V_1t = V'|s^+\}, \{V\backslash z^+V'\}\cup\sigma_J$	$(V' \in \Phi)$
R10.	$\{Vs = z|Xt\}, \sigma_J$	$\rightarrow$	$\{Vs = zX|t\}, \sigma_J$	$(s{=}\varepsilon$ or $t{\neq}\varepsilon$ or $X \in \Psi)$

Example 12. In the first step/connection of the proof for F_1 we need to unify the prefixes $s = a_2Z_3'Z_4'$ and $t = a_2a_5$. This is done as follows: $(\{a_2Z_3'Z_4'{=}\varepsilon|a_2a_5\}, \{\})$ $\xrightarrow{\text{R3}}$ $(\{Z_3'Z_4'{=}\varepsilon|a_5\}, \{\})$ $\xrightarrow{\text{R6}}$ $(\{Z_4'{=}\varepsilon|a_5\}, \{Z_3'\backslash\varepsilon\})$ $\xrightarrow{\text{R10}}$ $(\{Z_4'{=}a_5|\varepsilon\}, \{Z_3'\backslash\varepsilon\})$ $\xrightarrow{\text{R5}}$ $(\{\varepsilon{=}\varepsilon|\varepsilon\}, \{Z_3'\backslash\varepsilon, Z_4'\backslash a_5\})$ $\xrightarrow{\text{R1}}$ $(\{\}, \{Z_3'\backslash\varepsilon, Z_4'\backslash a_5\})$. Note that the second (most general) unifier can be calculated by using rule R10 instead of R6.

4 An Implementation: ileanCoP

The calculus presented in Section 3 was implemented in Prolog. We will present details of the implementation and performance results. The source code (and more information) can be obtained at **http://www.leancop.de/ileancop/**.

4.1 The Code

leanCoP is an implementation of the clausal connection calculus presented in Section 3 for *classical* first-order logic [19]. The reduction rule is applied before extension rules are applied and open branches are selected in a depth-first way. We adapt leanCoP to intuitionistic logic by adding

a. prefixes to atomic formulae and a prefix unification procedure,
b. to each clause the set of term variables contained in it and an admissibility check in which these sets are used to check condition (2) of Lemma 1.

The main part of the ileanCoP code is depicted in Figure 5. The underlined text was added to leanCoP; no other changes were done. A prefix **Pre** is added to

```
(1)     prove(Mat,PathLim) :-
(2)         append(MatA,[FV:Cla|MatB],Mat), \+ member(-(_):_,Cla),
(3)         append(MatA,MatB,Mat1),
(4)         prove([!:[]],[FV:[-(!):-(-[])|Cla]|Mat1],[],PathLim,[PreSet,FreeV]),
(5)         check_addco(FreeV), prefix_unify(PreSet).
(6)     prove(Mat,PathLim) :-
(7)         \+ ground(Mat), PathLim1 is PathLim+1, prove(Mat,PathLim1).

(8)     prove([],_,_,_,[[],[]]).
(9)     prove([Lit:Pre|Cla],Mat,Path,PathLim,[PreSet,FreeV]) :-
(10)        (-NegLit=Lit;-Lit=NegLit) ->
(11)           ( member(NegL:PreN,Path), unify_with_occurs_check(NegL,NegLit),
(12)             \+ \+ prefix_unify([Pre=PreN]), PreSet1=[], FreeV3=[]
(13)           ;
(14)             append(MatA,[Cla1|MatB],Mat), copy_term(Cla1,FV:Cla2),
(15)             append(ClaA,[NegL:PreN|ClaB],Cla2),
(16)             unify_with_occurs_check(NegL,NegLit),
(17)             \+ \+ prefix_unify([Pre=PreN]),
(18)             append(ClaA,ClaB,Cla3),
(19)             ( Cla1==FV:Cla2 ->
(20)                 append(MatB,MatA,Mat1)
(21)             ;
(22)                 length(Path,K), K<PathLim,
(23)                 append(MatB,[Cla1|MatA],Mat1)
(24)             ),
(25)             prove(Cla3,Mat1,[Lit:Pre|Path],PathLim,[PreSet1,FreeV1]),
(26)             append(FreeV1,FV,FreeV3)
(27)           ),
(28)           prove(Cla,Mat,Path,PathLim,[PreSet2,FreeV2]),
(29)           append([Pre=PreN|PreSet1],PreSet2,PreSet),
(30)           append(FreeV2,FreeV3,FreeV).
```

Fig. 5. Main part of the ileanCoP source code

atomic formulae q:Pre in which Pre is a list of prefix variables and constants. The set of term variables Var is added to each clause Var:Cla in which Var is a list containing pairs [V,Pre] in which V is a term variable and Pre its prefix.

Condition (2) of the admissibility check, check_addco, and the unification of prefixes, prefix_unify, are done after a classical proof, i.e. a set of connections, has been found. Therefore the set of prefix equations PreSet and the clause-variable sets FreeV are collected in an additional argument of prove. check_addco and prefix_unify require 5 and 26 more lines of code, respectively (see [20,22]). Condition (1) of the admissibility check of Lemma 1 is realized by the occurs-check of Prolog during term unification. Two prefix constants are considered equal if they can be unified by term unification. Note that we perform a *weak prefix unification* (line 12 and 17) for each connection already during the path checking process (double negation prevents any variable bindings).

To prove the formula F_1 from Section 2 and 3 we call prove(M,1). in which M= [[[X,[1^[],Z]]]:[-(p(X)): -([1^[],Z,Y])], []:[p(b):\[1^[],2^[]],p(c): [1^[],3^[]]]] is the prefixed matrix of F_1 (with clause-variable sets added).

4.2 Refuting Some First-Order Formulae

In order to obtain completeness ileanCoP performs iterative deepening on the proof depth, i.e. the size of the active path. The limit for this size, PathLim, is increased after the proof search for a given path limit has failed (line 7 in Figure 5). If the matrix contains no variables, the given matrix is refuted.

We will integrate a more restrictive method in which the path limit will only be increased if the current path limit was actually reached. In this case the predicate `pathlim` is written into Prolog's database indicating the need to increase the path limit if the proof search fails. It can be realized by modifying the following lines of the code in Figure 5:

```
(7)          retract(pathlim), PathLim1 is PathLim+1, prove(Mat,PathLim1).

(22)         length(Path,K), ( K<PathLim ->
(23a)        append(MatB,[Cla1|MatA],Mat1) ;
(23b)        \+ pathlim -> assert(pathlim), fail )
```

The resulting implementation is able to refute a large set of first-order formulae (but it does not result in a decision procedure for propositional formulae).

4.3 Performance Results

We have tested ileanCoP on all 1337 non-clausal (so-called FOF) problems in the TPTP library [27] of version 2.7.0 that are classically valid or are not known to be valid/invalid. The tests were performed on a 3 GHz Xeon system running Linux and ECLiPSe Prolog version 5.7 ("nodbgcomp." was used to generate code without debug information). The time limit for all proof attempts was 300 s.

The results of ileanCoP are listed in the last column of Table 4. We have also included the results of the following five theorem provers for intuitionistic first-order logic: JProver [25] (implemented in ML using a non-clausal connection calculus and prefixes), the Prolog and C versions of ft [24] (using an intuitionistic tableau calculus with many additional optimisation techniques and a contraction-free calculus [6] for propositional formulae), ileanSeP (using an intuitionistic tableau calculus; see `http://www.leancop.de/ileansep`) and ileanTAP [22] (using a tableau calculus and prefixes). The timings of the classical provers Otter [17], lean*TAP* [1] and leanCoP [19] are provided as well.

Of the 234 problems solved by ileanCoP 112 problems could not be solved by any other intuitionistic prover. The rating (see rows eight to twelve) expresses the relative difficulty of the problems from 0.0 (easy) to 1.0 (very difficult). The intuitionistic rating is taken from the ILTP library [23].

4.4 Analysing Test Runs

The proof search process of ileanCoP can be divided into the following four sections: transformation into clausal form, search for a "classical" proof (using weak prefix unification), admissibility check and prefix unification. We mark the transition to these sections with the letters a, b and c as follows:

	a		b		c	
clausal transformation		"classical"-proof		admissibility check		prefix unification

Table 4. Performance results of ileanCoP and other provers on the TPTP library

	Otter	lean*TAP*	leanCoP	ileanSeP	JProver	ft$_{Prolog}$	ft$_C$	ileanTAP	ileanCoP
solved	509	220	335	90	96	112	125	116	**234**
proved	509	220	335	87	94	99	112	113	**188**
refuted	0	0	0	3	2	13	13	3	**46**
0 to <1s	431	202	280	76	80	106	122	110	**192**
to <10s	54	9	23	8	9	5	1	2	**16**
to <100s	22	8	29	4	7	1	2	2	**17**
to <300s	2	1	3	2	0	0	0	2	**9**
intui. 0.0	-	-	-	60	76	74	76	76	**76**
to ≤0.7	-	-	-	21	20	33	40	34	**39**
to ≤1.0	-	-	-	9	0	5	9	6	**119**
class. 0.0	256	163	227	75	85	92	100	88	**175**
to ≤1.0	253	57	108	15	11	20	25	28	**59**
time out	360	1012	944	1173	1237	1171	667	1165	**1041**
error	468	105	58	74	4	54	545	56	**62**

Table 5. Analysis of the TPTP test runs

	$a=b=c=1$	$a=b=1; c>1$	$a=1; b,c>1$	$a=1; b\geq1; c=0$	$a=1; b=c=0$	$a=b=c=0$
proved	159	19	10	-	-	-
var(ø,max)	(13,216)	(21,62)	(11,23)	-	-	-
pre(ø,max)	(29,1191)	(22,119)	(13,26)	-	-	-
refuted	1	-	3	2	40	-
var(ø,max)	(2,2)	-	(3,5)	(5,5)	-	-
pre(ø,max)	(3,3)	-	(4,4)	-	-	-
time out	3	31	17	2	987	62
var(ø,max)	(26,34)	(25,106)	(41,146)	(47,70)	-	-
pre(ø,max)	(15,19)	(21,99)	(40,112)	-	-	-

Table 5 contains information about how often each section is entered when running ileanCoP on the 1337 TPTP problems. The numbers assigned to the letters express the number of times this section is entered. For example the column $a=b=c=1$ means that each section is entered just once. The rows var(ø,max) contain the average and maximum number of variables collected when entering section b, whereas the rows pre(ø,max) contain the average and maximum number of prefix equations collected when entering section c.

Most of the problems (178 formulae) have been proved without backtracking in sections a/b (1st column) or section a (2nd column). This means that the corresponding classical sequent proofs can be converted into intuitionistic sequent proofs by simply reordering the rules. Most of the refuted problems (40 formulae) failed because of weak prefix unification in section a (5th column).

For 987 problems no "classical" proof could be found (5th column). For 62 problems (6th column) the size of the clausal form is too big and the transformation results in a stack overflow ("error" in Table 4). For 31 problems sections b and c did not finish after a "classical" proof has been found (2nd column).

5   Conclusion

We have presented a clausal matrix characterisation for intuitionistic logic based on the standard characterisation [30] (see also [29]). Encoding the tree-ordering by an additional substitution replaces the reflexivity test of the reduction ordering by checking if the (extended) combined substitution is idempotent. Due to this Skolemization technique and the transformation into clausal form, multiplicities can be restricted to clauses and clause instances can be generated by just renaming the term and prefix variables of the original clauses. This allows the use of existing clausal calculi and implementations for classical logic (in contrast to specialised calculi [28]), without the redundancy caused by (relational) translations [18] into classical first-order logic. Instead an additional prefix unification is added, which captures the specific restrictions of intuitionistic logic, i.e.

$$
\begin{array}{lll}
\text{first-order logic} & = & \text{propositional logic} + \text{term unification ,} \\
\text{intuitionistic logic} & = & \text{classical logic} \quad\quad + \text{prefix unification .}
\end{array}
$$

We have adapted a clausal connection calculus and transformed the classical prover leanCoP into the intuitionistic prover ileanCoP with only minor additions. Experimental results on the TPTP library showed that the performance of ileanCoP is significantly better than any other (published) automated theorem prover for intuitionistic first-order logic available today. The correctness proof provides a way to convert the clausal matrix proofs back into non-clausal matrix proofs, which can then be converted into sequent-style proofs [26].

While a clausal form technically simplifies the proof search procedure, it has a negative effect on the size of the resulting clausal form and the search space. A non-clausal connection-based proof search, which increases multiplicities in a demand-driven way, would not suffer from this weakness. Whereas the used prefix unification algorithm is very straightforward to implement, it can be improved by solving the set of prefix equations in a more simultaneous way. The clausal characterisation, the clausal connection calculus and the implementation can be adapted to all modal logics considered within the matrix characterisation framework [30,21,11] by changing the unification rules [20]. By adding tests for the linear and relevant constraints within the complementary condition it can even be adapted to fragments of linear logic [10,11,15].

Acknowledgements

The author would like to thank Thomas Raths for providing the performance results for the listed classical and intuitionistic theorem proving systems.

References

1. B. BECKERT, J. POSEGGA. leanT^AP: lean, tableau-based theorem proving. 12th CADE, LNAI 814, pp. 793–797, Springer, 1994.
2. Y. BERTOT, P. CASTÉRAN. *Interactive Theorem Proving and Program Development*. Texts in Theoretical Computer Science, Springer, 2004.
3. W. BIBEL. Matings in matrices. *Communications of the ACM*, 26:844–852, 1983.
4. W. BIBEL. *Automated Theorem Proving*. Vieweg, second edition, 1987.
5. R. L. CONSTABLE ET. AL. *Implementing Mathematics with the NuPRL proof development system*. Prentice Hall, 1986.
6. R. DYCKHOFF. Contraction-free sequent calculi for intuitionistic logic. *Journal of Symbolic Logic*, 57:795–807, 1992.
7. M. C. FITTING. *First-Order Logic and Automated Theorem Proving*. Springer, 1990.
8. G. GENTZEN. Untersuchungen über das logische Schließen. *Mathematische Zeitschrift*, 39:176–210, 405–431, 1935.
9. R. HÄHNLE, P. SCHMITT The Liberalized δ-rule in free variable semantic tableaux. *Journal of Automated Reasoning*, 13:211–221, 1994.
10. C. KREITZ, H. MANTEL, J. OTTEN, S. SCHMITT Connection-based proof construction in linear logic. 14th CADE, LNAI 1249, pp. 207–221, Springer, 1997.
11. C. KREITZ, J. OTTEN. Connection-based theorem proving in classical and nonclassical logics. *Journal of Universal Computer Science*, 5:88–112, Springer, 1999.
12. R. LETZ, J. SCHUMANN, S. BAYERL, W. BIBEL. SETHEO: A high-performance theorem prover. *Journal of Automated Reasoning*, 8:183–212, 1992.
13. R. LETZ, G. STENZ Model elimination and connection tableau procedures. *Handbook of Automated Reasoning*, pp. 2015–2114, Elsevier, 2001.
14. D. LOVELAND. Mechanical theorem proving by model elimination. *JACM*, 15:236–251, 1968.
15. H. MANTEL, J. OTTEN. linTAP: A tableau prover for linear logic. 8th TABLEAUX Conference, LNAI 1617, pp. 217–231, Springer, 1999.
16. A. MARTELLI, U. MONTANARI. An efficient unification algorithm. *ACM Transactions on Programming Languages and Systems (TOPLAS)*, 4:258–282, 1982.
17. W. MCCUNE. OTTER 3.0 reference manual and guide. Technical Report ANL-94/6, Argonne National Laboratory, 1994.
18. H. J. OHLBACH ET AL. Encoding two-valued nonclassical logics in classical logic. *Handbook of Automated Reasoning*, pp. 1403–1486, Elsevier, 2001.
19. J. OTTEN, W. BIBEL. leanCoP: lean connection-based theorem proving. *Journal of Symbolic Computation*, 36:139–161, 2003.
20. J. OTTEN, C. KREITZ. T-string-unification: unifying prefixes in non-classical proof methods. 5th TABLEAUX Workshop, LNAI 1071, pp. 244–260, Springer, 1996.
21. J. OTTEN, C. KREITZ. A uniform proof procedure for classical and non-classical logics. *KI-96: Advances in Artificial Intelligence*, LNAI 1137, pp. 307–319, Springer, 1996.
22. J. OTTEN. ileanTAP: An intuitionistic theorem prover. 6th TABLEAUX Conference, LNAI 1227, pp. 307–312, Springer, 1997.
23. T. RATHS, J. OTTEN, C. KREITZ. The ILTP library: benchmarking automated theorem provers for intuitionistic logic. *TABLEAUX 2005*, this volume, 2005. (http://www.iltp.de)
24. D. SAHLIN, T. FRANZEN, S. HARIDI. An intuitionistic predicate logic theorem prover. *Journal of Logic and Computation*, 2:619–656, 1992.

25. S. SCHMITT ET. AL. JProver: Integrating connection-based theorem proving into interactive proof assistants. *IJCAR-2001*, LNAI 2083, pp. 421–426, Springer, 2001.

26. S. SCHMITT, C. KREITZ Converting non-classical matrix proofs into sequent-style systems. *13th CADE*, LNAI 1104, pp. 2–16, Springer, 1996.

27. G. Sutcliffe, C. Suttner. The TPTP problem library - CNF release v1.2.1. *Journal of Automated Reasoning*, 21: 177–203, 1998. (http://www.cs.miami.edu/~tptp)

28. T. TAMMET. A resolution theorem prover for intuitionistic logic. *13th CADE*, LNAI 1104, pp. 2–16, Springer, 1996.

29. A. WAALER Connections in nonclassical logics. *Handbook of Automated Reasoning*, pp. 1487–1578, Elsevier, 2001.

30. L. WALLEN. *Automated deduction in nonclassical logic*. MIT Press, 1990.

Automatic 'Descente Infinie' Induction Reasoning

Sorin Stratulat

LITA, Université Paul Verlaine-Metz, 57000, France
stratulat@univ-metz.fr

Abstract. We present a framework and a methodology to build and analyse automatic provers using the 'Descente Infinie' induction principle. A stronger connection between different proof techniques like those based on implicit induction and saturation is established by uniformly and explicitly representing them as applications of this principle. The framework offers a clear separation between logic and computation, by the means of i) an abstract inference system that defines the maximal sets of induction hypotheses available at every step of a proof, and ii) reasoning modules that perform the computation and allow for modular design of the concrete inference rules. The methodology is applied to define a concrete implicit induction prover and analyse an existing saturation-based inference system.

1 Introduction

Induction is a widely employed term with different meanings, by which the mathematicians and, in our case, the theorem proving community understand a proof technique that allows to use 'not yet' certified information in terms of *induction hypotheses*. Nowadays, it is considered as a standard method to validate potentially infinite sets of formulas, for example when reasoning on unbounded data structures like integers and lists.

The proof-by-induction procedures are implementations of *induction principles*. Starting from college, students usually get acquainted with the mathematical (Peano) induction principle: if $P(x)$ is a formula over the naturals x, to prove that $P(0)$, $P(1)$, ... are true, it is enough to prove that i) $P(0)$ is true, and ii) for an arbitrary natural $i \geq 0$, $P(i)$ implies $P(i+1)$, i.e. the induction hypothesis $P(i)$ can be soundly used to prove $P(i+1)$.

The soundness argument of any induction principle is based on the existence of a *Noetherian* (well-founded) order: defined on a set of elements, it would make impossible the creation of any infinite strictly decreasing sequence of elements. For example, 'less-than' over the naturals is the Noetherian order used by the Peano principle. Another induction principle, the 'Descente Infinie', was firstly described explicitly by Fermat in 1659 and captures inherently a proof by contradiction. In its most general form, a statement is true if the assumption of its negation induces a contradiction presented under the form of an infinite strictly decreasing sequence of elements over which a Noetherian order is defined.

B. Beckert (Ed): TABLEAUX 2005, LNAI 3702, pp. 262–276, 2005.

Usually, the elements are either terms (like the naturals from the Peano principle) or formulas. In the paper, we will discuss the applications of a formula-based variant of the 'Descente Infinie' induction principle. It can certify that a set of formulas $\mathcal{P}$ are true if for any false formula there is another smaller false formula (wrt a Noetherian order over $\mathcal{P}$). If, by absurdity, the existence of a false formula ϕ in $\mathcal{P}$ is assumed, then it exists a smaller one for which it exists an even smaller formula, and so on. Iteratively, an infinite strictly descending sequence of false formulas starting with ϕ can be built.

The 'Descente Infinie' principle contributes to the achievement of one of the main goals of the theorem proving community: given a set of formulas Ax, called *axioms*, and a formula ϕ called *conjecture*, to prove that ϕ is a consequence of Ax. In the paper, it will be applied to reason on *inductive* consequence relations, denoted generically by $\models$, which satisfy the following *characterisation property*: ϕ is an inductive consequence of Ax, written as $Ax \models \phi$, iff $Ax \models \psi$, for any ψ ground (no variable) instance of ϕ. Examples of inductive consequences are the *initial* consequence (by Proposition 3.1 of [7]), the *parameterised* consequence (by Lemma 3.4 of [6]), the *observational* consequence (by Theorem 1 of [5]) and the *positive/negative* consequence (by Corollary 3.12 of [2]). Usually, the proof of $Ax \models \psi$ consists in testing a finite set of deductive (logical) consequence relations. So, the difficulty of the approach is represented by the test of a potentially infinite set of ground instances. To link with the 'Descente Infinie' setting, a formula ϕ is *true* if $Ax \models \phi$, otherwise it is *false* when it exists a ground instance ψ of it, called *counterexample*, such that $Ax \not\models \psi$. The set $\mathcal{P}$ corresponds to the ground instances of all the conjectures from a derivation, defined as follows.

Testing a set of conjectures consists in a successive application of 'Descente Infinie' *inference rules*. They represent transition relations $\vdash$ between states made of sets of conjectures that process one conjecture at once: $E \cup \{\phi\} \vdash E \cup \Phi$. Here, the processed conjecture ϕ is transformed into a (potentially empty) set of new conjectures Φ. If E^0 is the set of initial conjectures, a *derivation* is a sequence of such transitions $E^0 \vdash \ldots \vdash E^n \vdash \ldots$. It may not terminate (diverges), or terminate with i) success, ii) error, by providing a counterexample, or iii) failure, when the derivation stops without reporting a success or an error.

Any Noetherian order defined on a non-empty set implies the existence of minimal elements, for which there is no smaller element. The *soundness* property of the 'Descente Infinie' inference systems guarantees the persistence of minimal counterexamples in any derivation containing false conjectures. The *implicit induction* inference systems certify that the conjectures are true whenever the derivations end with an empty set. The *saturation-based* inference systems detect the minimal counterexamples in the last (saturated) state of finite derivations. To satisfy the soundness property, any 'Descente Infinie' inference rule should satisfy the following requirement: whenever the processed conjecture is false, then for any of its counterexamples either i) somewhere in the derivation, there exists a conjecture with a smaller counterexample, or ii) in the future derivation states, there exists a conjecture with an equivalent (w.r.t. order) counterexample. The conjectures not containing minimal counterexamples are *redundant* and can be eliminated or used as *induction hypotheses* in further computations.

The rest of the paper is organised in 3 sections. Section 2 presents a framework and a methodology to design and analyse 'Descente Infinie' theorem provers that automatically verify inductive consequence relations. The framework was firstly adapted for automatic implicit induction provers in [18] is extended to represent also saturation-based procedures. It offers a clear separation between logic and computation by the means of i) a unique sound *abstract inference system* that defines maximal sets of induction hypotheses at every proof step, and ii) *reasoning modules* that perform the computation and allow for modular design of the concrete inference rules. To witness that the above proof techniques share the same logic, the methodology is applied in Section 3 to build a concrete implicit induction prover and analyse an existing saturation-based inference system, basically by instantiating the abstract system with specific reasoning modules. Their soundness is a direct consequence of the soundness of the abstract system and the instantiation results. The conclusions, some potential extensions of the framework that deal with more powerful but less automatic inference rules and the related works are presented in the last section.

2   The Framework

Basic Notions For the sake of simplicity, we assume that the conjectures and the axioms are quantifier-free first-order formulas. A (ground) *substitution* maps variables to (ground) terms. The *instance* of a formula ϕ obtained from the application of a substitution σ is denoted by $\phi\sigma$. A binary relation R is *stable* if whenever $\phi_1 R \phi_2$ then $(\phi_1\sigma) R (\phi_2\sigma)$ for any substitution σ.

Let $\mathcal{A}$ be a non-empty set of elements and $\leq$ a reflexive and transitive binary relation over the elements from $\mathcal{A}$, called *quasi-order*. The *equivalence* part $\sim$ of $\leq$ is defined by $x \sim y$ iff $x \leq y$ and $y \leq x$ and its *strict* part $<$, called *order*, by $x < y$ iff $x \leq y$ and $y \not\leq x$. An order is *total* if any two distinct elements are comparable. $\leq$ is *Noetherian* if there is no infinite strictly descending sequence $x_1 > x_2 > \ldots$ of elements of $\mathcal{A}$. Moreover, it is stable if its strict and equivalent part are stable. $\Psi_{\leq\phi}$ (resp. $\Psi_{<\phi}$ and $\Psi_{\sim\phi}$) denotes all the instances of formulas from Ψ that are 'smaller or equal' to (resp. 'smaller than' and 'equivalent' to) ϕ.

2.1   Contextual Cover Sets

In the following, it is assumed that $\leq$ is a stable Noetherian quasi-order over formulas. To guarantee their soundness, the automatic 'Descente Infinie' inference systems require that a counterexample equivalent to $\phi\tau$ be in the next state $E \cup \Phi$ of a proof step $E \cup \{\phi\} \vdash E \cup \Phi$ whenever ϕ contains a minimal counterexample $\phi\tau$. More precisely, the condition $\Gamma \models \phi\tau$ should be satisfied, where Γ can contain both $(E \cup \Phi)_{\sim\phi\tau}$ and true formulas: i) the axioms Ax, ii) $\mathsf{C}^1_{\leq\phi\tau}$, where C^1 has elements of E (but not of Φ) and other conjectures with no minimal counterexamples, iii) $\mathsf{C}^2_{<\phi\tau}$, where C^2 can be any set of conjectures from the derivation, and iv) $\Phi_{\leq\phi\tau}$. Usually, the minimal counterexample $\phi\tau$ is hard to be identified among the other ground instances of ϕ. To be sure that this condition is satisfied by $\phi\tau$, it is sufficient to generalise it for any ground instance of ϕ. To sum up, the condition becomes

$$Ax \cup C^1_{\leq\phi\gamma} \cup C^2_{<\phi\gamma} \cup \Phi_{\leq\phi\gamma} \models \phi\gamma, \text{ for any ground instance } \phi\gamma, \qquad (1)$$

i.e. Φ is a (general) *contextual cover set* (CCS) of ϕ in the *context* $C = (C^1, C^2)$.

Several kinds of CCSs are distinguished: i) *cover set* if $C^1 = C^2 = \emptyset$, ii) *strict* if $\Phi_{\leq\phi\gamma}$ is replaced by $\Phi_{<\phi\gamma}$, iii) *empty* if $\Phi = \emptyset$, and iv) *universal* if $Ax = \emptyset$. They are not mutually exclusive; for example, any empty CCS is also strict. The notion of cover set as a particular CCS corresponds to that from [8], which is a generalisation of [17]. Variants of it can be found in different completion-based induction methods [3,21,12,11]. In the definition (1), the formulas from $C^1_{\leq\phi\gamma}$, $C^2_{<\phi\gamma}$ and $\Phi_{\leq\phi\gamma}$ can be used to deduce $\phi\gamma$ even if they are not true or not yet proved to be true. They play the role of *induction hypotheses*.

In the rest of the paper, the 'contextually cover' relation is generalised to sets of formulas: $\Psi \sqsubseteq_C (\sqsubseteq_C)\Phi$ iff Φ is a (strict) CCS of any $\phi \in \Psi$ in the context C.

Properties As shown in [18], the 'contextually covers' relation is a quasi-order: i) (reflexivity) $\Phi \sqsubseteq_C \Phi$, for any set of formulas Φ, and ii) (transitivity) for any set of formulas Φ, Ψ and Γ, if $\Phi \sqsubseteq_C \Psi$ and $\Psi \sqsubseteq_C \Gamma$, then $\Phi \sqsubseteq_C \Gamma$. Due to the transitivity property of $\sqsubseteq_C$, new 'contextually cover' relations can be obtained by composition operations.

- *horizontal composition.* Given the chain of 'contextually cover' relations $\Phi_1 \sqsubseteq_C \ldots \sqsubseteq_C \Phi_i \sqsubseteq_C \Phi_{i+1} \sqsubseteq_C \ldots \Phi_n$, then i) $\Phi_i \sqsubseteq_C \Phi_j$, for all $i, j \in [1..n]$ with $i \leq j$, and ii) if $\Phi_i \sqsubset_C \Phi_{i+1}$ then $\Phi_k \sqsubset_C \Phi_j$, for all $k \leq i$ and $j > i$.
- *vertical composition.* Given the set of formulas $\Phi = \{\phi_1, \ldots, \phi_n\}$ such that $\forall i \in [1..n]$, $\{\phi_i\} \sqsubseteq_C$ (resp. $\sqsubset_C$)Ψ_i then $\Phi \sqsubseteq_C$ (resp. $\sqsubset_C$)$\bigcup_{j=1}^n \Psi_j$.

Our framework is able to provide maximal sets of induction hypotheses at any step of a derivation by the means of the inference system A.

2.2   The A Inference System

Generally, the CCS contexts may have conjectures from the whole derivation to be computed at any inference step. To *automate* their computation, a new set of particular formulas will join the set of conjectures such that the context has only formulas from the current state of the derivation. They are called *premises* and represent processed conjectures that do not contain minimal ground formulas. The inference rules have now the form:

$$\text{NAME} \quad (E \cup \{\phi\}, \mathbf{H}) \vdash (E \cup \Phi, \mathbf{H}') \text{ [if } Conditions]$$

where $\mathbf{H}$ and $\mathbf{H}'$ are premises. We propose an inference system, denoted by A, that replaces ϕ by Φ if Φ is a CCS of ϕ, whose context has only formulas from

1-ADDPREMISE	1-SIMPLIFY
$(E \cup \{\phi\},\ H) \ \vdash^{\mathsf{A}}\ (E \cup \Phi,\ H \cup \{\phi\})$	$(E \cup \{\phi\},\ H)\ \vdash^{\mathsf{A}}\ (E \cup \Phi,\ H)$
if $\{\phi\}\ \sqsubset_{(H,\ E)}\ \Phi$	if $\{\phi\}\ \sqsubseteq_{(E \cup H,\ \emptyset)}\ \Phi$

Fig. 1. The one-step inference rules

E and H. In its simplest form, it consists of two inference rules, presented in Fig. 1, that build the new set of conjectures in just one step. The 1-ADDPREMISE rule firstly computes Φ as a strict CCS of ϕ, then adds it to the current set of premises to participate to further computations. 1-SIMPLIFY does not make such addition, but it is less restrictive: Φ can be a general CCS and instances of E equivalent to ϕ are allowed. The induction hypotheses from the contexts do not affect the soundness of A.

Theorem 1. *(A-soundness) The minimal counterexamples are persistent in any A-derivation starting with an empty set of premises.*

Proof. Firstly, 1-ADDPREMISE cannot be applied on conjectures containing minimal counterexamples. By absurdity, ϕ is assumed to contain a minimal counterexample $\phi\tau$ when applying 1-ADDPREMISE from Fig. 1. Since $\{\phi\} \sqsubset_{(H,\ E)} \Phi$, it holds that $Ax \cup H_{\leq \phi\tau} \cup E_{<\phi\tau} \cup \Phi_{<\phi\tau} \models \phi\tau$. Therefore, it should exist a counterexample in the set $H_{\leq\phi\tau} \cup E_{<\phi\tau} \cup \Phi_{<\phi\tau}$. On the other hand, $(H \cup E \cup \Phi)_{<\phi\tau}$ has no counterexample since $\phi\tau$ is minimal and $<$ is stable. It means that H is not empty and should contain a premise γ having a counterexample equivalent to $\phi\tau$ (by the stability of $\sim$), so minimal. Then, 1-ADDPREMISE should have been applied on γ in a previous derivation step with fewer premises than H, since it does not contain γ. The same reasoning can be repeated for γ as for ϕ, and so on. At each iteration, there should always exist a previous step with a smaller non-empty set of premises having a minimal counterexample. On the other hand, this reasoning should finish because the cardinality of the set of premises decreases with each iteration and the initial set of premises is empty. Contradiction.

Assuming now that 1-SIMPLIFY is applied as in Fig. 1, then $Ax \cup (H \cup E \cup \Phi)_{\leq\phi\tau} \models \phi\tau$. As previously, $(H \cup E \cup \Phi)_{<\phi\tau}$ and $H_{\sim\phi\tau}$ have no counterexamples. Therefore, $E \cup \Phi$ should contain a counterexample equivalent to $\phi\tau$. ∎

More complex inference rules can be designed by using the composition properties of CCSs. The inference rules from Fig. 2 build the new set of conjectures in two steps. At the step (a) of 2-ADDPREMISE, an intermediate CCS of ϕ is created, denoted by Ψ. Then, for any formula from Ψ a strict CCS is built and added as new conjectures. By vertical composition at the step (b), Φ is a strict CCS of Ψ. Φ is also a strict CCS of ϕ, by horizontal composition. Similarly, 2-SIMPLIFY builds Φ as a CCS of ϕ. It can be shown that the two-step inference system is sound as in [18] and following the idea of the proof of Theorem 1. The one-step inference system is an instance of it by considering $\{\phi\}$ as the (trivial) CCS for $\{\phi\}$ at the step (a) of the corresponding two-step inference rules.

2-ADDPREMISE

$(E \cup \{\phi\}, H) \vdash^{\mathsf{A}} (E \cup \{\phi\} \cup \underbrace{\Phi_1 \cup \ldots \cup \Phi_p}_{\Phi}, H \cup \{\phi\})$ if

(a) $\{\phi\} \sqsubseteq_{(H, E \cup \Phi)} \bigcup_{j=1}^{p} \{\psi_j\}$ **and**
(b) $\{\psi_j\} \sqsubseteq_{(H, E \cup \{\phi\} \cup (\Phi \setminus \Phi_j))} \Phi_j$, for each $j \in [1..p]$

2-SIMPLIFY

$(E \cup \{\phi\}, H) \vdash^{\mathsf{A}} (E \cup \{\phi\} \cup \underbrace{\Phi_1 \cup \ldots \cup \Phi_p}_{\Phi}, H)$ if

(a) $\{\phi\} \sqsubseteq_{(E \cup H \cup \Phi, \emptyset)} \bigcup_{j=1}^{p} \{\psi_j\}$ **and**
(b) $\{\psi_j\} \sqsubseteq_{(E \cup H \cup (\Phi \setminus \Phi_j), \{\phi\})} \Phi_j$, for each $j \in [1..p]$

Fig. 2. The two-step inference rules

2-ADDPREMISE is asymmetric in the construction of Φ. To be complete, the two-step inference system should include a variant of 2-ADDPREMISE that creates the strict CCS at the step (a) instead of (b).

Proving by Implicit Induction The implicit induction proving technique guarantees that a set of conjectures E^0 is true if it exists a finite 'Descente Infinie' proof of the form $(E^0, \emptyset) \vdash^{\mathsf{A}} \ldots \vdash^{\mathsf{A}} (\emptyset, H^n)$. By Theorem 1, if the conjectures contain a counterexample, then the minimal counterexamples are persistent. But this is not the case because the last set of conjectures is empty.

Proving by Saturation The proofs by saturation have an important place in the automated reasoning domain, for instance the Handbook of Automated Reasoning contains at least three chapters devoted to this topic [4,9,16]. They provide the means to detect the minimal counterexamples in the last state of any finite 'Descente Infinie' derivation. It is required that i) the formulas from the last state be *saturated*, i.e. no new information can be produced by the exhaustive application of the inference rules, and ii) the inference system be *refutationally complete*, i.e. the minimal counterexamples from any saturated formulas are 'easily detectable'. Such a minimal counterexample is usually the formula that is false in any model, denoted here by $\square$, which is generated if and only if the saturated formulas are unsatisfiable.

To show the flexibility of our approach, we will prove the refutational completeness of a variant of A, denoted by A', for which the CCSs are universal. Indeed, the set of axioms is empty when applying the inference rules, but the argumentation for its refutational completeness is based on the existence of a *candidate model* [4] (or generated model [16]) for any saturated set of formulas, which plays the role of the axioms when defining counterexamples. We assume that it satisfies the following property:

Property 1. Let $(E^0, \emptyset) \vdash^{\mathsf{A'}} \cdots \vdash^{\mathsf{A'}} (E^n, H^n)$ be a saturated A'-derivation such that I_{E^n} be a model for E^n. Then, an A'-inference rule is applicable on any formula from E^n containing a minimal counterexample other than $\square$.

Theorem 2. *(A′-refutational completeness) Let* $(E^0, \emptyset) \vdash^{A'} \cdots \vdash^{A'} (E^n, H^n)$ *be a saturated* A′*-derivation with* $\square \notin E^n$ *and* $I_{E^n} \models E^n$. *If* A′ *satisfies Property 1, for any ground formula* ϕ *from* E^n, $I_{E^n} \models \phi$.

Proof. By contradiction, we assume that there is a counterexample in E^n. Theorem 1 still holds for A′ as it is an instance of A. Therefore, there is a minimal counterexample γ in the derivation, other than $\square$, that should be in E^n. On the other hand, Property 1 guarantees the applicability of an A′-rule on the formula containing γ, which contradicts the fact that the derivation is saturated. ∎

2.3   Reasoning Modules

The inference systems A and A′ are abstract because they ignore how the CCSs from its inference rules are built. The last components of our framework, the *reasoning modules*, are in charge to compute the *elementary* CCSs, i.e. that are not built by composition operations. They represent implementations of reasoning techniques adequate to the nature of the employed consequence relation. Most of them are deductive, like rewriting and subsumption, and can reason on any kind of consequence relation. Some others are more specific, as the replacement of a natural variable with 0, 1, … for the initial consequences, or work only for particular reasoning domains, like the decision procedures.

A reasoning module is defined by two functions that take as arguments a context and a formula for which a CCS is built: i) the *generation* function g, and ii) the *condition* function *cond*, such that whenever $g(\phi, \mathsf{C}) = \Phi$ then $\{\phi\} \sqsubseteq_{\mathsf{C}} \Phi$ under the condition $cond(\phi, \mathsf{C})$. All the reasoning modules presented in the paper have the trivial condition function that returns true, but in general this is not the case.

An inference system is *recursive* if the validation process of the conditions is performed by the prover itself. An integration schema of reasoning modules in recursive implicit induction inference systems has been proposed in [18], which presents the advantage that elements of the context given as argument to the condition function can contribute as initial premises to the proof of the conditions.

2.4   Methodology for Designing and Analysing Provers

The framework can be used to design new provers and to analyse, improve and extend existing inference systems. To build a new prover that reasons on a given inductive consequence relation, the following steps can be followed:

1. provide a stable Noetherian order over the formulas,
2. provide a set of reasoning techniques adequate for the consequence relation,
3. define reasoning modules based on the reasoning techniques,
4. instantiate abstract inference rules by showing how their CCSs are created by the reasoning modules.

The soundness of the new inference system is guaranteed if each inference rule is an instance of an A-rule by showing i) the *compatibility* and ii) the *context inclusion* of its CCSs w.r.t. the corresponding CCSs defined by the A-rule. The compatibility predicate is defined in Table 1. Any kind of generated CCS is compatible with general CCSs (third line) and only the strict and the empty CCSs are compatible with strict CCSs (last line).

If the generation function builds a CCS of several kinds, for example a strict cover set, then the result is the disjunction of the compatibility results for each kind. A context (C^{11}, C^{12}) is *included* in (C^{21}, C^{22}) if $(C^{11}_{\leq \phi} \cup C^{12}_{< \phi}) \subseteq (C^{21}_{\leq \phi} \cup C^{22}_{< \phi})$, for any ground formula ϕ.

Analysing existing systems is an easier task since the order and the reasoning techniques are already provided. The most difficult steps are the representation

Table 1. The compatibility predicate

Kind of CCS produced by the generation function				
general	strict	empty	cover set	universal
$\sqsubseteq$ true	true	true	true	true
$\sqsubseteq$ false	true	true	false	false

of the derivation states under the form of (E, H) and that of the rules as instances of the abstract rules. Then, the rules can be soundly improved, for example, by expanding the contexts to the maximal values allowed by the abstract rules. Existing systems can be expanded modularly with new rules created in a flexible way, as described in the next section.

3 Applications

The methodology presented in the previous section will be used to i) design a simple implicit induction prover that reasons on naturals, and ii) analyse an existing saturation-based inference system.

3.1 Designing an Implicit Induction Prover

An implicit induction inference system will be designed to generate proofs that validate properties over the naturals, with axioms of the form $s = t$, representing equalities between terms.

⸦ *Order* ⸧ Defined over equalities, the stable Noetherian order of our choice is syntactic. Starting from an ordering (precedence) over the function symbols $\leq_F$, it requires an intermediate order $<_t$ over the terms defined by recurrence on their structure, similar to recursive path orderings [10]:
$$t \equiv g(t_1, \ldots, t_n) \quad <_t \quad f(s_1, \ldots, s_m) \equiv s$$
if either i) $g \sim_F f$ and $\{t_1, \ldots, t_n\} \ll_t \{s_1, \ldots, s_m\}$, or ii) $g <_F f$ and $t_i <_t s$, for all $i \in [1..n]$, or iii) $t <_t s_i$, for an index $i \in [1..m]$, or iv) $t \equiv s_i$, for an index $i \in [1..m]$, where $\equiv$ is the syntactic equality. $<_t$ is mutually dependent on its multiset extension $\ll_t$. Given two multisets of terms A_1 and A_2, then $A_1 \ll_t A_2$

if $\forall a \in A_1 \backslash A_2$, $\exists b \in A_2 \backslash A_1$ such that $a <_t b$, where $\backslash$ is the difference operation over the multisets.

The order over equalities $<_e$ can be defined as a multiset extension of $<_t$ because any equality $s = t$ can be represented as the multiset $\{s, t\}$:

$$s = t \quad <_e \quad l = r \quad \text{if } max(s, t) \ll_t max(l, r),$$

where $max(s, t)$ is the singleton containing the maximal term between s and t (wrt $<_t$) when it exists, otherwise $\{s, t\}$. It can be shown that $<_e$ and $<_t$ are Noetherian and stable if $<_\mathsf{F}$ is Noetherian.

For example, given the precedence $0 <_\mathsf{F} S$ (the successor function) $<_\mathsf{F} +$, then i) $x <_t 0 + x$, ii) $S(x + y) <_t S(x) + y$ and iii) $S(x) + 0 = S(x) <_e S(S(x) + 0) = S(S(x))$.

$\boxed{Reasoning\ Techniques}$ An inductive consequence relation appropriate for equalities is the initial consequence, denoted by $\models_{ini}$, and a typical reasoning technique for it is the replacement of a variable with finite descriptions of its domain. For the domain of the naturals, the most common are $\{0, S(x)\}$ and $\{0, S(0), S(S(x))\}$.

Two more general, deductive, reasoning techniques will be used: the elimination of identities of the form $s = s$ because they cannot contain counterexamples, and rewriting. Rewriting is an implementation of the Leibniz's substitution of equals for equals. To become operational, any equality $a = b$ is oriented into a rewrite rule $a \rightarrow b$ if $a >_t b$. By rewriting, a new equality is obtained by replacing a by b in an equality containing a. More formally, by $e[t]_p$ is unambiguously indicated that the equality (or term) e contains the term t at the position p. Then, given a set of rewrite rules ρ, the rewrite relation $\leadsto_\rho$ is defined as $e'[a'\sigma]'_p \leadsto_\rho e'[b'\sigma]'_p$ if $a' \rightarrow b' \in \rho$ and σ is a substitution.

$\boxed{Reasoning\ Modules}$ The previously presented reasoning techniques will serve to build new reasoning modules:

- D. Its generation function is $g_D(t = t, (\emptyset, \emptyset)) = \emptyset$ and returns an empty CCS.
- R, with $g_R(e, (\mathsf{C}^1, \mathsf{C}^2)) = \{e'\}$, where $e \leadsto_{Ax \cup C^1_{\leq_e} \cup C^2_{\leq_e}} e'$. It builds a strict CCS.
- Ex, with $g_{Ex}(e[x]_p, (\emptyset, \emptyset)) = \{e\sigma', e\sigma''\}$ that generates a cover set. σ' (resp. σ'') is the substitution that replaces x by 0 (resp. $S(x')$ and x' is a fresh variable).

DELETE IDENTITY
$(E \cup \{t = t\}, H) \vdash^\mathsf{P} (E \cup \Phi, H)$ if $\Phi = g_D(t = t, (\emptyset, \emptyset))$
REWRITE
$(E \cup \{e\}, H) \vdash^\mathsf{P} (E \cup \Phi, H \cup \{e\})$ if $\Phi = g_R(e, (H, E))$
EXPAND
$(E \cup \{e\}, H) \vdash^\mathsf{P} (E \cup \Phi, H)$ if $\Phi = g_{Ex}(e, (\emptyset, \emptyset))$

Fig. 3. The inference system P

Inference Rules Our inference system P will contain the inference rules from
Fig. 3. For example, given the set of orientable axioms that define the addition
over the naturals, $0 + x \rightarrow x$ and $S(x) + y \rightarrow S(x + y)$, P can prove that
$Ax \models_{ini} x + 0 = x$:

$$(\{x + 0 = x\}, \emptyset) \vdash^{\mathsf{P}}_{Ex} (\{\underline{0 + 0} = 0, \underline{S(x') + 0} = S(x')\}, \emptyset) \vdash^{\mathsf{P}}_{R} (twice)$$
$$(\{0 = 0, S(x' + 0) = S(x')\}, \{0 + 0 = 0, S(x') + 0 = S(x')\}) \vdash^{\mathsf{P}}_{D}$$
$$(\{S(x' + 0) = S(x')\}, \{0 + 0 = 0, S(x') + 0 = S(x')\}) \vdash^{\mathsf{P}}_{Ex}$$
$$(\{S(\underline{0+0}) = S(0), S(S(x'') + 0) = S(S(x''))\}, \{0+0=0, S(x') + 0 = S(x')\}) \vdash^{\mathsf{P}}_{R}$$
$$(\{S(0) = S(0), S(\underline{S(x'') + 0}) = S(S(x''))\}, \{\ldots, S(x') + 0 = S(x')\}) \vdash^{\mathsf{P}}_{R}$$
$$(\{S(0) = S(0), S(S(x'')) = S(S(x''))\}, \{\ldots\}) \vdash^{\mathsf{P}}_{D} (twice)(\emptyset, \{\ldots\}),$$

where by $\vdash^{\mathsf{P}}_{D}$ (resp. $\vdash^{\mathsf{P}}_{R}$ and $\vdash^{\mathsf{P}}_{Ex}$) is denoted the application of DELETE IDENTITY
(resp. REWRITE and EXPAND). The rewrite operations transform the underlined
subterms with rewrite rules from the axioms, excepting the last operation that
uses the instance $S(x'') + 0 \rightarrow S(x'')$ from the premises. This proves $x + 0 = x$
because the last set of conjectures becomes empty.

Instantiation Result P is an instance of A, as shown in Table 2.

Each column presents respectively for each P-rule the instantiated A-rule,
together with the name of the reasoning module, the context and kind of the
generated CCS. It is
easy to observe from
Table 1 and Fig. 1
that the generated
CCSs are compatible
with and the contexts
are included in the
contexts of the corre-

Table 2. P-rules as instances of A-rules

P-rule	A-rule	RM, context, kind of CCS
DELETE	1-SIMPLIFY	D, $(\emptyset, \emptyset)$, empty
REWRITE	1-ADDPREMISE	I, (H, E), strict
EXPAND	1-SIMPLIFY	Ex, $(\emptyset, \emptyset)$, cover set

sponding A-rules. Therefore, P is sound and $Ax \models_{ini} x+0 = x$ can be concluded.

Simpler proofs of $x + 0 = x$ can be obtained from instances of two-step
A-rules, like EXPANDREWRITE in Fig. 4. As an instance of 2-ADDPREMISE, it
allows the addition of $x+0 = x$ to the set of premises directly from the first step
of the proof for a later use as induction hypothesis in the subsequent REWRITE
operations.

$$(E \cup \{e\}, H) \vdash^{\mathsf{P}} (E \cup \Phi, H \cup \{e\})$$
$$\text{if } \Psi = g_{Ex}(e, (\emptyset, \emptyset)) \text{ and } \Phi = \bigcup_{e' \in \Psi} g_R(e', (H, E))$$

Fig. 4. The EXPANDREWRITE rule

3.2   Analysing a Saturation-Based Prover

The inference system to analyse, denoted by G in Fig. 5 and by $\mathcal{G}$ in [16], reasons
on ground Horn equational clauses of the form $e_1 \wedge \ldots \wedge e_n \Rightarrow$ or $e_1 \wedge \ldots \wedge e_n \Rightarrow e$,
where $e, e_1, \ldots, e_n$ are equalities and $s \succ \tau(\Gamma)$ (resp. $s \succeq \tau(\Gamma)$) means that

---

SUPERPOSITION RIGHT $E \vdash^G E \cup \{\Gamma', \Gamma \Rightarrow s[r]_p = t\}$

if $\Gamma' \Rightarrow l = r \in E$ s.t. $l \succ r$ and $l \succ \tau(\Gamma')$

and $\Gamma \Rightarrow s[l]_p = t \in E$ s.t. $s \succ t$ and $s \succ \tau(\Gamma)$

SUPERPOSITION LEFT $E \vdash^G E \cup \{\Gamma', \Gamma, s[r]_p = t \Rightarrow \Delta\}$

if $\Gamma' \Rightarrow l = r \in E$ s.t. $l \succ r$ and $l \succ \tau(\Gamma')$

and $\Gamma, s[l]_p = t \Rightarrow \Delta \in E$ s.t. $s \succ t$ and $s \succeq \tau(\Gamma \wedge \Delta)$

EQUALITY RESOLUTION $E \vdash^G E \cup \{\Gamma \Rightarrow \Delta\}$

if $\Gamma, s = s \Rightarrow \Delta \in E$ s.t. $s \succ \tau(\Gamma \wedge \Delta)$

Fig. 5. The inference system G

$s \succ u$ (resp. $s \succeq u$), for any term u from the equalities occurring in Γ. Based on ordered paramodulation, the rule SUPERPOSITION LEFT (resp. SUPERPOSITION RIGHT) transforms (resp. strictly) maximal literals in the processed conjecture, while EQUALITY RESOLUTION eliminates from the negative part of a clause the identities involving its maximal term. It can be noticed that all these rules preserve the processed conjecture in the next derivation. The orders over terms, $\prec$, and clauses, $\prec_c$, are Noetherian and total. The stability of $\prec_c$ is guaranteed because the formulas are ground. It has been shown in [16] that G is refutationally complete and that Property 1 is satisfied. Moreover, it is (trivially) sound because all the processed conjectures, and therefore the minimal counterexamples, are preserved in the derivation.

G-Variant. G is not adapted for the automated reasoning on long derivations. The time required to apply any of the superposition rules is directly proportional to the size of the current set of conjectures E, size that increases with every inference step and therefore reduces the prover's performances. We propose a G-variant, denoted by G', that reduces considerably the size of E at the expense of less powerful rules by applying our methodology.

$\boxed{A'\text{-representation}}$ Firstly, an empty set of premises is appended to any G' state and the processed conjecture is no longer added to the new set of conjectures.

$\boxed{Reasoning\ Modules}$ The new set of conjectures generated by any G'-rule is a strict CCS of the processed conjecture built by one of the following reasoning modules:

1. SR, used by the SUPERPOSITION RIGHT-variant, has the generation function, g_{SR}, defined as $g_{SR}(\Gamma \Rightarrow s[l]_p = t, (\{\Gamma' \Rightarrow l = r\}, \emptyset)) = \{\Gamma', \Gamma \Rightarrow s[r]_p = t\}$ since $\Gamma \Rightarrow s[l]_p = t \succeq_c \Gamma' \Rightarrow l = r$. Moreover, it builds a strict CCS because $\Gamma \Rightarrow s[l]_p = t \succ_c \Gamma', \Gamma \Rightarrow s[r]_p = t$, according to the definition of $\succ_c$ from [16].

2. SL intervenes in the definition of the SUPERPOSITION LEFT-variant. $g_{LR}((\Gamma, s[l]_p = t \Rightarrow \Delta), (\{\Gamma' \Rightarrow l = r\}, \emptyset)) = \{\Gamma', \Gamma, s[r]_p = t \Rightarrow \Delta\}$ is a strict CCS for similar reasons as above.

3. ER is used by EQUALITY REASONING-variant and has $g_{ER}((\Gamma, s = s \Rightarrow \Delta), (\emptyset, \emptyset)) = \{\Gamma \Rightarrow \Delta\}$ that generates a strict CCS because s is maximal and $\Gamma, s = s \Rightarrow \Delta \succ_c \Gamma \Rightarrow \Delta$.

| Instantiation Result | Any G'-rule is an instance of the 1-SIMPLIFY rule of A', as shown in Table 3. Its columns have the same meaning as for Table 2.

G'-*Variants* In our setting, the order $\succ_c$ is not required to be total. Such non-total (partial) orders can be built, for example, by generalising $<_e$ from equalities to conditional clauses. Representing $s_1 = t_1 \wedge \ldots \wedge s_n = t_n \Rightarrow l_1 = r_1 \vee \ldots \vee l_m = r_m$ as the multiset $\{s_1, t_1, \ldots, s_n, t_n, l_1, r_1, \ldots, l_m, r_m\}$ allows for the generation of smaller or equal clauses if:

i) non-maximal terms of the processed conjecture are reduced by SUPERPOSITION LEFT

Table 3. G'-rules as instances of A'-rules

G'-rule	A'-rule	RM, context, kind of CCS
SUPERPOSITION RIGHT	1-SIMPLIFY	SR, $(E, \emptyset)$, strict
SUPERPOSITION LEFT	1-SIMPLIFY	SL, $(E, \emptyset)$, strict
EQUALITY REASONING	1-SIMPLIFY	ER, $(\emptyset, \emptyset)$, strict

and SUPER-POSITION RIGHT, and ii) identities of the form $s = s$ are eliminated by EQUALITY REASONING even if s is not maximal. These weakened versions of the G'-rules are still instances of 1-SIMPLIFY because it does not require strict CCSs.

Our methodology allows for a fine tuning for powerful and automated inference systems by considering two new rules for each G'-rule, with contexts defined by the corresponding one-step A'-inference rules:

1. As instances of 1-ADDPREMISE, with respect to their original specifications, SUPERPOSITION LEFT and SUPERPOSITION RIGHT are constrained to forbid the reduction of the maximal terms at the root position in order to build a strict CCS. There is no restriction for EQUALITY REASONING.
2. As instances of 1-SIMPLIFY, the rules are not constrained to reduce maximal terms.

An inference system containing only 1-ADDPREMISE instances is as powerful as G: the processed conjecture is not saved in the new state as a conjecture, but as a premise. Notice that the G' and its variants satisfy Property 1 for similar reasons as G.

Theorem 3. *The inference system* G' *and its variants are sound and refutationally complete.*

Proof. By Theorem 2 and the instantiation results.   ∎

The saturation process always finishes on ground Horn clauses with any of G, G' and G'-variants. This is not the case for (Horn) clauses with variables, where the G'-variants are expected to be more effective. Techniques for lifting inference systems from the ground to the non-ground case are presented in [16,4].

4   Conclusions, Future and Related Works

The presented work has both applicative and theoretical interests. We have proposed a framework and a methodology to design and analyse automatic provers based on a 'Descente Infinie' induction principle. A stronger connection between two important fields of the automated deduction, the implicit induction and saturation-based theorem proving, is established. Their logic is captured by an abstract inference system: to witness, the inference systems from the given examples have been shown to be instantiations of it. The orders over formulas need only to be Noetherian and stable to satisfy the soundness property, as can be observed from the proof of Theorem 1. Thanks to the clear separation between the logic and computation provided by the framework, it can be concluded that other order properties encountered in the literature, as the totality, are required either in the computation part or to satisfy additional properties like Property 1.

We believe that the abstract system is a landmark for the completely automated 'Descente Infinie' inference systems. [18] compares a simplified version of A with other implicit induction systems and shows that it provides the maximal set of induction hypotheses. We expect the same results for the saturation-based provers. Usually, the one- and two-step inference rules are sufficient to apply the methodology, but the framework is able to provide 'more than two'-step rules when necessary. The computation process has the advantage to be parameterisable; it is performed by reasoning modules that allow for the modular design of concrete inference systems.

The methodology indicates how to instantiate the abstract system with specific reasoning techniques without affecting its soundness. The soundness proofs of concrete procedures become advantageously simpler because the soundness of the abstract systems was established once for all. They are mainly reduced to show that any rule is the instance of an abstract inference rule.

Future Works. We intend to implement a highly parameterised 'Descente Infinie' proving environment to build modular implicit induction and saturation-based provers having user-configurable reasoning techniques that benefit of the most general set of induction hypotheses. Today, many of these provers are highly automated and able to perform multiple (but easy) proof steps. However, in order to successfully treat industrial-size applications, the user should intervene and combine automatic with interactive proofs. In this paper, the abstract inference rules are adapted to run in *automatic* mode but at the expense of limiting the search of the induction hypotheses only to the formulas from the current state. Extending the scope of the search beyond the current state would increase the potential (power) of the provers; in *interactive* mode, the user would be allowed to build the appropriate CCSs with formulas from previous states, while in a *proof-check* mode, these formulas can be taken from the whole derivation to check. The environment would guarantee the soundness of these operations by using the composition properties of CCSs and the compatibility constraints imposed by the abstract rules. Its computational power will be represented by a library of reasoning techniques and orders, classified on the type of consequence relation and formulas. They can be "imported" from different 'Descente Infinie' provers, especially from saturation-based systems like those using inductionless induction [9], ordered paramodulation [16], ordered resolution [4] and different 'Knuth-Bendix completion'-like procedures. An integration schema of the conditional reasoning modules in recursive systems is part of our perspectives.

Thanks to the strong connection presented above, the environment would perform different implicit induction and saturation-based proofs during the same session. We can imagine an application where i) the properties of a specification are tested by saturation-based procedures, like the ground convergence for conditional specifications [13], ii) the (implicit induction) proof of specification's properties, and iii) specific saturation-based reasoning techniques, like the combination of different decision procedures [1], are all performed by the same 'Descente Infinie' inference system.

Related Works. Even if the 'Descente Infinie' principle traces back to the ancient Greeks, there are very few papers that refer to its explicit application in theorem proving. The most representative, [19], gives a thorough description of it and its applications to interactive rather than automatic proof techniques, as the sequent and tableau calculi. No clear connection with the saturation-based techniques is provided.

The orders used by the current saturation-based and the most of the implicit induction procedures are defined on formulas. Other implicit induction orders are more flexible, as [15] that uses pairs (formula, substitution), or semantic and more powerful [20,19].

The saturation procedures are known to be 'induction-based' for more than twenty years [14]. In spite of this, to our knowledge, no existing framework separates their logic from computation. For instance, the abstract resolution system RP from [4] does not abstract the subsumption operations. On the other hand, there are several abstract systems for implicit induction similar to A (see [18] for a comparison). For example, the 'switched frame' system from [20] is as powerful as the one-step A. In the same paper, the (non-switched) frame system is more powerful, thanks to a memorising rule which collects all the processed conjectures. However, this makes it more suited for provers that run in interactive mode rather than in automated and proof-check modes, for the reasons shown in the *Future Works* paragraph.

Acknowledgements. We thank the anonymous referees and Claus-Peter Wirth for useful comments on previous versions of this paper.

References

1. A. Armando, S. Ranise, and M. Rusinowitch. Uniform derivation of decision procedures by superposition. In *CSL2001*, number 2142 in LNCS, pages 513–527, 2001.
2. J. Avenhaus and K. Madlener. Theorem proving in hierarchical clausal specifications. *Advances in Algorithms, Languages, and Complexity*, pages 1–51, 1997.

3. L. Bachmair. Proof by consistency in equational theories. In *Proceedings 3rd IEEE Symposium on Logic in Computer Science, Edinburgh (UK)*, pages 228–233, 1988.

4. L. Bachmair and H. Ganzinger. *Resolution Theorem Proving*, pages 19–99. Handbook of Automated Reasoning. 2001.

5. N. Berregeb, A. Bouhoula, and M. Rusinowitch. Observational proofs with critical contexts. In *Fundamental Approaches to Software Engineering (FASE'98)*, volume 1382 of *LNCS*, pages 38–53. Springer Verlag, March–April 1998.

6. A. Bouhoula. Using induction and rewriting to verify and complete parameterized specifications. *Theoretical Computer Science*, 1-2(170):245–276, 1996.

7. A. Bouhoula. Automated theorem proving by test set induction. *Journal of Symbolic Computation*, 23:47–77, 1997.

8. F. Bronsard, U. Reddy, and R. W. Hasker. Induction using term orders. *Journal of Symbolic Computation*, 16:3–37, 1996.

9. H. Comon. *Inductionless induction*, pages 913–970. Handbook of Automated Reasoning. 2001.

10. N. Dershowitz. Orderings for term-rewriting systems. *Theoretical Computer Science*, 17(3):279–301, 1982.

11. D. Kapur, P. Narendran, and H. Zhang. Automating inductionless induction using test sets. *Journal of symbolic Computation*, 11:83–112, 1991.

12. E. Kounalis and M. Rusinowitch. Mechanizing inductive reasoning. *Bulletin of European Association for Theoretical Computer Science*, 41:216–226, June 1990.

13. E. Kounalis and M. Rusinowitch. Studies on the ground convergence property of conditional theories. In *AMAST'1991*, pages 363–376, 1991.

14. D.S. Lankford. Some remarks on inductionless induction. Technical Report MTP-11, Louisiana Tech University, Ruston, LA, 1980.

15. D. Naidich. On generic representation of implicit induction procedures. Technical Report CS-R9620, CWI, 1996.

16. R. Nieuwenhuis and A. Rubio. *Paramodulation-Based Theorem Proving*, pages 371–443. Handbook of Automated Reasoning. 2001.

17. U. Reddy. Term rewriting induction. In *10th International Conference on Automated Deduction*, volume 814 of *LNCS*, pages 162–177, 1990.

18. S. Stratulat. A general framework to build contextual cover set induction provers. *Journal of Symbolic Computation*, 32(4):403–445, 2001.

19. C.-P. Wirth. Descente Infinie + Deduction. *Logic Journal of the IGPL*, 12(1):1–96, 2004.

20. C.-P. Wirth and K. Becker. Abstract notions and inference systems for proofs by mathematical induction. In *CTRS'94*, number 968 in LNCS, pages 353–373, 1994.

21. H. Zhang, D. Kapur, and M. S. Krishnamoorthy. A mechanizable induction principle for equational specifications. In *CADE'09*, number 310 in LNCS, pages 162–181. Springer-Verlag, 1988.

A Decision Procedure for the Alternation-Free Two-Way Modal μ-Calculus*,**

Yoshinori Tanabe[1,2], Koichi Takahashi[2], Mitsuharu Yamamoto[3],
Akihiko Tozawa[4], and Masami Hagiya[5,6]

[1] CREST, Japan Science and Technology Agency
[2] Research Center for Verification and Semantics,
National Institute of Advanced Industrial Science and Technology (AIST)
[3] Faculty of Science, Chiba University
[4] IBM Research, Tokyo Research Laboratory, IBM Japan ltd., Japan
[5] Graduate School of Information Science and Technology, University of Tokyo
[6] NTT Communication Science Laboratories

Abstract. The satisfiability checking problem is known to be decidable for a variety of modal/temporal logics such as the modal μ-calculus, but effective implementation has not necessarily been developed for all such logics. In this paper, we propose a decision procedure using the tableau method for the alternation-free two-way modal μ-calculus. Although the size of the tableau set maintained in the method might be large for complex formulas, the set and the operations on it can be expressed using BDD and therefore we can implement the method in an effective way.

1 Introduction

There are various applications of satisfiability checking in modal logics. For example it has been used to synthesize a concurrent program from a specification expressed as a temporal formula by checking satisfiability of the formula [3, 10]. The authors have proposed the abstraction method [16] that can be applied to verification problems of graph rewriting systems using satisfiability checking of temporal formulas, and applied the method to analysis of cellular automata [6]. We also applied satisfiability checking of temporal formulas to verification and analysis of programs that process XML documents, whose tree structures are naturally expressible using branching time temporal logic formulas [17].

In applications mentioned above, we regard a graph with labelled edges as a Kripke structure, and describe properties of the graph by formulas in temporal

---

* This research was supported by Core Research for Evolutional Science and Technology (CREST) Program "New High-performance Information Processing Technology Supporting Information-oriented Society" of Japan Science and Technology Agency (JST).
** This research was partially supported by the Ministry of Education, Science, Sports and Culture, Grant-in-Aid for Scientific Research on Priority Areas, 16016211, 2004.

B. Beckert (Ed): TABLEAUX 2005, LNAI 3702, pp. 277–291, 2005.
© Springer-Verlag Berlin Heidelberg 2005

logics. For the purpose of the applications, we often need to follow edges not only in the forward direction but also in the backward direction. Therefore we need to use temporal logics that can handle both directions of edges as modalities. From this point of view the two-way modal μ-calculus[1] is one of the logics we regard as powerful enough [19].

It is known that for the general two-way modal μ-calculus the satisfiability problem is decidable and its complexity is EXPTIME-complete [19]. The decision procedure is constructed by converting the problem into the emptiness problem of the language recognized by a certain alternating tree automaton on infinite trees. In order to solve the emptiness problem, complex operations are required including determinization of parity automata [15]. The algorithm described in [5] has the time complexity $2^{O(n^4 \log n)}$ with regard to the length n of the given formula and actual implementation has not been reported. In possible applications of satisfiability checking mentioned above, on the other hand, the decision procedure need to be called repeatedly. From this point finding a decision procedure that can be implemented effectively is meaningful even if it can only be applied to a subset of the logic as long as the subset is reasonably powerful for expressing problems in such applications.

In this research we focus on the alternation-free two-way modal μ-calculus as such a subset. We propose a decision procedure for checking satisfiability in the logic and describe experimental implementation of the procedure. The time complexity of the decision procedure is $2^{O(n \log n)}$, which is faster than the time complexity for the whole two-way modal μ-calculus as expected. The experimental implementation uses BDD [1]. Effectiveness of the use of BDD in verification tools are well-known. Classical examples are model checking tools such as SMV [11]. It is also applied to satisfiability checking in the experimental procedure for the basic modal logic K [13, 14], and in Mona [7], a verification tool on the monadic second-order logic WS2S, which is closely related to temporal logics and allows an effective decision procedure. Our decision procedure is also suitable to implement with BDD since it is built with iteration of set operations on finite sets such as tableaux. The experimental implementation shows that it finishes in reasonable time for formulas in modest sizes.

As described in detail in a later section, the key point for deriving the time complexity $2^{O(n \log n)}$ is to effectively express the condition concerning loops that include both forward and backward modalities. This point is specific to two-way logics.

For analysis of programs that process XML documents as mentioned above, we need to restrict models to be binary-branching finite trees. We show that our decision procedure can also be used for this purpose, though the performance may not be very attractive. We reported a more effective decision procedure specialized to binary-branching finite tree models [18]. In this case the problem

---

[1] The two-way modal μ-calculus is also known as the full modal μ-calculus. In this paper we use the word "two-way" because the word "full" usually means the entire system of a logic while we mainly deal with a subsystem of the μ-calculus.

can be converted into WS2S, so we could use Mona for satisfiability checking, but our experimental implementation shows that the specialized procedure performs better.

The rest of the paper is organized as follows: Section 2 reviews basic definitions and properties of the alternation-free two-way modal μ-calculus and Kripke structures. In Section 3 we describe the decision procedure for the calculus. In Section 4 we specialize the procedure for binary-branching finite tree models. A correctness proof of the decision procedure is given in Section 5. In Section 6 we describe our experimental implementation and Section 7 concludes.

2   Preliminaries

2.1   The Alternation-Free Two-Way Modal μ-Calculus

We denote the set of propositional constants by PC, the set of propositional variables by PV and the set of modalities by Mod. We assume that a function $^-$: Mod $\rightarrow$ Mod is defined and that $\overline{\overline{a}} = a$ for each $a \in$ Mod.

Definition 1. *We define the set* L_μ *of formulas of the two-way modal μ-calculus.*

- *For propositional constant* $P \in$ PC, $P, \neg P \in L_\mu$. *For propositional variable* $X \in$ PV, $X \in L_\mu$.
- *If* $\varphi_1, \varphi_2 \in L_\mu$ *then* $\varphi_1 \vee \varphi_2, \varphi_1 \wedge \varphi_2 \in L_\mu$. *We call* $\vee$ *and* $\wedge$ *the principal operator of* $\varphi_1 \vee \varphi_2$ *and* $\varphi_1 \wedge \varphi_2$, *respectively.*
- *If* $\varphi \in L_\mu$ *and* $a \in$ Mod *then* $[a]\varphi, \langle a \rangle \varphi \in L_\mu$. *We call* $[]$ *and* $\langle \rangle$ *the principal operator of* $[a]\varphi$ *and* $\langle a \rangle \varphi$, *respectively.*
- *If* $\varphi \in L_\mu$ *and* $X \in$ PV *then* $\mu X \varphi, \nu X \varphi \in L_\mu$. *We call* μ *and* ν *the principal operator of* $\mu X \varphi$ *and* $\nu X \varphi$, *respectively.*

An occurrence of $X \in PC$ *in* $\varphi \in L_\mu$ *is bound if it is contained in a subformula of the form of* $\mu X \varphi'$ *or* $\nu X \varphi'$, *otherwise it is free. We call* $\varphi \in L_\mu$ *a formula of the alternation-free two-way modal μ-calculus if the following conditions are satisfied:*

- *If* $\mu X \psi$ *is a subformula of* φ *and* $\nu Y \chi$ *is a subformula of* ψ *then* X *does not occur freely in* χ.
- *If* $\nu X \psi$ *is a subformula of* φ *and* $\mu Y \chi$ *is a subformula of* ψ *then* X *does not occur freely in* χ.

We denote by L_μ^{af} *the set of formulas of the alternation-free two-way modal μ-calculus.*

For example, $\mu X \nu Y (P \wedge [a]Y \wedge [\overline{a}]X)$ is a formula of the two-way modal μ-calculus but it is not alternation-free since X appears in the scope of νY. On the other hand, $\mu X (\nu Y (P \wedge [b]Y) \vee \langle a \rangle X \vee \langle \overline{a} \rangle X)$ is alternation-free.

Let $F \subseteq L_\mu^{af}$ and S be a sequence of $\vee$, $\wedge$, $[]$, $\langle \rangle$, μ and ν. We denote by PO($F; S$) the set of elements of F whose principal operators are listed in S. For example PO($\{P, P \wedge Q, \langle a \rangle P, [b]Q\}; \langle \rangle, []$) = $\{\langle a \rangle P, [b]Q\}$.

2.2 Kripke Structures

Definition 2. *A Kripke structure is a triple $\mathcal{M} = (M, R, \lambda)$ that satisfies the following conditions: (1) M is a set. (2) R is a function whose domain is* Mod *and for $a \in$ Mod, $R(a) \subseteq M \times M$ and $R(\bar{a}) = R(a)^{-1}$. (3) λ is a function whose domain is* PC *and for $P \in$ PC, $\lambda(P) \subseteq M$.*
A valuation v is a function whose domain is PV *and for $X \in$ PV, $v(X) \subseteq M$. For $S \subseteq M$ we denote by $v[X \mapsto S]$ the valuation obtained from v by replacing the value of $v(X)$ with S.*

Definition 3. *For $\varphi \in L_\mu^{af}$, Kripke structure $\mathcal{M} = (M, R, \lambda)$, valuation v, and $m \in M$, we define relation $\mathcal{M}, v, m \models \varphi$, or $m \models \varphi$ for short.*

- *For $P \in$ PC, $m \models P \iff m \in \lambda(P)$.*
- *For $P \in$ PC, $m \models \neg P \iff m \notin \lambda(P)$.*
- *For $X \in$ PV, $m \models X \iff m \in v(X)$.*
- *$m \models \varphi_1 \vee \varphi_2 \iff m \models \varphi_1$ or $m \models \varphi_2$.*
- *$m \models \varphi_1 \wedge \varphi_2 \iff m \models \varphi_1$ and $m \models \varphi_2$.*
- *$m \models \langle a \rangle \varphi \iff$ There exists $m' \in M$ such that $(m, m') \in R(a)$ and $m' \models \varphi$.*
- *$m \models [a]\varphi \iff$ For all $m' \in M$ if $(m, m') \in R(a)$ then $m' \models \varphi$.*
- *$m \models \mu X \varphi \iff m \in \bigcap \{S \subseteq M \mid S \supseteq \{m \in M \mid \mathcal{M}, v[X \mapsto S], m \models \varphi\}\}$.*
- *$m \models \nu X \varphi \iff m \in \bigcup \{S \subseteq M \mid S \subseteq \{m \in M \mid \mathcal{M}, v[X \mapsto S], m \models \varphi\}\}$.*

If φ is a sentence, that is all propositional variables occurring in φ are bound, the relation does not depend on the valuation v so we write $\mathcal{M}, m \models \varphi$ instead of $\mathcal{M}, v, m \models \varphi$.

We will introduce a few more concepts to describe a decision procedure.

First, for $\varphi_1, \varphi_2 \in L_\mu^{af}$, if $\mathcal{M}, m \models \varphi_1 \iff \mathcal{M}, m \models \varphi_2$ holds for any Kripke structure $\mathcal{M} = (M, R, \lambda)$ and $m \in M$, we say φ_1 and φ_2 are *equivalent* and write $\varphi_1 \equiv \varphi_2$.

Next we define formulas in normal form.

Definition 4. *Formula $\varphi \in L_\mu^{af}$ is said to be in normal form if it is well-named and guarded, that is,*

- *For any propositional variable X, either all the occurrences of X are free or there is a unique subformula ψ of φ in the form of $\mu X \varphi'$ or $\nu X \varphi'$ and X does not occur outside of ψ.*
- *If $\mu X \varphi'$ or $\nu X \varphi'$ is a subformula of φ, for any occurrence of X in φ' there is a subformula of φ' that contains the occurrence and is in the form of $\langle a \rangle \varphi''$ or $[a]\varphi''$.*

It is known [12, Lemma 2.2] that for every $\varphi \in L_\mu^{af}$ there exists a formula $\psi \in L_\mu^{af}$ in normal form that is equivalent to φ. So in the rest of the paper we assume that all formulas in consideration are in normal form.

We will then define the closure of a formula and related concepts. For $\varphi = \mu X \psi$ or $\varphi = \nu X \psi$, we denote by $\exp(\varphi)$ the formula ψ with replacing all the occurrence of X with φ. We define relation $\to_e$ on L_μ^{af} to be the least relation that satisfies the following:

$$- \varphi_1 \vee \varphi_2 \to_e \varphi_i \ (i = 1, 2). \qquad - [a]\varphi \to_e \varphi.$$
$$- \varphi_1 \wedge \varphi_2 \to_e \varphi_i \ (i = 1, 2). \qquad - \mu X \varphi \to_e \exp(\mu X \varphi).$$
$$- \langle a \rangle \varphi \to_e \varphi. \qquad\qquad\quad - \nu X \varphi \to_e \exp(\nu X \varphi).$$

The *closure* of formula $\varphi \in L_\mu^{af}$ is defined as the smallest set S that contains φ and closed under relation $\to_e$, that is, if $\psi \in S$ and $\psi \to_e \psi'$ then $\psi' \in S$. We denote the closure of φ by $cl(\varphi)$. Let $\mathcal{D}$ be the set of strongly connected components of $cl(\varphi)$ with respect to the relation $\to_e$, and we define $\mathcal{D}_\mu = \{D \in \mathcal{D} \mid PO(D; \mu) \neq \varnothing\}$ and $\mathcal{D}_\nu = \{D \in \mathcal{D} \mid PO(D; \nu) \neq \varnothing\}$. Following are some basic properties of the concepts defined so far.

Proposition 1.
(1) $\varphi \equiv \exp(\varphi)$.
(2) $cl(\varphi)$ *is a finite set and its size is linear to the length of φ, that is the number of operators and propositional variables appearing in φ.*
(3) $\mathcal{D}_\mu \cap \mathcal{D}_\nu = \varnothing$.

Proof. Refer to [9] for claims (1) and (2). Claim (3) follows from the fact that φ is alternation-free. □

For example, consider formula $\varphi = \mu X([a]X \wedge \nu Y \langle b \rangle Y)$. If we denote the subformula $\nu Y \langle b \rangle Y$ by ψ then φ can be written as $\mu X([a]X \wedge \psi)$. With these, $\exp(\varphi)$ is $[a]\varphi \wedge \psi$, and the closure $cl(\varphi)$ is $\{\varphi, [a]\varphi \wedge \psi, [a]\varphi, \psi, \langle b \rangle \psi\}$. The relation $\to_e$ contains $\varphi \to_e [a]\varphi \wedge \psi \to_e [a]\varphi \to_e \varphi$. There are two strongly connected components, namely $D_1 = \{\varphi, [a]\varphi \wedge \psi, [a]\varphi\}$ and $D_2 = \{\psi, \langle b \rangle \psi\}$. D_1 belongs to $\mathcal{D}_\mu$ while D_2 belongs to $\mathcal{D}_\nu$.

Let Lit be the set of propositional constants and their negations that are members of $cl(\varphi)$. We call the set $\text{Lit} \cup PO(cl(\varphi); \langle \rangle, [])$ the *lean* of φ after Pan et al [13]. Since we only consider formulas in normal form, every "path" in $(cl(\varphi), \to_e)$ reaches some element of the lean after finitely many steps and on the path we see only $\vee$, $\wedge$, μ and ν as principal operators.

2.3   Properties of Kripke Structures

In this subsection we state two basic properties of Kripke structures and formulas in the alternation-free two-way modal μ-calculus without proof. They will be used to prove correctness of the decision procedure in the following subsections. Let $\mathcal{M} = (M, R, \lambda)$ be a Kripke structure. We denote the set $\{(\varphi, m) \in cl(\varphi_I) \times M \mid m \models \varphi\}$ by $\text{Sat}(\mathcal{M})$.

Let $Q \subseteq cl(\varphi_I) \times M$. We call a function F a *choice function* on Q if its domain is the set $\{(\varphi, m) \in Q \mid \varphi \in PO(cl(\varphi_I); \vee, \langle \rangle)\}$ and satisfies the following:

$$- F(\varphi_1 \vee \varphi_2, m) \in \{\varphi_1, \varphi_2\}$$
$$- F(\langle a \rangle \varphi, m) \in M, \ (m, F(\langle a \rangle \varphi, m)) \in R(a),$$

We define a relation R_F on $cl(\varphi_I) \times M$ to be the least one that satisfies the following:

- If $\varphi = \varphi_1 \vee \varphi_2$ and $(\varphi, m) \in Q$ then $(\varphi, m) \, R_F \, (F(\varphi, m), m)$.
- $(\varphi_1 \wedge \varphi_2, m) \, R_F \, (\varphi_i, m) \quad (i = 1, 2)$.
- $(\mu X \varphi, m) \, R_F \, (\exp(\mu X \varphi), m)$.
- $(\nu X \varphi, m) \, R_F \, (\exp(\nu X \varphi), m)$.
- If $(\langle a \rangle \varphi, m) \in Q$ then $(\langle a \rangle \varphi, m) \, R_F \, (\varphi, F(\langle a \rangle \varphi, m))$.
- If $(m, n) \in R(a)$ then $([a]\varphi, m) \, R_F \, (\varphi, n)$.

Proposition 2. *If $Q \subseteq \mathrm{cl}(\varphi_I) \times M$ and a choice function F on Q satisfies the following three conditions, then $Q \subseteq \mathrm{Sat}(\mathcal{M})$ holds.*
(1)   For $P \in \mathrm{PC}$, if $(P, m) \in Q$ then $m \in \lambda(P)$. If $(\neg P, m) \in Q$ then $m \notin \lambda(P)$.
(2) Q is closed under R_F, that is, $(\varphi, m) \in Q$ and $(\varphi, m) \, R_F \, (\psi, n)$ implies $(\psi, n) \in Q$.
(3)   If $D \in \mathcal{D}_\mu$ and $Q_D = (D \times M) \cap Q$ then $R_F^{-1} \upharpoonright (Q_D \times Q_D)$ is a well-founded relation on Q_D, that is, there is no sequence $(q_i \mid i < \omega)$ such that $q_i \in Q_D$ and $q_i R_F q_{i+1}$ holds for all $i < \omega$.

We fix $D \in \mathcal{D}_\mu$ until the end of this subsection. Let us denote the class of ordinals by On. For $\alpha \in \mathrm{On}$ we define $U_\alpha \subseteq \mathrm{Sat}(\mathcal{M})$ as the least set that satisfies the following:

- $\{(\varphi, m) \in \mathrm{Sat}(\mathcal{M}) \mid \varphi \notin D\} \subseteq U_0$
- If $(\varphi_1, m) \in U_\alpha$ or $(\varphi_2, m) \in U_\alpha$ then $(\varphi_1 \vee \varphi_2, m) \in U_\alpha$.
- If $(\varphi_1, m) \in U_\alpha$ and $(\varphi_2, m) \in U_\alpha$ then $(\varphi_1 \wedge \varphi_2, m) \in U_\alpha$.
- If $(\exp(\mu X \varphi), m) \in U_\alpha$ then $(\mu X \varphi, m) \in U_\alpha$.
- If there is $m' \in M$ and $\beta < \alpha$ such that $(\varphi, m') \in U_\beta$ and $(m, m') \in R(a)$ then $(\langle a \rangle \varphi, m) \in U_\alpha$.
- If for all $m' \in M$ such that $(m, m') \in R(a)$ there exists $\beta < \alpha$ with $(\varphi, m') \in U_\beta$ then $([a]\varphi, m) \in U_\alpha$.

Proposition 3. $\mathrm{Sat}(\mathcal{M}) = \bigcup_{\alpha \in \mathrm{On}} U_\alpha$

For $(\varphi, m) \in \mathrm{Sat}(\mathcal{M})$ we denote by $\mathrm{rank}_D(\varphi, m)$ the least ordinal α such that $(\varphi, m) \in U_\alpha$.

3   The Decision Procedure

In this section we give a decision procedure to judge whether there is a Kripke structure $\mathcal{M} = (M, R, \lambda)$ and $m \in M$ such that $\mathcal{M}, m \models \varphi_I$ for a given sentence $\varphi_I \in \mathrm{L}_\mu^{\mathrm{af}}$ using the tableau method. We assume $\mathcal{D}_\mu \neq \varnothing$ without loss of generality, since if it is not the case we can consider $\varphi_I \vee \mu X \langle a \rangle X$ instead of φ_I. Note that $\mu X \langle a \rangle X$ is equivalent to $\bot$.

We first define a φ_I-type, which will be a node of the tableau. A φ_I-type is usually defined as a set of formulas in $\mathrm{cl}(\varphi_I)$ that are true at the corresponding node in the tableau, but our definition is based on the following two considerations. First, we do not need to handle all elements of $\mathrm{cl}(\varphi_I)$ since every $\varphi \in \mathrm{cl}(\varphi_I)$ is equivalent to a boolean combination of elements of a set called Lean, which is defined as $(\text{PC} \cap \text{cl}(\varphi_{\text{I}})) \cup \text{PO}(\text{cl}(\varphi_{\text{I}}); \langle\rangle, [])$. Second, we want to store information on the rank in a φ_{I}-type. More specifically, for φ_1 and φ_2 that are true at a node, we want to know which has a higher rank (or they are of the same rank) if they belong to the same $D \in \mathcal{D}_\mu$ and there are $\psi_1, \psi_2 \in D$ such that $\text{mod}(\varphi_1) = \overline{\text{mod}(\psi_1)}$ and $\text{mod}(\varphi_2) = \overline{\text{mod}(\psi_2)}$, where we define $\text{mod}(\langle a\rangle\varphi) = a$ and $\text{mod}([a]\varphi) = a$. According to these considerations we define a φ_{I}-type as a function t from Lean to $\omega + 1$, intending that the value ω for $t(\varphi)$ means that φ is "false" and all other values mean "true." And $t(\varphi_1) < t(\varphi_2)$ means that φ_1 has a lower rank than that of φ_2.

Definition 5. *A φ_{I}-type is a function $t : \text{Lean} \to \omega + 1 = \{0, 1, \ldots, \omega\}$ satisfying the following conditions: for $\varphi \in \text{Lean}$,*

- *If $\varphi \in D \cap \text{Lean}$ for $D \in \mathcal{D}_\mu$, $t(\varphi) \le |\text{BForm}_D|$ or $t(\varphi) = \omega$*
- *otherwise, $t(\varphi) \in \{0, \omega\}$*

where

- $\text{BForm}_D = \{\varphi \in \text{PO}(D; \langle\rangle, []) \mid \text{mod}(\varphi) \in \text{BMod}_D\}$.
- $\text{BMod}_D = \{a \in \text{Mod} \mid \text{There exist } \varphi, \psi \in \text{PO}(D; \langle\rangle, []) \text{ such that } \text{mod}(\varphi) = a \text{ and } \text{mod}(\psi) = \overline{a}.\}$.

and $|A|$ is the size of a finite set A.
  The set of all φ_{I}-types is denoted by T_{I}. We also define $\text{sat}(t) = \{\varphi \in \text{Lean} \mid t(\varphi) < \omega\}$ for $t \in T_{\text{I}}$.

As mentioned above, every formula $\varphi \in \text{cl}(\varphi_{\text{I}})$ is equivalent to a boolean combination of elements of Lean. Thus we can define a function $\overline{t} : \text{cl}(\varphi_{\text{I}}) \to \omega + 1$ that naturally extends t:

- For $\varphi \in \text{Lean}$, $\overline{t}(\varphi) = t(\varphi)$.
- For $P \in \text{PC} \cap \text{cl}(\varphi_{\text{I}})$, $\overline{t}(\neg P) = 0$ if $t(P) = \omega$ and $\overline{t}(\neg P) = \omega$ if $t(P) = 0$
- $\overline{t}(\varphi_1 \vee \varphi_2) = \min\{\overline{t}(\varphi_1), \overline{t}(\varphi_2)\}$.
- $\overline{t}(\varphi_1 \wedge \varphi_2) = \max\{\overline{t}(\varphi_1), \overline{t}(\varphi_2)\}$.
- $\overline{t}(\mu X\varphi) = \overline{t}(\exp(\mu X\varphi))$.
- $\overline{t}(\nu X\varphi) = \overline{t}(\exp(\nu X\varphi))$.

For $t \in T_{\text{I}}$ and $\varphi \in \text{cl}(\varphi_{\text{I}})$ we define relation $t \Vdash \varphi$ as $\overline{t}(\varphi) < \omega$, which intuitively means φ is true at t.

Next we want to define a relation $\xrightarrow{a}$ on T_{I} that corresponds to $R(a)$ in a Kripke structure (M, R, λ). What conditions should the relation satisfy? If a formula $[a]\varphi$ is true at t and $t \xrightarrow{a} t'$ holds, then φ must be true at t'. A similar condition is needed for $[\overline{a}]\varphi$. There is another condition that is more complicated. We take the formula $\psi = \mu X([a]X \wedge [\overline{a}]X)$ as an example to describe it. Suppose all of $t([a]\psi), t([\overline{a}]\psi), t'([a]\psi), t'([\overline{a}]\psi)$ are less than ω. It can easily be checked that $t \Vdash \psi$ and $t' \Vdash \psi$ hold, thus the above two conditions do not rule out $t \xrightarrow{a} t'$. But if $t \xrightarrow{a} t'$ held there would be an infinite loop of "expansion" corresponding to the relation R_F in a Kripke structure: $(\psi, t) \to ([a]\psi \wedge [\overline{a}]\psi, t) \to$

$([a]\psi, t) \to (\psi, t') \to ([a]\psi \wedge [\bar{a}]\psi, t') \to ([\bar{a}]\psi, t') \to (\psi, t) \to \cdots$. Such a loop is not acceptable as the principal operator of ψ is μ.

We use the information on the rank to prohibit this. Suppose $t \xrightarrow{a} t'$, $t \Vdash [a]\varphi_1$, $t' \Vdash \varphi_1$, $t' \Vdash [\bar{a}]\varphi_2$ and $t \Vdash \varphi_2$ hold. According to the definition, $\text{rank}_D(\varphi_2, t) < \text{rank}_D([\bar{a}]\varphi_2, t')$ and $\text{rank}_D(\varphi_1, t') < \text{rank}_D([a]\varphi_1, t)$. Therefore when $t'([\bar{a}]\varphi_2) \leq \bar{t'}(\varphi_1)$ holds, which intuitively means $\text{rank}_D([\bar{a}]\varphi_2, t') \leq \text{rank}_D(\varphi_1, \bar{t'})$, we should have $\bar{t}(\varphi_2) < t([a]\varphi_1)$. A similar condition is imposed by exchanging the roles of t and t'.

By introducing these conditions, we can avoid the loop caused by the formula ψ above. We can assume $t'([\bar{a}]\psi) = \bar{t'}(\psi)$ without loss of generality. But $\bar{t}(\psi) = \max\{t([a]\psi), t([\bar{a}]\psi)\} \geq t([a]\psi)$, therefore $t \xrightarrow{a} t'$ cannot hold.

To define the conditions precisely, the relation $\text{LoopSafe}(t, t', \varphi_1, \varphi_2)$ for $t, t' \in T_I$ and $\varphi_1, \varphi_2 \in \text{PO}(\text{cl}(\varphi_1), \langle\rangle, [])$ is defined as follows:

- $\bar{t'}(\varphi_2) \leq \bar{t'}(\vec{\varphi_1}) < \omega \implies \bar{t}(\vec{\varphi_2}) < \bar{t}(\varphi_1)$
- $\bar{t}(\varphi_1) \leq \bar{t}(\vec{\varphi_2}) < \omega \implies \bar{t'}(\vec{\varphi_1}) < \bar{t'}(\varphi_2)$

where $\vec{\varphi} = \varphi'$ if $\varphi = \langle a \rangle \varphi'$ or $\varphi = [a]\varphi'$.

Possible infinite loops may contain a formula $\langle a \rangle \varphi$. To prohibit such loops we should define the relation $\xrightarrow{\langle a \rangle \varphi}$ rather than $\xrightarrow{a}$. We will regard the union of $\{\xrightarrow{\langle a \rangle \varphi} \mid \langle a \rangle \varphi \in \text{cl}(\varphi_I)\}$ as the counterpart of $R(a)$ of a Kripke structure (M, R, λ).

Definition 6. *For $\langle a \rangle \varphi \in \text{cl}(\varphi_I)$ the relation $\xrightarrow{\langle a \rangle \varphi}$ is defined so that for $t, t' \in T_I$ $t \xrightarrow{\langle a \rangle \varphi} t'$ holds if and only if the following conditions are satisfied:*

- *$t' \Vdash \varphi$.*
- *For $[a]\psi \in \text{cl}(\varphi_I)$ if $t \Vdash [a]\psi$ then $t' \Vdash \psi$.*
- *For $[\bar{a}]\psi \in \text{cl}(\varphi_I)$ if $t' \Vdash [\bar{a}]\psi$ then $t \Vdash \psi$.*
- *$\text{LoopSafe}(t, t', [a]\psi_1, [\bar{a}]\psi_2)$ holds for all $D \in \mathcal{D}_\mu$ and $[a]\psi_1, [\bar{a}]\psi_2 \in \text{BForm}_D$.*
- *$\text{LoopSafe}(t, t', \langle a \rangle \varphi, [\bar{a}]\psi_2)$ holds for all $[\bar{a}]\psi_2 \in \text{BForm}_D$, provided that $\langle a \rangle \varphi \in \text{BForm}_D$ holds.*

The decision procedure starts with the tableau of all φ_I-types, and repeatedly disposes of φ_I-types that are "inconsistent," until all remaining φ_I-types are consistent. There are two types of inconsistency, which we call $\Diamond$-inconsistency and μ-inconsistency, respectively. The former is simple: for $T \subseteq T_I$ we say that $t \in T$ is $\Diamond$-consistent in T if for all $\langle a \rangle \varphi \in \text{sat}(t)$ there exists $t' \in T$ such that $t \xrightarrow{\langle a \rangle \varphi} t'$. We denote by $\text{Con}_\Diamond(T)$ the set of $t \in T$ that is $\Diamond$-consistent in T.

An example of the μ-consistency is as follows: suppose that $\psi = \mu X([a]X)$, $T = \{t\} \subseteq T_I$, $t([a]\psi) = 0$, and $t \xrightarrow{\langle a \rangle \xi} t$ holds. Although $t \Vdash \psi$ holds, there is an infinite sequence of expansion: $(\psi, t) \to ([a]\psi, t) \to (\psi, t) \to \cdots$. Our strategy to identify such an inconsistency is to mark those formulas in φ_I-types which are guaranteed as consistent, i.e. which have finite sequences of expansion. The check is done independently for each $D \in \mathcal{D}_\mu$ and formulas that do not belong to D are always marked. Marking should obey the rank information stored in φ_I-types: if $t \in T$ and S is the set of formulas in $D \cap \text{sat}(t)$ that are marked at a moment, the pair (t, S) should be a member of the set $\text{TS}(D, T) = \{(t, S) \in T \times \mathcal{P}(D \cap \text{sat}(t)) \mid$ for any $\varphi \in S$ and $\psi \in D \cap \text{Lean}$ if $t(\psi) < t(\varphi)$ then $\psi \in S\}$.

With this observation we define a sequence of subsets $V_0, V_1, \ldots$ of $\text{TS}(D, T)$ so that $(t, S) \in V_j$ if S is the set of formulas in $D \cap \text{sat}(t)$ marked at stage j. We sometimes write $V_j(D, T)$ for V_j to make D and T explicit. At stage 0, as nothing has been marked in D, we define $V_0 = \{(t, \varnothing) \mid t \in T\}$. We will define $\text{Step}_D(V, T) \subseteq \text{TS}(D, T)$, the set of formulas that can be marked if formulas in V have already been marked. Then we can define $V_{j+1} = V_j \cup \text{Step}_D(V, T)$.

To define $\text{Step}_D(V, T)$, we need to decide whether $\varphi \in \text{cl}(\varphi_I)$ is marked or not, even if φ is not a member of Lean. For $(t, S) \in \text{TS}(D, T)$ and $\varphi \in \text{cl}(\varphi_I)$, we say $(t, S) \Vdash \varphi$ if $t' \Vdash \varphi$ holds, where $t' \in T_I$ is defined as $t'(\varphi) = \omega$ if $\varphi \in D \backslash S$ and $t'(\varphi) = t(\varphi)$ otherwise. Then we define $\text{Step}_D(V, T)$ so that $(t, S) \in \text{Step}_D(V, T)$ holds if and only if for any $\langle a \rangle \psi \in \text{sat}(t)$ there exists $(t', S') \in V$ such that:

- $t \xrightarrow{\langle a \rangle \psi} t'$ holds.
- For any $\varphi \in \text{PO}(S, [])$ with $\text{mod}(\varphi) = a$, $(t', S') \Vdash \overrightarrow{\varphi}$.
- If $\langle a \rangle \psi \in S$, $(t', S') \Vdash \psi$.

Each V_j is a subset of a finite set T_I and $(V_j \mid j < \omega)$ is an increasing sequence with respect to the inclusion relation. Therefore there exists $J = J(D, T)$ such that if $j \geq J$ then $V_j = V_J$.

We finally define the μ-consistency: for $T \subseteq T_I$ and $t \in T$, t is said to be μ-consistent if $(t, D \cap \text{sat}(t)) \in V_{J(D,T)}(D, T)$ holds for all $D \in \mathcal{D}_\mu$. We denote by $\text{Con}_\mu(T)$ the set of $t \in T$ that is μ-consistent in T. With two types of consistency defined, we can describe the main loop of the decision procedure: for $k \in \omega$, we define $T_k \subseteq T_I$: $T_0 = T_I$ and $T_{k+1} = \text{Con}_\Diamond(T_k) \cap \text{Con}_\mu(T_k)$. Since $(T_k \mid k < \omega)$ is a decreasing sequence with respect to the inclusion relation, there exists $K < \omega$ such that if $k \geq K$ then $T_k = T_K$.

Theorem 1. *For a sentence $\varphi_I \in L_\mu^{\text{af}}$ the following two conditions are equivalent.*
(1) φ_I *is satisfiable, that is, there is a Kripke structure $\mathcal{M} = (M, R, \lambda)$ and $m \in M$ such that $\mathcal{M}, m \models \varphi_I$ holds.*
(2) *There is $t \in T_K$ such that $t \Vdash \varphi_I$.*

Proof. Refer to Sections 5.1 and 5.2. □

Theorem 1 gives a decision procedure for checking the satisfiability of a sentence in the alternation-free two-way modal μ-calculus. For given sentence φ_I we can effectively calculate the sets appearing in this subsection such as $\text{cl}(\varphi_I)$, $\mathcal{D}_\mu, T_0, T_1, \ldots, V_0(T_j, D), \ldots$, and so on. Each set can be calculated with finitely many operations and the two loops (calculating T_k and V_j) are guaranteed to terminate.

The complexity of the decision procedure can be estimated as follows. Let n be the length of φ_I. The size of $\text{cl}(\varphi_I)$ is proportional to n. The size of T_I is bounded by the number of functions from $\mathrm{cl}(\varphi_{\mathrm{I}})$ to $\mathrm{cl}(\varphi_{\mathrm{I}})$, which is $2^{\mathcal{O}(n \log n)}$. Therefore the number of iterations of the main loop is $2^{\mathcal{O}(n \log n)}$. Each iteration of the main loop consists of two consistency checks for each $t \in T_k$. The μ-consistency of t can be judged by checking whether $t \xrightarrow{\langle a \rangle \varphi} t'$ holds or not for all $(\langle a \rangle \varphi, t') \in \mathrm{cl}(\varphi_{\mathrm{I}}) \times T_k$, and the size of $\mathrm{cl}(\varphi_{\mathrm{I}}) \times T_k$ is also $2^{\mathcal{O}(n \log n)}$. Similarly we can see that the sizes of all sets that are to be iterated over to calculate the relations $\xrightarrow{\langle a \rangle \varphi}$, the sets V_j, the relation LoopSafe, and so on, do not exceed $2^{\mathcal{O}(n \log n)}$. Therefore the time complexity of the whole decision procedure is $2^{\mathcal{O}(n \log n)}$.

4   Finite Binary Tree Model

When we apply the satisfiability checking procedure to XML problems as mentioned in the introduction, we usually pose the following restrictions:

- The set of modality is fixed: $\mathrm{Mod} = \{1, 2, \overline{1}, \overline{2}\}$.
- A model of a formula must form a finite binary tree. More precisely, (M, R, λ) is a finite binary tree model (or FBTM for short) if M is finite and $(M, R(1) \cup R(2))$ is a tree and for all $m \in M$ and $j \in \{1, 2\}$, there is at most one $m_j \in M$ such that $(m, m_j) \in R(j)$.

Definition 7.
(1)   An FBTM $\mathcal{M} = (M, R, \lambda)$ satisfies φ if $\mathcal{M}, r \models \varphi$ where r is the root of the tree M.
(2)   We denote by rootAx the formula $\mu X([1]X \wedge [2]X) \wedge [\overline{1}]\bot \wedge [\overline{2}]\bot$.
(3)   For a formula φ, we denote by φ^{FBT} the formula φ with replacing all occurrences of $\langle a \rangle \psi$ by $\langle a \rangle \top \wedge [a](\psi^{\mathrm{FBT}})$, where a is either 1, 2, $\overline{1}$ or $\overline{2}$.

Note that the formula $\mu X([1]X \wedge [2]X)$ means that all descending branches from the node are finite, $[\overline{1}]\bot \wedge [\overline{2}]\bot$ means the node has no parent, and $\langle a \rangle \top \wedge [a]\psi$ means that all children satisfy ψ and there is at least one child. We can use the decision procedure in Section 3 for deciding whether a formula is satisfied by an FBTM or not:

Proposition 4. *A formula φ is satisfied by some FBTM if and only if $\mathrm{rootAx} \wedge \varphi^{\mathrm{FBT}}$ is satisfiable.*

Proof. The "only if" part is almost immediate. To prove the "if" part, we assume $(M, R, \lambda), m_0 \models \mathrm{rootAx} \wedge \varphi^{\mathrm{FBT}}$. We construct a tree model (T, S, λ') and a function $f : T \to M$. Add the root t_0 of T with $f(t_0) = m_0$. Suppose we have constructed a node t of T. For $j \in \{1, 2\}$, if there is $m' \in M$ with $(f(t), m') \in R(j)$ then pick one such m' and add a node t' to T with $f(t') = m'$. Finally we define $\lambda'(P) = \{t \in T \mid f(t) \in \lambda(P)\}$.

There is no infinite branch in T since if there were such B then $\{f(t) \mid t \in B\}$ would be a counterexample for $\mu X([1]X \wedge [2]X)$ at m_0. By König's lemma, T is finite. To show $(T, S, \lambda') \models \varphi'$, it is enough to show that for all formula ψ and $t \in T$, $(T, S, \lambda'), t \models \psi$ holds if $(M, R, \lambda), f(t) \models \psi^{\mathrm{FBT}}$ holds, which can easily be shown by induction on the construction of ψ.   □

5 Correctness

5.1 Completeness

In this subsection we prove that the decision procedure is complete, that is, $(1) \Longrightarrow (2)$ of Theorem 1. Take a Kripke structure $\mathcal{M} = (M, R, \lambda)$ with $m_I \in M$ such that $m_I \models \varphi_I$. For $m \in M$ we define $g(m) \in T_I$ as follows. For $\varphi \in$ Lean, $g(m)(\varphi) = \omega$ if $m \not\models \varphi$, $g(m)(\varphi) = 0$ if $m \models \varphi$ and $\varphi \notin \bigcup \mathcal{D}_\mu$, and $g(m)(\varphi) = k_\varphi$ if $m \models \varphi$ and $\varphi \in \bigcup \mathcal{D}_\mu$, where $k_\varphi = |\{\mathrm{rank}_D(\psi, m) \mid \psi \in \mathrm{BForm}_D, \mathrm{rank}_D(\psi, m) \le \mathrm{rank}_D(\varphi, m)\}|$ with $\varphi \in D \in \mathcal{D}_\mu$. It is clear from the definition that $g(m) \in T_I$ and that $m \models \varphi \iff g(m) \Vdash \varphi$. We have three propositions. Proposition 5 can be proved using the definition of g and Proposition 6 can be shown using Proposition 5 twice. We leave the details to the reader.

Proposition 5. *Suppose $D \in \mathcal{D}_\mu$, $\varphi \in \mathrm{BForm}_D$, $\psi \in D$, $m \models \varphi$ and $m \models \psi$. Then $\mathrm{rank}_D(\psi, m) < \mathrm{rank}_D(\varphi, m)$ implies $\overline{g(m)}(\psi) < g(m)(\varphi)$.*

Proposition 6. *Suppose $m, m' \in M$, $(m, m') \in R(a)$, $m \models \langle a \rangle \varphi$, $m' \models \varphi$ and $\mathrm{rank}_D(\langle a \rangle \varphi, m) > \mathrm{rank}_D(\varphi, m')$. Then $g(m) \xrightarrow{\langle a \rangle \varphi} g(m')$ holds.*

Proposition 7. *$g(m) \in T_k$ holds for any $m \in M$ and $k < \omega$.*

Proof. We prove the proposition by induction on k. Since it trivially holds for $k = 0$, we assume it holds for k and show it also holds for $k + 1$.

By the induction hypothesis $g(m) \in T_k$. It can easily been checked that $g(m) \in \mathrm{Con}_\Diamond(T_k)$ using Proposition 6.

In order to show $g(m) \in \mathrm{Con}_\mu(T_k)$, take $D \in \mathcal{D}_\mu$. For $\alpha \in \mathrm{On}$ let $S(\alpha, m) = \{\varphi \in \mathrm{Lean} \cap D \mid m \models \varphi$ and $\mathrm{rank}_D(\varphi, m) < \alpha\}$. It is clear by the definition of $g(m)$ that $(g(m), S(\alpha, m)) \in \mathrm{TS}(D, T)$ holds. We can see that for any $\alpha \in \mathrm{On}$, $(g(m), S(\alpha, m)) \in V_{J(D, T_k)}$ by induction on α. The details are left to the reader. Then we take α large enough (for example larger than the cardinality of M), and we have $S(\alpha, m) = D \cap \mathrm{sat}(g(m))$.   □

Using this proposition, completeness is almost clear: since $m_I \models \varphi_I$, we have $g(m_I) \Vdash \varphi_I$ as we have already noticed. From Proposition 7, $g(m_I) \in T_K$ holds.

5.2 Soundness

In this subsection we prove that the decision procedure is sound, that is, $(2) \Longrightarrow (1)$ of Theorem 1.

Take $t_I \in T_K$ with $t_I \Vdash \varphi_I$ in the condition (2) of Theorem 1. Let $L = \{(t, S, D) \mid D \in \mathcal{D}_\mu, (t, S) \in V_J = V_{J(D, T_K)}(D, T_K)\}$ and let $\tau : \mathcal{D}_\mu \to \mathcal{D}_\mu$ be a cyclic permutation on $\mathcal{D}_\mu$. Also pick $D_0 \in \mathcal{D}_\mu$. We construct a finite branching tree (but possibly infinite) $\hat{T}$ with an element of L as the label of each node as follows: Prepare a node n_I labelled by $(t_I, D_0 \cap \mathrm{sat}(t_I), D_0)$ and make it the root of tree $\hat{T}$. This node has not yet been "processed."

Then pick an unprocessed node n whose position is one of the shallowest. Let (t, S, D) be the label of n. For each $\varphi \in \text{PO}(\text{cl}(\varphi_I); \langle\rangle)$ such that $t \Vdash \varphi$, we create a node n' as a child of node n and denote it by $\text{Succ}(n, \varphi)$. If $S \neq \varnothing$, let $j = \min\{j \leq J \mid (t, S) \in V_{j+1}\}$. Since $(t, S) \in \text{Step}_D(V_j, T_K)$, we can take $(t', S') \in V_j$ that satisfies the three conditions in the definition of $\text{Step}_D(V_j, T_K)$ and make (t', S', D) the label of node n'. If $S = \varnothing$, take any $t' \in T_K$ such that $t \xrightarrow{\varphi} t'$ (such one exists since $t \in \text{Con}_\Diamond(T_K)$) and make $(t', \tau(D) \cap \text{sat}(t'), \tau(D))$ the label of node n'. That completes the construction of $\hat{T}$.

We denote the label of node n by $l(n)$. When $l(n) = (t, S, D)$, we denote t, S and D by $t(n)$, $S(n)$ and $D(n)$ respectively. Let $j(n) = \min\{j < \omega \mid (t(n), S(n)) \in V_j(D(n), T_K)\}$. The following proposition is clear from the construction of $\hat{T}$.

Proposition 8. *Suppose $(n_k \mid k < \omega)$ is an infinite path in $\hat{T}$ beginning with n_0 such that for any $k < \omega$ there exists φ such that $n_{k+1} = \text{Succ}(n_k, \varphi)$.*
(1)   For any $k < \omega$ there is $k' > k$ such that $S(n_{k'}) = \varnothing$.
(2)   For any $D \in \mathcal{D}_\mu$ there exists $k < \omega$ such that $D(n_k) = D$ and $S(n_k) = D \cap \text{sat}(t(n_k))$

We define a Kripke structure $\mathcal{M} = (M, R, \lambda)$ as follows: M is the set of all nodes of tree $\hat{T}$. $R(a) = \{(n_1, n_2) \mid n_2 = \text{Succ}(n_1, \langle a \rangle \varphi)$ for some $\langle a \rangle \varphi$ or $n_1 = \text{Succ}(n_2, \langle \overline{a} \rangle \varphi)$ for some $\langle \overline{a} \rangle \varphi\}$. And for $P \in \text{PC}$, $\lambda(P) = \{n \in M \mid t(n) \Vdash P\}$. We will show $\mathcal{M}, n_I \models \varphi_I$. Let $Q = \{(\varphi, n) \in \text{cl}(\varphi_I) \times M \mid t(n) \Vdash \varphi\}$. We define a choice function F as follows: for $\langle a \rangle \varphi$, simply let $F(\langle a \rangle \varphi, n) = \text{Succ}(n, \langle a \rangle \varphi)$. For $F(\varphi_1 \vee \varphi_2, n)$, we first compare $t(n)(\varphi_1)$ and $t(n)(\varphi_2)$ and pick the one with a smaller value. If the values are the same, we pick the one that satisfies $(t(n), S(n)) \Vdash \varphi_j$ if only one of them satisfies it, otherwise we pick either of them. We want to show $Q \subseteq \text{Sat}(\mathcal{M})$ by checking the three conditions in Proposition 2. Conditions (1) and (2) can easily be checked.

In the rest of this subsection we prove condition (3). Suppose, on the contrary, it does not hold. There are $D \in \mathcal{D}_\mu$ and sequence $((\varphi_i, n_i) \mid i < \omega)$ satisfying $\varphi_i \in D$, $n_i \in M$, $(\varphi_i, n_i) \in Q$ and $(\varphi_i, n_i) R_F (\varphi_{i+1}, n_{i+1})$ $(i < \omega)$.

Proposition 9. *Suppose $i \leq j$, $n_i = n_j$, $t = t(n_i) = t(n_j)$. Then $\overline{t}(\varphi_j) \leq \overline{t}(\varphi_i)$ holds. Furthermore if there exists k such that $i < k < j$ and $n_i \neq n_k$, $\overline{t}(\varphi_j) < \overline{t}(\varphi_i)$ holds.*

This proposition can be proved by induction on $j - i$ from the fact that adjacent nodes of the tree has the relation LoopSafe. We leave the details to the reader.

Since the formulas are in normal form the sequence cannot stay at a node infinitely long: for each $i < \omega$ there exists $j > i$ such that $n_j \neq n_i$. By Proposition 9, for each node $n \in M$ there are only finitely many i such that $n_i = n$. Therefore there is a subsequence $(n_{i(k)} \mid k < \omega)$ that forms a postfix of an infinite branch of tree $\hat{T}$. When there are two or more suffixes that are suitable for $i(k)$, we pick the largest suffix so that for all $i > i(k)$ we have $n_i \neq n_{i(k)}$.

Take $k < \omega$ such that $D(n_{i(k)}) = D$ and $S(n_{i(k)}) = D \cap \text{sat}(n_{i(k)})$ by Proposition 8(2). Clearly $(t(n_{i(k)}), S(n_{i(k)})) \Vdash \varphi_{i(k)}$. We can check that for any $k' > k$,

$(t(n_{i(k')}), S(n_{i(k')})) \Vdash \varphi_{i(k')}$ by induction on k'. But this is impossible for k' with $S(n_{i(k')}) = \varnothing$, which exists by Proposition 8(1), since $\varphi_{i(k')} \in D \cap \mathrm{Lean}$. That completes the proof of soundness of the decision procedure.

6   An Experimental Implementation

We have implemented the decision procedure proposed in Section 3. Since the main part of the procedure repeatedly computes subsets of some fixed set, implementation using BDD [1] is suitable.

Table 1 shows the results for a few formulas. The column v is the number of the BDD variables, n is the number of the BDD nodes, and t is the execution (elapsed) time in milliseconds.

The formulas are defined in Figure 1. The intention of lap$n$, for example, is "p_0 holds at the start point and if p_{i-1} holds one can find a point where p_i holds by following arrows labelled by a. But one cannot find a point where p_n holds and connected to a point where p_0 holds with reverse arrows labelled by a." No model satisfies this formula. There is a $D \in \mathcal{D}_\mu$ such that the size of BForm$_D$ is $2n$ in the formula loop$n$, while BForm$_D$ is an empty set in the other formulas for all $D \in \mathcal{D}_\mu$. The formula tree$n$ are satisfiable but they do not have finite models, while the other formulas are unsatisfiable.

The experimental implementation is written in Java with JavaBDD 1.0 [8], which calls BuDDy [2]. Values for the tuning parameters are: nodenum $= 2^{20}$, maxincrease $= 2^{18}$, cachesize $=$ variable, cacheratio $= 1/64.0$. Our experiments have been performed on a Pentium 4 2.4 GHz machine with 512 megabytes of RAM, running Microsoft Windows XP, Sun Java Development Kit version 1.5.0.

Table 1. Experiments

lap$n$				loop$n$				tree$n$			
n	v	b	t	n	v	b	t	n	v	b	t
2	15	1.7×10^3	560	1	21	4.2×10^3	352	4	16	1.4×10^3	313
6	35	1.7×10^4	610	3	49	6.7×10^4	453	8	32	6.9×10^3	343
10	55	2.0×10^5	1020	5	73	5.9×10^5	1047	12	48	9.8×10^4	453
14	75	2.0×10^6	7140	7	113	4.1×10^6	43016	16	64	1.4×10^6	5875

$$\mathtt{lap}n = p_0 \wedge \bigwedge_{i=1}^{n} \nu X ((p_{i-1} \to \mu Y (p_i \vee \langle a \rangle Y)) \wedge [a]X) \wedge \neg \mu Z ((p_n \wedge \mu W (p_0 \vee \langle \overline{a} \rangle W)) \vee \langle a \rangle Z)))$$

$$\alpha_0 = p \vee [a_0](p \vee [\overline{a_0}]x_1) \quad \alpha_n = p \vee [a_n] \mu X_n ([\overline{a_n}]X_{n+1} \vee \alpha_{n-1})$$

$$\beta_0 = \neg p \quad \beta_n = \neg p \wedge \langle a_{n-1} \rangle \beta_{n-1}$$

$$\mathtt{loop}n = \mu X_{n+1}(\alpha_n) \wedge \beta_n$$

$$\mathtt{tree}0 = \top \quad \mathtt{tree}n = \nu Y_n (\langle a_n \rangle (\mu X_n ([\overline{a_n}]X_n) \wedge Y_n \wedge \mathtt{tree}(n-1)))$$

Fig. 1. Definitions of formulas

7   Conclusion and Future Work

In this study we proposed an effective decision procedure for checking satisfiability of formulas in the alternation-free two-way modal μ-calculus and reported an experimental implementation of the procedure.

In this section we describe two directions for future work. The first direction is to expand the range of logics where the method is applicable. A natural extension of the two-way modal μ-calculus is μ-LGF, which extends the decidable subsystem, LGF (Loosely Guarded Fragment), of the first-order predicate logic with fixed point operators. Although a general decision procedure is already known for μ-LGF [4], it is a complex procedure using alternating tree automata as in the two-way modal μ-calculus. Therefore it is meaningful to find an appropriate subsystem of μ-LGF to which the method in this study can be applied. An obvious attempt would be to define the alternation-free μ-LGF, in which we restrict fixed operators not to alternate. It does not work well, however, since in the tree model constructed from the tableau the expansion of (φ, n) with the relation R_F may range over an infinite set for $\varphi \in D$ where D has the least fixed point operator, while its finiteness is a key property of our decision procedure for L_μ^{af}.

The second direction is to apply the decision procedure to verification problems. As mentioned in the introduction, we already employed a decision procedure for judging satisfiability of two-way CTL formulas for analyzing properties of cellular automata. Replacing it with the decision procedure in this study will make more properties verifiable. We also have a plan to apply it to software model checking [20]. Temporal logics such as the two-way modal μ-calculus has enough power to describe properties of a heap when we regard a heap as a Kripke structure with the "points-to" relation as the transition relation. Therefore the decision procedure developed in this study may be useful to build an abstract state space for properties on a heap.

Acknowledgments

The authors thank the anonymous reviewers for their careful reading and helpful comments.

References

[1] R. E. Bryant. Symbolic boolean manipulation with ordered binary-decision diagrams. *ACM Computing Surveys*, 24(3):293–318, 1992.

[2] BuDDy. http://sourceforge.net/projects/buddy.

[3] E. A. Emerson and E. M. Clarke. Using branching-time temporal logic to synthesize synchronization skeletons. *Science of Computer Programming*, 2(3):241–266, 1982.

[4] E. Grädel. Guarded fixed point logics and the monadic theory of countable trees. *Theoretical Computer Science*, 288:129–152, 2002.

[5] E. Grädel, W. Thomas, and T. Wilke, editors. *Automata, Logics, and Infinite Games: A Guide to Current Research*, volume 2500 of *Lecture Notes in Computer Science*. Springer-Verlag, 2002.

[6] M. Hagiya, K. Takahashi, M. Yamamoto, and T. Sato. Analysis of synchronous and asynchronous cellular automata using abstraction by temporal logic. In *FLOPS2004: The Seventh Functional and Logic Programming Symposium*, volume 2998 of *Lecture Notes in Computer Science*, pages 7–21. Springer-Verlag, 2004.

[7] J. G. Henriksen, J. L. Jensen, M. E. Jørgensen, N. Klarlund, R. Paige, T. Rauhe, and A. Sandholm. Mona: Monadic second-order logic in practice. In *Proceedings of the First International Workshop on Tools and Algorithms for Construction and Analysis of Systems*, volume 1019 of *Lecture Notes in Computer Science*, pages 89–110. Springer-Verlag, 1995.

[8] JavaBDD. http://javabdd.sourceforge.net/.

[9] D. Kozen. Results on the propositional μ-calculus. *Theoretical Computer Science*, 27:333–354, 1983.

[10] Z. Manna and P. Wolper. Synthesis of communicating processes from temporal logic specifications. *ACM Transactions on Programming Languages and Systems*, 6(1):68–93, 1984.

[11] K. McMillan. *Symbolic Model Checking*. Kluwer Academic Publ., 1993.

[12] D. Niwinski and I. Walukiewicz. Games for the μ-calculus. *Theoretical Computer Science*, 163(1,2):99–116, 1996.

[13] G. Pan, U. Sattler, and M. Y. Vardi. BDD-based decision procedures for K. In *Proceedings of the Conference on Automated Deduction*, volume 2392 of *Lecture Notes in Artificial Intelligence*, pages 16–30. Springer-Verlag, 2002.

[14] G. Pan and M. Y. Vardi. Optimizing a BDD-based modal solver. In *19th International Conference on Automated Deduction*, volume 2741 of *Lecture Notes in Computer Science*, pages 75–89. Springer-Verlag, 2003.

[15] S. Safra. On the complexity of omega-automata. In *Proceedings of the 29th Annual Symposium on Foundations of Computer Science, FoCS '88*, pages 319–327. IEEE Computer Society Press, 1988.

[16] K. Takahashi and M. Hagiya. Abstraction of graph transformation using temporal formulas. In *Supplemental Volume of the 2003 International Conference on Dependable Systems and Networks (DSN-2003)*, pages W–65 to W–66, 2003.

[17] A. Tozawa. On binary tree logic for XML and its satisifiability test. In *Sixth Workshop on Programming and Programming Language (PPL2004)*, 2004.

[18] A. Tozawa, Y. Tanabe, and M. Hagiya. Experiments on global type checking and termination checking for XML transducer. *IBM Research Report RT0614*, 2005.

[19] M. Y. Vardi. Reasoning about the past with two-way automata. In *Automata, Languages and Programming, 25th International Colloquium, ICALP'98*, volume 1443 of *Lecture Notes in Computer Science*, pages 628–641. Springer-Verlag, 1998.

[20] M. Yamamoto, Y. Tanabe, K. Takahashi, and M. Hagiya. Abstraction of graph transformation systems by temporal logic and its verification. In *IFIP Working Conference on Verified Software: Tools, Techniques, and Experiments*. (to appear).

On the Partial Respects in Which a Real Valued Arithmetic System Can Verify Its Tableaux Consistency

Dan E. Willard*

State University of New York at Albany
dew@cs.albany.edu

Abstract. Gödel's Second Incompleteness Theorem states axiom systems of sufficient strength are unable to verify their own consistency. We will show this theorem does not preclude axiomizations for a computer's floating point arithmetic from recognizing their own consistency, in certain well defined partial respects.

1 Introduction

Let $A(x, y, z)$ and $M(x, y, z)$ denote two 3-way predicates indicating $x + y = z$ and $x * y = z$. An axiom system α will be said to **recognize** successor, addition and multiplication as **Total Functions** iff it can prove (1), (2) and (3).

$$\forall x \, \exists z \; A(x, 1, z) \tag{1}$$

$$\forall x \, \forall y \, \exists z \; A(x, y, z) \tag{2}$$

$$\forall x \, \forall y \, \exists z \; M(x, y, z) \tag{3}$$

It is known that Equations (1) – (3) are related to both generalizations of Gödel's Second Incompleteness Theorem and to its boundary-case exceptions. For instance, Equation (1) will enable the Second Incompleteness Theorem to apply to Hilbert deduction. Also, [36,38] showed that the semantic tableaux version of the Second Incompleteness Theorem generalizes for essentially all axiom systems that can prove the validity of Equations (1) – (3) for integer arithmetic. On the other hand, [35,38,39,41] showed exceptions to the semantic tableaux version of the Second Incompleteness Theorem do exist when an axiom system fails to support Equation (3).

The preceding research naturally raises the question whether or not an analogous phenomenon holds when one changes the venue of application from integer arithmetic to the addition and multiplication operations of a computer's floating point arithmetic set. Throughout this paper, we will use the term *simulated real-arithmetic* to refer to an instruction set that is slightly more general and powerful than the common floating point instructions on a digital computer's hardware. We will prove that simulated real arithmetic is *quite unlike* integer arithmetic — insofar as an axiom system can simultaneously recognize its semantic tableaux consistency and the validity of Equations (1) – (3) *for simulated real arithmetic*.

---
* Supported partially by NSF Grant CCR 99-02726.

B. Beckert (Ed): TABLEAUX 2005, LNAI 3702, pp. 292–306, 2005.

This result is significant because a computer's floating point instruction set has essentially as many practical applications as an integer arithmetic. Moreover, Section 5 will formalize another very unusual aspect of simulated arithmetic. It will be that our partial exceptions to the Second Incompleteness Theorem actually house a novel type of limited Gentzen-style deductive cut rule for simulated real arithmetic, *whose analog for integer-based arithmetics is infeasible.* Thus, our main contribution will be the demonstration that simulated real-valued arithmetic *in several respects* supports more *robust forms* of tableaux-like exceptions to Gödel's Second Incompleteness Theorem than an integer arithmetic can feasibly do.

2   Literature Survey

Gödel's 1931 paper on incompleteness [10] contained two major results, called the "First" and "Second" Incompleteness Theorems. The former theorem (after Rosser [23] strengthened it) established that it was impossible to construct a consistent r.e. axiomatic extension of Peano Arithmetic that could prove or disprove every logical sentence.

Gödel's "Second Incompleteness Theorem" showed that neither Peano Arithmetic nor any extension of it can confirm its own consistency. During the last 75 years, there has been a substantial effort to determine under exactly what circumstances, Gödel's Second Incompleteness Theorem is applicable to axiom systems weaker than Peano Arithmetic. In our summary of this topic, it is it is useful to classify an *"integer" arithmetic* axiom system α as being of:

a. **Type-M** iff α can formally prove the assertions in Equations (1) – (3), indicating that "integer" multiplication, addition and successor are "total" functions.

b. **Type-A** iff α can prove that integer addition and successor satisfy Equations (1) and (2), but α cannot formally prove integer multiplication satisfies (3).

c. **Type-S** iff α can prove successor is a total function (as specified by Equation (1)), but it is uncertain about whether integer arithmetic satisfies (2) and (3).

d. **Type-NS** iff α is unable to prove any of Equations (1), (2) or (3).

The research into generalizations of the Incompleteness Theorem for weak axiom systems began with Tarski-Mostowski-Robinson [30]'s observation that a Type-M axiom system, which they called Q, had the property that Q (but none of its proper subsets) satisfied a condition called "essential undecidability". Also, Bezboruah-Shepherdson [6] proved a version of the Second Incompleteness Theorem for Q which stated that Q was unable to prove some types of particularized theorems affirming its own Hilbert consistency. The subsequent research can be roughly summarized as follows:

I. Pudlák (1985) made the seminal observation [20] that the Bezboruah-Shepherdson [6] result could be generalized for all extensions of Q, all encodings of its Hilbert-proof predicate, and also for all its local extensions in the range of a specified "Definable Cut". Wilkie-Paris [34] further noted that the axiom system $I\Sigma_0 + Exp$ is unable to prove the Hilbert consistency of Q. Nelson [17] observed that many of the properties of the Tarski-Mostowski-Robinson [30] axiom system Q can be extended to also hold for Type-S axiom systems. In 1994, Solovay [26] had combined these three formalisms to obtain the following result:

Theorem 1. (Solovay's modification [26] of Pudlák's Theorem 2.3 from [20] using the additional mathematical methodologies of Nelson and Wilkie-Paris [17,34].) *No consistent Type-S axiom system, formalizing integer addition and multiplication as 3-way relations A(x,y,z) and M(x,y,z), can prove the non-existence of a Hilbert-proof of* $0 = 1$ *from itself.*

II. An open question that Paris-Wilkie [19] posed in 1981 was whether an analog of Theorem 1 held for cut-free methods of deduction, such as semantic tableaux. Adamowicz-Zbierski [1,4] answered this question in the positive for the axiom system $I\Sigma_0 + \Omega_1$, and Willard extended their result for $I\Sigma_0$ and yet weaker derivatives of this system. The main result of [36,38] was to construct a Π_1 sentence V such that every consistent Type-M system $\alpha \supset V$ is necessarily unable to prove a theorem affirming its semantic tableaux consistency. (Here and elsewhere in this paper, our definition of semantic tableaux is similar to that in Fitting's textbook [9].)

A more detailed literature review about the properties of the Second Incompleteness Theorem is provided in [41]. Since the sundry generalizations of the Second Incompleteness Theorem are quite powerful, we have *cautiously* called our partial evasions of the Second Incompleteness Effect *"boundary-case exceptions"*. These exceptions have consisted mostly of Type-NS systems [37] that can verify their Hilbert consistency and Type-A systems [37,39] that can verify their tableaux-oriented consistencies. These results are near-maximal because the Type-NS evasions closely interface against Item I's generalization of the Second Incompleteness Theorem for Type-S systems, and similarly Item II's variant of the Second Incompleteness Theorem for Type-M systems complements [35,37,39]'s Type-A evasions.

Our new results in the current article will differ from the prior work by *changing the venue of application* from integer arithmetic to a computer's simulated real-valued arithmetic instruction set, so that our tableaux-style partial exceptions to the Second Incompleteness Theorem shall be able to recognize *both* addition and multiplication as total functions for simulated real arithmetic. Aside from directly eschewing Item II's integer-based form of the Second Incompleteness Theorem (given above), this approach will also help clarify the nature of numerical analysis's logical foundations.

Let us now assume α is an axiom system and D a deduction method. The pair (α, D) will be called an **Introspectively Unified Logic** iff

A) one of α's formal theorems will state that the deduction method D, applied to the axiom system α, will produce a consistent set of theorems,

B) and the axiom system α is in fact consistent.

Also, an axiom system α will be called **Self-Justifying** iff there exists some frequently employed deduction method D, such as perhaps the tableaux, Hilbert, Herbrand or sequent-calculus formalisms, where (α, D) is an introspectively unified logic.

Kleene's Fixed Point Theorem implies that every r.e. axiom system α can be easily extended into a broader system α^* which satisfies Part-A of the definition for introspective unification. For a fixed deduction method D, Kleene essentially proposed [14] to let α^* contain all α's axioms plus one additional axiom sentence stating :

+   There is no proof (using deduction method D) of a prototypical absurdity-sentence (such as for example "0=1") from the union of the axiom system α with *this* sentence "+"(looking at itself).

Kleene explained how to apply the Fixed Point Theorem to encode a self-referencing statement, similar to the axiom (above). However, he pointed out that the catch is that α^* may be inconsistent even while its added axiom formally asserts α^* 's consistency.

For this reason, Kleene, Rogers and Jeroslow [13,14,22] each warned their readers that most axiom systems, similar to α^*, were useless on account of their inconsistency, *although they were technically well-defined.* In the notation used in this section, the difficulty is that (α^*, D) may satisfy Part-A but not the equally important Part-B of the definition of "introspective unification".

This problem arises in settings considerably more general than Gödel's original paradigm, where α was an extension of Peano Arithmetic. Thus, [1–6, 8, 12, 18–21, 24–26, 29–31, 34, 36, 38, 40, 42] discuss a variety of generalizations of the incompleteness theorem where a similar paradigm applies to weak axiom systems.

Our interest in this topic began [35,37,39] with the observation that there are certain well-defined settings where it is feasible to construct introspectively unified logics (α , D) using the Kleene-like "I am consistent" axioms, analogous to the sentence $+$. In particular, boundary-case exceptions to the Second Incompleteness Theorem do exist when either 1) α is a Type-NS axiom system and D represents Hilbert deduction, or when 2) α is a Type-A integer arithmetic and D represents some cut-free deductive method, such as a tableaux formalism. We again remind the reader that these two results are near maximal on account of [20,26,36,38]'s Type-S and Type-M generalizations of the Second Incompleteness Effect, summarized earlier by Items I and II of this section.

Prior to our research, the proof-theoretic literature has sought to partially evade the Second Incompleteness Theorem largely by studying what perhaps can be called the localized (α, D, φ)−consistency statements. In particular, let $\lceil \Psi \rceil$ denote a formula Ψ's Gödel number, and $\mathrm{Prf}_\alpha^D (t, p)$ denote p is a proof of the theorem t from the axiom system α using the deduction method D . Let us say α can recognize its **localized (α, D, φ)−consistency property** if α can prove the theorem:

$$\forall p \quad \{ \quad \varphi(p) \quad \rightarrow \quad \neg \, \mathrm{Prf}_\alpha^D \, (\lceil 0 = 1 \rceil , \, p) \quad \} \tag{4}$$

In order to analyze Equation (4)'s meaning, let **Sequence**(φ) denote the formula:

$$\varphi(0) \quad \text{AND} \quad \forall x \quad \varphi(x) \quad \rightarrow \quad \varphi(x + 1) \tag{5}$$

Let **Induction**(φ) denote (6)'s statement about the validity of the principle of induction:

$$\text{Sequence}(\varphi) \quad \rightarrow \quad \forall x \, \varphi(x) \tag{6}$$

It is clear that any conventional axiom system that can simultaneously prove Equations (4) – (6) will be able to combine these results to infer its own global consistency (and thereby evade the Second Incompleteness Theorem). To avoid these effect, [11,12,15,16,17,18,19,20,21,25,26,27,31,32,33,34] have considered localized

(α, D, φ)−consistency statements that come tantalizingly close to such an evasion via their axiom systems retaining an ability to prove Equations (4) and (5) *but not the also-needed* Equation (6). The strongest results about this subject matter were derived by Pudlák [20], who showed that if α represents essentially any axiom system of finite cardinality and D denotes either the Herbrand or tableaux deduction method, then there exists formulae $\varphi(x)$ where α has the tantalizing property that it can prove the first two of the preceding three wffs.

It is difficult to make detailed comparisons between our self-justifying formalisms, which employ the Statement $+$'s "I am consistent axiom", with the literature about localized consistency because each approach has its own separate objectives, advantages and difficulties when it seeks to evade the Second Incompleteness Theorem. Thus, self-justifying systems are unable to recognize *integer* multiplication as a total function, whereas the comparable sacrifice in the literature concerning localized consistency is that sentences, similar to Equation (4), have their meaning diluted when α is unable to prove Induction(φ). In both cases, it is of theoretic interest to categorize the maximal types of partial evasions of the Second Incompleteness Theorem that are feasible.

3   General Self-justification Framework

Define a mapping $F(a_1, a_2...a_j)$ to be a **Non-Growth** function iff it satisfies the invariant $F(a_1, a_2, ...a_j) \leq Maximum(a_1, a_2, ...a_j)$. Six examples of non-growth functions are *Integer Subtraction* (where $x - y$ is defined to equal zero when $x \leq y$), *Integer Division* (where $x \div y$ is defined to equal x when $y = 0$, and it equals $\lfloor x/y \rfloor$ otherwise), $Maximum(x, y)$, $Logarithm(x)$, $Root(x, y) = \lceil x^{1/y} \rceil$ and $Count(x, j)$ designating the number of "1" bits among x's rightmost j bits. These function are called the **Grounding Functions**.

The term **U-Grounding Function** will refer to a set of eight operations, which includes the six non-growth "Grounding" functions plus the *growth operations* of addition and $Double(x) = x + x$. For simplicity, we will use a U-Grounding language. Its notation is technically unnecessary — because a system that uses Equation (7) (which specifies addition is a total function) along with our first six *non-growth* Grounding operations would have properties similar to the U-Grounding language.

$$\forall x \, \forall y \, \exists z \quad x = z - y \tag{7}$$

However, the U-Grounded notation makes it easier to present our results.

Our formal analogs for Logic's Π_n and Σ_m sentences in the U-Grounding language will be called Π_n^* and Σ_m^*. Here, a *term* t is defined to be a constant, variable or a U-Grounding function symbol (whose input arguments are recursively defined terms). Also, the quantifiers in the wffs $\forall \, v \leq t \; \Psi(v)$ and $\exists \, v \leq t \; \Psi(v)$ are called *bounded integer quantifiers*. Any formula in the U-Grounding language, all of whose quantifiers are bounded, will be called Δ_0^*. Following conventional notation, every Δ_0^* formula will be considered to satisfy the " Π_0^* " and " Σ_0^* " conditions. For $n \geq 1$, a formula Υ shall be called Π_n^* iff it is written in the form $\forall v_1 \, \forall v_2 \, ... \, \forall v_k \quad \Phi$, where Φ is Σ_{n-1}^* . Likewise, Υ is called Σ_n^* iff it is written in the form $\exists v_1 \, \exists v_2 \, ... \, \exists v_k \quad \Phi$, where Φ is Π_{n-1}^* . Henceforth, we will also use the following definitions:

1. A **Level(n) Definition** of an axiom system α's tableaux consistency is the declaration that there exists no Π_n^* sentence Υ supporting simultaneous semantic tableaux proofs from α of both Υ and its negation.

2. A **Level(0−) Definition** of a system α's tableaux consistency is the statement that there exists no proof of 0=1 from α.

We will also sometimes use a notation referring to Π_n^- and Σ_n^- sentences. These sentences will have the same definitions as Π_n^* and Σ_n^* except that they will use a language employing the Grounding (instead of U-Grounding) functions.

In essence, both notations are useful. The U-Grounding based notation is preferable when an axiom system uses Equation (7)'s axiom, declaring addition is a total function. It then leads to more succinct proofs. On the other hand, the slightly more cumbersome Π_n^- and Σ_n^- notation is essential for axiom systems that do not recognize addition as a total function. In essence, we will employ the U-Grounding notation in Sections 3 and 5 and the Grounding language notation in Section 4.

Theorem 2. (A summary of the central results published in [35,37,39]) *Let A denote an arbitrary consistent axiom system employing the U-Grounding language. Then it is possible to construct two further consistent axiom systems, called IS(A) and IS-1(A) [35,37,39], having the following properties.*

1. *Both IS(A) and IS-1(A) will retain a capacity to prove all the Π_1^* theorems of A ,*
2. *Both IS(A) and IS-1(A) will recognize addition as a total function,*
3. *IS(A) will retain an ability to recognize its own Level(0-) consistency, and IS-1(A) will have a stronger ability to also recognize its Level(1) consistency.*

The underlying technique used in [35,37,39] was that IS(A) and IS-1(A) axiom systems would apply an essentially 2-part formalism to achieve the above conditions. Their first part would achieve the conditions (1) and (2) by essentially simulating the actions of the axiom system A in a relatively straightforward manner. These axiom systems will satisfy the property (3) by using an analog of a Kleene-like "I am consistent" axiom, similar to the statement $+$ (defined in Section 2).

The main challenge in [35,37,39] was to demonstrate that the axiom systems IS(A) and IS-1(A) did not become inconsistent on account of the presence of their final axiom — which declared their own consistencies. In the nomenclature of Section 2, the challenge was to show that the presence of the axiom $+$ did not cause either IS(A) or IS-1(A) to violate Part-B of Section 2's definition of introspective unification.

In particular, let us recall that Section 2 had noted that if an axiom system α is sufficiently strong and we add to it a Kleene-like "I am consistent" axiom, then the resulting new system α^* will be rendered inconsistent because of the presence of a Gödel-like diagonalization proof. The main theorems in [37,39] showed that this difficulty did not pertain to either IS(A) or the stronger IS-1(A) system because these formalisms treated integer multiplication as a 3-way relation, rather than as a total function.

Our articles [35,37,39] of course, raised almost as many as questions as they had settled because an axiom system that fails to recognize multiplication as total is inherently weak. *How useful is it to employ such axiom systems which gain a knowledge about their own consistency only by sacrificing the common axiom that multiplication is a total function?* This question is especially pressing because [36,38] demonstrated it is impossible to construct a consistent axiom system that simultaneously recognizes integer-multiplication as a total function and its own consistency — even when one uses the quite weak Level(0-) definition of tableaux consistency !

Within such a context, we will now show that there is an analog of a computer's floating point instruction set, *called simulated-real arithmetic,* where the axiom systems IS(A) and IS-1(A) from [37,39] can prove that addition, multiplication, subtraction and division *among simulated real numbers* are total functions.

Definition 1. We will use two formalizations of an integer, called NN and IPN, in this paper. The first definition "NN" will represent the set of non-negative integers. (This is the usual definition employed when investigating the Incompleteness Theorem.) Our alternate definition "IPN" will regard an integer as being any *positive or negative* whole number, as well as reserve a special symbol for representing ∞ .

Definition 2. The symbol F will denote a 1-1 function F that maps the set of NN integers onto IPN integers. In particular, let $\text{Even}(x)$ denote a function that equals 1 if x is an even number and -1 if x is odd. Let $\text{Half}(x)$ denote the integer-truncated quantity $\lfloor x \div 2 \rfloor$. Then $F(x)$ is defined by the convention that:

$$F(x) = \text{Even}(x) \cdot \text{Half}(x) \text{ when } x \neq 1 \quad \text{AND} \quad F(1) = \infty$$

Lower case letters x will henceforth denote NN-integers, and upper case letters X will denote IPN integers.

Definition 3. Let i denote an arbitrary indexing integer. Then the i-th **Simulated Real-Number** will be defined to be an ordered pair (M_i, E_i) where M_i is an IPN number storing the mantissa, and E_i is a second IPN integer storing the exponent. The bold-face symbol $\mathbf{R}_i$ will denote this simulated real-number. It is defined as follows:

1. If $E_i \neq \infty$ and $0 \neq M_i \neq \infty$ then $\mathbf{R}_i = M_i \cdot 2^{-\lfloor Log_2(|M_i|) \rfloor} \cdot 2^{E_i}$.

2. If $E_i = \infty$ and M_i is a power of 2 , then $\mathbf{R}_i$ represents the real number 0 written in a binary notation with $\text{Log}(M_i)$ digits to the right of the decimal point.

3. Otherwise, $\mathbf{R}_i$ will represent an "overflow" symbol following division by zero.

Important Comment: We will often use the NN notation (m_i, e_i) to denote a simulated real number, instead of IPN. In this case, Definition 2's function F will map (m_i, e_i) onto its IPN counterpart (M_i, E_i) , so as to calculate $\mathbf{R}_i$'s value.

Definition 4. Let $\mathbf{R}_1$, $\mathbf{R}_2$ and $\mathbf{R}_3$ denote three simulated real-numbers that are encoded by the respective ordered pairs (m_1, e_1), (m_2, e_2) and (m_3, e_3) when written in the NN-integer notation. Let S denote one of the four arithmetic symbols of $+$, $\times$, $-$ or $\div$. Then $\Theta_S(m_1, e_1, m_2, e_2, m_3, e_3)$ will henceforth denote a formula which states that the two real numbers $\mathbf{R}_1$ and $\mathbf{R}_2$, combined under the arithmetic operation of S , will produce a third simulated real of $\mathbf{R}_3$. More precisely, $\Theta_S(m_1, e_1, m_2, e_2, m_3, e_3)$'s definition will employ the usual computerized floating point hardware rounding convention that $\mathbf{R}_3$'s computed mantissa has a bit-length L equal to the maximum of the lengths for the two input mantissas of $\mathbf{R}_1$ and $\mathbf{R}_2$. It will thus specify $\mathbf{R}_3$ represents the *closest approximation of the combination* of $\mathbf{R}_1$ and $\mathbf{R}_2$ under the operation S that is feasible for a number which has a mantissa-length of L and which uses Definition 3's formula for defining $\mathbf{R}_3$'s value.

Lemma 1. *For each of the four cases where S denotes one of the symbols of $+$, $\times$, $-$ or $\div$, the predicate $\Theta_S(m_1, e_1, m_2, e_2, m_3, e_3)$ has a Δ_0^* encoding.*

Proof Sketch: For the cases where S denotes the $+$ or $-$ symbols, it is easy to encode $\Theta_S(m_1, e_1, m_2, e_2, m_3, e_3)$ as a Δ_0^* formula. It is also reasonably routine to encode $\Theta_S(m_1, e_1, m_2, e_2, m_3, e_3)$ as a Σ_1^* formula when S denotes the $\times$ or $\div$ symbol. A much more meticulous analysis in a longer version of this paper will show that this formal encoding for multiplication and division can, in fact, be compressed into a more ideally terse Δ_0^* form. The intuition behind this compression is for one to break the initial mantissas m_1 and m_2 into two substrings of essentially equal length, each, so that they are sufficiently short so that the Σ_1^* formula's *unbounded existential quantifiers* can then be made bounded. $\square$.

Lemma 2. *Let S again denote one of the four arithmetic symbols of $+$, $\times$, $-$ or $\div$ under simulated real arithmetic. In each of these four cases using the NN-integer notation, the statement that S is a total function can be encoded as a Π_1^* sentence.*

Proof: It is clear that if the statement of Lemma 2 was changed so that the totality of S was expressed as a Π_2^* (rather than Π_1^*) sentence, then Lemma 2 would be an immediate consequence of Lemma 1. This is because in each of the four cases where S denotes the symbol of $+$, $\times$, $-$ or $\div$, Equation (8) is a formal statement declaring the totality of the operation of S :

$$\forall m_1 \, \forall e_1 \, \forall m_2 \, \forall e_2 \, \exists m_3 \, \exists e_3 \quad \Theta_S(m_1, e_1, m_2, e_2, m_3, e_3) \tag{8}$$

In order to construct a Π_1^* sentence that is equivalent to (8) under the Standard Model, we will use the fact that in each of the four cases where S denotes the symbol of $+$, $\times$, $-$ or $\div$, a 6-tuple will satisfy $\Theta_S(m_1, e_1, m_2, e_2, m_3, e_3)$ only when:

$$** \quad m_3 \leq \text{Double}(\text{Max}(m_1, m_2)) \quad \text{AND} \quad e_3 \leq \text{Double}(\text{Double}(\text{Max}(e_1, e_2)))$$

Let t denote the term of $\text{Double}(\text{Double}(\text{Max}(m_1, m_2, e_1, e_2)))$. Item $**$ then implies that Equation (9) is a Π_1^* sentence that logically implies the validity of Equation (8) under the Standard Model of the Natural Numbers:

$$\forall m_1 \, \forall e_1 \, \forall m_2 \, \forall e_2 \, \exists m_3 \leq t \, \exists e_3 \leq t \quad \Theta_S(m_1, e_1, m_2, e_2, m_3, e_3) \tag{9}$$

Definition 5. In a context where $\mathbf{R}$ denotes a simulated real number, the term Expand $(\mathbf{R})$ will denote a second simulated real whose value is identical to that of $\mathbf{R}$ except that the mantissa for Expand$(\mathbf{R})$ will have one extra bit of precision. Thus, if $\mathbf{R}_1 = (m_1, e_1)$ and $\mathbf{R}_2 = (m_2, e_2)$ then Equation (10) is a Δ_0^* formula, denoted formally as $\Theta^*(m_1, e_1, m_2, e_2)$, which indicates that $\mathbf{R}_2 = \text{Expand}(\mathbf{R}_1)$.

$$m_2 = \text{Double}(m_1) - \text{Count}(m_1, 1) \quad \text{AND} \quad e_1 = e_2 \tag{10}$$

Lemma 3. *There exists a Π_1^* formula indicating that Expand($\mathbf{R}$) is a total function.*

Proof: Obvious because the Equation (11) is the needed formula:

$$\forall\, m_1 \quad \forall\, e_1 \quad \exists\, m_2 \leq \text{Double}(m_1) \quad \exists\, e_2 \leq e_1 \qquad \Theta^*(m_1, e_1, m_2, e_2) \qquad (11)$$

Theorem 3. (An initial result that will be made much more robust in Section 5): *For every consistent axiom system A employing the U-Grounding functions, there exists a consistent axiom system α that can* 1) *prove all A's Π_1^* theorems,* 2) *recognize integer-addition as a total function,* 3) *confirm that simulated real arithmetic operations of addition, multiplication, subtraction, division and Expand are each total functions, and* 4) *recognize its own Level(1) semantic tableaux consistency.*

Proof: Let Ψ_1, Ψ_2, Ψ_3, Ψ_4 and Ψ_5 denote the five Π_1^* sentences, defined by Lemmas 2 and 3 that indicate the operations of addition, multiplication, subtraction, division and Expand are each total functions. Let A' denote the union of the axiom system A with these five added Π_1^* sentences of Ψ_1, Ψ_2, Ψ_3, Ψ_4 and Ψ_5. It follows from the combination of Theorem 2 and Lemmas 2 and 3 that the axiom system IS-1(A') satisfies Theorem 3's four requirements   □

The remainder of this article will have two parts. Section 4 will introduce two new versions of the Second Incompleteness Theorem, whose negative results *tightly contrast against* Theorem 3's positive result. Section 5 will introduce a stronger form of Theorem 3, which replaces the notion of Level(1) tableaux consistency with the significantly stronger construct of "$Tier(1)$" tableaux consistency. It will also contemplate potential applications to numerical analysis. It will arrive at a pleasing mathematical observation about the foundational nature of Gentzen-style deductive cuts.

4   Generalizations of the Second Incompleteness Theorem

The term **"Base"** substructure will refer to the starting axiom system A which Theorem 2 used to construct its more elaborate self-justifying axiom systems of IS(A) and IS-1(A). It is apparent that Theorem 3's partial exception to the Second Incompleteness Theorem possesses some type of non-trivial quality because its axiom systems of IS(A) and IS-1(A) have a capacity to prove all the Π_1^* theorems of Peano Arithmetic (PA) whenever $A \supset PA$. Since the Second Incompleteness Theorem precludes a self-justifying axiom system α from becoming excessively strong, it is thus helpful to remind ourselves about the exact properties which α *is precluded from possessing*. For example, Items I and II from the literature survey chapter indicated Type-S axiomizations *of integer arithmetic* cannot recognize their own Hilbert consistency [20,26], and similarly Type-M axiomizations *of integer arithmetic* cannot recognize their Level(0-) tableaux consistency [38]. Also in another recent article [40], we established Type-A axiomizations *of integer arithmetic* are unable to recognize their own Level(2) tableaux consistency. In this section, we will develop two variants of the Second Incompleteness Theorem that apply uniquely to *simulated real arithmetics*.

Definition 6. Let **AddComp**$(m_1, e_1, m_2, e_2, m_3, e_3)$ denote a Δ_0^- *predicate formula* indicating that (m_3, e_3) represents the additive sum of the simulated real numbers of (m_1, e_1) and (m_2, e_2) that satisfies the usual computerized hardware-rounding convention that the mantissa m_3 has a bit-length equal to the maximum of the bit-lengths for m_1 and m_2. Also, let **LongMult**$(m_1, e_1, m_2, e_2, m_3, e_3)$ denote a Δ_0^- predicate formula indicating that (m_3, e_3) represents the *untruncated* multiplicative product of the simulated real numbers of (m_1, e_1) and (m_2, e_2). (This is the variant of multiplication where the floating-point truncate-and-round operation is now absent.)

The first of our two new versions of the Second Incompleteness Theorem is Theorem 4. It will show that Theorem 3's partial evasion of the Second Incompleteness Theorem has no analog when one tries to generalize it from semantic tableaux styled definitions of consistency to stronger definitions of consistency focused around Hilbert deduction. Theorem 4's result is surprising because semantic tableaux and Hilbert deduction produce the same set of theorems in first order logic. The intuitive reason that Theorems 3 and 4 provide such a contrasting pair of results is because these two formalisms produce sharply different proof lengths.

Theorem 4. *There exists a Π_1^- sentence W (with a relatively simple structure) such that every consistent Grounding-language based axiom system $\alpha \supset W$ is **unable to both** prove its own consistency (under the Hilbert-styled method of deduction) and to also prove Equation (12)'s Π_2^- sentence.* (The latter states AddComp formalizes a total function among simulated real numbers).

$$\forall m_1 \ \forall e_1 \ \forall m_2 \ \forall e_2 \ \exists m_3 \ \exists e_3 \quad \text{AddComp}(m_1, e_1, m_2, e_2, m_3, e_3) \qquad (12)$$

Proof Sketch: We shall employ the Theorem 1 by Pudlák-Solovay, which had indicated that *no axiom system can simultaneously* recognize successor as a total function and prove a theorem affirming its own Hilbert consistency. Our current discussion cannot assume *apriori* that Theorem 4's axiom system α will recognize successor as a total function — because α is employing a Grounding (rather than U-Grounding) language *without growth functions*. Thus to prove Theorem 4, we will need to show that α can infer from Equation (12) that successor is a total function.

Let us first recall that our notation convention has the upper case symbol X replacing the lower case symbol x when we are viewing an integer object in the IPN rather than the NN notation. Also, let $\bar{C}$ be a representation for the constant " $+1$ " written in IPN notation. Using this notation to simplify (12) and letting $\bar{C}$ substitute for m_1 and m_2, it follows that the axiom system α can infer (13) from (12).

$$\forall E_1 \ \exists M_3 \ \exists E_3 \quad \text{AddComp}(\bar{C}, E_1, \bar{C}, E_1, M_3, E_3) \qquad (13)$$

A longer version of this paper will use Equation (13) to derive that if our threshold Π_1^* sentence W contains more than some tiny amount of strength then any $\alpha \supset W$ which proves Equation (12)'s validity shall be able to infer from (13) that successor is a total function among the set of NN integers. Also assuming that W contains more than a tiny amount of initial strength, α will then satisfy Theorem 1's hypothesis. Thus α cannot verify its own Hilbert consistency. $\square$

Theorem 5. *There exists some particular Π_1^- sentence W such that no consistent axiom system $\alpha \supset W$ can simultaneously prove its own Level(0-) tableaux consistency and the validity of Equation (14)'s Π_2^- sentence.* (Equation (14) indicates that

LongMult formalizes a total function among simulated real numbers).

$$\forall \, m_1 \; \forall \, e_1 \; \forall \, m_2 \; \forall \, e_2 \; \exists \, m_3 \; \exists \, e_3 \quad \text{LongMult}(m_1, e_1, m_2, e_2, m_3, e_3) \qquad (14)$$

Proof Sketch: Our justification of Theorem 5 will be brief because its proof is similar to Theorem 4's proof. After one changes the underlying notation from NN to IPN, it is apparent that LongMult($M_1, E_1, M_2, E_2, M_3, E_3$) can be satisfied only when $M_3 = M_1 \cdot M_2$. Hence, (14) implies multiplication is a total function among IPN integers. Also, there clearly exists a Π_1^- sentence Φ that can infer multiplication is a total function *among NN integers* from its totality *among IPN integers*.

Let us next recall that our Tableaux-2000 conference paper [36] (and/or its journal counterpart [38]) constructed a Π_1^- sentence V where no Type-M *integer-based* system $\alpha \supset V$ can prove its Level(0-) tableaux consistency. The analog of this result for simulated real arithmetics thus follows by setting $W = V \cup \Phi$. $\qquad \square$

On the Significance of Theorems 4 and 5: We suspect that Theorem 4 is the more important of the two restrictions upon the capacities of simulated real systems. One reason Theorem 5 is less significant is because digital computers simply do not process floating point numbers with the LongMult instruction set. (If they did, then the bit-length of a floating point number would double each time a multiplication was performed: It would exceed the number of atoms in the universe after roughly 100 consecutive multiplications.) It is thus reassuring that numerical analysts, starting with Leibnitz and Newton, have demonstrated most mathematical computations are *only infinitesimally changed* when one one throws away their low-order bits.

On the other hand, Theorem 4 formalizes a *more serious* constraint on self-justifying systems. It signals a Gentzen-style deductive cut rule will usher in the force of the Second Incompleteness Theorem. Thus, we are left to inquire: *In what respects, beyond those delineated by Theorem 3, can Theorem 4's actual generalization of the Second Incompleteness Theorem be evaded? Also how robust may such evasions be?* The next section will contain our chief results. It will bring this article to a surprising conclusion.

5   On the Two Surprising Facets of Hybrid Deduction Formalisms

Let us say a U-Grounding sentence belongs to the Tier(k) class if it is either Π_k^* or Σ_k^*. Also, let H denote a sequence of ordered pairs (t_1, p_1), (t_2, p_2), ... (t_n, p_n), where p_i is a semantic tableaux proof of the theorem t_i. In a context where $\Re$ designates some class of sentences, such as possibly Tier(k), Π_k^* or Σ_k^*, define H to be a Tab$-\Re$ proof of a theorem T from the axiom system α iff $T = t_n$ and also:

1. Each axiom in p_i's proof is either one of $t_1, t_2, ...t_{i-1}$ or comes from α.
2. Each of the "intermediate results" $t_1, t_2, ...t_{n-1}$ lie in the pre-specified class $\Re$.

Thus, Tab$-\Re$ deduction is stronger than classic tableaux by allowing for a type of Gentzen-like deductive cut rule for sentences that belong to the intermediate class, that is formalized by $\Re$. From Gentzen's Cut Elimination Theorem [28] it is known that the set of the theorems that can be proven by a Tab$-\Re$ proof *are the same* as those that can be derived by a purely cut-free proof formalism, similar to semantic tableaux. However, the proof length efficiencies of these two alternate approaches *can be quite different.*

If $\Re$ denotes the universal class of all possible sentences then Tab$-\Re$ deduction will thus be essentially equivalent to the Hilbert-style deductive method in its proof-efficiency. On the other hand, if $\Re$ denotes a class of sentences that is a strict subset of the universal class, such as for example Tier(k), then Tab$-\Re$ will be a methodology lying properly between classic tableaux and Hilbert deduction.

This terminology was introduced in [40]. It proved Type-A axiom systems are unable to verify their self-consistency under either Tab$-\Sigma_2^*$ or Tab$-\Pi_2^*$ deduction.

The acronym Tab-1 will refer to a version of Tab$-\Re$ deduction with $\Re = $ Tier(1). An expanded version of our Tableaux-2002 paper [39], which will soon appear in the *Journal of Symbolic Logic* [41], will explore a stronger variant of [39]'s IS-1(A) formalism, called $\text{IS}_D(A)$. Their difference is that $\text{IS}_D(A)$ can recognize its consistency under Tab-1 deduction, while IS-1(A) recognized its consistency under only Section 3's Level(1) definition. (Both formalisms prove all A 's Π_1^* theorems.) The last page of our Tableaux-2002 paper actually defined $\text{IS}_D(A)$ and called it "IS-1*(A)". The longer JSL article [41] will finish this topic by formally proving $\text{IS}_D(A)$ is consistent.

Theorem 6. (A stronger version of Theorem 3) *For every consistent axiom system A employing the U-Grounding functions, there exists a consistent axiom system α that can* 1) *prove all A's Π_1^* theorems,* 2) *recognize integer-addition as a total function,* 3) *confirm that the five basic operations of simulated real arithmetic are total functions,* **and** 4) **recognize its own consistency under Tab-1 deduction.**

Proof Sketch: The proof for Theorem 6 is identical to Theorem 3's proof, except that all references to the prior paper [39]'s IS-1 system should be replaced by the forthcoming paper [41]'s IS_D system to achieve Feature (4)'s added functionality.   □

Important Comment. Since Theorem 6 is a direct generalization of Theorem 3, it is tempting to suspect both results have similar implications. *To the contrary,* we will now show Theorem 6 has a special added *"Tier(1)$^\oplus$ floating-point"* property, which is actually central for achieving our main goals and purposes.

Notation. The symbol $\langle m, e \rangle$ will denote the simulated real number with mantissa m and exponent e. Also $|\langle m, e \rangle|^J$ will denote the quantity begotten by taking the absolute value of this simulated real number and raising it to the J−th power. Assuming that $J \neq 0$, $\langle n, f \rangle \geq 1$ and that Length(m) denotes m's bit-length, the formal expression of $\langle m, e \rangle \ll_L^J \langle n, e \rangle$ will have the following meaning:

1. Length(m) $\leq$ Length(n) $+ L$ and if $J \geq 1$ then $|\langle n, f \rangle|^J \geq |\langle m, e \rangle| \geq 1$.
2. Length(m) $\leq$ Length(n) $+ L$ and if $J \leq -1$ then $|\langle n, f \rangle|^J \leq |\langle m, e \rangle| \leq 1$.

Assuming that **R** is a term that specifies the value of a simulated real number whose value is greater than 1, we will also use the preceding notation to define **Bounded Real Quantifiers** of the form $\exists \langle m, e \rangle \ll_L^J$ **R** and $\forall \langle m, e \rangle \ll_L^J$ **R** . The term **Bounded Integer Quantifiers** will refer to the expressions of the form $\forall x \leq t$ and $\exists x \leq t$ that were defined in Section 3. A wff will be called $\Delta_0^\oplus$ if it is built in any arbitrary manner out of these four forms of bounded quantifiers, together with the usual U-Grounding function symbols, the equality and greater-than predicates and the standard Boolean connectives. If Ψ is a $\Delta_0^{\oplus}$ formula then the formal expressions of $\forall v_1 \ \forall v_2 \ ... \ \forall v_k \ \Psi$ and $\exists v_1 \ \exists v_2 \ ... \ \exists v_k \ \Psi$ will be called $\Pi_1^{\oplus}$ and $\Sigma_1^{\oplus}$ formulae. Also a sentence will be called Tier(1)$^{\oplus}$ when it is either $\Pi_1^{\oplus}$ or $\Sigma_1^{\oplus}$.

Theorem 7. *There exists a computable function F that maps each Tier$(1)^{\oplus}$ formula ϕ onto a Tier(1) formula Φ such that ϕ and Φ are logically equivalent to each other.*

The theory of LinH functions [43] can be used to prove Theorem 7. This proof is rather lengthy. It will thus appear in a longer version of this article.

On the Surprising Combined Implications of Theorems 3, 4, 6 and 7: Theorem 4 had showed that Theorem 3's evasion of the Second Incompleteness Effect for tableaux deduction does not generalize for Hilbert deduction. In this context, Theorem 6 had offered an alternate hybrid approach, called Tab-1 deduction, for at least partially extending the prior tableaux results. One subsequent question is therefore to ask: *How much more robust is Tab-1 deduction than classic tableaux ?*

In order to answer this question, let Φ denote a Tier(1) sentence of either the form of $\forall v_1 \ \forall v_2 \ .. \ \forall v_k \ \Psi$ or $\exists v_1 \ \exists v_2 \ .. \ \exists v_k \ \Psi$, and let us call Ψ the Δ_0^* **Stem** of Φ. Since by definition, all the quantifiers in the stem are bounded integer quantifiers of the form $\forall x \leq t$ or $\exists x \leq t$, it follows that the size of the integer x is greatly limited by the size of the input integers $p_1, p_2 ... p_j$ for the term t. In particular since the only growth functions available in our U-Grounding language are addition and doubling, the size of x may exceed $\text{Max}(p_1, p_2 ... p_j)$ by only a scalar constant of k — whose value depends on the number of function symbols in t . The latter fact shows that the $\text{IS}_D(A)$ self-justifying formalism can actually prove only a *limited range* of theorems about *"integer"* arithmetics because $\text{IS}_D(A)$ is able to include only the Π_1^* theorems of A as its generating set of axioms, and it can apply a Gentzen-style deductive cut rule *only to the quite limited Tier(1) class of formulae.*

However if one *shifts the venue of application* from integer arithmetic to *real-value arithmetic*, then the range of permissible uses of a Gentzen-style deductive cut rule becomes much more robust. This is because Theorem 7 stated that every Tier$(1)^{\oplus}$ formula ϕ can be translated into an equivalent Tier(1) formula Φ. Moreover, the bounded real quantifiers in a Tier$(1)^{\oplus}$ formula of the form $\exists \ \langle m, e \rangle \ \ll_L^J \ \mathbf{R}$ and $\forall \ \langle m, e \rangle \ \ll_L^J \ \mathbf{R}$ allow $\langle m, e \rangle$'s real number to attain values essentially as large as $\mathbf{R}^J$ (quite unlike the narrower range of values associated with integer-based bounded quantifiers). Thus *from the perspective of real valued arithmetic*, the $\text{IS}_D(A)$ axiom has a good deal of flexibility, since any logically valid $\Pi_1^{\oplus}$ sentence can essentially be made into an axiom (by choosing an initial broad enough base axiom system of A) , *and the Tab-1 deductive cut rule can then be applied to any Tier$(1)^{\oplus}$ formula.*

Hence, while self-justifying systems may be viewed as primarily a *theoretical-only* device for exploring *integer arithmetics*, their significance is broader for real-valued arithmetics *because of the permissible use of bounded real quantifiers.*

To reinforce this point, we observe that all the major algorithms in numerical analysis [7] can be simulated by proofs under simulated real arithmetic. The basic purpose of numerical analysis is to produce sequences of real numbers $\mathbf{R}_1$, $\mathbf{R}_2$, $\mathbf{R}_3$, ... that converge upon target answers with a decreasing error rate $\epsilon_1, \epsilon_2, \epsilon_3, \ldots$. By choosing to use mantissas with sufficiently large lengths, it is easy to formalize such algorithms under simulated real arithmetic. Each of the axiom systems of IS(A), IS-1(A) and $\text{IS}_D(A)$ can accomplish this task. Their distinction is that they house the desired simulations under three increasingly broad definitions of tableaux self-consistency.

The seminal nature of Gödel's Second Incompleteness Theorem is, of course, impossible to overestimate. It has had many fascinating generalizations. Our exceptions to Gödel's Second Incompleteness Theorem illustrate that there are certain *limited-but-tangible* respects where a formalism *not using the integer version of multiplication,* can possess a *partial-but-not-full* knowledge of its own consistency.

Acknowledgments. I thank the referees for their careful reading of this article and for several suggestions on how to improve the presentation.

References

1. Z. Adamowicz, "Herbrand Consistency and Bounded Arithmetic", *Fundamenta Mathematica* 171 (2002) pp. 279-292.
2. Z. Adamowicz and T. Bigorajska, "Existentially Closed Structures and Gödel's Second Incompleteness Theorem", *Journal of Symbolic Logic* 66 (2001), pp. 349-356.
3. Z. Adamowicz and P. Zbierski, *The Logic of Mathematics*, John Wiley and Sons, 1997.
4. Z. Adamowicz and P. Zbierski, "On Herbrand consistency in weak theories", *Archive for Mathematical Logic* 40 (2001) pp. 399-413.
5. T. Arai, "Derivability Conditions on Rosser's Proof Predicates", *Notre Dame Journal on Formal Logic* 31 (1990) pp. 487-497.
6. A. Bezboruah and J. Shepherdson, "Gödel's Second Incompleteness Theorem for Q", *Journal of Symb Logic* 41 (1976) 503-512.
7. R. Burden and J. Faires, *Numerical Methods*, Brookes-Cole (2003)
8. S. Buss and A. Ignjatovic, "Unprovability of Consistency Statements in Fragments of Bounded Arithmetic", *Annals of Pure and Applied Logic* 74 (1995) pp. 221-244.
9. M. Fitting, *First Order Logic & Automated Theorem Proving*, SpringerVerlag 1996.
10. K. Gödel, " Über formal unentscheidbare Sätse der Principia Mathematica und Verwandte Systeme I", *Monatshefte für Math. Phys.* 37 (1931) pp. 349-360.
11. P. Hájek, "On Interpretability in Set Theory" Part I in *Com. Math. Univ. Carol* 12 (1971) pp. 73–79 and Part II in *Comm. Math. Univ. Carol* 13 (1972) pp 445-455.
12. P. Hájek and P. Pudlák, *Metamathematics of First Order Arithmetic*, Springer Verlag 1991.
13. R. Jeroslow, "Consistency Statements in Formalal Mathematics.", *Fundamenta Mathemtaica* 72 (1971) pp. 17-40.
14. S. Kleene, "On the Notation of Ordinal Numbers", *Journal Symb Logic* 3 (1938), 150-156.
15. J. Krajícek, "A Note on Proofs of Falsehood", *Archive for Math Logic* 26 (1987) 169-176.
16. G. Kreisel and G. Takeuti, "Formally Self-Referential Propositions for Cut-Free Classical Analysis", *Dissertationes Mathematicae* 118 (1974) pp. 1–55
17. E. Nelson, *Predicative Arithmetic*, Princeton Math Notes, 1986.
18. J. Paris and C. Dimitracopoulos, "A Note on the Undefinability of Cuts", *Journal of Symbolic Logic* 48 (1983) pp. 564-569.
19. J. Paris and A. Wilkie, "Δ_0 Sets and Induction", *1981 Jadswin Conf Proc*, pp. 237-248.
20. P. Pudlák, "Cuts, Consistency Statements and Interpretations", *Journal of Symbolic Logic* 50 (1985) 423-442.

21. P. Pudlák, "On the Lengths of Proofs of Consistency", in *Collegium Logicum: 1996 Annals of the Kurt Gödel Society* (Volume 2), Springer-Wien-NewYork, pp 65-86.

22. H. Rogers Jr. , *Recursive Functions and Effective Compatibility,* McGrawHill 1967.

23. J. Rosser, "Extensions of Theorems by Gödel and Church", *Jour Symb Logic* 1 (1936) 87-91.

24. S. Salehi, Ph D thesis for the Polish Academy of Sciences, Oct. 2001.

25. C. Smoryński, "Non-standard Models and Related Developments in the Work of Harvey Friedman", in *Harvey Friedman's Research in Foundations of Math.*, North Holland 1985.

26. R. Solovay, Private telephone communications during 1994 describing Solovay's generalization of one of Pudlák's theorems [20], using the additional formalisms of Nelson and Wilkie-Paris [17,34]. Solovay never published this result (called Theorem 1 in Section 2) or his other observations that several logicians [12,17,18,19,20,21,34] have attributed to his private communications. The Appendix A of [37] offers a 4-page summary of Solovay's theorem.

27. V. Švejdar, "Modal Analysis of Generalized Rosser Sentences", *Journal of Symbolic Logic* 48 (1983) 986-999.

28. G. Takeuti, *Proof Theory,* Studies in Logic Volume 81, North Holland, 1987.

29. G. Takeuti, "Gödel Sentences of Bounded Arithmetic", *Journal of Symbolic Logic* 65 (2000) pp. 1338-1346

30. A. Tarski, A. Mostowski and R. M. Robinson, *Undecidable Theories*, North Holland, 1953.

31. A. Visser, "The Unprovability of Small Inconsistency", *Archive for Mathematical Logic* 32 (1993) pp. 275-298.

32. A. Visser, "Faith and Falsity", to appear in *Annals of Pure and Applied Logic.*

33. P. Vopěnka and P. Hájek, "Existence of a Generalized Semantic Model of Gödel-Bernays Set Theory", *Bulletin de l'Academie Polonaise des Sciences* 12 (1973) pp.1079-1086.

34. A. Wilkie and J. Paris, "On the Scheme of Induction for Bounded Arithmetic", *Annals of Pure and Applied Logic* (35) 1987, 261-302

35. D. Willard, "Self-Verifying Axiom Systems", *Proceedings of the Third Kurt Gödel Colloquium* (1993), Springer-Verlag LNCS#713, pp. 325-336.

36. D. Willard, "The Semantic Tableaux Version of the Second Incompleteness Theorem Extends Almost to Robinson's Arithmetic Q", *Proceedings of the Tableaux 2000 Conference*, SpringerVerlag LNAI#1847, pp. 415-430.

37. D. Willard, "Self-Verifying Systems, the Incompleteness Theorem and the Tangibility Principle", in *Journal of Symbolic Logic* 66 (2001) pp. 536-596.

38. D. Willard, "How to Extend The Semantic Tableaux And Cut-Free Versions of the Second Incompleteness Theorem Almost to Robinson's Arithmetic Q", in *Journal of Symbolic Logic* 67 (2002) pp. 465–496.

39. D. Willard, "Some Exceptions for the Semantic Tableaux Version of the Second Incompleteness Theorem", *Proceedings of the Tableaux 2002 Conference*, SpringerVerlag LNAI#2381, pp. 281–297.

40. D. Willard, "A Version of the Second Incompleteness Theorem For Axiom Systems that Recognize Addition But Not Multiplication as a Total Function", *First Order Logic Revisited, (Year 2003 Proceedings FOL-75 Conference),* Logos Verlag (Berlin) 2004, pp. 337–368.

41. D. Willard, "An Exploration of the Partial Respects in which an Axiom System Recognizing Solely Addition as a Total Function Can Verify Its Own Consistency", to appear in the *Journal of Symbolic Logic.*

42. D. Willard, "A New Variant of Hilbert Styled Generalization of the Second Incompleteness Theorem and Some Exceptions to It", to appear in the *Annals of Pure and Applied Logic.*

43. C. Wrathall, "Rudimentary Predicates and Relative Computation", *Siam Journal on Computing* 7 (1978), 194-209

Pdk: The System and Its Language

Marta Cialdea Mayer, Carla Limongelli,
Andrea Orlandini, and Valentina Poggioni

Università di Roma Tre, Dipartimento di Informatica e Automazione
{cialdea, limongel, orlandin, poggioni}@dia.uniroma3.it

Abstract. This paper presents the planning system Pdk (Planning with
Domain Knowledge), based on the translation of planning problems into
Linear Time Logic theories, in such a way that finding solution plans
is reduced to model search. The model search mechanism is based on
temporal tableaux. The planning language accepted by the system allows
one to specify extra problem dependent information, that can be of help
both in reducing the search space and finding plans of better quality.

1 Introduction

Artificial Intelligence planning is concerned with the automatic synthesis of sequences of actions that, when executed, lead from a given initial state to a state
satisfying some logics have been used in different perspectives in this context,
either in the deductive view [10,12], or by model checking [5], or as a language
to add control knowledge to a specialised planner [1,6]. The system presented in
this work is a planner fully based on Linear Time Logic (LTL), planning problem
into LTL in such a way that planning is reduced to model search. This corresponds to the idea of *executing temporal logics*, where executing a formula means
building a model of it [7] (the application to planning was already sketchily proposed in [2]).

The system Pdk (Planning with Domain Knowledge) accepts the description
of a planning problem, given in the planning language PDDL-K (Planning Domain Description Language with control Knowledge), that offers the possibility
of specifying extra problem dependent and control information, that can be of
help both in reducing the search space and finding plans of better quality. The
specification of the planning problem is translated into a set S of LTL formulae
in such a way that any model of S represents a plan solving the problem. The
reduction consists of a "linear encoding" [3], that recalls the classical Situation
Calculus representation of planning [11].

The small example that follows illustrates how a planning problem is encoded
into LTL. In the *initial state*, a robot is in room A, its *goal* is to be in room
B and the only *actions* it can perform is going from a location to another,
specifically either from A to B (go_A_B), or from B to A (go_B_A). This problem
is represented by the following set of LTL formulae:

B. Beckert (Ed): TABLEAUX 2005, LNAI 3702, pp. 307–311, 2005.

$$S = \{ \; at\_A \wedge \neg at\_B, \Diamond at\_B,$$
$$\Box(go\_A\_B \rightarrow at\_A), \; \Box(go\_B\_A \rightarrow at\_B),$$
$$\Box(\bigcirc at\_A \equiv (at\_A \wedge \neg go\_A\_B) \vee go\_B\_A),$$
$$\Box(\bigcirc at\_B \equiv (at\_B \wedge \neg go\_B\_A) \vee go\_A\_B)\}$$

The first two formulae represent the initial state and goal of the problem (sometimes in the future the goal will be achieved). The two formulae in the second line above represent the preconditions for the executability of the two actions (the agent can move from a place only if it is there). The last two formulae are the LTL reformulation of Reiter's "successor state axioms" [11]. A model of S is the sequence of states $s_0, s_1, \dots$ such that the only true atoms at s_0 are at_A and go_A_B, and the only true atom at s_i, for $i > 0$, is at_B. The plan corresponding to such a model is the sequence of actions $\langle go\_A\_B \rangle$.

Pdk, which is available at http://pdk.dia.uniroma3.it/, is implemented in Objective Caml (http://caml.inria.fr/) and its model search mechanism relies on the system ptl, an efficient implementation of proof search in LTL by means of tableaux techniques, developed in C by G. Janssen at Eindhoven University [9].

A first version of the planner was already presented in [4]. The experiments with that system raised the need for tools supporting the domain expert in the specification task, especially when stating domain specific and control knowledge. In fact, the domain expert charged of the description of the problem cannot be assumed to be a logician, and it is therefore important to have a simple, compact and easy-to-use planning language, similar to the special purpose formalisms widely adopted in the planning community.

The new release of the planner, briefly described in this paper, beyond being a more efficient implementation, accepts problem specifications written in a planning language where heuristic knowledge can not only be provided explicitly as a set of LTL formulae, but also as instances of a set of specific control schemata. Moreover, the system provides meta-level tools that can be of help in debugging the problem specification (synthetically described at the end of the next section). Such tools can be developed thanks to the fact that the whole planning domain is encoded into a logical theory.

2 The Planning Language and Off-Line Checks

The planning language PDDL-K can be viewed as an extension of the ADL-subset of classical PDDL [8]. A full description of the language and its semantics, given by means of the translation into LTL, can be found at the above cited URL, together with a set of sample domains. Due to space restrictions, only a brief overview can be given here. The PDDL-K description of a planning problem follows a multi-sorted first-order syntax. However, as usual in classical planning, domains are finite and fixed, and (typed) quantification is actually an abbreviation for propositional formulae. For instance, $\forall x : t \, A(x)$ stands for $A(c_1) \wedge \dots \wedge A(c_n)$, where $c_1, \dots, c_n$ are all the constants of type t.

The specification of the problem contains the definition of the signature: type declarations, with associated sets of constants, and predicate declarations. In PDDL-K predicates are distinguished into *static* predicates (denoting properties that do not change over time) and *fluents*. The value of static predicates is declared as a *background theory*, consisting of a set of classical formulae. The background theory is completed with respect to static atoms, according to the closed world assumption: what is not classically derivable from the background theory is false. Therefore, each ground static atom is either true or false at each time point. All literals built up from fluents which are derivable from the background theory are also added to its completion. After completion, the background theory consists of a set of literals. It is used to simplify the encoding of the planning problem: each static atom is replaced by either True or False, and the same happens with fluent literals occurring in the completed theory. The theory is also used to filter out operator instances, by elimination of those actions whose preconditions or effects are inconsistent.

The other fundamental declarations in the description of the problem are the specification of the initial state, goal and operators. Since, like in classical planning, the knowledge about the initial state is assumed to be complete, the initial state is completed wrt fluents. The description of the initial state may also contain temporal operators. Non-classical formulae, that will just be added to the encoding, may be used to specify intermediate goals that must be achieved or actions that must be performed. For instance, the description of the initial state can contain $\Diamond(\exists x : location\ go(x, bank) \land \Diamond \exists x : location\ go(x, post\_office))$: during plan execution, the agent must sooner or later go (from some place x) to the bank and afterwards to the post office.

Control knowledge can be specified either explicitly, as a set of temporal formulae, or according to a predefined set of *control schemata* in the description of operators. Such schemata allow one, for instance, to specify that a given action should be performed only if given conditions hold, or that it should be performed whenever possible (see the example given below).

A peculiarity of the language is that it is possible to refer to actions in formulae. Beyond fluents and static predicates, in fact, atoms can be build out of predicates denoting actions. This allows one to specify for instance that the execution of a given action must be either accompanied or followed by others, or that the effect of performing two actions together is different from what can be obtained by executing each of them separately.

In the specification of control information, the possibility to refer to the goal to be achieved, as well as to what is true in the initial state can be of use, especially to eliminate useless actions. To this aim, the syntax of PDDL-K formulae is extended by means of the unary modal operators *goal* (which can dominate only literals) and *initially* (which can dominate only atoms): *goal* ℓ means that ℓ is a goal of the problem, *initially* p means that p is true in the initial state.

In order to give the flavour of the language, let us consider the simple domain where a robotic agent has to move some objects from/to different locations by means of a briefcase. Constants are either of sort *location* or *object*. The fluents are *atRobby*(x) (the robot is at location x), *in*(x) (the object x is in the robot's briefcase), *at*(x, y) (the object x is at location y). The complete specification of this class of problems can be found at the above cited URL. Here, we only show how the *take* action can be specified using some of the available control schemata.

```
(:action take :parameters (x - object y - location)
              :precondition (atRobby y) (at x y)
              :effect (in x) (not (at x y))
              :only-if (not (goal (at x y))) (initially (at x y))
              :s-asap )
```

The parameters of the action are the object x and the place y where both the object and the robot are (this is specified in the action precondition). After the execution of the action, the object is no more at y, but inside the briefcase (the effect of the action). The :only-if field specifies additional (heuristic) restrictions on the action execution: take x from place y only if x is not already at its destination and x is at y in the initial state. The :s-asap field (*strong* asap) requires all instances of the action to be performed "as soon as possible". If $O = \neg goal(at(x, y)) \wedge initially(at(x, y))$ and $P = atRobby(y) \wedge at(x, y)$, the encoding of the two control fields above (:only-if and :s-asap) are (the propositional counterparts of) the following formulae, respectively:

$$\Box \forall x : object \, \forall y : location(take(x, y) \rightarrow O)$$
$$\Box \forall x : object \, \forall y : location(P \wedge O \rightarrow take(x, y))$$

Following the guidelines given by action oriented control schemata, the specification task becomes easier, but surely not free from errors. It often happens in fact that the addition of too strong control requirements makes the problem unsolvable. In order to help the domain expert, the system Pdk provides some meta-level tools that can be of help in debugging the specification. The system allows one to check for the consistency of the theory encoding the kernel problem (i.e. excluding control knowledge and goal), and to check whether such a theory becomes inconsistent with the addition of control knowledge. Sometimes, moreover, it happens that, although the theory is consistent, the goal cannot be achieved because some important actions can never be performed, since they would violate some too strong control requirement. The system allows one to check actions, one by one, to verify their executability.

3   Conclusion

Some experiments have been carried out in order to verify whether the expressive power of the language compensates for the loss in time efficiency that often derives from the use of general logical procedures (especially if based on exhaustive search) and/or allows the planner to find solutions of better quality. The performances of Pdk have been compared with some planners that showed best performances in the last International Planning Competitions (IPC 2002 and IPC 2004). The experiments – whose results cannot be reported in detail here for space reasons – show that Pdk outperforms optimal planners, such as SAT-PLAN_2004. With respect to suboptimal planners, such as LPG, the relative performances of the planners in terms of execution times vary in dependence of the planning domains: in some cases LPG is much faster, in other domains the execution times are comparable and sometimes Pdk is faster than LPG. In nearly all experimented cases, however, Pdk finds shorter plans.

Comparing Pdk with other planners that allow for the employment of heuristic knowledge, such as TLPLAN [1], involves, beyond efficiency, also expressiveness considerations. Although TLPLAN is faster than Pdk, the specification of heuristic knowledge is often quite long and cumbersome. One of the reasons is that all statements must be made in terms of fluents. The fact that, in PDDL-K, actions are represented by atoms allows for much simpler control formulae, that, in turn, can often be reduced to the addition of even simpler control fields in operators specifications. We believe that this is not a secondary issue, since the statement of correct control knowledge is often a subtle and difficult task.

References

1. F. Bacchus and F. Kabanza. Using temporal logics to express search control knowledge for planning. *Artificial Intelligence*, 116:123–191, 2000.
2. H. Barringer, M. Fisher, D. Gabbay, and A. Hunter. Meta-reasoning in executable temporal logic. In *Proc. of KR'91*, pages 40–49, 1991.
3. S. Cerrito and M. Cialdea Mayer. Using linear temporal logic to model and solve planning problems. In *Proc. of AIMSA'98*, pages 141–152, 1998.
4. M. Cialdea Mayer, A. Orlandini, G. Balestreri, and C. Limongelli. A planner fully based on linear time logic. In *Proc. of AIPS-2000*, pages 347–354, 2000.
5. A. Cimatti, E. Giunchiglia, F. Giunchiglia, and P. Traverso. Planning via model checking: a decision procedure for $\mathcal{AR}$. In *Proc. ECP-97*, pages 130–142, 1997.
6. P. Doherty and J. Kvarnström. TALplanner: A temporal logic based planner. *AI Magazine*, 22:95–102, 2001.
7. M. Fisher and R. Owens. An introduction to executable modal and temporal logics. In *Executable modal and temporal logics (Proc. of the IJCAI'93 Workshop)*, pages 1–20, 1995.
8. M. Fox and D. Long. PDDL2.1: An extension to PDDL for expressing temporal planning domains. *Journal of Artificial Intelligence Research*, 20:61–124, 2003.
9. G. L. J. M. Janssen. *Logics for Digital Circuit Verification. Theory, Algorithms and Applications*. CIP-DATA Library Technische Universiteit Eindhoven, 1999.
10. J. Koehler and R. Treinen. Constraint deduction in an interval-based temporal logic. In *Executable Modal and Temporal Logics, (Proc. of the IJCAI'93 Workshop)*, pages 103–117, 1995.
11. R. Reiter. *Knowledge in Action: logical foundations for describing and implementing dynamical systems*. MIT Press, 2001.
12. B. Stephan and S. Biundo. Deduction based refinement planning. In *Proc. of AIPS-96*, pages 213–220, 1996.

Proof Output and Transformation for Disconnection Tableaux[*]

Philipp Correll and Gernot Stenz

Institut für Informatik,
Technische Universität München,
D-85748 Garching, Germany
{correll, stenzg}@in.tum.de

Abstract. For applications of first-order automated theorem provers in a wider verification context it is essential to provide a means of presenting and checking automatically found proofs. In this paper we present a new method of transforming disconnection tableau proofs found by the prover system DCTP into a series of resolution inferences representing a resolution refutation of the proof problem.

1 Introduction

Recent years have seen an increased interest in the use of formal methods, not only in scientific environments but also in industrial applications. First-order theorem provers have also been used in such contexts but usually just as proof oracles returning *yes* or *no* wrt. a given proof task. However, the *pervasive* use of formal methods requires a more detailed form of output. Not only are automated theorem provers required to return proofs but those proofs should be presented in a form that allows them to be mechanically checked. Here, we present an algorithm that transforms disconnection tableau proofs found by the automated theorem prover DCTP into ordinary resolution proofs. Resolution has been chosen because available proof checking tools are mostly built for checking resolution proofs; also switching to a different (preferably simpler) paradigm for proof checking simplifies the task of finding errors in the original proof method.

This paper is organised as follows: First, the disconnection calculus is briefly revisited. Then in Section 3 the transformation procedure is outlined. Section 4 explains how certain calculus refinements can be integrated into the transformation procedure. Finally, the concluding Section 5 describes possible extensions of the proof transformation procedure.

2 The Disconnection Tableau Calculus

Detailed descriptions of the disconnection calculus can be found in [2,3,5]. Here it is sufficient to recapture the most significant aspects of this calculus. For

---

[*] This work was partially funded by the German Federal Ministry of Education, Science, Research and Technology (BMBF) in the framework of the Verisoft project under grant 01 IS C38. The responsibility for this article lies with the author(s).

B. Beckert (Ed): TABLEAUX 2005, LNAI 3702, pp. 312–317, 2005.

this we use the standard terminology for clausal tableaux. The disconnection tableau calculus consists of a single complex inference rule, the so-called *linking rule*. This linking rule is used to systematically expand a clausal tableau with suitable instances of the input clauses. The guidance for the application of the linking rule is provided by *links*, pairs of potentially complementary subgoals on the tableau path of the open subgoal to be extended. All variables on the tableau are considered to be implicitly universally quantified. In order to provide an initial set of links to start the tableau construction, an arbitrary *initial active path* is selected, marking one subgoal from each of the input clauses. A branch in a disconnection tableau is called ∀-*closed* if it contains two literals K and L such that $K\sigma$ is the complement of $L\sigma$ where σ is a substitution identifying all variables in the tableau. The disconnection tableau calculus is refutation complete for first-order clause logic, provided the linking rule is applied in a fair manner. Additionally, it is sufficient to use each link only once on each branch. If the set of links on an open branch is exhausted, the literals on the branch describe a model for the clause set.

3   The Basic Procedure

The aim of the basic proof transformation procedure is to transform its input, a closed disconnection tableau for a clause set S, into a resolution refutation of S.

Prover Output. In our system, proof search and proof output and transformation are kept separate. While searching for a proof, DCTP simply outputs its actions in a sequential manner. For the most basic proof output, there are three types of output statements:

$$\texttt{initial}(clause\_id, literals).$$
$$\texttt{instance}(clause\_id, parent\_literal, initial\_id, literals).$$
$$\texttt{close}(leaf\_literal, branch\_literal).$$

First, the `initial` statements list the set of input clauses. The `instance` statements denote which instances of the input clauses have been placed where on the tableau. Finally, the `close` statements denote where branch closures have occurred and which branch literals were used for these closures. The parameters of all statements contain the necessary information to enable the complete reconstruction of the proof tableau. Except for the `initial` statements, DCTP will only output ground clauses. In accordance with the closure condition given in Section 2, all variables are replaced by the special constant $.

The Transformation Procedure. The main problem encountered here is that proof tableaux represent dual clause forms (a tableau branch is the conjunction of its literals, while the entire tableau is the disjunction of its branches) whereas resolution proofs are clause sets (conjunctions of the input clauses, the resolvents and the empty clause). Also, the disconnection calculus is (in its pure form) purely analytic; clauses are not recombined as in a resolution step. So proof tableaux cannot be translated into resolution derivations in a straightforward manner. We solve this problem by replacing the actual tableau structure by the series of branch closures indicated by the close statements. Basically each one of these closures is translated into a resolution step. To obtain the proper input clauses for these resolution steps we have introduced the concept of the *annotated resolvent*. An annotated resolvent r_c is assigned to each tableau clause c and is initialised to the set of literals of c. The resolution steps generated by our procedure operate solely on these annotated resolvents, not on the original tableau clauses. Every time a subgoal $L \in c$ is solved, r_c is updated to the resolvent resulting from the solution of L. As the number of unresolved literals is never greater than the number of unsolved tableau subgoals, running the procedure on a closed tableau will finally produce the empty clause.

```
    Input  : a clause set S and a closed disconnection tableau T for S
    Output: a resolution proof for S
 1  foreach clause s ∈ S do print(s)
 2  foreach clause c ∈ T do setAnnotatedRes(c,c)
 3  repeat
 4      L ← getOpenLeaf(T)
 5      ∼ L ← getComplementOnBranch(L T)
 6      r_new ← resolve(getAnnotatedRes(L) getAnnotatedRes(∼ L))
 7      markLiteralSolved(L)
 8      currentClause ← getClause(L)
 9      while (allNodesMarkedInClause(currentClause)) do
10          markLiteralSolved(parentNodeOf(currentClause))
11          currentClause ← getClause(parentNodeOf(currentClause))
12      end
13      setAnnotatedRes(currentClause,r_new)
14      printResolutionStep(r_new)
15  until (r_new = □)
```

Fig. 1. The basic transformation algorithm

The transformation algorithm starts by reconstructing the proof tableau based on the prover output. Then, the input clause set is written to output and the annotated resolvents of the tableau clauses are initialised. This is followed by an iteration over all branch closures. For each closure, the annotated resolvents of the clauses containing the closing branch literals L and $\sim L$ are used as the input for a resolution step producing a new resolvent r_{new}. The leaf L is marked as solved and r_{new} is passed on as the new annotated resolvent to the clause containing the next unsolved subgoal in the tableau in clause currentClause. Finally the resolution step producing r_{new} is written to output.

It should be noted that in the algorithm of Figure 1, the function getOpenLeaf and the variable currentClause identify different concepts. getOpenLeaf always returns the next *leaf node* in a depth-first manner, while currentClause holds the lowermost clause of the leaf node's branch containing an unsolved subgoal. Also, the depth-first processing order of function `getOpenLeaf` is a basic requirement for securing the correctness of the algorithm depicted in Figure 1.

The transformation output consists of two kinds of statements: the `initial` statements, displaying the input clause set and the `resolvent` statements specifying the resolution inferences.

$$\texttt{initial}(\mathit{clause\_id}, \mathit{literals}).$$
$$\vdots$$
$$\texttt{resolvent}(\mathit{resolvent\_id}, [\mathit{annotatedResolvent1\_id}, \mathit{annotatedResolvent2\_id}]).$$

An example showing how the transformation procedure works is presented in Figure 2. The example depicts a sequence of two transformation steps of a partial tableau containing the clauses $c_1, \ldots, c_3$. The boxed tableau nodes indicate the complementary literals to be resolved. An annotated resolvent r_{c_i} highlighted in grey is assigned to each clause c_i. In subfigure (a) the annotated resolvents have been initialised to the respective clause literals. The annotated resolvents r_{c_2} and r_{c_3} are resolved and the resulting resolvent replaces r_{c_3} in subfigure (b). There, the second literal L_1 of c_3 is solved by a closure with $\sim L_1$ of c_1. Again, the annotated resolvents are used for a resolution step. Now all literals of c_3 have been solved (and with them $\sim L_3$ of c_2), so the resulting resolvent $L_2 \vee L_4$ is passed up to replace r_{c_2} in subfigure (c).

Factorisation Effects. Analytic tableau calculi do not allow certain optimisations across proof clauses such as factorisation. This means that in a tableau proof a subgoal has to be solved several times, whereas in the corresponding resolution refutation the same subgoal has to be resolved only once. Although a partial remedy to this problem will be discussed in Section 4, it is still possible that the necessity arises to repeatedly solve the same subgoal. These multiple solutions can be ignored in the resolution context. Therefore, a closure statement will produce a resolution step only if the closing leaf $K \in d$ is contained in the current annotated resolvent r_d of d. This may also lead to a premature termination of the transformation algorithm when the empty clause is produced before all closure statements have been processed.

Tautological Resolvents. The disconnection calculus can delete tautological clauses, yet it must allow linking steps producing the tableau equivalent of tautological resolvents in order to remain complete [5]. Therefore, the transformed tableau proof can contain tautological resolvents. The resolution proofs thus generated are sound but not optimal. We have extended the transformation algorithm by a markup technique allowing the removal of tautological resolvents and their descendents.

4   Extensions

The algorithm of Figure 1 covers only the most basic version of the disconnection tableau calculus. The calculus of DCTP makes use of many advanced

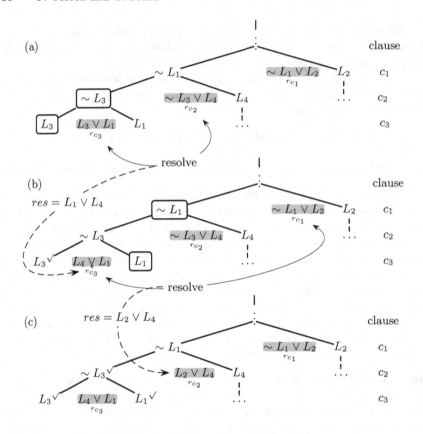

Fig. 2. Basic transformation of branch closures

refinements, so the transformation procedure has to be enhanced in order to handle these techniques.

Pruning. The *proof confluence* property of the disconnection calculus allows a straightforward depth-first proof search, but it also causes the finished tableau proof to contain a large number of redundant clauses. *Tableau pruning* [4], also known as *justification based backtracking*, is a vital redundancy elimination technique that allows to remove unnecessary clauses or subproofs from tableau proofs. A clause c is considered redundant if one of its subgoals $L \in c$ can be solved without using L for a closure. As other subgoals $K_1, \ldots, K_n$ from c might have been solved before L was selected for solving, the subproofs of the K_i become redundant, too. This means that all resolvents produced in the subproofs of the K_i and their generating resolution steps must be removed from the transformed proof. In order to do this, the entire hierarchy of resolvents (and the corresponding resolution steps) must be cached and can only be written to output after the entire tableau proof transformation has been completed.

Folding Up. The DCTP prover system makes heavy use of the *folding-up* [1] of solved subgoals. Folding-up is used as a way to both simulate factorisation in a tableau framework and to integrate atomic cuts in a controlled fashion. This also indicates how folding-up is transformed into resolution steps: below the c-point of the subgoal folded up, an atomic cut is inserted and becomes part of the transformation output.

Unit Theory. In order to obtain a more efficient proof procedure, it is vital to make use of *unit clauses* via *unit subsumption* and *unit simplification* [5]. As subsumption is only used to guide the proof search it does not affect the proof transformation. *Unit simplification* can be directly transformed into unit resolution.

5   Conclusion and Future Work

In this paper we have introduced a method to transform and output disconnection tableau proofs in resolution style. As most of the complications of the disconnection calculus pertain only to the proof search and not to the proofs themselves, the transformation can be realised in an elegant way. In addition, the procedure can produce redundancy-free tableau proof trees in PSTricks format. The transformation procedure described here has been implemented in Scheme as a separate converter tool communicating with the DCTP prover. Both the converter and DCTP are available from the authors. Two important topics not addressed in this paper will be mentioned briefly.

Equality. The transformation of disconnection proofs for problems with equality is not part of our current work. The main problem here is that the built-in theory equality reasoning of DCTP [3] is not analytical due to the addition of new *expansion literals* to clause instances. Therefore, a translation of these *eq-linking* steps into paramodulation steps is far more difficult.

CNF Conversion. As most proof problems are stated in full first-order logic, the prover output should include the clause form transformation. However, the resulting output would be checkable only by a higher order prover system.

References

1. R. Letz, K. Mayr, and C. Goller. Controlled integration of the cut rule into connection tableau calculi. *Journal of Automated Reasoning*, 13(3):297–338, 12/1994.
2. R. Letz and G. Stenz. Proof and Model Generation with Disconnection Tableaux. In *Proceedings, LPAR 2001*, pages 142–156. Springer, Berlin, 12/2001.
3. R. Letz and G. Stenz. Integration of Equality Reasoning into the Disconnection Calculus. In *Proceedings, TABLEAUX-2002, LNAI* 2381, pages 176–190. Springer, Berlin, 7/2002.
4. F. Oppacher and E. Suen. HARP: A tableau-based theorem prover. *Journal of Automated Reasoning*, 4:69–100, 1988.
5. G. Stenz. *The Disconnection Calculus*. Logos Verlag, Berlin, 2002. Dissertation, Fakultät für Informatik, Technische Universität München.

LoTREC: Logical Tableaux Research Engineering Companion

Olivier Gasquet, Andreas Herzig, Dominique Longin, and Mohamad Sahade

IRIT - Université Paul Sabatier — Toulouse (France)
{gasquet, herzig, longin, sahade}@irit.fr
www.irit.fr/recherches/LILaC/Lotrec/

Abstract. In this paper we describe a generic tableaux system for building models or counter-models and testing satisfiability of formulas in
modal and description logics. This system is called LoTREC2.0. It is
characterized by a high-level language for tableau rules and strategies.
It aims at covering all Kripke-semantic based logics. It is implemented
in Java and characterized by a user-friendly graphical interface. It can
be used as a learning system for possible worlds semantics and tableaux
based proof methods.

1 Introduction

In general, most tableaux-based theorem provers [7,6] and DPLL-based implementations [10] use many different optimization techniques to speed up proof
search in a particular logic. But they are limited in the sense that they treat a
fixed set of logics: we cannot use them to treat other logics directly.

In general any user of a tableau based prover needs (1) high-level languages
for tableau rules and strategy definition, (2) user-friendly interfaces and (3)
flexibility and portability of the implementation. LoTREC2.0 is a such generic
tableau prover. It is designed for researchers specialized in modal logic and
for students concerned by learning modal logic. It aims at covering all logics
having possible worlds semantics, in particular modal and description logics.
LoTREC2.0 not only allows to test satisfiability, but it also builds models. In
that perspective LoTREC2.0 traverses the whole search space and keeps all the
models in memory. This is also the price to pay for a generic tool, whose efficiency cannot be compared with other more specialized provers such as FaCT
[14], MSPASS [11] or LWB [8].

In its genericity LoTREC2.0 is similar to the higher-order provers Isabelle
[12] and PVS [13]. In its aims LoTREC2.0 is similar to the Tableaux Workbench
TWB [9]. With TWB, one can go beyond traditional tableaux systems, too, and
handle for example modal tableaux with back/forth rules [3]. The differences are
as follows: while LoTREC2.0 is semantics-driven and tries to build models, TWB
is rather syntax-driven and closer to sequent calculi. With TWB, it is neither
possible to implement calculi that require two or more passages of the tableaux
algorithms (as required for example for Linear Temporal Logic LTL), nor is

B. Beckert (Ed): TABLEAUX 2005, LNAI 3702, pp. 318–322, 2005.

it possible to test loops or to handle relational properties like weak-directness. Moreover, TWB does not offer proof editing capabilities.

LoTREC2.0 is implemented in JAVA, a flexible and portable object oriented programming language. A first version was implemented by D. Fauthoux [1]. This version neither allowed to implement semantic properties like linearity, confluence, density, nor to test for example if a formula $\Diamond A$ is realized in a node. Recently, a new version of LoTREC2.0 has been implemented by M. Sahade [2], with a lot of modifications in its high-level language and its capacities. This motivated us to write this paper.

The theoretical basis of our system LoTREC2.0 is presented in [15] that is submitted as research paper at TABLEAUX'05 and that is available from our website.

In the next section, we will present the general architecture of LoTREC2.0, and in section 3 we present the main extensions we have made w.r.t. the preceding version.

2   LoTREC2.0 in a Nutshell

Tableaux are usually presented in tree form. In LoTREC2.0, they are generalized to RDAGs (Rooted Directed Acyclic Graphs) in order to enable complex logics such as the modal logic of confluence, or modal logics with complex interaction between knowledge and action. Graphs also allow to visualize possible worlds models. Graph nodes are labelled by sets of formulae, and edges by relation expressions.

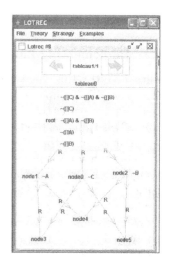

LoTREC2.0 graphically presents the tableaux it has generated (Fig 1), and allows the user to restructure them by "drag-and-drop". It can also output the results in a file for reuse. (In particular we plan to interact with other provers.)

LoTREC2.0 allows the user to define his logic and search strategy via textual files (as in Isabelle and PVS), or via a graphical interface. The user has to define the logical connectors, the semantic tableaux rules and the search strategy.

Fig. 1. A model for the formula: $\Diamond \neg A \wedge \Diamond \neg B \wedge \Diamond \neg C$

A tableau rule is interpreted as mapping a pattern to a pattern, where patterns are connected fragments of a given RDAGs. Every rule has a *condition* part, e.g. `if node hasElement (variable A)`, `if node1 isLinked node2 R`, ..., and an *actions* part, e.g. `do add A to node0`, `do link node0 node1 R`, ...The condition part contains the conditions to be verified to apply the rule, and the action part contains the actions to apply if all the conditions are satisfied. Very briefly, LoTREC2.0 strategies iteratively do the following: basically, each entity that has just been modified or created (formula, node, link,...) calls for application all rules whose condition part matches some subgraph containing this entity. These calls are managed by the Java event machine.

For more details about how to define connectors, rules and strategies see [2].

3 Improvements w.r.t. the Previous Version

The language of LoTREC2.0 allows for a very declarative way of programming, but this simplicity induced a lot of redundancies in the LoTREC1.0.

3.1 Duplication

As we have said, at every iteration step, each entity that has just been modified or created calls for application all rules that may be concerned. In the LoTREC1.0 if we have k conditions in the condition part of a rule, each rule was applied k times (one to each elementary component of the pattern under concern).

Such redundancies quickly turned out to be prohibitive for more complex logics: too much time and memory is wasted. This has been fixed in the new version, and now rules apply only once to each applicable pattern. This permits to treat logics like LTL or product logics in a more efficient manner.

3.2 Implementing Linearity and Seriality

With LoTREC1.0, there was no way of enforcing that a node has only one successor. For example in LTL, for a node which contains **next A** and **next B**, LoTREC1.0 creates two different successors (while what we want is one successor only). This is due to the fact that the rule finds two instances of the conditions, and so it executes the action part for every instance.

Due to this problem we were not able to treat in a satisfactory manner logics where the accessibility relation is linear or serial, like for LTL or the logics with the axiom of seriality D because LoTREC1.0 had to do redundant work and thus lost time and memory.

Our solution was to create a new action called **createOneSuccessor** which creates only one successor to a given node if it hasn't a successor yet, otherwise the rule continues to execute the other actions. And so now, we can build serial and linear models, and treat the corresponding logics in a satisfactory way.

3.3 Implementing Confluence

Normally if we want to apply the confluence rule to the pattern in (Fig 2) we will get one successor, for example called D and we link B and C to D, but LoTREC1.0 does not stop!

The problem is: in the first step it gives 4 successors because (1) we didn't specify that the two successor nodes must be different, (2) LoTREC1.0 does not consider that two patterns like {A,B,C} and {A,C,B} are equivalent modulo commutativity and must be considered as one pattern.

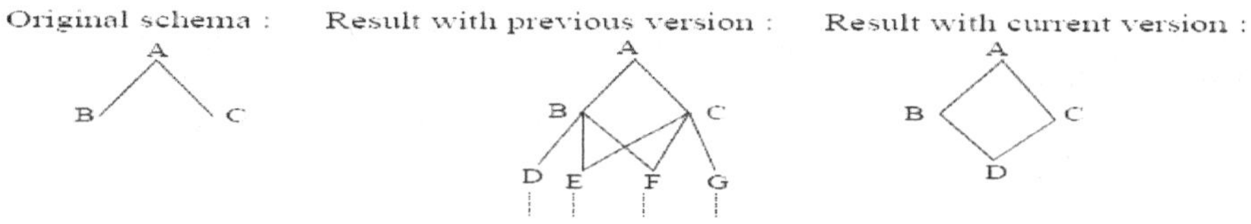

Original schema :   Result with previous version :   Result with current version :

Fig. 2. The confluence problem

For the first problem we add two new conditions which test if two given nodes are identical or not. For the second one we had to modify the internal operation of the rules to be able to express if a rule must consider two patterns equivalent modulo commutativity. In other words, we have now two kinds of rules: normal rules and commutative rules.

3.4 Test If an Expression Is Realized

In some logics where we stop by inclusion test (like in LTL and PDL), in order to know if we can build a model we have to do some postprocessing: we have to test whether a formula of the form $\Diamond A$ is realized or not. So we need to traverse the graph and to mark those $\Diamond$-formulas that are realized. With the LoTREC1.0 we could not perform such tasks, and thus couldn't treat LTL and PDL.

Our solution for this problem was to create two new actions: markExpression and unMarkExpression, which mark or unmark expressions in a specified node and two new conditions: isMarkedExpression and isNotMarkedExpression, to test if an expression is marked or not.

3.5 Extending the Strategy Language

The strategy language of the LoTREC1.0 had only two operators: repeat and firstRule. The first one repeats the application of a list of rules in the order of their appearance in the list. firstRule applies only the first applicable rule among a list of rules.

So with this restrictive language we couldn't build strategies where we want to apply a sequence of rules only once. For example, in model checking one may have a rule which builds the model and other rules to do the remaining work. The model building rule must be applied once and after it works we apply the other rules.

In the new version, we have extended the strategy language by a new operator allRules which applies the applicable rules among a sequence of rules only once.

3.6 Graphical User Interface and Execution via Internet

In the new version, the programming interface is far more user-friendly: The user can define connectors, rules, strategy and the formula to test in the corresponding space. He can choose his connectors and rules by selecting them from a list. He can save his logics and strategies in files and the reuse them in a user-friendlier way, and choose between printing the result of the computation on the screen, or saving it in a text file. Finally, one can run LoTREC2.0 now via Internet from the LoTREC2.0 home page.

4   Conclusion and Acknowledgements

We believe LoTREC2.0 is a useful tool to learn and play with modal and description logics and tableaux systems for them, through an improved graphic interface which allows the user to easily define his logic and strategy.

We have implemented with LoTREC2.0 several theorem provers for the logics: K, KB, KD, KT, K4, K5, S4, S5, KD45, PLTL, PDL, K4+confluence, K4+density and several logics of knowledge and action as well as intuitionistic logics. One can find a library of all defined logics at the LoTREC2.0 web site.

References

1. F.Del Cerro, D.Fauthoux, O.Gasquet, A.Herzig and D. Longin, Lotrec: the generic tableau prover for modal and description logics. In International Joint Conference on Automated Reasoning, LNCS, 2001, 453-458.
2. M.Sahade. LoTREC: User Manuel available at:LoTREC2.0 home page.
3. F. Massacci. Single step tableaux for modal logics: methodology, computations, algorithms. *JAR*, 24(3):319–364, 2000.
4. M. Castilho, L. Fariñas del Cerro, O. Gasquet, and A. Herzig. Modal tableaux with propagation rules and structural rules. *Fund. Inf.*, 32(3):281–297, 1997.
5. L. Fariñas del Cerro, O. Gasquet. Tableaux Based Decision Procedures for Modal Logics of Confluence and Density. *Fund. Inf.*, 40(4): 317-333 (1999).
6. I. Horrocks, U. Sattler, and S. Tobies. Practical reasoning for expressive description logics. *Log. J. of the IGPL*, 8(3):239–263, 2000.
7. I. Horrocks and P. F. Patel-Schneider. Optimizing description logic subsumption. *JLC*, 9(3):267–293, 199
8. A Heuerding. LWB theory http://www.lwb.unibe.ch
9. P. Abate and R. Gore. System Description: The Tableaux Work Bench
10. E. Giunchiglia, F. Giunchiglia, R. Sebastiani, and A. Tacchella. Sat vs. translation based decision procedures for modal logics: a comparative evaluation. *JANCL*, 10(2):145–173, 2000.
11. U. Hustadt and R. A. Schmidt (2000) In Dyckhoff, R. (eds), Automated Reasoning with Analytic Tableaux and Related Methods (TABLEAUX 2000). Lecture Notes in Artificial Intelligence, Vol. 1847, Springer, 67-71
12. L. C. Paulson. *Isabelle: A Generic Theorem Prover*, LNCS. Springer-Verlag, 1994.
13. S. Owre, J. M. Rushby, and N. Shankar. PVS: A prototype verification system. In *Proc. of CADE'92*, LNAI, pp. 748–752, 1992.
14. I.Horrocks and P.K. Patel-Schnieder. Optimising propositional modal satisfiability for discription logic subsumption. In LNCS 1476,1998.
15. O. Gasquet, A. Herzig and M. Sahade: Programming Modal Tableaux Systems. Submitted to Tableaux05.

A Tableau-Based Explainer for DL Subsumption

Thorsten Liebig and Michael Halfmann

Dept. of AI, University of Ulm, D-89069 Ulm, Germany
{liebig, michael.halfmann}@informatik.uni-ulm.de

Abstract. This paper describes the implementation of a tableau-based reasoning component which is capable of providing quasi natural language explanations for subsumptions within $\mathcal{ALEHF}_{R^+}$ TBoxes.

1 Motivation

W3C's recently recommended ontology language OWL is expected to be used even by non-sophisticated end users. However, the Description Logics (DLs) underlying OWL Lite ($\mathcal{SHIF}$) and OWL DL ($\mathcal{SHOIN}$) are quite expressive [1]. Authoring ontologies presumably is not possible without a basic understanding of the underlying reasoning services. Our explainer MEX[1] aims at supporting a deeper comprehension of subsumption, the core inference service, by providing an on-demand step by step quasi-natural language explanation.

MEX is capable of explaining subsumptions within a significant fraction of the DL underlying OWL Lite, namely definitorial $\mathcal{ALEHF}_{R^+}$ TBoxes with global domain and range restrictions. Such TBoxes require all axioms to be of the form $A \sqsubseteq D$ or $A \equiv D$, with A atomic and unique. The language $\mathcal{ALE}$ covers the top and bottom concept ($\top, \bot$), conjunction ($C \sqcap D$), qualified existential ($\exists r.C$) and universal quantification ($\forall r.C$), and atomic negation ($\neg A$). $\mathcal{ALEHF}_{R^+}$ extends $\mathcal{ALE}$ with role hierarchies ($p \sqsubseteq r$), a limited form of cardinality restrictions ($\leq n\ r$) with n either 0 or 1, and transitive as well as functional roles. Our system implements and extends the theoretical approach for explaining $\mathcal{ALC}$ subsumptions described in [2] which proposes to utilize a sequent proof derived from a tableau style algorithm. Although not explicitly present in our language, disjunction comes implicitly on rhs due to the refutation strategy of the tableau algorithm.

2 Implementation

Tableaux systems implement a refutation proof strategy. However, to prove a conclusion by showing its opposite to be contradictory doesn't seem to be intuitive. In our opinion, humans usually try to reduce a complex problem into sensible pieces which are more easily comprehensible. In consideration of the latter, our system explains a subsumption by breaking it down into sub-subsumptions

---

[1] Acronym for My EXplainer.

B. Beckert (Ed): TABLEAUX 2005, LNAI 3702, pp. 323–327, 2005.
© Springer-Verlag Berlin Heidelberg 2005

until those can be explained by simple statements. This course of action is triggered by a tableau-based approach, where tableaux rules are applied until a terminating clash occurs. However, for the task of explaining how a proof has been derived, the refutation strategy has to be hidden to the user. In order to achieve this we use a technique called tagging [2]. Tagging allows to reconstruct the original query for illustration of the derivation steps at any time of tableau processing. To be concrete, we distinguish between the right-hand side (rhs) and the left-hand side (lhs) of a subsumption query by tagging the rhs.

Our explainer MEX is implemented in Lisp. Its syntax for defining concepts and roles follows the KRSS standard [3], which easily allows to transfer existing ontologies from other systems, most notably the widely used RACER [4] reasoner. The internal data-structure is based on Lisp `structure` data-types. A node of a tableau proof tree is represented as a `structure` object, storing the subsumee (lhs) and subsumer (rhs) of the corresponding query, its role successor nodes, a list of disjunctive alternative nodes in case of a disjunction on rhs, a reference to its parent node, and some further parameters used for blocking and optimization. MEX also implements lazy unfolding, a well known optimization technique of DL tableaux algorithms. Lazy unfolding delays the unfolding of a concept to its given definition until it is required by the proof algorithm to proceed. This also helps to maximize performance in typical ontologies since only those axioms are taken into account which are actually proof relevant.

While successively creating the tableau tree MEX generates a corresponding proof explanation in parallel. Such an explanation is a list of atomic explanation steps. For each relevant tableau rule application or unfolding one or more explanation steps are added. An explanation step consists of its corresponding type, its depth in the tableau tree, a list of additional information (relevant concepts, nodes, etc.) and a textual explanation. The type and the additional parameters stored within each step should enable an external component to layout explanations with respect to the given application context (e. g. degree of detail, language, hierarchical structure vs. flat ordering).

As mentioned before, each tableau transformation will result in one or more explanation steps. A tableau transformation consists of unfolding steps as well as tableaux rules. For example, a role-successor node will be explained by introducing the new node and a description of the constraints which follow from the qualified quantifications on that role. Explanations will be given only to those nodes which are clash relevant. Since we extend previous work about explaining $\mathcal{ALC}$ we had to add explanations for cardinality restrictions, role hierarchies, merging of role-successors, and domain and range restrictions.

The expansion of a tableau branch terminates as soon as a clash is found. For each type of clash we have formulated a quasi-natural language statement, explaining the subsumption relationship of the original query. For example, due to the fact that we have to take the origin of the clash relevant constructs into account we have to distinguish between four types of cardinality clashes. Other language constructs like transitive roles and domain as well as range restrictions do not add extra clash types but require additional steps explaining the source of the expressions they append to existing tableaux nodes. The elements of a disjunction (which only can occur on rhs) within a node are stored in an or-branch slot of the **structure** object. They all have to clash in order to close the node they occur in. From the viewpoint of explaining they correspond to a conjunction and will therefore be explained with help of an enumeration of sub-explanations. For the lack of space we refer to [5] for a more detailed description about explaining $\mathcal{ALEHF}_{R^+}$ TBoxes.

When processing a conjunction of different expansions (e. g. role successors) the first one in which a clash is found will be explained by MEX. A future optimization could select the most intuitive explanation from the set of alternatives.

To generate concise and simple explanations we have implemented several optimizations which condense the explanation in specific situations. The two most important ones, filtering and mode switching, are described in the following.

Filtering is a simple method for pruning disjunction on rhs with help of a structural comparison. Consider the subsumption $A \sqcap B \sqcap C \sqsubseteq A \sqcap C$. The standard approach would split up the explanation into two sub-subsumptions according to the internal disjunction on rhs. Very likely this is unnecessary because this subsumption obviously holds, since the subsumee is a direct specialization of the subsumer. Therefore MEX structurally compares the lhs and rhs after lazy unfolding and prior to any further processing. If the rhs is a syntactical subset of the lhs an obvious subsumption is found and explained accordingly. A distantly related technique, called normalizing and naming is found in the DL literature.

Another optimization method is called mode switching and applies to situations where at some stage of tableau processing either the rhs or lhs is unsatisfiable on its own (i. e. independently from each other). This corresponds to either the subsumee being equivalent to $\bot$ or the subsumer being equivalent to $\top$ on explanation side. In such a case our explainer will switch to either unsatisfiability or tautology explaining while disregarding the other side. Since our explanations are generated on-the-fly while building the tableau tree this requires to check each side concerning unsatisfiability in advance. We use the RACER reasoner as an external component for this test. An additional benefit of the optional RACER connection is the increased performance when using the dual-reasoner architecture as proposed in [6]. This optimization feature uses RACER to determine the effectively clash relevant expressions within each side. This will prune the explainer tableau by leaving out irrelevant node expansions.

3   OntoTrack Integration

Explaining an inference service can improve the authoring process of an ontology to a large degree. Consequently, explanations are most powerful when combined with an ontology editing tool. Therefore we made MEX accessible to our interactive ontology authoring tool ONTOTRACK [7] via the offered plug-in interface. Figure 1 displays the most important system components with respect to this integration. As an on-demand functionality ONTOTRACK converts the current

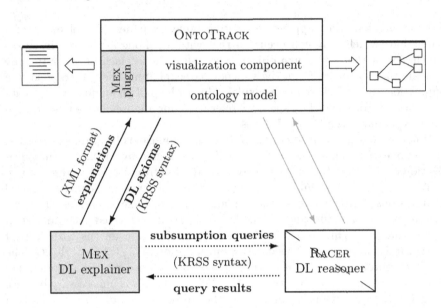

Fig. 1. Architecture of MEX as an ONTOTRACK plugin with optional RACER link-up

internal ontology model into KRSS [3] syntax and sends it together with an explanation request to MEX via TCP. An explanation is generated and transmitted as an XML serialization back to ONTOTRACK. This enables the ONTOTRACK plug-in to display the explanation in a graphical way as an expandable tree list.

The RACER link-up is optional, but leads to optimized explanations (with mode switching) and an increased performance. ONTOTRACK itself also uses RACER for instant reasoning feedback to user interactions. As indicated in Figure 1, ONTOTRACK and MEX currently use two different RACER TBox instances because of lack of a standardized mapping of ontology axioms into RACER TBoxes. Future work will be concerned with a standardized RACER connection, so that MEX is able to access the ONTOTRACK model serialization directly.

Figure 2 shows a screen-shot of ONTOTRACK displaying a sample ontology and an explanation for A subsuming B provided by MEX. To invoke an explanation a user only has to select "Explain" within the mouse enabled context menu on edges representing a subsumption relationship.

References

1. Horrocks, I., Patel-Schneider, P.: Reducing OWL entailment to description logic satisfiability. Journal of Web Semantics **1** (2004) 345–357
2. Borgida, A., Franconi, E., Horrocks, I., McGuinness, D., Patel-Schneider, P.: Explaining $\mathcal{ALC}$ subsumption. In: Proc. of the Int. Workshop on Description Logics (DL99), Linköping, Sweden (1999) 37–40
3. Patel-Schneider, P., Swartout, B.: Description-Logic Knowledge Representation System Specification (1993) Working Version (Draft).

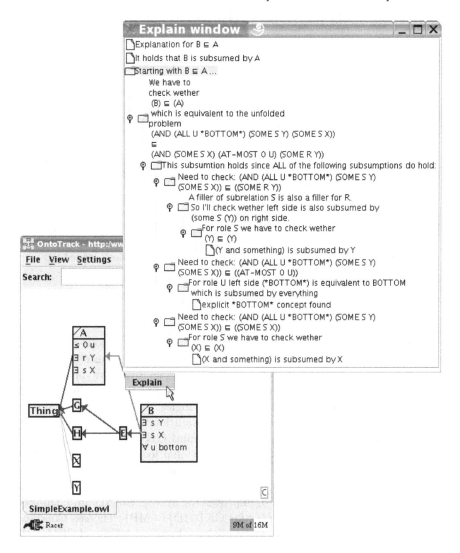

Fig. 2. OntoTrack with an explanation

4. Haarslev, V., Möller, R.: Description of the RACER System and its Applications. In: Proc. ot the Int. Workshop on Description Logics (DL01), Stanford, CA, USA (2001)

5. Liebig, T., Halfmann, M.: Explaining Subsumtion in $\mathcal{ALEHF}_{R+}$ TBoxes. In: Int. Workshop on Description Logics (DL05). (2005) to appear.

6. Kwong, F.: Explaining Description Logic Reasoning. In: Proc. of the 2004 Int. Workshop on Description Logics (DL04), Whistler, BC, Canada (2004) 210

7. Liebig, T., Noppens, O.: ONTOTRACK: Combining Browsing and Editing with Reasoning and Explaining for OWL Lite Ontologies. In: Proc. of the Int. Semantic Web Conference (ISWC 2004), Hiroshima, Japan (2004) 244–258

CondLean 3.0: Improving CondLean for Stronger Conditional Logics

Nicola Olivetti and Gian Luca Pozzato

Dipartimento di Informatica, Università degli Studi di Torino,
Corso Svizzera 185, 10149 Torino, Italy
{olivetti, pozzato}@di.unito.it

Abstract. In this paper we present CondLean 3.0, a theorem prover
for propositional conditional logics CK, CK+ID, CK+MP, CK+CS,
CK+CEM and some of their combinations. CondLean 3.0 implements
sequent calculi for these logics. CondLean 3.0 improves CondLean
and is developed following the methodology of leanTAP. It is implemented in SICStus Prolog and also comprises a graphical user interface implemented in JAVA. CondLean 3.0 can be downloaded at
www.di.unito.it/~pozzato/condlean3/

1 Introduction

Conditional logics have a long history, and recently they have found interesting
applications in several areas of computer science and artificial intelligence, such
as knowledge representation, non-monotonic reasoning, deductive databases, belief revision and natural language semantics. In spite of their significance, very
few proof systems have been proposed for these logics. In [1] labelled sequent calculi SeqS are introduced for minimal normal conditional logic **CK** and for three
extensions of it, namely **CK+ID, CK+MP** and **CK+MP+ID**. In order to
develop a reasoning system for stronger conditional logics, in [2] SeqS systems
have been extended to well-known conditional logics **CS** and **CEM**.

In this work we describe an implementation of SeqS calculi in SICStus Prolog.
The program, called CondLean 3.0, improves the system CondLean, introduced
in [3]. CondLean is a theorem prover for CK{+ID}{+MP}, whereas CondLean
3.0 gives a decision procedure also for conditional logics with CS and CEM; as far
as we know this is the first theorem prover for these logics. For weaker systems
CK{+ID}, CondLean 3.0 inherits the three different versions of CondLean: 1.
a simple version, where Prolog *constants* are used to represent SeqS's labels; 2.
a more efficient one, where labels are represented by Prolog *variables*, inspired
by the free-variable tableaux presented in [4]; 3. a "two-phase" theorem prover,
which first attempts to prove a sequent by using an incomplete, but fast, proof
procedure (phase 1), and then it calls the free-variable proof procedure (phase
2) in case of failure. For systems allowing (MP), (CEM) and (CS), CondLean
3.0 implements calculi for these logics where the crucial ($\Rightarrow$ L) rule is invertible.
For these systems, we only present the constant label version. CondLean 3.0 also
comprises a graphical interface implemented in Java.

B. Beckert (Ed): TABLEAUX 2005, LNAI 3702, pp. 328–332, 2005.

2   Conditional Logics and Their Sequent Calculi

We consider a propositional language $\mathcal{L}$ over a set ATM of propositional variables. Formulas of $\mathcal{L}$ are built from propositional variables by means of the boolean operators $\to, \perp$ and the conditional operator $\Rightarrow$. We adopt the so-called propositional selection function semantics [5]. A selection function model for $\mathcal{L}$ is a triple $\mathcal{M} = \langle \mathcal{W}, f, [\ ] \rangle$, where $\mathcal{W}$ is a non-empty set of items called *worlds*, f is a function of type $f : \mathcal{W} \times 2^{\mathcal{W}} \longrightarrow 2^{\mathcal{W}}$, called the *selection function* and $[\ ]$ is an evaluation function of type $ATM \longrightarrow 2^{\mathcal{W}}$. $[\ ]$ assigns to an atom p the set of worlds where p is true. The evaluation function $[\ ]$ can be extended to every formula by means of the following inductive clauses: 1. $[\perp] = \emptyset$; 2. $[A \to B] = (\mathcal{W} - [A]) \cup [B]$; 3. $[A \Rightarrow B] = \{w \in \mathcal{W} \mid f(w, [A]) \subseteq [B]\}$. A formula A is *valid* in a model $\mathcal{M}$ as above if $[A] = \mathcal{W}$. A formula A is *valid* (denoted by $\models A$) if it is valid in every model $\mathcal{M}$.

The above one is the semantics of the basic conditional logic **CK**, where no specific properties of the selection function f are assumed. Moreover, we consider extensions of **CK** obtained by imposing one or more of the following conditions on the selection function: **(ID)** $f(x, [A]) \subseteq [A]$; **(MP)** if $x \in [A]$ then $x \in f(x, [A])$; **(CS)** if $x \in [A]$, then $f(x, [A]) \subseteq \{x\}$; **(CEM)** $|f(x, [A])| \leq 1$. The above semantic conditions correspond to the axiom schemata $A \Rightarrow A$, $(A \Rightarrow B) \to (A \to B)$, $(A \wedge B) \to (A \Rightarrow B)$ and $(A \Rightarrow B) \vee (A \Rightarrow \neg B)$, respectively. In **Figure 1** we present the calculi for CK and its mentioned extensions ([1] and [2]); the calculi make use of *labelled formulas*, where the labels are drawn from a denumerable set $\mathcal{A}$; there are two kinds of formulas: 1. *world formulas*, denoted by $x : A$, where $x \in \mathcal{A}$ and $A \in \mathcal{L}$; 2. *transition formulas*, denoted by $x \xrightarrow{A} y$, where $x, y \in \mathcal{A}$ and $A \in \mathcal{L}$. A transition formula $x \xrightarrow{A} y$ represents that $y \in f(x, [A])$. In systems CK{+ID} the ($\Rightarrow$ L) does not copy the principal formula into its premises, i.e. it is not invertible. Systems combining two or more semantic conditions are characterized by all rules capturing those conditions; systems allowing *both* (CEM) and (MP) axioms are not presented in this work.

Definition (Sequent validity). Given a model $\mathcal{M} = \langle \mathcal{W}, f, [\ ] \rangle$ for $\mathcal{L}$, and a label alphabet $\mathcal{A}$, we consider any *mapping* $I : \mathcal{A} \to \mathcal{W}$. Let F be a labelled formula, we define $\mathcal{M} \models_I F$ as follows: $\mathcal{M} \models_I x: A$ iff $I(x) \in [A]$ and $\mathcal{M} \models_I x \xrightarrow{A} y$ iff $I(y) \in f(I(x), [A])$. We say that $\Gamma \vdash \Delta$ is *valid* in $\mathcal{M}$ if for every mapping $I : \mathcal{A} \to \mathcal{W}$, if $\mathcal{M} \models_I F$ for every $F \in \Gamma$, then $\mathcal{M} \models_I G$ for some $G \in \Delta$. We say that $\Gamma \vdash \Delta$ is valid in a system (CK or one of its extensions) if it is valid in every $\mathcal{M}$ satisfying the specific conditions for that system (if any).

Theorem 1 (Soundness and completeness [2]). A sequent $\Gamma \vdash \Delta$ is valid if and only if $\Gamma \vdash \Delta$ is derivable in SeqS.

In order to obtain a decision procedure we have to control the application of rules ($\Rightarrow$ L) and (CEM) in a backward proof search of a sequent derivation, since their premises have a higher complexity than the conclusion. To this regard we have the following results:

$$(\mathbf{AX}) \ \Gamma, F \vdash \Delta, F \qquad\qquad (\mathbf{A}\bot) \ \Gamma, x : \bot \vdash \Delta$$

$$(\to \mathbf{L}) \ \frac{\Gamma \vdash x : A, \Delta \quad \Gamma, x : B \vdash \Delta}{\Gamma, x : A \to B \vdash \Delta} \qquad (\to \mathbf{R}) \ \frac{\Gamma, x : A \vdash x : B, \Delta}{\Gamma \vdash x : A \to B, \Delta}$$

$$(\mathbf{EQ}) \ \frac{u : A \vdash u : B \quad u : B \vdash u : A}{\Gamma, x \xrightarrow{A} y \vdash x \xrightarrow{B} y, \Delta}$$

$$(\Rightarrow \mathbf{L}) \ \frac{\Gamma, x : A \Rightarrow B \vdash x \xrightarrow{A} y, \Delta \quad \Gamma, x : A \Rightarrow B, y : B \vdash \Delta}{\Gamma, x : A \Rightarrow B \vdash \Delta} \qquad (\Rightarrow \mathbf{R}) \ \frac{\Gamma, x \xrightarrow{A} y \vdash y : B, \Delta}{\Gamma \vdash x : A \Rightarrow B, \Delta} \ (y \notin \Gamma, \Delta)$$

$$(\mathbf{ID}) \ \frac{\Gamma, y : A \vdash \Delta}{\Gamma, x \xrightarrow{A} y \vdash \Delta} \qquad (\mathbf{CEM}) \ \frac{\Gamma, x \xrightarrow{A} y \vdash \Delta, x \xrightarrow{A} z}{(\Gamma, x \xrightarrow{A} y \vdash \Delta)[y, z/u]}{\Gamma, x \xrightarrow{A} y \vdash \Delta} \ \begin{array}{l}(y \neq z, \\ u \notin \Gamma, \Delta)\end{array}$$

$$(\mathbf{CS}) \ \frac{\Gamma \vdash \Delta, x : A \quad \Gamma[x, y/u], u \xrightarrow{A} u \vdash \Delta[x, y/u]}{\Gamma, x \xrightarrow{A} y \vdash \Delta} \ \begin{array}{l}(x \neq y, \\ u \notin \Gamma, \Delta)\end{array} \qquad (\mathbf{MP}) \ \frac{\Gamma \vdash x : A, \Delta}{\Gamma \vdash x \xrightarrow{A} x, \Delta}$$

Fig. 1. Sequent calculi SeqS. $\Sigma[x, y/u]$ denotes the substitution of labels x and y with the label u wherever they occur in Σ. Rules (ID), (MP), (CS) and (CEM) are only used in corresponding extensions of the basic system SeqCK.

Theorem 2 ([2]). *SeqS systems are sound and complete even if the $(\Rightarrow L)$ rule is applied to $\Gamma, x : A \Rightarrow B \vdash \Delta$ with the following restrictions: 1. $(\Rightarrow L)$ is applied at most once by using the same transition $x \xrightarrow{A} y$ on the same conditional formula $x : A \Rightarrow B$; 2. $(\Rightarrow L)$ is applied by using a transition formula $x \xrightarrow{A} y$ such that there exists $x \xrightarrow{C} y \in \Gamma$ or $x = y$. In systems $SeqCEM\{+CS\}\{+ID\}$ one can control the application of (CEM) by means of the same restrictions.*

Theorem 3 ([1], [2]). *SeqCK and SeqID are sound and complete even if $(\Rightarrow L)$ is formulated as follows:*

$$\frac{x \xrightarrow{C} y \vdash x \xrightarrow{A} y \quad \Gamma, x \xrightarrow{C} y, y : A \vdash \Delta}{\Gamma, x \xrightarrow{C} y, x : A \Rightarrow B \vdash \Delta} \ (\Rightarrow L)$$

These results give a constructive proof of decidability of the respective systems, alternative to the semantic proof based on the finite model property.

3   Design of CondLean 3.0

In this section we present an implementation of the sequent calculi SeqS; it is a SICStus Prolog program inspired by leanTAP ([6], [7]) and extends CondLean

[3] to systems CS, CEM and their mentioned extensions. The program comprises a set of clauses, each one of them represents a sequent rule or axiom. The proof search is provided for free by the mere depth-first search mechanism of Prolog, without any additional ad hoc mechanism. We represent each component of a sequent (antecedent and consequent) by a **list** of formulas, partitioned into three sub-lists: atomic formulas, transitions and complex formulas. Atomic and complex formulas are represented by a list like [x,a], where x is a Prolog constant and a is a formula. A transition $x \xrightarrow{A} y$ is represented by [x,a,y]. SeqS's labels are represented by Prolog's constants. The sequent calculi are implemented by the predicate **prove(Cond, Sigma, Delta, Labels).** which succeeds if and only if $\Sigma \vdash \Delta$ is derivable in SeqS, where Sigma and Delta are the lists representing the multisets Σ and Δ, respectively and Labels is the list of labels introduced in that branch. Cond is a list of couples of kind $[F, Used]$, where F is a conditional formula [X,A => B] and $Used$ is a list of transitions $[[X, A_1, Y_1], ..., [X, A_n, Y_n]]$ such that $(\Rightarrow$ L) has already been applied to $x : A \Rightarrow B$ by using transitions $x \xrightarrow{A_i} y_i$. $(\Rightarrow$ L) is applied to $x : A \Rightarrow B$ by choosing $x \xrightarrow{A} y$ such that $x \xrightarrow{A_y} y$ belongs to Sigma and $[x, a_y, y]$ *does not belong to* Used. For instance, to prove $x : B \vdash x : A \Rightarrow A$ in CK+ID, one queries CondLean 3.0 with the goal prove([],[[[x,b]],[],[]], [[],[],[[x,a=>a]]], [x]). Each clause of prove implements one axiom or rule of SeqS; for example, the clause[1] implementing $(\Rightarrow$ L) is as follows:

prove(Cond,[LitSigma,TransSigma,ComplexSigma],
    [LitDelta,TransDelta,ComplexDelta], Labels):-
  member([X,A => B],ComplexSigma),select([[X,A => B],Used],Cond,TempCond),
  member([X,C,Y],TransSigma), \+member([X,C,Y],Used),!,
  put([Y,B],LitSigma,ComplexSigma,NewLitSigma,NewComplexSigma),
  prove([[[X, A => B],[[X,C,Y] | Used]] | TempCond],
    [LitSigma,TransSigma,ComplexSigma],
    [LitDelta,[[X,A,Y]|TransDelta],ComplexDelta],Labels),
  prove([[[X, A => B],[[X,C,Y] | Used]] | TempCond],
    [NewLitSigma,TransSigma,NewComplexSigma],
    [LitDelta,TransDelta,ComplexDelta],Labels).

The predicate put is used to put [Y,B] in the proper sub-list of the antecedent. To search a derivation of a sequent $\Sigma \vdash \Delta$, CondLean 3.0 proceeds as follows. First of all, if $\Sigma \vdash \Delta$ is an axiom, the goal will succeed immediately by using the clauses for the axioms. If it is not, then the first applicable rule will be chosen, e.g. if ComplexDelta contains a formula [X,A -> B], then the clause for $(\rightarrow$ R) rule will be used, invoking prove on the unique premise of $(\rightarrow$ R). CondLean 3.0 proceeds in a similar way for the other rules. The ordering of the clauses is such that the application of the branching rules is postponed as much as possible.

---

[1] There are other clauses implementing $(\Rightarrow$ L), taking care of cases when Cond is empty and when $x \xrightarrow{A} x$ is used (systems allowing MP only), which we cannot present here due to space limitation.

In systems SeqCEM{+CS}{+ID}, in order to control the application of (CEM) another parameter Tr is added to the predicate prove.

In systems SeqCK and SeqID CondLean 3.0 implements the reformulated version of ($\Rightarrow$ L) stated by Theorem 3. Since ($\Rightarrow$ L) is not invertible, the auxiliary parameter Cond is removed from prove. When the ($\Rightarrow$ L) clause is used to prove $\Sigma \vdash \Delta$, a backtracking point is introduced by the choice of a label Y occurring in the two premises of the rule; choosing, sooner or later, the right label to apply ($\Rightarrow$ L) may strongly affect the theorem prover's efficiency. In order to reduce this potential loss of efficiency for CK{+ID}, CondLean 3.0 implements the same *free-variables* and *heuristic* versions given by its predecessor CondLean [3]. These versions have not been implemented for the other systems.

CondLean 3.0 has also a graphical interface (GUI) implemented in Java. The GUI interacts with the SICStus Prolog implementation by means of the package se.sics.jasper. Thanks to the GUI, one does not need to know how to call the predicate prove: one just introduces a sequent in a text box and searches a derivation by clicking a button; moreover, one can choose the intended system of conditional logic. When the submitted sequent is valid, CondLean offers these options: display a proof tree of the sequent in a special window, build a latex file containing the proof tree, and view some statistics of the proof.

4   Statistics and Conclusions

The performances of CondLean 3.0 are promising. For instance, considering the sequent-degree (defined as the maximum level of nesting of the $\Rightarrow$ operator) as a parameter, it succeeds in less than 2 seconds for all sequents of degree 15 used in our tests, 700 ms for degree 10, and 5 ms for degree 2. In future research we intend to extend it to other conditional logics and we intend to develop a free variable version for stronger conditional systems in which the ($\Rightarrow$ L) is invertible.

References

1. N. Olivetti and C. B. Schwind. A calculus and complexity bound for minimal conditional logic. *Proc. ICTCS01, vol LNCS 2202*, pages 384–404, 2001.
2. N. Olivetti, G. L. Pozzato, and C. B. Schwind. A sequent calculus and a theorem prover for standard conditional logics: Extended version. *Technical Report 87/05, Dip. di Informatica, Università degli Studi di Torino, available at www.di.unito.it/~pozzato/condlean3*, 2005.
3. N. Olivetti and G. L. Pozzato. Condlean: A theorem prover for conditional logics. *In Proc. of TABLEAUX 2003 (Automated Reasoning with Analytic Tableaux and Related Methods), volume 2796 of LNAI, Springer*, pages 264–270, 2003.
4. B. Beckert and R. Gorè. Free variable tableaux for propositional modal logics. *Tableaux-97, LNCS 1227, Springer*, pages 91–106, 1997.
5. D. Nute. Topics in conditional logic. *Reidel*, 1980.
6. B. Beckert and J. Posegga. leantap: Lean tableau-based deduction. *Journal of Automated Reasoning, 15(3)*, pages 339–358, 1995.
7. M. Fitting. leantap revisited. *J. of Logic and Computation, 8(1)*, pages 33–47, 1998.

The ILTP Library: Benchmarking Automated Theorem Provers for Intuitionistic Logic

Thomas Raths*, Jens Otten, and Christoph Kreitz

Institut für Informatik, University of Potsdam,
August-Bebel-Str. 89, 14482 Potsdam-Babelsberg, Germany
{raths, jeotten, kreitz}@cs.uni-potsdam.de

Abstract. The Intuitionistic Logic Theorem Proving (ILTP) Library provides a platfom for testing and benchmarking theorem provers for first-order intuitionistic logic. It includes a collection of benchmark problems in a standardised syntax and performance results obtained by a comprehensive test of currently available intuitionistic theorem proving systems. These results are used to provide information about the status and the difficulty rating of the benchmark problems.

1 Introduction

Benchmarking automated theorem proving (ATP) systems using standardised problem sets is a well-established method for measuring their performance. The TPTP library [10] is the largest collection of problems (currently more than 7000 formulas) for testing and benchmarking ATP systems for classical logic. Other problem libraries for, e.g., termination and induction problems have been developed as well.[1]

Unfortunately the availability of such libraries for non-classical logics is very limited. For intuitionistic logic several small collections of formulas have been published and used for testing ATP systems. Sahlin et al. [8] compiled one of the first collections of first-order formulas for testing their intuitionistic ATP system ft. The same collection was also used for benchmarking other intuitionistic theorem provers [11,5]. A second collection of first-order formulas was used to test the intuitionistic ATP system JProver [9], which has been integrated into the constructive interactive proof assistants NuPRL [3] and Coq [2].

Another collection of propositional formulas was compiled by Dyckhoff.[2] It introduces six classes of scalable formulas following the methodology of the Logics Workbench [1]. The advantage of this approach is the possibility to study the time complexity behaviour of an ATP system on a specific generic formula

---

* The first author's research is sponsored by DARPA under agreement number FA8750-04-2-0216.

[1] For the termination problem library see http://www.lri.fr/~marche/tpdb/, for the induction problem libraries see http://dream.dai.ed.ac.uk/dc/lib.html and http://www.cs.nott.ac.uk/~lad/research/challenges/.

[2] See http://www.dcs.st-and.ac.uk/~rd/logic/marks.html .

B. Beckert (Ed): TABLEAUX 2005, LNAI 3702, pp. 333–337, 2005.

as its size increases. But in order to achieve more meaningful benchmark results the number of generic formulas would have to be increased significantly. Most of the formulas in the collection have a rather syntactical nature, often specifically designed with the presence (or absence) of a specific search strategy in mind. To provide a better view of the usefulness of intuitionistic ATP systems on problems arising in practice, like in program synthesis [3], a benchmark collection should cover a broader range of more realistic problems. These kind of problems are typically presented in a first-order logic (as already mentioned in Dyckhoff's benchmark collection).

The ILTP library was developed for exactly that purpose. In the following we will describe the content of the ILTP library, which contains two major problem sets, some benchmark tools, and a database of currently available intuitionistic ATP systems with performance results. We will also present information about the intuitionistic status and difficulty rating of the problems in the ILTP library based on comprehensive tests with existing intuitionistic ATP systems.

2   The Content of the ILTP Library

The ILTP library contains two main set of problems: the first one is taken from the TPTP library and the second one from three problem collections, which have been used previously for testing and benchmarking intuitionistic ATP systems.

2.1   The TPTP Problem Set

Whereas the semantics of classical and intuitionistic logic differs, they share the same syntax. This allows in principle the use of classical benchmark libraries like the TPTP library [10] for benchmarking intuitionistic ATP systems as well. Starting mainly as a library of first-order formulas in clausal form, today the TPTP library contains a large number of formulas in non-clausal form as well. Problems in clausal (i.e. disjunctive or conjunctive normal) form are intuitionistically invalid and therefore useless for intuitionistic reasoning. Furthermore, the conversion of formulas to clausal form does not preserve intuitionistic validity, because it involves (intuitionistically invalid) laws like $\neg(A \wedge B) \Rightarrow (\neg A \vee \neg B)$ and $\neg\neg A \Rightarrow A$. Adding double negation to classically valid formulas in order to generate intuitionistically valid formulas is of less interest as well since the resulting problems are just encodings of the classical ones.

1745 of the problems in the TPTP library version 2.7.0 are in non-clausal form, so called "first-order formulas" (FOF). Of these formulas 408 are classically invalid. Since every intuitionistically valid formula needs to be classically valid as well, it is straightforward to refute these formulas with a classical ATP system. Therefore we will focus on the remaining 1337 formulas whose classical status is either valid or unknown.

These 1337 formulas form the first part, the TPTP problem set, of the ILTP library. The status (i.e. Theorem, Non-Theorem, Unknown) and the difficulty rating of the problems have been adapted to the intuitionistic case (see Section 3) and are provided separately.

2.2   The ILTP Problem Set

The second part of the ILTP library, which we call the ILTP problem set, contains 108 formulas from three benchmark collections. Their syntax was standardised and adapted to the TPTP input format. Each problem file was given a header with useful information, like references, as done in the TPTP library. The intuitionistic status and difficulty rating (see Section 3) was included as well.

The first collection contains 39 intuitionistically valid first-order formulas originally used to test the intuitionistic ATP system ft [8]. Five of the problems are already part of the TPTP problem set and therefore excluded. These are problems ft3.1 to ft3.5 which are identical with Pelletier's problems no. 39 to 43.

The second collection contains 36 propositional formulas from Dyckhoff's benchmark collection. From each of the six problem classes three (intuitionistically) valid and three invalid formulas are included. These six formula instances have been chosen according to their difficulty relative to current intuitionistic ATP systems.

The third collection contains 33 propositional and first-order formulas from the problem set used to test the intuitionistic ATP system JProver [9]. The type information, which was used to test JProver within the NuPRL environment, is removed. Three problems are left out because they are already classically invalid or cannot be represented in pure first-order logic.

2.3   Tools and Prover Database

In addition to the two problems sets, we provide so-called format files, which can be used to convert the problems in the ILTP library into the input syntax of the ATP systems listed in the prover database. These format files are used together with the TPTP2X utility, which is part of the TPTP library. The ILTP library also contains a small database with information about published intuitionistic ATP systems. For each prover we provide some basic information (like author, homepage, short description, references) and a test run on two example formulas. A summary and a detailed list of the performance results on running each system on the problems in the ILTP library are given as well.

3   Rating the Difficulty of Problems in the ILTP Library

In the TPTP library the difficulty of every problem is rated according to the performance of current (classical) state-of-the-art ATP systems. It expresses the ratio of systems which can solve a problem. For example a rating of 0.0 indicates that every state-of-the-art prover can solve the problem, a rating of 0.5 indicates that half of the systems were able to solve it, and a problem with rating 1.0 was not solved by any ATP system.

We adapt this notation to the problems in the ILTP library. To this end we need to specify a set of intuitionistic state-of-the-art ATP systems. We performed comprehensive tests of all currently available systems on the problems in the ILTP library and analysed the performance results [7]. We have selected four first-order and one purely propositional ATP system, which solved the highest number of problems: the first-order systems ft (C-version) [8], JProver [9], ileanTAP[5], ileanCoP[6], and the propositional system STRIP [4].

Each problem is assigned its status. The status can be Theorem, Non-Theorem or Unknown. We did not perform any theoretical investigations into the intuitionistic validity of the formulas in the TPTP problem set. We mark the status of a problem as Theorem or Non-Theorem if any ATP system was able to show that the given problem is valid or invalid, respectively. All other TPTP problems were given the status Unknown.

3.1   The TPTP Problem Set

Table 1 shows a summary of the rating and status information of the TPTP problem set. The rating and status information refers to intuitionistic logic. Only the last line shows the (original TPTP) classical rating of the problem set.

Table 1. Rating of the TPTP problem set

Rating	0.0	0.01–0.25	0.26–0.50	0.51–0.75	0.76–0.99	1.0	Σ
Theorem	74	21	28	97	0	0	220
Non-Theorem	2	0	5	45	1	0	53
Unknown	0	0	0	0	0	1064	1064
Classical	286	245	102	256	265	183	1337

Domain	AGT	ALG	COM	GEO	LCL	MGT	NLP	PUZ	SET	SWV	SYN
Theorem	14	7	3	7	1	25	11	2	75	1	74
Non-Theorem	0	1	0	0	2	0	0	0	0	0	50
Unknown	38	137	0	65	0	42	11	2	244	1	92
intuit. 0.0	0	0	0	0	1	5	3	0	14	1	52
>0.0	52	145	3	72	2	62	19	4	305	1	164
classic. 0.0	43	6	1	0	3	29	7	3	40	1	123
>0.0	9	139	2	72	0	38	15	1	279	1	93

The lower part of Table 1 contains information with respect to the TPTP problem domain (e.g. SET contains problems from set theory). All problems of the domains GRP, HAL and SWC are Unknown (and not included). From the 1337 problems 220 have been proven intuitionistically valid, 53 invalid.

Table 2. Rating of the ILTP problem set

Rating	0.0	0.01–0.25	0.26–0.50	0.51–0.75	0.76–0.99	1.0	Σ
Theorem	59	11	10	8	1	1	90
Non-Theorem	0	0	5	9	3	1	18
Propositional	14	3	8	17	4	2	48
First-order	45	8	7	0	0	0	60

3.2   The ILTP Problem Set

Table 2 shows a summary of the rating and status information of the ILTP problem set. Again the rating and status information refers to intuitionistic logic. From the 108 problems 90 are intuitionistically valid, 18 are invalid.

4   Conclusion

Like the TPTP library for classical logic, the main motivation for the ILTP library is to put the testing and evaluation of intuitionistic ATP systems onto a firm basis. It is the first systematic attempt to assemble a benchmark library for intuitionistic theorem provers. This will help to ensure that published results reflect the actual performance of an ATP system and make meaningful system evaluations and comparisons possible. We expect that such a library will be fruitful for the development of novel, more efficient calculi and implementations for intuitionistic first-order logic, which — compared to classical logic — is still in its infancy. We have mainly focused on first-order logic which is practically more relevant than the propositional fragment. Future work includes adding more formulas which occur during the practical use of interactive proof assistants like NuPRL [3]. Extending the library to other non-classical logics like first-order modal logics or fragments of linear logic is under consideration as well.

The ILTP library is available at `http://www.iltp.de`. We welcome submissions of new intuitionistic benchmark problems and intuitionistic ATP systems.

References

1. P. Balsiger, A. Heuerding, S. Schwendimann.   Logics workbench 1.0. 7th TABLEAUX Conference, LNCS 1397, Springer, pp. 35–37, 1998.
2. Y. Bertot, P. Castéran. *Interactive theorem proving and program development.* Texts in Theoretical Computer Science, Springer, 2004.
3. R. L. Constable et. al. *Implementing mathematics with the NuPRL proof development system.* Prentice Hall, 1986.
4. D. Larchey-Wendling, D. Méry, D. Galmiche. STRIP: Structural sharing for efficient proof-search. *IJCAR-2001*, LNAI 2083, pp. 696–700, Springer, 2001.
5. J. Otten. ileanTAP: An intuitionistic theorem prover. 6th TABLEAUX Conference, LNAI 1227, pp. 307–312, Springer, 1997.
6. J. Otten. Clausal connection-based theorem proving in intuitionistic first-order logic. *TABLEAUX 2005*, this volume, 2005. (`http://www.leancop.de`)
7. T. Raths. Evaluating intuitionistic automated theorem provers. Technical Report, University of Potsdam, 2005.
8. D. Sahlin, T. Franzen, S. Haridi. An intuitionistic predicate logic theorem prover. *Journal of Logic and Computation*, 2:619–656, 1992.
9. S. Schmitt et al. JProver: Integrating connection-based theorem proving into interactive proof assistants. *IJCAR-2001*, LNAI 2083, pp. 421–426, Springer, 2001.
10. G. Sutcliffe, C. Suttner. The TPTP problem library - CNF release v1.2.1. *Journal of Automated Reasoning*, 21: 177–203, 1998. (`http://www.cs.miami.edu/~tptp`)
11. T. Tammet. A resolution theorem prover for intuitionistic logic. 13th CADE, LNAI 1104, pp. 2–16, Springer, 1996.

Unit Propagation in a Tableau Framework[*]

Gernot Stenz

Institut für Informatik,
Technische Universität München,
D-85748 Garching, Germany
stenzg@in.tum.de

Abstract. Unit propagation is one of the most important techniques of efficient SAT solvers. Unfortunately, this technique is not directly applicable to first-order clausal tableaux. We show a way of integrating a variant of unit propagation into the disconnection calculus and present some results obtained with an implementation of unit propagation in the DCTP theorem prover that show the usefulness of our new method.

1 Introduction

Over the last years, SAT solvers based on the Davis-Putnam-Loveland-Logeman (DPLL) method have become increasingly popular and successful. One of the single most important features of the DPLL procedure is the use of *unit propagation* to guide the proof search. In fact, when DPLL was used as the basis for a new first-order proof method, the *model evolution calculus* [1], the ability to use unit propagation for guiding the proof search was named as one of the prime motivations. Therefore it would be of great interest to integrate unit propagation into classic clausal tableau calculi as well. However, clausal tableau and semantic trees are of a different nature, as are the respective ways of searching for a proof. In [4], a form of unit propagation is introduced for propositional sequent calculi, but without considering the first-order case. In this paper, we present a method of adapting unit propagation to a first-order clausal tableau calculus, the disconnection calculus implemented in the DCTP prover system.

This paper is organised as follows: First, the disconnection calculus is briefly revisited. Section 3 describes the SAT version of unit propagation. Section 4 shows how unit propagation is integrated into the disconnection calculus. Finally, the implementation is briefly described in Section 5.

2 The Disconnection Tableau Calculus

Detailed descriptions of the disconnection calculus can be found in [7]. Here it is sufficient to recapture the most significant aspects of this calculus. For

---

[*] This work was partially funded by the German Federal Ministry of Education, Science, Research and Technology (BMBF) in the framework of the Verisoft project under grant 01 IS C38. The responsibility for this article lies with the author(s).

B. Beckert (Ed): TABLEAUX 2005, LNAI 3702, pp. 338–342, 2005.

this we use the standard terminology for clausal tableaux. The disconnection tableau calculus consists of a single complex inference rule, the so-called *linking rule*, shown in Figure 1. The guidance for the application of the linking rule is provided by *links*, pairs of potentially complementary subgoals on the tableau path of the open subgoal to be extended. The linking rule is used to expand the current clausal tableau with suitable *linking instances* of the input clauses. In order to provide an initial set of links to start the tableau construction, an arbitrary *initial active path* is selected, marking one subgoal from each of the input clauses. A branch in a disconnection tableau is called ∀-*closed* if it contains two literals K and L such that $K\sigma$ is the complement of $L\sigma$ where σ is a substitution identifying all variables in the tableau. The disconnection tableau calculus is refutation complete for first-order clause logic, provided the linking rule is applied in a fair manner[1]. Additionally, it is sufficient to use each link only once on each branch. If the set of links on an open branch is exhausted, the literals on the branch describe a model for the clause set. In order to increase efficiency, *folding up* and *unit simplification*, a tableau version of unit resolution, are employed [3].

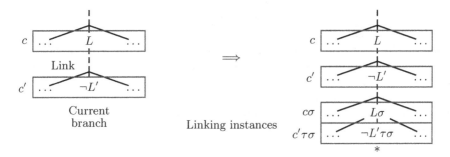

Fig. 1. Illustration of a linking step

3   Unit Propagation for Semantic Trees

DPLL procedures [2] operate on a data structure called a *semantic tree*. A semantic tree is a succession of atomic cuts arranged in a binary tree. The cuts that form the semantic tree are made over the propositional variables occurring in the input clauses of the proof problem. A branch B of the semantic tree is closed if in its context an input clause becomes false, i.e. there is an input clause c such that for every literal $l_1, \ldots, l_m \in c$ the complement $\sim l_i$ of l_i is on B. A semantic tree whose branches are all closed constitutes a proof for the input formula. An un-extendable open branch indicates a model for the input formula. Unit propagation [8] is a method for finding the locally optimal choice of cut variable. For every input clause the *context length* of that clause is recorded,

---

[1] That is, if every available link is used for a linking step after a finite number of inferences.

that is the number of clause literals not complemented by the current branch. The branch literals l_j are used to successively decrement the context length of all input clauses containing $\sim l_j$. When an input clause has a context length of 1, the remaining un-complemented clause variable will be used for the next cut. Thus, one of the new branches will immediately be closed and the proof search can continue without an actual split of the branch and with an extended branch context that can again be used to decrement the context lengths.

4   Unit Propagation for Disconnection Tableaux

Disconnection tableaux are fundamentally different from semantic trees. Here, the branches are labeled with the subgoals of linking instances of clauses. In semantic trees, unit propagation helps to decide over which variable to split next. In disconnection tableaux, unit propagation is to provide heuristic information on which link to use next for a linking step. So our unit propagation procedure does not work on the tableau clauses, but on the *potential linking instances* of the links. As shown in Figure 1, every linking step produces at least one closed branch as well as a number of new open branches. Some of the new branches, however, might already be closed by context literals on the branch. On the left side of Figure 2, a link is depicted, whose use in a linking step would create two open branches. If a path subgoal like $\neg R(v, b)$ has been placed on the branch in the meanwhile as shown on the right side, using the link would create only one open branch. The unit propagation procedure keeps track of the number of open branches to be created by every available link dynamically changing with the context of the current branch. In addition, the simplification of linking instances by unit clauses can also permanently decrease the number of potential new open branches.

Depending on the open branch count of each link, the unit propagation procedure can act in several ways:

Ordinary Unit Propagation. When the current path context ensures that for a link ℓ all but one of the linking instance subgoals can immediately be closed, applying ℓ will augment the path context without causing actual branching. This is the equivalent of the propositional unit propagation described in Section 3.

Lemma Generating Propagation. When all but one of the linking instance subgoals of a link can be simplified, the remaining subgoal will become a unit lemma. This is a special case of ordinary unit propagation, but preferable as unit lemmas are more useful and versatile than path context subgoals.

Branch Closure Propagation. When the current path context allows the immediate ∀-closure of all the linking instance subgoals of a link ℓ, then an application of ℓ can be used to close the current branch. In a DPLL context, branch closure propagation has no equivalent, since an according branch would be closed without further action.

Complementary Unit Propagation. When all linking subgoals of a link ℓ are simplified by the current set of unit lemmas, an application of ℓ will finish the proof (this being equivalent to the creation of the empty clause in a resolution context).

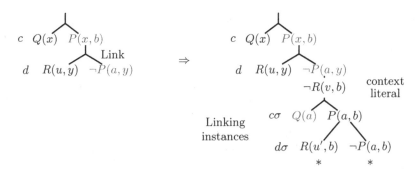

Fig. 2. Reduction of open instance subgoals by the path context

It must be pointed out that, unlike in propositional semantic trees, where unit propagation can be used as a deciding strategy, in first-order tableaux unit propagation serves as a mere *heuristic* to guide the proof search. It can easily be seen that, given a unit lemma $P(a)$ and a path clause $\neg P(x) \lor P(f(x))$, the uncontrolled use of unit propagation would merely produce new unit lemmas $P(f(a)), P(f(f(a))), P(f(f(f(a)))), \ldots$ that would not necessarily contribute to the proof search, thus destroying the completeness of the calculus. Therefore, in order to maintain the fairness condition required in Section 2, ordinary unit propagation and lemma generating propagation are used only in an intermittent fashion, alternating at a fixed rate with regular linking steps. This way, the application of each link will still be delayed for a finite number of inferences only, and thus the completeness of the calculus is preserved. On the other hand, branch closure propagation (and also, trivially, complementary unit propagation) is not subject to such an alternation scheme, since the fairness requirement is restricted only to the current (and subsequently closed) branch.

5   Implementation and Results

Unit propagation is implemented in the DCTP prover system[6] by means of a *propagation index*. Upon the generation of a new link, all subgoals of the potential linking instances are entered into the propagation index. Also the number of open subgoals created by the application of the link is stored within the link. Whenever the branch context is extended by a new subgoal, that subgoal is used to search the propagation index for matching entries. All matching entries not marked by another branch subgoal[2] are marked as closed and the open subgoal counts of their respective links are decremented. Upon backtracking, the markings caused by the branch subgoal are undone and the link counts corrected. Simplifying units will prune the propagation index destructively, erasing all matching branches. All links classified according to the propagation categories of Section 4 are kept in four lists. When selecting a new link, DCTP first

---

[2] A branch may contain multiple occurrences of the same non-ground subgoal and only the topmost occurrence may be used to decrement link counts.

checks these lists for valid entries, for ordinary unit propagation and lemma generating propagation an additional alternation scheme is operated, so that a link of that kind can only be used for every n-th inference. Due to the obvious usefulness of unit propagation, only a very small number of tests has been run and the results obtained substantiate our claims. With the use of unit propagation, the first-order theorem prover DCTP has been able to solve a number of SAT solver benchmark problems (ssa2670-130, ssa2670-141) that previously were simply beyond the abilities of DCTP (and which also cannot be solved be the E prover[5]). In the TPTP SYN class of syntactic ALC problems, unit propagation also brought significant speedups. Finally, TPTP problem PUZ037-3, the Rubik's Cube problem where three rotations of the cube are needed for a solution, had hitherto been out of reach for DCTP. Using unit propagation PUZ037-3 can be solved in less than three seconds.

6   Conclusion

In this paper we described the adaptation of the unit propagation SAT technique to a clausal tableau framework and the successful integration of this technique into the DCTP theorem prover. This serves to show that it is worthwhile looking beyond first-order theorem proving for new techniques. The DCTP theorem prover is under constant development,but a stable snapshot incorporating unit propagation is available from the author.

References

1. P. Baumgartner and C. Tinelli. The Model Evolution Calculus. In Franz Baader, editor, *Automated Deduction - CADE-19, LNAI 2741*. Springer, 2003.
2. M. Davis, G. Logemann, and D. Loveland. A machine program for theorem proving. *Communications of the Association for Computing Machinery*, pages 394–397, 1962.
3. R. Letz and G. Stenz. Proof and Model Generation with Disconnection Tableaux. In Andrei Voronkov, editor, *Proceedings of LPAR 2001*, pages 142–156. Springer, Berlin, December 2001.
4. Fabio Massacci. Simplification: A general constraint propagation technique for propositional and modal tableaux. In Harrie de Swart, editor, *Proceedings, TABLEAUX-1998, LNAI 1397*, pages 217–231. Springer, Berlin, May 1998.
5. S. Schulz. System Abstract: E 0.61. In *Proceedings, IJCAR-2001, LNAI 2083*, pages 370–375. Springer, Berlin, June 2001.
6. G. Stenz. DCTP 1.2 – System Abstract. In *Proceedings, TABLEAUX-2002, LNAI 2381*, pages 335–340. Springer, Berlin, July 2002.
7. G. Stenz. *The Disconnection Calculus*. Logos Verlag, Berlin, 2002. Dissertation, Fakultät für Informatik, Technische Universität München.
8. H. Zhang and M. E. Stickel. An efficient algorithm for unit propagation. In *Proceedings, AI-MATH'96*, December 1996.

Author Index

Lecture Notes in Artificial Intelligence (LNAI)

Vol. 3501: B. Kégl, G. Lapalme (Eds.), Advances in Artificial Intelligence. XV, 458 pages. 2005.

Vol. 3492: P. Blache, E. Stabler, J. Busquets, R. Moot (Eds.), Logical Aspects of Computational Linguistics. X, 363 pages. 2005.

Vol. 3488: M.-S. Hacid, N.V. Murray, Z.W. Raś, S. Tsumoto (Eds.), Foundations of Intelligent Systems. XIII, 700 pages. 2005.

Vol. 3487: J. Leite, P. Torroni (Eds.), Computational Logic in Multi-Agent Systems. XII, 281 pages. 2005.

Vol. 3476: J. Leite, A. Omicini, P. Torroni, P. Yolum (Eds.), Declarative Agent Languages and Technologies II. XII, 289 pages. 2005.

Vol. 3464: S.A. Brueckner, G.D.M. Serugendo, A. Karageorgos, R. Nagpal (Eds.), Engineering Self-Organising Systems. XIII, 299 pages. 2005.

Vol. 3452: F. Baader, A. Voronkov (Eds.), Logic for Programming, Artificial Intelligence, and Reasoning. XI, 562 pages. 2005.

Vol. 3451: M.-P. Gleizes, A. Omicini, F. Zambonelli (Eds.), Engineering Societies in the Agents World V. XIII, 349 pages. 2005.

Vol. 3446: T. Ishida, L. Gasser, H. Nakashima (Eds.), Massively Multi-Agent Systems I. XI, 349 pages. 2005.

Vol. 3445: G. Chollet, A. Esposito, M. Faundez-Zanuy, M. Marinaro (Eds.), Nonlinear Speech Modeling and Applications. XIII, 433 pages. 2005.

Vol. 3438: H. Christiansen, P.R. Skadhauge, J. Villadsen (Eds.), Constraint Solving and Language Processing. VIII, 205 pages. 2005.

Vol. 3430: S. Tsumoto, T. Yamaguchi, M. Numao, H. Motoda (Eds.), Active Mining. XII, 349 pages. 2005.

Vol. 3419: B. Faltings, A. Petcu, F. Fages, F. Rossi (Eds.), Constraint Satisfaction and Constraint Logic Programming. X, 217 pages. 2005.

Vol. 3416: M. Böhlen, J. Gamper, W. Polasek, M.A. Wimmer (Eds.), E-Government: Towards Electronic Democracy. XIII, 311 pages. 2005.

Vol. 3415: P. Davidsson, B. Logan, K. Takadama (Eds.), Multi-Agent and Multi-Agent-Based Simulation. X, 265 pages. 2005.

Vol. 3403: B. Ganter, R. Godin (Eds.), Formal Concept Analysis. XI, 419 pages. 2005.

Vol. 3398: D.-K. Baik (Ed.), Systems Modeling and Simulation: Theory and Applications. XIV, 733 pages. 2005.

Vol. 3397: T.G. Kim (Ed.), Artificial Intelligence and Simulation. XV, 711 pages. 2005.

Vol. 3396: R.M. van Eijk, M.-P. Huget, F. Dignum (Eds.), Agent Communication. X, 261 pages. 2005.

Vol. 3394: D. Kudenko, D. Kazakov, E. Alonso (Eds.), Adaptive Agents and Multi-Agent Systems II. VIII, 313 pages. 2005.

Vol. 3392: D. Seipel, M. Hanus, U. Geske, O. Bartenstein (Eds.), Applications of Declarative Programming and Knowledge Management. X, 309 pages. 2005.

Vol. 3374: D. Weyns, H. V.D. Parunak, F. Michel (Eds.), Environments for Multi-Agent Systems. X, 279 pages. 2005.

Vol. 3371: M.W. Barley, N. Kasabov (Eds.), Intelligent Agents and Multi-Agent Systems. X, 329 pages. 2005.

Vol. 3369: V. R. Benjamins, P. Casanovas, J. Breuker, A. Gangemi (Eds.), Law and the Semantic Web. XII, 249 pages. 2005.

Vol. 3366: I. Rahwan, P. Moraitis, C. Reed (Eds.), Argumentation in Multi-Agent Systems. XII, 263 pages. 2005.

Vol. 3359: G. Grieser, Y. Tanaka (Eds.), Intuitive Human Interfaces for Organizing and Accessing Intellectual Assets. XIV, 257 pages. 2005.

Vol. 3346: R.H. Bordini, M. Dastani, J. Dix, A.E.F. Seghrouchni (Eds.), Programming Multi-Agent Systems. XIV, 249 pages. 2005.

Vol. 3345: Y. Cai (Ed.), Ambient Intelligence for Scientific Discovery. XII, 311 pages. 2005.

Vol. 3343: C. Freksa, M. Knauff, B. Krieg-Brückner, B. Nebel, T. Barkowsky (Eds.), Spatial Cognition IV. XIII, 519 pages. 2005.

Vol. 3339: G.I. Webb, X. Yu (Eds.), AI 2004: Advances in Artificial Intelligence. XXII, 1272 pages. 2004.

Vol. 3336: D. Karagiannis, U. Reimer (Eds.), Practical Aspects of Knowledge Management. X, 523 pages. 2004.

Vol. 3327: Y. Shi, W. Xu, Z. Chen (Eds.), Data Mining and Knowledge Management. XIII, 263 pages. 2005.

Vol. 3315: C. Lemaître, C.A. Reyes, J.A. González (Eds.), Advances in Artificial Intelligence – IBERAMIA 2004. XX, 987 pages. 2004.

Vol. 3303: J.A. López, E. Benfenati, W. Dubitzky (Eds.), Knowledge Exploration in Life Science Informatics. X, 249 pages. 2004.

Vol. 3301: G. Kern-Isberner, W. Rödder, F. Kulmann (Eds.), Conditionals, Information, and Inference. XII, 219 pages. 2005.

Vol. 3276: D. Nardi, M. Riedmiller, C. Sammut, J. Santos-Victor (Eds.), RoboCup 2004: Robot Soccer World Cup VIII. XVIII, 678 pages. 2005.

Vol. 3275: P. Perner (Ed.), Advances in Data Mining. VIII, 173 pages. 2004.

Vol. 3265: R.E. Frederking, K.B. Taylor (Eds.), Machine Translation: From Real Users to Research. XI, 392 pages. 2004.

Vol. 3264: G. Paliouras, Y. Sakakibara (Eds.), Grammatical Inference: Algorithms and Applications. XI, 291 pages. 2004.

Vol. 3259: J. Dix, J. Leite (Eds.), Computational Logic in Multi-Agent Systems. XII, 251 pages. 2004.

Vol. 3257: E. Motta, N.R. Shadbolt, A. Stutt, N. Gibbins (Eds.), Engineering Knowledge in the Age of the Semantic Web. XVII, 517 pages. 2004.

Vol. 3249: B. Buchberger, J.A. Campbell (Eds.), Artificial Intelligence and Symbolic Computation. X, 285 pages. 2004.

Vol. 3248: K.-Y. Su, J. Tsujii, J.-H. Lee, O.Y. Kwong (Eds.), Natural Language Processing – IJCNLP 2004. XVIII, 817 pages. 2005.

Vol. 3245: E. Suzuki, S. Arikawa (Eds.), Discovery Science. XIV, 430 pages. 2004.

Vol. 3244: S. Ben-David, J. Case, A. Maruoka (Eds.), Algorithmic Learning Theory. XIV, 505 pages. 2004.